International Symposium on Current Progress in Functional Materials (ISCPFM 2016)

IOP Conference Series: Materials Science and Engineering
Volume 188

Bali, Indonesia
26 – 27 July 2016

ISBN: 978-1-5108-4027-0
ISSN: 1757-8981

Printed from e-media with permission by:

Curran Associates, Inc.
57 Morehouse Lane
Red Hook, NY 12571

Some format issues inherent in the e-media version may also appear in this print version.

Printed by Curran Associates, Inc. (2017)

For permission requests, please contact the Institute of Physics
at the address below.

Institute of Physics
Dirac House, Temple Back
Bristol BS1 6BE UK

Phone: 44 1 17 929 7481
Fax: 44 1 17 920 0979

techtracking@iop.org

Additional copies of this publication are available from:

Curran Associates, Inc.
57 Morehouse Lane
Red Hook, NY 12571 USA
Phone: 845-758-0400
Fax: 845-758-2633
Email: curran@proceedings.com
Web: www.proceedings.com

Table of contents

Volume 188

International Symposium on Current Progress in Functional Materials

26–27 July 2016, Bali, Indonesia

Accepted papers received: 31 March 2017
Published online: 2 May 2017

Preface

Papers

IOP Conf. Series: Materials Science and Engineering **188** (2017) 011001 doi:10.1088/1757-899X/188/1/011001

Preface from the editors

The energy requirement increases exponentially and becomes an important issue for the entire world due to its significant impact on environmental damage and human health. Therefore, it is an urgent priority for the researchers to design new materials that can fulfill the energy requirements and to reduce the environmental risks and health problems in the same time. Functionalization of materials can be a potential response to the growing challenges in the energy, environment and health applications. Materials with improved functional has not only attracted more attention from researchers, but also be required in a wide range of advanced technology fields. This demand leads to the rapid global development of the new and novel materials.

The International Conference on Current Progress in Functional Materials 2016 (ISCPFM 2016) is the first international conference in functional materials organized by the Faculty of Mathematics and Natural Sciences (FMIPA), Universitas Indonesia. It was held in Bali (Indonesia) and accommodated two days of intense discussion regarding the current progress in the functional material for energy, environment and health applications. Three plenary speakers from Japan and UK have featured the fascinating results in experimental works, while the other three from US, Japan, and China gave compelling presentations from the theoretical point of views. Concurrent sessions were covered a more wide range of topics and issues, including low-dimensional, nano and 2D materials, metals, metal oxides, organometals and coordination chemistry, polymers, ceramics and composite materials, biomaterials as well as computation and simulation of materials. After a peer review process, 62 papers have been accepted to be published in the conference proceeding.

We would like to thank all of the reviewers for their time and effort to review and select the papers and also express our heartfelt appreciation to the faculty's conference proceeding teams to finalize the articles to be included in the proceeding. It would be difficult for us to finish the proceeding without the commitment and the willpower from the researchers and technical teams to involve in all processes. We have enjoyed working hand in hand with the reviewers and the faculty's proceedings teams. It has been an honor for us to have the chance to become the editors of the proceeding.

We are very grateful and highly appreciate the contribution of various parties who have been involved directly and indirectly- the international advisory board, the steering committee, the organizing committee, the scientific committee, the program committee and financial support from Universitas Indonesia- for helping to execute the conference and the proceedings. Finally, we thank all participants and authors for their contributions.

Depok, March 2017

Editors
Ivandini Tribidasari A.
Djoko Triyono

International Advisory Board

Rosari Saleh (Universitas Indonesia, Indonesia)

Russell F Howe (The University of Aberdeen, United Kingdom)

Wei Ku (Shanghai Jiao Tong University, China)

Yasuaki Einaga (Keio University, Japan)

Scientific Committee

Jarnuzi Gunlazuardi (Chair)

Muhammad Aziz Majidi (Co-chair)

Teruna J Siahaan (The University of Kansas, United States)

Hisao Yoshida (Kyoto University, Japan)

Mochamad Chalid

Dede Djuhana

Faculty's Conference Proceeding Committee

Ridla Bakri

Rokhmatuloh

Lusitra Munisa

Yustin S Dewi

Hesni A Shabrany

Lisna Arumsari

IOP Conf. Series: Materials Science and Engineering **188** (2017) 011002 doi:10.1088/1757-899X/188/1/011002

Organizing Committee

Advisor	:	Prof. Dr. Ir. Muhammad Anis, M.Met
		Prof. Dr.rer.nat. Rosari Saleh
Honorary	:	Dr.rer.nat. Abdul Haris
Chairman		Prof. Dr. Ridla Bakri, M.Phil
		Dr. Djoko Triyono
Conference Chair	:	Dr. Ivandini Tribidasari A
Co-chair	:	Dr. Muhammad Aziz Majidi
Programme	:	Dr. Yuni K Krisnandi
Committee	:	Lusitra Munisa, M.Si
		Yustin S Dewi, S.Si
		Devvi Sarwinda, M.Kom
		Lisna Arumsari, A.Md
		Asri Nurlaeli D H, A.Md
		Dedy Kurniawan, AMd
		Caya Mikail, A.Md

IOP Conf. Series: Materials Science and Engineering **188** (2017) 011002 doi:10.1088/1757-899X/188/1/011002

ISCPFM 2016

The International Symposium on Current Progress in Functional Materials 2016 (ISCPFM 2016)

held by

FMIPA

Faculty of Mathematics and Natural Sciences
Universitas Indonesia

Supported by

UNIVERSITAS INDONESIA
Veritas, Probitas, Iustitia

Universitas Indonesia

Peer review statement

All papers published in this volume of *IOP Conference Series: Materials Science and Engineering* have been peer reviewed through processes administered by the proceedings Editors. Reviews were conducted by expert referees to the professional and scientific standards expected of a proceedings journal published by IOP Publishing.

The potency of electrical energy production from urine by microbial fuel cell using boron-doped diamond electrode

I Rahmawati, T A Ivandini and E Saepudin

Department of Chemistry, Faculty of Mathematics and Natural Sciences Universitas Indonesia, Kampus UI Depok, Depok 16424, Indonesia

Corresponding author's e-mail: ivandini.tri@sci.ui.ac.id

Abstract. Microbial fuel cell was developed since it is one of the prospective alternative energy and eco-friendly, using urine as the fuel and *Candida fukuyamaensis* as a biocatalyst. Boron-doped diamond was used as the electrode. At pH 7, maximum power and current densities of 109.6 mW/m^2 and 970 mA/m^2 can be obtained, respectively. The results indicated the potency of the system to produce an alternative energy. Furthermore, glucose and creatinine in urine are proposed to be responsible as the carbon sources for the metabolism of *C. fukuyamaensis*.

Keywords: MFC, urine, Candida fukuyamaensis, boron-doped diamond, glucose, creatinin

1. Introduction

In the last decades, many researches focused on finding an environmentally friendly alternative energy to replace fossil energy and batteries. Microbial fuel cell (MFC) is one of the prospective alternative energy, which is also eco-friendly. MFC converts organic material into electrical energy by using microorganisms as biocatalysts [1].

Urine is an abundant waste. Normal person can produce urine approximately 1-2 L/day, which usually useless and wasted. Urine contains many dissolved organic compounds [2] that can contaminate the environment, especially water, causing an unpleasant odor and eutrophication. On the other hand, the organic substances contained in urine are potential to be applied as fuels in MFC.

In this study, urine is used as fuel in MFC. As a biocatalyst in the system, a local yeast *Candida fukuyamaensis* UICC Y-247 is employed. This yeast was selected based on its properties, such as easinest to obtain, aerobic property, as well as relatively stable and easily grow property [3]. Boron-doped diamond (BDD) was selected as the electrode due to its superior properties among other conventional solid electrodes, including wide potential window, low background current, and high stability of the surface [4, 5]. Whereas there are many reports of BDD applications for sensors and biosensors [6, 7], a very limited work were reported to discuss its potency to produce energy. This work shows the MFC system with BDD electrodes can successfully produce electrical energy with urine as the fuel.

2. Materials and methods

The urine samples used in this study were from three women, aged 20-25 years old. KH_2PO_4, K_2HPO_4, Nafion 117, NaOH, $K_3Fe(CN)_6$, H_2SO_4, glucose, urea, creatinine, H_2O_2 3%, and other compounds were from Sigma or Sigma-Aldrich. *Candida fukuyamaensis* UICC Y-247 were obtained from Universitas Indonesia Culture Collection (UICC).

The urine was characterized for its pH as well as the content of glucose, protein and creatinine. *C. fukuyamaensis* was regenerated by transfer it into YMA medium aseptically followed by 24-h incubation. Then, the yeast from YMA were transferred into YMB aseptically and incubated for 24 h. This suspension in YMB was used as the biocatalyst in MFC experiments.

A two compartments reactor was used in this study with a total volume of 100 mL. These compartments were separated by Nafion-117 membrane. The anode compartment contained suspension of *C. fukuyamaensis* and urine, while cathode compartment contained a solution of 0.1 M phosphate buffer and 0.1 M potassium ferri cyanide solution. BDD was used as the anode with a surface area of 1 cm^2, while a spiral-shaped platinum was used as the cathode and an Ag/AgCl system was used as the reference electrode. The anaerobic anode compartment was set up by closing the compartment firmly with nitrogen gas flown to the system to prevent oxygen entrance. On the opposite, the aerobic cathode compartment was set up with an open-air compartment.

3. Results and discussion

Prior to the application, urine for the fuel of the MFC fuel was characterized. The urine pH was around 6, whereas the content of glucose, other proteins, and creatinine were 0.5%, 10 mg/dL, and 0.6 mg/dL, respectively, which was in the normal range of the human urine.

Various pHs from pH 5-8 were examined to observe the effect of pH on the electrical energy produced. The pH changes in anode compartment may affect the microbial activity in substrate metabolism, which indirectly affects the produced electrical energy. The data of current and voltage produced is presented in figure 1.

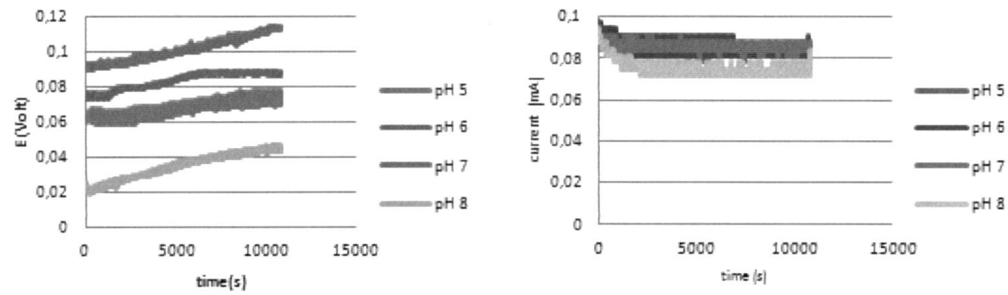

(a) (b)

Figure 1. Graph of current (a) and potential (b) measurements at various pH.

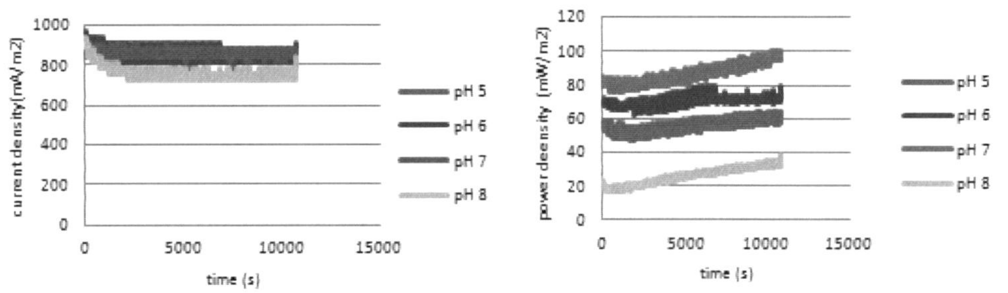

(a) (b)

Figure 2. Graphs of current (a) and power (b) densities at various pHs.
The data was transformed from figure 1.

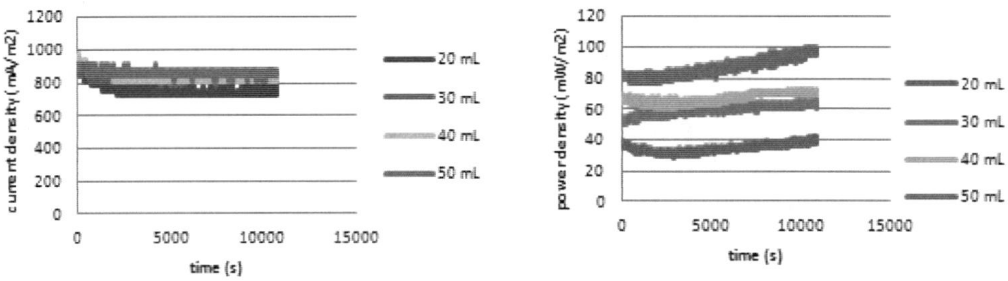

(a) (b)

Figure 3. Graph of current (a) and power (b) densities at various volume of *C. fukuyamaensis* suspension.

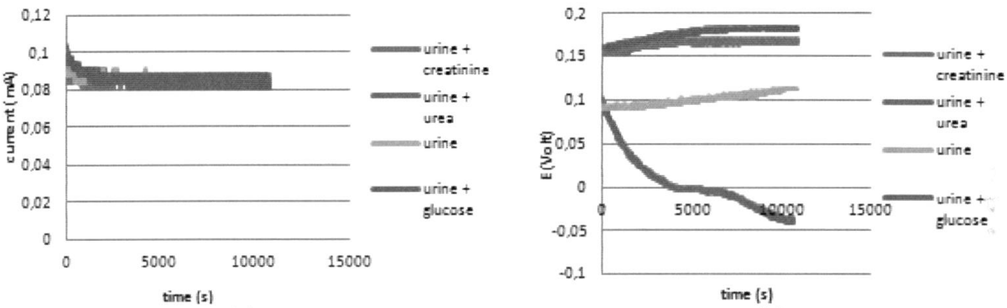

(a) (b)

Figure 4. Graph of current (a) and potential (b) mesaurements at various subtrates.

Conversion to the current density and power density is shown in figure 2. The figures show that the change in pH affects the electrical energy produced. At pH 7, the optimum current density and power was generated, whereas at pH 8, the generated energy significantly decreased. In this study, the measurement of energy was performed without the addition of external mediator. Therefore, it can be assumed that the generated electrons are transferred directly from microbial cells to the electrode surface or commonly called the direct electron transfer [8].

Furthermore, various volumes of *C. fukuyamaensis* suspension (20-50 mL) were examined to study the correlation between the number of microbes and the production of electrical energy. As expected the more *C. fukuyamaensis* the higher energy will be generated. The data of the current and power densities generated by the developed MFC in various amount of microbes is displayed in figure 3.

The influences of glucose, creatinine, and urea in the MFC system were also studied to confirm the main carbon sources used for the metabolisms of *C. fukuyamaensis*. Figure 4 shows the data of current and potential measurements in the system with the presence of glucose, creatinine, and urea. The data shows that the addition of glucose and creatinine in urine increase the current and power density. Meanwhile, the addition of urea into urine decreases the potential significantly as *C. fukuyamaensis* cannot metabolize urea to produce protons and electrons. In addition, the urea induced the system to be highly alkaline for *C. fukuyamaensis*. Therefore, the potential different between the cathode and the anode becomes very high, hence the potential generated become dropped dramatically.

The energy produced by the MFC was stable for 3 h. Until the second day the system was applied, the generated current and power density was stable. However, after the third day the current and power density drastically decreased to zero. The reasons may probably due to the limited substrate in urine as well as the lifetime of *C. fukuyamaensis*. Therefore, it is necessary to consider the method to input the fuel continously.

4. Conclusions

Boron-doped diamond electrodes were successfully employed in urine-based MFC with *C. fukuyamaensis* as the biocatalyst with the optimum pH of 7. Glucose and creatinine containing in urine play the role in the production of the MFC systems.

References

[1] Lu N, Zhou S G, Zhuang L , Zhang J T and Ni J R 2009 *Biochem. Eng. J.* **43** 246-51
[2] Putnam D F 1971 *Composition and Concentrative Properties of Human Urine* (Washington DC: NASA-Langley Reserach Center) 112
[3] Anggraningrum I T, Zulfia F T and Saepudin E 2011 *Makara Sains* **15** 33-7
[4] Fujishima A, Einaga Y, Rao T N and Tryk D 2005 *Diamond Electrochemistry* (Tokyo: BKC; Amsterdam: Elsevier)
[5] Ishii Y, Ivandini T A, Murata K and Einaga Y 2013 *Anal. Chem.* **85** 4284-88
[6] Ivandini T A, Saepudin E, Wardah H, Dewangga N and Einaga Y 2012 *Anal. Chem.* **84** 9825-32
[7] Pino F, Ivandini T A, Nakata K, Fujishima A, Merkoci A and Einaga Y 2015 *Anal. Sci.* **31** 1061-8
[8] Schroder U 2007 *Phys. Chem. Chem. Phys.* **9** 2619-29

IOP Conf. Series: Materials Science and Engineering **188** (2017) 012002 doi:10.1088/1757-899X/188/1/012002

Effect of Sr-doped on thermal properties of nanoparticle (La,Sr)FeO_3

H Laysandra and D Triyono

Department of Physics, Faculty of Mathematics and Natural Sciences Universitas Indonesia, Kampus UI Depok, Depok 16424, Indonesia

Corresponding author's e-mail: djoko.triyono@sci.ui.ac.id

Abstract. Structure and thermal properties of $La_{1-x}Sr_xFeO_3$ have been studied. (La,Sr)FeO_3 was synthesized by sol-gel method result in powder form. Structural study was carried out using X-Ray Diffraction (XRD) and TEM. XRD characterization for all samples show orthorhombic perovskite structure. The crystallite size of the materials lie in the range of 20-40 nm as calculated by using Schrerrer's formula which is also supported by TEM images. The thermal properties were characterized by using DSC. DSC curves show increasing of Neel temperature with addition of Stronsium. Average specific heat capacity was also calculated.

1. Introduction

$LaFeO_3$, one of perovskite oxide, has been paid attention for many applications, such as cathode materials for fuel cells and alcohol sensor [1]. $LaFeO_3$ is antiferromagnetic material with Neel temperature of 740 K [2]. The A-site substituted $LaFeO_3$ can be used in various applications because their stable structure, high thermal stability, high dielectric constant, low dielectric loss, and high performances are significantly affected by the substitution [3]. Thermal properties of $LaFeO_3$ is dependent on crystallite size [1], in which crystallite size is also dependent on doped on A-site. However, the thermal properties study for this compound has not been well studied. Parida *et al.* [4] have measured the enthalpy and standard molar Gibbs energies of $LaFeO_3$ using high temperature Calvet micro calorimeter and solid oxide galvanic cell and found that -1334 kJ/mol and 128.9 J/mol.K, respectively.

In this work, we have synthesized (La,Sr)FeO_3 with x = 0.0, 0.1, 0.2, 0.3, and 0.4 at.% by sol-gel method to form material in nano-scale. We focused on the effect of Sr-doped on the structural and thermal properties of (La,Sr)FeO_3 nanoparticles materials.

2. Materials and methods

$La_{1-x}Sr_xFeO_3$ samples were prepared by sol-gel method. Briefly, La_2O_3 (Merck), $Fe(NO_3)_3.9H_2O$ (Merck), $Sr(NO_3)_2$ (Merck), and citric acid were dissolved in 150 mL water. The resulting suspension was continuously stirred on a heating plate until it turned to a highly viscous gel. The gel was heated at 200°C until formed brown powder. Then, it was calcined at 900°C for 6 h. The phase identification was checked by using X-Ray Diffractometer (XRD) with Cu Kα (λ = 1.54060 Å) radiation in the 2θ range of 10°- 90° with a step size of 0.02°. Rietveld refinement of the diffraction pattern was carried out using the HighScore Plus software. The morphology of the samples was checked by TEM. Thermal properties of $La_{1-x}Sr_xFeO_3$ were characterized by *DSC* with range temperature of 30-500°C in nitrogen atmosphere.

3. Results and discussion

3.1. Structure analysis

Figure 1 shows XRD pattern of the nanoparticles of $La_{1-x}Sr_xFeO_3$ with x = 0.1, 0.2, 0.3, and 0.4 at.% at room temperature. High score plus software was employed to analyze all XRD pattern. All samples exhibit single phase orthorhombic perovskite structure. The refined structural parameters are tabulated in table 1. There are no significantly shifting in 2θ indicates that there are no significantly change in lattice parameters. Broadening peaks are also observed clearly by increasing of Sr concentration. As shown in the Table 1, the crystallite size decrease as Sr-doped is increased. It can be caused the ionic radii of Sr^{2+} (144 pm) is larger than La^{3+} (136 pm), then the substitution of Sr effected the oxidation of Fe^{3+} ions to Fe^{4+} ions. While the ionic radii of Fe^{4+} (58.5 pm) is smaller than Fe^{3+} (64.5 pm) that lead to the decrease of crystallite size [5].

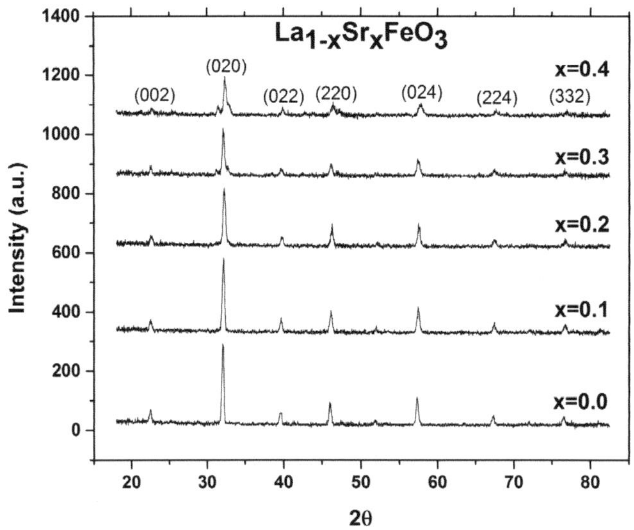

Figure 1. The XRD patterns of $(La,Sr)FeO_3$ nanoparticles calcined at 900°C for 6 h in air.

Table 1. Refined lattice parameters and crystallite size of $La_{1-x}Sr_xFeO_3$ nanoparticles calcined at 900°C for 6 h in air.

Samples	$La_{1-x}Sr_xFeO_3$				
	x=0	x=0.1	x=0.2	x=0.3	x=0.4
Lattice parameters of $La_{1-x}Sr_xFeO_3$ (Å)					
a	5.5724	5.5558	5.5610	5.5575	5.5608
b	7.8765	7.8384	7.8499	7.8617	7.8676
c	5.5696	5.5434	5.5547	5.5275	5.4998
Crystallite size (nm)	29.5	25.7	24.6	23.9	23.0

3.2 TEM

The morphology and structure of $La_{1-x}Sr_xFeO_3$ nanoparticles were characterized by TEM, as shown in figure 2. The TEM characterization were prepared by placing microdrops of solution on a carbon film which supported by copper grid. It can be seen from figure 2 that the average particle size of the samples decreased with increasing of Sr-doping ((a) x = 0.0 and (b) x = 0.4) content which is consistent with XRD results.

3.3 Thermal analysis

Figure 3 shows DSC measurements of $La_{1-x}Sr_xFeO_3$ nanoparticles with x = 0.0, 0.3 and 0.4. The measurement were performed with heating rate of 10°C/min from 30°C to 500°C using 5 mg powder sample encapsulated in standard Al crucibles under nitrogen atmosphere. A phase transition which are Neel temperatures were observed at about 420°C in DSC curve of x = 0.0, 432°C of x = 0.3, and 435°C of x = 0.4. It can be seen that Neel temperature increases as Sr-doped increased. The similar phenomenon is found by Caranoni *et al.* [6] for double perovskite Pb_2ScTaO_6 and Pb_2ScNbO_6 at lower temperature. These mean antifferomagnetic phase spread in high temperature with increase of Sr-doped. The average specific heat (C_p) calculations were also analyzed using the corresponding DSC curves.

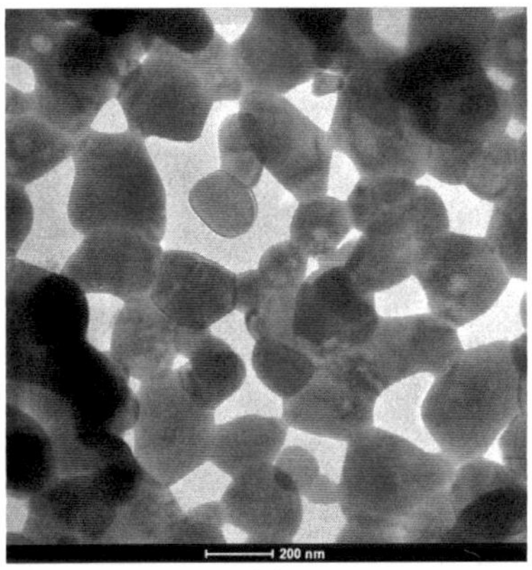

(a)

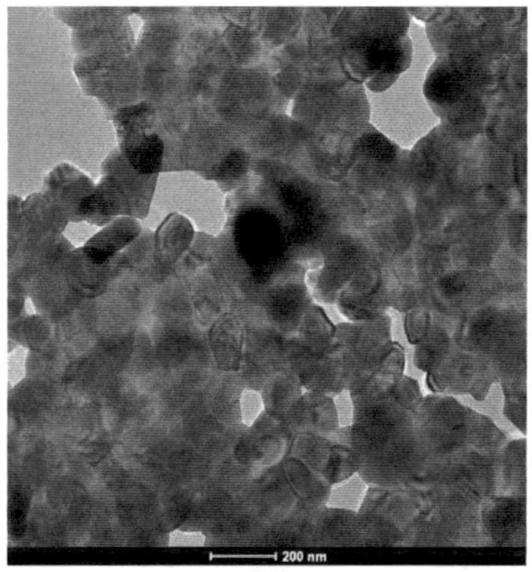

(b)

Figure 2. TEM images of $La_{1-x}Sr_xFeO_3$ nanoparticles for the samples with (a) x = 0.0, (b) x = 0.4.

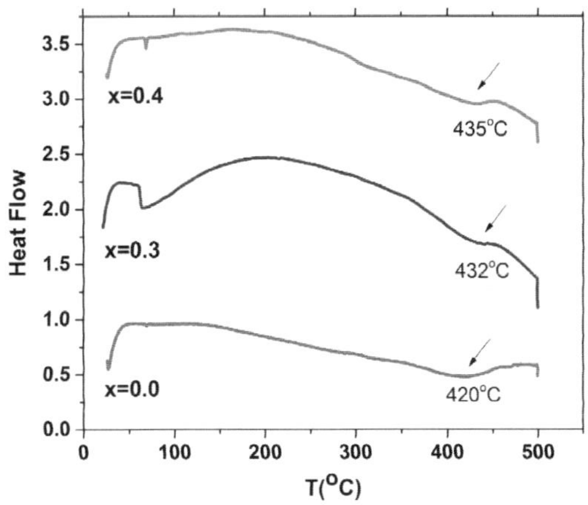

Figure 3. DSC curve of $La_{1-x}Sr_xFeO_3$ nanoparticles with x = 0.0, 0.3 and 0.4 between 30°C and 500°C with rate of 10°C/min in nitrogen atmosphere.

Table 2. Average specific heat capacity (C_p) value calculation of $La_{1-x}Sr_xFeO_3$ nanoparticles with x = 0.0, 0.1, 0.2, 0.3 and 0.4.

$La_{1-x}Sr_xFeO_3$	C_p (10^{-4} J/g°C)
x = 0.0	4.2
x = 0.1	4.3
x = 0.2	4.4
x = 0.3	6.8
x = 0.4	6.9

Table 2 shows average specific heat capacity (C_p) value nanoparticles $La_{1-x}Sr_xFeO_3$ with x = 0.0, 0.1, 0.2, 0.3, and 0.4. The heat capacity was calculated from 300°C to 400°C. These results show that specific heat capacity of $La_{1-x}Sr_xFeO_3$ increase with the increasing of Sr-doped. Maximum value of specific heat capacity was found at x = 0.4.

4. Conclusions

$La_{1-x}Sr_xFeO_3$ with x = 0.0, 0.1, 0.2, 0.3, and 0.4 at.% nanoparticles has been successfully synthesized by sol-gel method. XRD characterization for all samples shows single phase orthorhombic perovskite structure. DSC curve shows the Neel temperature increases as Sr-doped increased.

Acknowledgements

The authors would like thanks to Universitas Indonesia for financial support from PITTA Grant No. 2007/UN2.R12/HKP.05.00/2016.

References

[1] Phokha S, Hunpratup S, Pinitsoontorn S, Putasaeng B, Rujirawat S and Maensiri S 2015 *Mater. Res. Bulletin* **67** 118-25
[2] Devi Chandrasekhar K, Mallesh S, Krishna Murthy J, Das A K and Venimadhav A 2014 *Physica* B **448** 304-11
[3] Cui X, Li S and Zhu X 2014 *Mater. Lett.* **130** 267-70
[4] Parida S C, Singh Z, Dash S, Prasad R and Venugopal V 1998 *J. Alloys and Comp.* **280** 94-8
[5] Zhao K *et al.* 2014 *Chinese J. Catal.* **35** 1196-205
[6] Caranoni C, Lampin P, Siny I, Zheng J G, Li Q, Kang Z C and Boulesteix C 1992 *Phys. Stat. Sol.* (a) **130** 25-37

IOP Conf. Series: Materials Science and Engineering **188** (2017) 012003 doi:10.1088/1757-899X/188/1/012003

Synthesis, characterization and trivalent arsenic sorption potential of Ce-Al nanostructured mixed oxide

S Bhattacharya[1], K Gupta[2] and U C Ghosh[2]

[1]School of Ecology and Environment Studies, Nalanda University, Rajgir, Bihar, India
[2]Department of Chemistry, Presidency University, Kolkata, India

Corresponding author's e-mail: sbhattacharya@nalandauniv.edu.in

Abstract. Arsenic contamination in the ground water has serious health consequences in many parts of the world. The surface sorption method for arsenic mitigation has been widely investigated due to its simple method, inexpensive operation, highly efficient and low content of by-products. In the present study, nanostructured hydrated cerium aluminum oxide (NHCAO) was synthesized and characterized and its arsenic (III) sorption behavior from the aqueous solution was studied. The material was characterized in SEM, FE-SEM, TEM, AFM, XRD, and FT-IR. Batch method was used for the kinetics of As (III) sorption on nanoparticles at 303 (± 1.6) K and at pH 7.0 (± 0.2). The experiments on isotherm subject were performed individually at 288K, 303K, 318K temperatures at pH 7.0 (± 0.2) using the batch sorption method. In the kinetics study of arsenic (III) sorption, the sorption percentage was observed to remain nearly unchanged up to pH 9.0, thereafter only slight reduction in sorption percentage. The equilibrium sorption results were tested using the models of Langmuir and the Freundlich isotherm. The Langmuir model is the most fitted model for the sorption reaction. NHCAO was highly efficient in As(III) removal out of the water in the extensive range of pH and could be used for arsenic removal from contaminated water.

1. Introduction

Arsenic contamination in ground water has been denounced in Bangladesh, India, China, Taiwan, Vietnam, USA, Argentina, Chile, and Mexico [1]. In many of these countries, the concentration has surpassed the allowed value of 10 ppb. The Bengal basin is considered to have the highest contamination of arsenic makes it the most affected geological place in the world [2]. Arsenic contamination of ground water and ingestion of contaminated water has long term consequences like social disorders, health hazards and socioeconomic dissolution. Long-term arsenic exposure can lead to cancer and many non-cancerous diseases [1]. In 1993, the World Health Organization (WHO) has amended the directive for the quality of drinking water in which the permissible limit of arsenic was lowered from 50 to 10 ppb so that to minimize the hazardous impact of arsenic on health [3].

Various technologies have been used for arsenic removal, and the effectivity of these techniques is dependent on the physical and chemical characteristics of the arsenic content in water. Commonly used techniques are (i) oxidation followed by precipitation, (ii) coagulation/electro-coagulation/co-precipitation, (iii) adsorption onto sorptive media, (iv) ion-exchange, and (v) membrane technique, (vi) foam floatation, (vii) solvent extraction, and (viii) bioremediation.

Nanomaterials are quickly came forth as strong candidates for water treatment in the countries with conventional technologies. Two interesting properties make nanoparticles highly usable as sorbents [4]. The nanoparticles have much larger surface areas on a mass basic in contrast to macro particles. They also can be extended using certain reactor groups to improve their chemical affinity against the target compounds. The researchers are exploiting these properties to promote exceptionally selective and effective sorbents to remove the organic and inorganic pollutants out of the contaminated water. Among the available methods, adsorption is highly useful because of its easy treatment, highly efficient, and can be applied to develop a filter for house hold purpose. The mixed oxides nano-agglomerates, namely iron–cerium, iron–manganese, iron–zirconium, iron–titanium, iron–chromium, cerium-manganese etc. were prepared, characterized and effectively used to remove the pollutant (i.e. arsenic, fluoride etc.) from aqueous solutions [5-8].

An attempt has been made to synthesize a mixed oxide nanoparticles for observing its capacity in arsenic adsorption. As many previous works have revealed immense potential in nanoparticles in pollutant adsorption, effort was given to develop a material which will be cheap, and at the same time will be very much efficient in arsenic adsorption, so that the material can be easily used for contaminated water purification for household purpose and in large scales as well. This manuscript reports herein the synthesis and characterization of Nanostructured Hydrous Cerium Aluminium Oxide (NHCAO) with arsenic (III) sorption properties from the aqueous solution.

2. Materials and methods

2.1. Synthesis of Ce(IV)-Al(III) binary oxide

Ammonium ceric nitrate and aluminum chloride solutions were mixed in mole ratio of 1:1. The mixed solution was heated at 80°C and 0.1M Na_2CO_3 solution was added to the mixture drop wise under vigorous stirring condition. With the addition of Na_2CO_3, pH of the solution mixture gradually increased. The final pH of the solution was maintained at 9.0. A yellowish precipitate was obtained which was then filtered, washed thoroughly with distilled water. The solid mass was then dried at 90°C. The Nanostructured Hydrous Cerium Aluminium Oxide (NHCAO) thus resulted was then meshed and applied for arsenic sorption study.

2.2. Arsenic solutions

Stock standard solution of As(III) (1000 mgL^{-1}) was made by dissolving 0.1320 gm. As_2O_3 (99.9% Aldrich, USA) in 10 mL 4% (w/v) NaOH, adjusted to acid condition with 2.0 mL concentrated HC1, and diluted with arsenic free de-ionized water to 100 mL. The working solutions of required As(III) concentrations were prepared by diluting the stock with 0.2% (v/v) HCl. The stock solution was freshly made every other three days and frozen to avoid oxidation.

2.3. Analytical methods

The arsenic in samples was determined by hydride generation atomic absorption spectrophotometer (Perkin Elmer-3100) and UV-VIS spectrophotometer (Systronics UV-VIS spectrophotometer 117) using the methods as described by the standard methods [9].

The X-ray diffraction (XRD) patterns of NHCAO were determined by power diffractometer (Philips Analytical PW-1710) utilizing Cu Kα radiation source. The voltage and applied potential current were 40 kV and 30 mA, respectively. The Atomic Force Microscopic (AFM) images for surface topography of the samples were carried out in non-contact mode using multimode scanning probe microscope (Agilent AFM 5500 series, USA) equipped with multipurpose small scanner using low coherence laser (1 mW power, 670 nm wavelength, coherence length (< 50 μm), scan range: XY: 0–10 μm; Z: 0–2 μm, noise level: XY < 0.1 nm RMS, Z < 0.02 nm RMS). The mixed oxide's Fourier Transform Infrared (FTIR) spectra and the associated pure oxides were individually noted by a Perkin Elmer (U.S.) system 2000 spectrophotometer in 2 cm^{-1} resolution. Scanning electron microscopic (SEM) images with EDS (Tescan Vega, U.K.; model LSU+) spectra were recorded spraying the sample on a carbon tape. Transmission electron micrograph (TEM) image was recorded on a H800 transmission electron micrograph (Hitachi, Japan) operating at 200 kV.

2.4. Batch experiments

Batch method was applied for the As(III) sorption kinetics on NHCAO at 303 (±1.6)K and at pH 7.0 (± 0.2). In this experiment, 500.0 mL of As(III) solution [concentrations (mgL^{-1}): 5.0 and 10.0] taking into one liter glass vessel, was positioned in a thermostat bath to achieve the desired temperature. 1 g NHCAO was added to that thermostatic solution and the reaction mixture was stirred (280 ± 5 rpm) using a speed adjustable agitator. A pH-meter electrode was fitted for frequent on line pH checking of the reaction mixture. Any change of pH noted was adjusted by the addition the small amount of 0.1 (M) NaOH or 0.1(M) HCl, as necessary. A certain volume of the reaction mixture, 2.0 mL at initial successive three steps and 5.0 mL in the later stages, was sampled time to time. The sample solutions were filtered by 0.45 μm-membrane filter and, the filtrates were tested for arsenic content.

The isotherm experiments were performed individually at temperatures 288K, 303K, 318K at pH 7.0 (± 0.2) by batch sorption method. Here, the As(III)-solutions of concentration (mgL^{-1}) ranged from 5.0 to 350.0 were used. The sorbent dose used was 2.0 g per L of solute solution. The agitation (speed: 280 ± 5 rpm) time applied was 2.5 h. The pH solution was adjusted twice throughout the experimental process (at 1.0 h and 2 h of agitation from zero time) with the addition of 0.1 M HCl and/or 0.1 M NaOH as necessary. Residual arsenic was analyzed in the filtered sample [9].

3. Results and discussion

3.1. XRD analysis
Figure 1 shows the NHCAO's XRD pattern. Presence of sharp peaks indicates the crystalline nature of the binary oxide. Peaks at 2θ values 28.6, 47.6, 56.4 and 69.5 correspond to the cubic antifluorite ceria phase. However peak characteristics of alumina phases are absent in the XRD pattern, indicating complete incorporation of Al^{3+} ions within the unit cells of ceria.

3.2. AFM analysis
Figure 2 shows the AFM image (surface topography) of NHCAO. The image shows small differences in contrast indicating presence of smooth surface. The difference between peak height and average baseline is 8.5 nm indicates the height of the particle.

3.3. SEM analysis
Figure 3a and 3b show SEM images of NHCAO in two different magnifications. From both the images it is clear that there is no particular surface morphology present in the material. The EDAX spectra indicate presence of both Ce and Al on the material surface (figure 4).

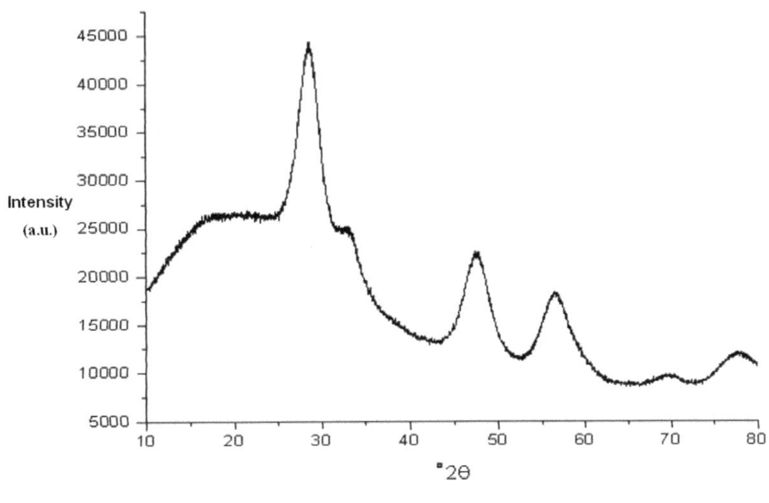

Figure 1. XRD pattern of NHCAO.

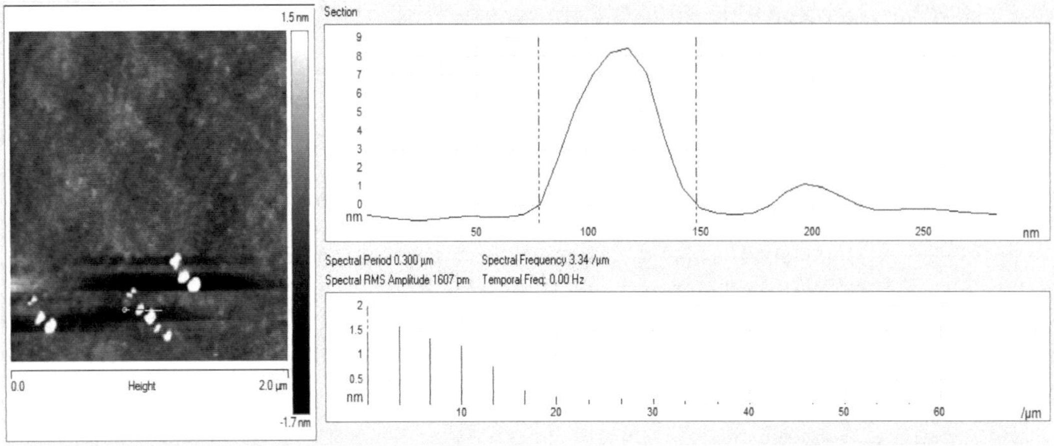

Figure 2. AFM analysis of NHCAO.

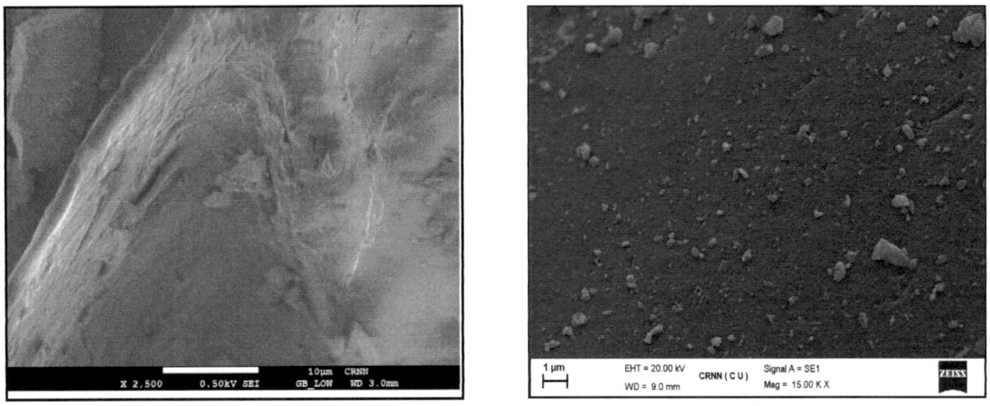

(a) (b)

Figure 3. SEM analysis of NHCAO.

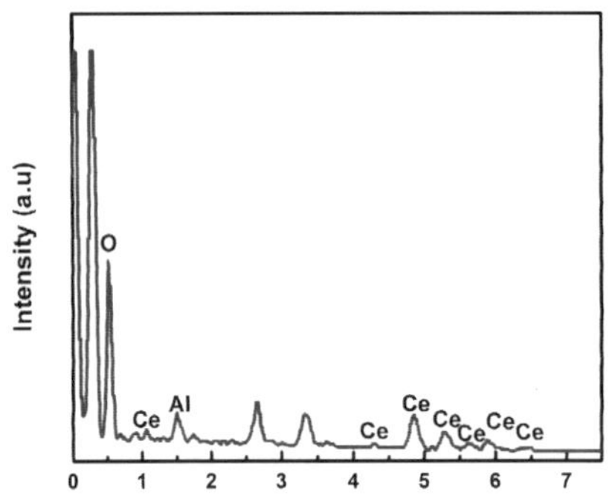

Figure 4. The EDAX spectra of cerium-aluminium binary oxide.

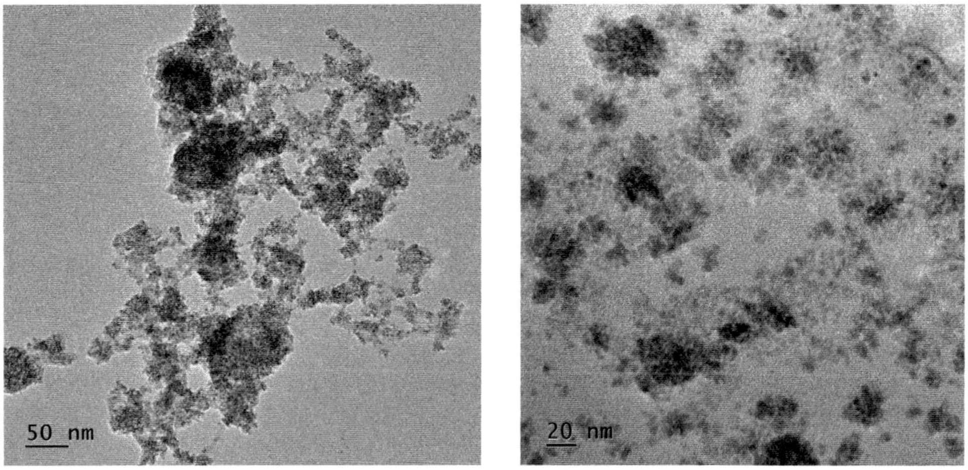

(a) (b)

Figure 5. TEM images of NHCAO

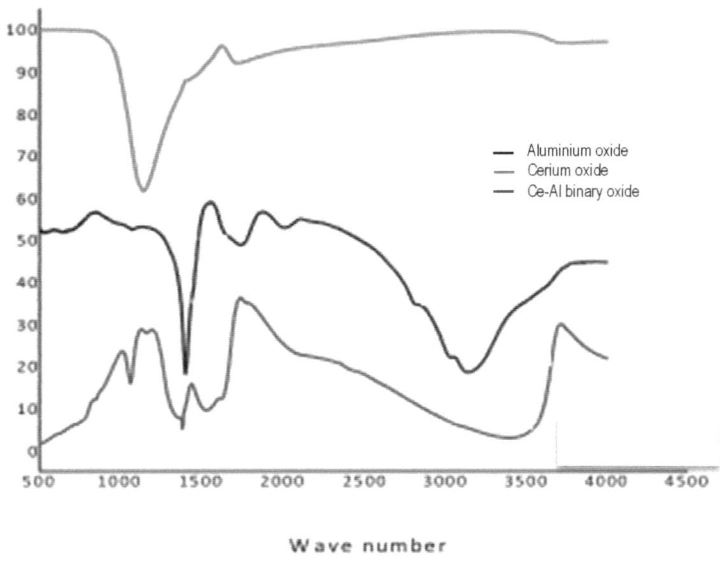

Figure 6. FT-IR analysis of NHCAO.

3.4. TEM analysis

Figure 5a and 5b show TEM images of NHCAO in two different magnifications. The black spots indicate presence of agglomerated particles. The particle size ranged between 8-10nm, which indicates that the synthesized mixed oxide is nanostructured in nature.

3.5. FT-IR analysis

Figure 6 shows FTIR spectra of all the three samples indicated. The FTIR spectra of cerium oxide contain a sharp peak at 1145 cm^{-1}, indicating the stretching vibration of Ce-O bond. The FTIR spectra of pure aluminum oxide contain sharp peaks at ~ 1415 cm^{-1} and 3150 cm^{-1} which represent the –OH stretching and bending mode of vibration of surface hydroxyl groups. The peak 660 cm^{-1} might be associated to the stretching vibration of Al-O bond. The FTIR spectrum of Ce-Al binary oxide shows peaks at 1390 and 3450 cm^{-1} which are due to the availability of surface –OH groups. The peak at 1065

cm^{-1} indicates presence of Ce-O bonds. The value of this wave number and also the peak intensity is somewhat lower compared to the wave number of pure cerium oxide (1145 cm^{-1}) indicating weakening of Ce-O bond due to the formation of Ce-O-Al bond.

3.6. Effect of pH on As(III) sorption

Figure 7 shows the variation of As (III) sorption percentage at initial pH solution. It was found from the plot that the sorption percentage stayed nearly unchanged up to pH = 9.0. Then, a slight reduction in sorption percentage was noted. The As(III) species contained in the solution as $As(OH)_3$ up to pH = 9.0, because the pK_a^1 value of $As(OH)_3$ is 9.2. As a result of the fact a small change in the sorption capacity was observed upto pH = 9.0 although the surface charge of the oxide gradually varied with the variation of the solution pH. Above pH 9.0 $As(OH)_3$ dissociated to produce $As(OH)_2O^-$ and the surface of the oxide also became predominantly negatively charged. As a result columbic repulsion was produce between the solute and sorbent which decreased down the sorption perntage.

3.7. Kinetics of arsenic (III) sorption

Figure 8 shows the variation of As(III) sorption by the increase contact time at T= 303 K and pH_i=7.0. The plot showed that a plateau was reached after the contact time of ~ 50 min., which means that the equilibrium was obtained in the sorption reaction within 50 min. of contact time. The obtained were analyzed by two well known kinetic equations the pseudo first order (1) and the pseudo second order model equations (2) as follows

$$q_t = q_e [1- \exp(-k_1 t)] \quad (1)$$
$$q_t = k_2 q_e^2 t/(1+k_2 q_e t) \quad (2)$$

where qe and qt are the amount (mg . g^{-1}) of arsenic(III) sorption at equilibrium and at time, t (min.), respectively; and k_1 ($min.^{-1}$) and k_2 (g. $mg^{-1}.min.^{-1}$) are the rate constants for the respective equations (2, 3). The parametres obtained are summarized in the table 1. Comparison of either χ^2 or R^2 values indicates that the sorption reaction followed the pseudo second order kinetics.

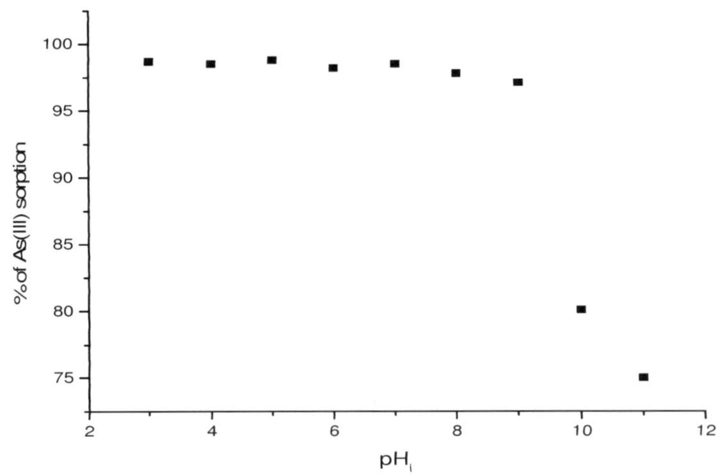

Figure 7. The plots of sorption percentage of arsenic (III) by NHCAO against pH_i at T = 303K.

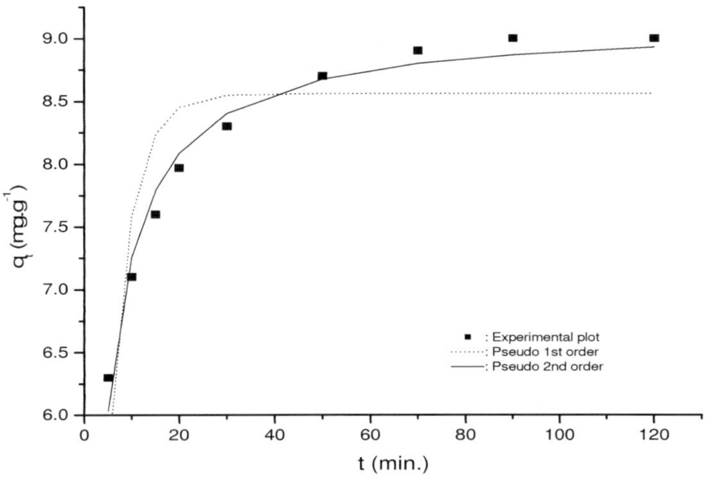

Figure 8. The plots of time dependent sorption capacity (q_t, mg.g^{-1}) of As(III) by NHCAO at T = (303 ± 1.0) K , pH_i = (7.0 ± 0.2)

Table 1. The parameters of kinetic model evaluated for the sorption of arsenic(III) on NHCAO at pH_i = (7.0 ± 0.1) and T = 303 (± 1.0K) for Arsenic (III) concentration 10.0 mgL^{-1}

Pseudo-first order parameters				Pseudo-second order parameters			
k_1	q_e	R^2	χ^2	k_2	q_e	R^2	χ^2
0.22 (±0.03)	8.56 (± 0.21)	0.74	0.26	0.04 (± 0.01)	9.12 (± 0.09)	0.97	0.03

3.8. Isotherm analysis of As(III) sorption

Figure 9 shows the equilibrium capacity (q_e, mg.g^{-1}) of As(III) sorption on NHCAO at 303K and pH=7.0 against equilibrium arsenic(III) concentrations, (C_e, mgL^{-1}). The equilibrium data were analyzed by the models of Langmuir (eq. 3) and the Freundlich (eq.4) isotherm [5], normally applied to express the equilibrium sorption result.

$$\text{Langmuir equation: } q_e = (q_m K_a C_e)/\ (1+K_a C_e) \quad (3)$$

$$\text{Freundlich equation: } q_e = K_F\, C_e^{(1/n)} \quad (4)$$

where q_m and K_a are the Langmuir constants associated to monolayer sorption capacity (mg.g^{-1}) and sorption equilibrium constant (L.g^{-1}), respectively. K_F and n are the Freundlich constants attributed to the sorption capacity (mg.g^{-1}) and sorption intensity, respectively.

The isotherm parameters obtained after non linear fit of these model equations to the experimental data were summarized in table 2. From the χ^2 or R^2 values it is discovered that the model of Langmuir is the most fitted model for this sorption reaction. The maximum monolayer sorption capacity value shows that NHCAO is very much effective in As(III) removal.

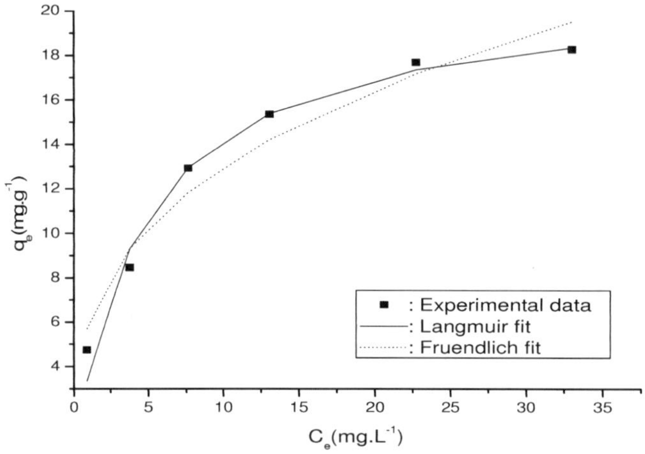

Figure 9. The plot of q_e (mg.g^{-1}) versus C_e (mg.L^{-1}) on As(III) sorption by NHCAO at T = 303 (± 1.0) K , pH_i = 7.0 (± 0.2).

Table 2. The isotherm parameters estimated using the non-linear analysis of equilibrium arsenic (III) sorption result on NHCAO at pH_i = (7.0 ± 0.1) and T = 303 ± 1.0K).

Langmuir				Freundlich			
q_m	K_a	R^2	χ^2	n	K_F	R^2	χ^2
20.99 ± 1.05	0.21 ± 0.04	0.98	0.70	2.93 ± 0.38	5.91 (± 0.75)	0.96	1.48

4. Conclusions

The Ce-Al binary oxide (NHCAO) synthesized by a simple precipitation method is highly efficient in As(III) removal out of the contaminated water in the wide range of pH. Kinetic study revealed that maximum adsorption happened in the first 50 minutes of contact time. The monolayer sorption capacity of NHCAO is rather proportional toward the other As(III) adsorbents reported earlier [5, 6, 8, 10-13]. Cerium aluminum mixed oxide particle prepared in this study can be regarded as the effective adsorbent for arsenic. Previous study reported the effectiveness of Aluminium-Cerium mixed oxide nanoparticles in deflourination of water [14]. Our study further shows the utility of NHCAO in inorganic remediation of arsenic from drinking water.

References

[1] Mondal B K and Suzuki K T 2002 *Talanta* **58** 201-35
[2] Bhattacharya S, Gupta K, Debnath S, Ghosh UC, Chattopadhyay D and Mukhopadhyay A 2012 *Toxicol. Environ. Chem.* **94** 429-41
[3] WHO 2001 *United Nations Synthesis Report on Arsenic Drinking Water* (World Health Organization)
[4] Bhattacharya S, Saha I, Mukhopadhyay A, Chattopadhyay D, Ghosh U C and Chatterjee D 2013 *Int. J. Chem. Sci. Tech.* **3** 59-64
[5] Gupta K, Bhattacharya S, Chattopadhyay D J, Mukhopadhyay A, Biswas H, Dutta J, Ray N R and Ghosh U C 2011 *Chem. Eng. J.* **172** 219-29
[6] Gupta K, Bhattacharya S, Nandi D, Dhar A, Maity A, Mukhopadhyay A, Chattopadhyay D J, Ray N R, Sen P and Ghosh U C 2012 *J. Colloid. Interf. Sci.* **377** 269-76
[7] Zhang Y, Yang M and Don X M 2005 *Environ. Sci. Technol.* **39** 7246-53
[8] Gupta K, Saha S and Ghosh U C 2008 *J. Nanopart. Res.* **10** 1361-68
[9] APHA 1985 *Standard Methods for the Examination of Water and Wastewater* (Washington, D.C.: American Public Health Association)

[10] Mohan D and Pittman Jr C U 2007 *J. Hazard. Mater.* **147** 1-53
[11] Gupta K, Biswas K and Ghosh U C 2008 *Ind. Eng. Chem. Res.* **47** 9903-12
[12] Gupta K, Maity A and Ghosh U C 2010 *J. Hazard. Mater.* **184** 832-42
[13] Deschamps E, Ciminelli V S T and Holl W H 2005 *Water Res.* **39** 5212-20
[14] Liu H, Deng S, Li Z, Yu G and Huang J 2010 *J. Hazard. Mater.* **179** 424-30

Biosynthesis and characterization of ZnO nanoparticles using the aqueous leaf extract of *Imperata cylindrica L.*

I S Saputra and Y Yulizar

Department of Chemistry, Faculty of Mathematics and Natural Sciences Universitas Indonesia, Kampus UI Depok, Depok 16424, Indonesia

Corresponding author's e-mail: yokiy@ui.ac.id

Abstract. ZnO nanoparticles (ZnO NPs) were biosynthesized.The growth was observed by a sol-gel method. ZnO were successfully formed through the reaction of zinc nitrate tetrahydrate $Zn(NO_3)_2.4H_2O$ precursor with aqueous leaf extract of *Imperata cylindrica L* (ICL). The structural and optical properties of ZnO were investigated. The as-synthesized products were characterized by UV-Visible (UV-Vis), UV diffuse reflectance spectroscopy (UV-DRS), Fourier transform infrared spectroscopy (FTIR), X-ray diffraction (XRD), scanning electron microscopy (SEM), and energy-dispersive X-ray spectroscopy (EDS). UV-Vis absorption data showed hydrolysis and characteristic of absorption peak at 300 nm of $Zn(OH)_2$. UV-DRS confirmed that ZnO NPs has the indirect band gap at 3.13 eV. FTIR spectrum revealed the functional groups and indicated the presence of protein as the capping and stabilizing agent on the ZnO surface. Powder XRD studies indicated the formation of pure wurtzite hexagonal structure with particle size of 11.9 nm. The detailed morphological and structural characterizations revealed that the synthesized products were hexagonal nanochip.

Keywords: ZnO nanoparticles, *Imperata cylindrica L.*, biosynthesis, sol-gel method and stabilizing agent.

1. Introduction

In recent years, metal nanoparticles are important concern for researchers [1]. Zinc oxide is an inorganic compound appears as white powder, and insoluble in water. The powder is widely used as an additive for numerous materials and products, including plastics, ceramics, glass, cement, batteries, etc. ZnO is found in the earth crust as zincite mineral. In materials science, ZnO is called II-VI semiconductor [2]. ZnO is a semiconductor and piezoelectric material with a direct wide band gap of 3.37 eV and a large exciton binding energy of 60 meV at room temperature [3].

There are several methods reported for the synthesis of ZnO nanoparticles (ZnO NPs), which include chemical vapor deposition, gas-phase method, spray pyrolysis, hydrothermal synthesis, micro emulsion, electrochemical method, pulsed laser deposition, microwave synthesis, and the sol-gel method [4]. The use of environmentally benign materials such as plant extract, fungi, and bacteria for generating metal or metal oxide NPs is considered as an eco-friendly approach [5-8]. ZnO is employed for a variety of important potential applications, such as antimicrobial activity [9], gas sensors [10], photocatalytic [11], and degradation of congo red organic dye [12].

The green synthesis of ZnO NPs using *Plectranthus amboinicus* leaf extract showed the morphology of rod shape with an average size of 88 nm [13]. The synthesis of ZnO NPs using *Aspalathus linearis* natural extract, as eco-friendly material, acted as reducing/oxidizing agent [14]. Characterization by

TEM image and XRD showed the ZnO NPs have an average diameter ranges from 1.0-8.5 nm. The crystal structures of ZnO exhibited a wurtzite, zinc blende, and rocksalt. *Agathosma betulina* natural extract was used in ZnO NPs formation as an effective chelating agent [15]. Based on the background in the present study, ZnO NPs formation is conducted using *Imperata cylindrica L.* leaf extract as a base source and stabilizing agent by sol-gel method.

2. Materials and methods

2.1. Preparation of the leaf extract
Imperata cylindrica L. (ICL) plant leaves were taken from Universitas Indonesia area. 5 g of dry powder ICL leaf was stirred with 100 mL of double distilled water for 1 h at 70°C until the colour of the colloid changes to light yellow. The extract was filtered with Whatman No. 1 paper filter.

2.2. Biosynthesis of ZnO NPs
25 ml of the extract was added with 3 g of $Zn(NO_3)_2.4H_2O$ under continuous stirring at 85°C for 24 h. This paste was heated in a furnace at 400°C for 4 h.

2.3. Characterization of ZnO NPs
The UV-Vis absorption spectrum was recorded using (Shimadzu UV-Vis 2600) spectrophotometer. UV-Vis diffuse reflectance spectra were observed by (UV-Vis DRS, Shimadzu 2450) in the wavelength range of 200-800 nm. The functional group of compound was examined using Fourier transform infrared spectroscopy (FTIR) (Shimadzu, Prestige 21). The crystal phase information was characterized from 20°-70° by XRD (Shimadzu 610) with Cu Kα (λ = 0.1546 nm) radiation. The surface morphology of ZnO NPs was examined by scanning electron microscope (SEM), and the chemical composition of the product was examined by energy dispersive X-ray spectroscopy (EDX, AMETEK).

3. Results and discussion

3.1. UV-Vis absorption and DRS spectroscopy analysis
Figure 1 shows the UV-Vis absorption spectra of ICL extract at maximum wavelength, λ_{max} 273 nm. The spectrum reveals a characteristic absorption peak of $Zn(OH)_2$ at λ_{max} of 300 nm wavelength. Hydrolysis of $Zn(NO_3)_2.4H_2O$ into $Zn(OH)_2$ produces a strong red shift. The DRS spectrum of ZnO NPs in figure 1b showed at the λ_{max} of 359 nm. The reflectance spectrum was analyzed using the Kubelka-Munk equations (equation 1):

$$F(R) = \frac{(1-R)^2}{2R} \qquad (1)$$

where R is the reflectance value, F (R) is the equivalent to the absorption coefficient. The band gap energy (Eg) of ZnO NPs was estimated from the variation of Kubelka-Munk function with photon energy [16]. Figure 2 shows the Kubelka-Munk plots of ZnO NPs that used to determine association of Eg with its indirect transitions. The ZnO NPs exhibits the indirect Eg of 3.13 eV.

3.2. FT-IR spectroscopy
The FT-IR spectrum of ZnO NPs (figure 3) shows the wavenumbers at 3375, 2349, 1375, 915 and 481 cm^{-1}, respectively. The broad peak at 3375 cm^{-1} indicated the -OH stretching vibrations. The peak at 2349 cm^{-1} indicated the free –C=O group. Stretching vibrations present at 1375 cm^{-1} showed C-O-H bending mode. The peak at 915 cm^{-1} showed the characteristic -NH of amine. The band at 481 cm^{-1} confirmed the stretching vibrations of ZnO NPs. The ICL leaf extract containing polyols, terpenoids, and proteins that has functional groups of amines, alcohols, ketones, and carboxylic acid contributed in hydrolysis and acted as stabilizing agent. The good stability of ZnO NPs is due to the free amino and

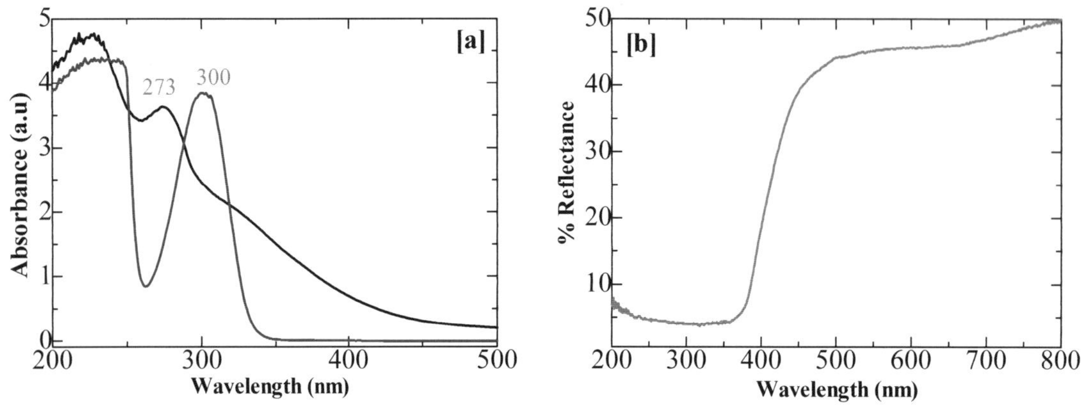

(a) (b)

Figure 1. (a) UV-Vis absorption and (b) DRS spectrum of biosynthesized ZnO NPs.

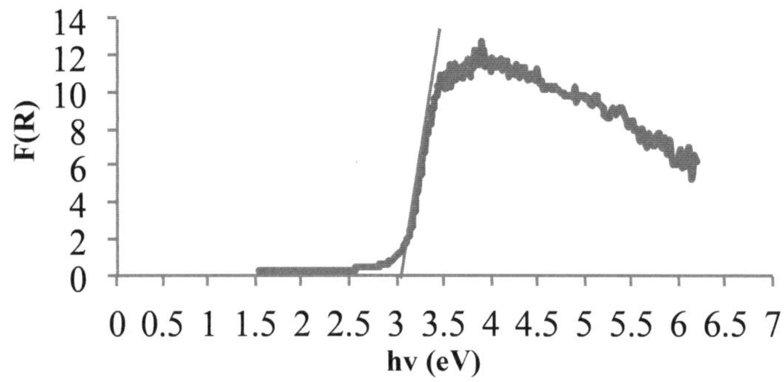

Figure 2. Plot of indirect band gap energy of ZnO NPs.

carboxylic groups interacted with the ZnO NPs surface. The bonds of functional groups such as –C-O-C-, C-O- and -C=C- were derived from heterocyclic compounds, and the amide bonds derived from the proteins in ICL leaf extract, and acted as capping agent of ligands of the nanoparticles[17].

3.3. XRD analysis

The XRD patterns of the calcined ZnO NPs 800°C are shown in figure 3b. XRD pattern confirmed the crystal phase and crystallinity of ZnO NPs. The XRD peaks are adjusted with the JCPDS data card 89-1397 of ZnO hexagonal [18]. The detected peaks corresponded to the hexagonal phase ZnO NPs are found in the lattice planes (h,k,l) of (1 0 0), (0 0 2), (1 0 1), (1 0 2), (1 1 0), (1 0 3), (1 1 2), and (2 0 1) in the 2θ value : 31.71°; 34.37°; 36.18°; 47.49°; 56.57°; 62.85°; 68.01° and 68.98° respectively. The average grain size of sample was calculated using the Scherer's (equation 2).

$$D = \frac{K\lambda}{\beta cos\theta} \text{Å} \quad (2)$$

where D is the average crystallite size in Å, K is the shape factor (0.9), K is the wavelength of X-ray (1.5406 Å) Cu Kα radiation, θ is the Bragg angle, and β is the corrected line broadening of NPs [16]. The average crystallite size of ZnO NPs is about 11.9 nm.

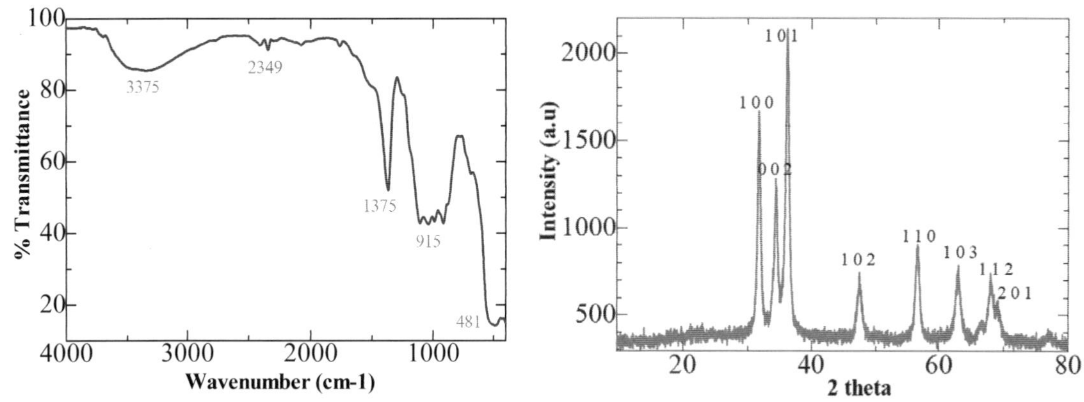

(a) (b)

Figure 3. (a) FT-IR spectra and (b) powder X-ray diffraction of biosynthesized ZnO NPs.

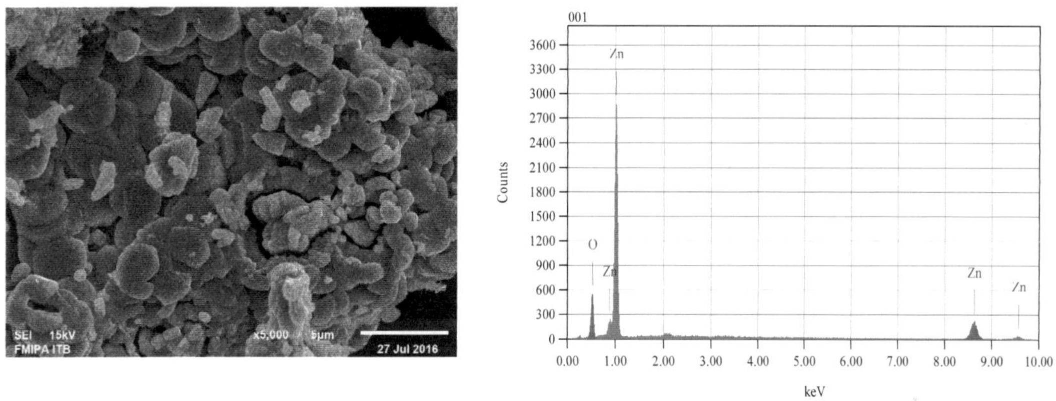

(a) (b)

Figure 4. Biosynthesized ZnO NPs (a) SEM image (b) EDS spectrum.

3.4. SEM-EDS analysis

The morphology of biosynthesized ZnO NPs was observed by SEM, as shown in figure 4a. The diameter of cluster ZnO NPs was 2 μm and in hexagonal nanochip shape with rough surface. The EDS analysis confirmed the chemical composition of ZnO NPs as shown in the figure 4b. The EDS spectrum shows the high value of zinc (80.3%) and oxygen (19.65%), respectively.

4. Conclusions

The ZnO NPs were successfully biosynthesized by sol-gel method using *Imperata cylindrica L.* leaf extract as base source and stabilizing agent. Hydrolysis process showed the characteristic absorption peak of $Zn(OH)_2$ at λ_{max} 300 nm and UV-Vis-DRS data confirmed the indirect band gap at 3.13 eV. FT-IR spectra revealed the functional groups of stretching bands of ZnO NPs around 600-400 cm^{-1}. The XRD result showed the sample was well crystallized in a hexagonal wurtzite structure with crystallite size of 11.9 nm. SEM-EDS analysis showed the morphology of hexagonal nanochip with the compositions of Zn and O elements.

Acknowledgements

The authors would like to thank Universitas Indonesia for funding this research through PITTA Grant Universitas Indonesia with contract No: 2039/UN2.R12/HKP.05.00/2016.

References

[1] Ariyanta H A and Yulizar Y 2016 *IOP Conf. Series: Materials Science and Engineering* **107** 012002-8
[2] Chen Y, Bagnall D M, Koh H, Park K, Hiraga K, Zhu Z and Yao T 1998 *J. Appl. Phys.* **84** 3912-8
[3] Ozgur U, Alivov Y I, Liu C, Teke A, Reschikov M A, Dogan S, Avrutin V, Cho S J and Morkoc H 2005 *J. Appl. Phys.* **98** 041301-103
[4] Darroudi M, Sabouri Z, Oskuee R K, Zake A K, Kargarc H and Hamid M H N A 2013 *Ceram. Int* **39** 9195-9
[5] Rathi Sre P R, Reka M, Poovazhagi R, Arul Kumar M and Murugesan K 2015 *Spectrochim. Acta A Mol Biomol Spectrosc.* **135** 1137-44
[6] Singh B N, Rawat A K S, Khan W, Naqvi A H and Singh B R 2014 *PloS ONE* **9** **e**106937-12
[7] Yuvakkumar R, Suresh J, Nathanael A J, Sundrarajan M and Hong S I 2014 *Mat. Sci. Eng. C* **41** 17-27
[8] Raliya R *et al.* 2014 *Int. J. Biol. Macromol* **65** 362-8
[9] Elumalai K, Velmurungan S, Ravi S, Kathiravan V, Ashokkumar S 2015 *Molecular and Biomolecular Spectroscopy* **143** 158-64
[10] Zhong Y Z, Wang M H and Liu T T 2015 *Materials Letters* **158** 274-7
[11] Udawatte N, Lee M, Kim J and Lee D 2011 *ACS Appl. Mater. Interfaces* **3** 4531-8
[12] Habibi M H and Rahmati M H 2015 *Spectrochim.Acta A* **137** 160-4
[13] Fu L and Fu Z 2015 *Ceramics International* **41** 2492-6
[14] Diallo A, Ngom B D, Park E and Maaza M 2015 *Journal of Alloys and Compounds* **646** 425-30
[15] Thema F T, Manikandan E, Dhlamini M S and Maaza M 2015 *Materials Latters* **161** 124-7
[16] Ramesh M, Anbuvannan M and Viruthagiri G 2015 *Molecular and Biomolecular Spectroscopy* **136** 864-70
[17] Peletiri C, Matur B M, Ihongbe J C and Okoye M 2012 *J. Med. Sci.* **2(1)** 013-7
[18] John R and Rajakumari R 2012 *Nano-Micro Lett.* **4(2)** 65-72

IOP Conf. Series: Materials Science and Engineering **188** (2017) 012005 doi:10.1088/1757-899X/188/1/012005

Adsorption and sonocatalytic performance of magnetite ZnO/CuO with NGP variation

A Taufik[1,2] **and R Saleh**[1,2]

[1]Department of Physics, Faculty of Mathematics and Natural Sciences Universitas Indonesia, Kampus UI Depok, Depok 16424, Indonesia
[2]Integrated Laboratory of Energy and Environment, Faculty of Mathematics and Natural Sciences Universitas Indonesia, Kampus UI Depok, Depok 16424, Indonesia

Corresponding author's e-mail: rosari.saleh@gmail.com

Abstract. A series of Fe_3O_4/ZnO/CuO/nanographene platelets (Fe_3O_4/ZnO/CuO/NGP) nanocomposites with various NGP weight percents were studied as catalysts for methylene blue removal under adsorption followed by sonocatalytic process. Weight percents (wt.%) of NGP in the nanocomposites were varied (5, 10, and 15 wt.%). The physicochemical properties of the samples were characterized using X-Ray diffraction (XRD), ultraviolet-visible (UV-VIS) spectroscopy, *Brunauer–Emmett–Teller (*BET) surface area analysis, and a vibrating sample magnetometer (VSM). The heterogeneous structure of all samples consisted of the cubic spinel structure of Fe_3O_4, hexagonal wurtzite structure of ZnO, monoclinic structure of CuO, and graphite-like structure of NGP. With increasing NGP weight percent, sample surface area increased from 14 m^2/g to 23 m^2/g. Adsorption and sonocatalytic activity were examined on degradation of methylene blue in alkaline conditions. The results show that the adsorption ability of samples increased with increasing NGP weight percent. However, in the sonocatalytic process, Fe_3O_4/ZnO/CuO/NGP with 10 wt.% NGP exhibited the maximum degradation. The effect of addition different radical scavenger was also examined to understand the sonocatalytic mechanism.

1. Introduction

Organic dyes used in industrial processes contaminate the environment and its surroundings [1-2]. For that reason, an effective method is needed to manage dye waste water before it is disposed of in the environment. Because it is cheap and easy, adsorption is the most commonly used method of organic dye removal [3-4].

The magnetic material is one of the potential adsorbent in adsorption process for organic dye removal. It has not only the good adsorption capacity but could make the separation process of material from organic dye solution became easier [5]. Among other magnetic material, Fe_3O_4 is the promising material because it has high adsorption capacity and unique magnetic properties. [6].

Aside from the adsorption process, advanced oxidation processes (AOPs) have been used for the removal of organic dye, as their observed degradation ability is better than adsorption process and could mineralize hazardous substances into simple molecules [7]. The AOPs method recently used is photocatalytic. However photocatalytic process has limitation when applied to degrade the high concentrate organic dye due to limited light penetration [8]. It is thought that ultrasonic waves may resolve that issue because of their high penetration of any liquid medium [8-9]. Applying ultrasonic waves in water causes acoustic cavitation, consisting of processes of nucleation, rapid growth, and Applying ultrasonic waves in water causes acoustic cavitation, consisting of processes of nucleation, rapid growth, and implosion, involving bubble collapse. This collapse produces high-temperature (5000 K) and high-pressure (1000 atm) "hot spots," followed by light emission [10-11], commonly known as sonoluminescence that could interacts with the catalyst to form electron-hole pairs that degrade the organic dye [12].

The semiconductor zinc oxide (ZnO) is commonly used in sonocatalytic process because it is chemically stable, cheap, and eco-friendly [13]. However, sonocatalytic activity of single ZnO is limited by the high rate of recombination of electrons and holes [14]. Combining ZnO with CuO and Fe_3O_4 is believed to hamper electron-hole combination, therefore increasing the efficiency of organic dye removal [15-16]. Because of its large surface area and good electron transport, the addition of graphene is also believed to increase adsorption capacity and sonocatalytic activity [17]. In our previous works, we have investigated the sonocatalytic activity using NGP materials on Fe_3O_4/ZnO/CuO nanocomposites [18]. However, the influence of NGP loading as well as adsorption performance of Fe_3O_4/ZnO/CuO/NGP are also need to be investigate. Therefore, in this research we try to investigate the influence of NGP loading in the adsorption ability and sonocatalytic performance of Fe_3O_4/ZnO/CuO/NGP nanocomposites.

2. Materials and methods

Nanoparticles of CuO and Fe_3O_4 and nanocomposite of Fe_3O_4/ZnO/CuO with molar ratio of Fe_3O_4:ZnO:CuO is 1:1:5 were synthesized as described in our previous study [19]. In synthesizing Fe_3O_4/ZnO/CuO/NGP, nanographene platelets (NGP) were purchased from Angstron Materials. In the typical process NGP were dispersed into a mixture of water and ethanol, using an ultrasonic bath for 2 hours. The synthesized 2 g of Fe_3O_4/ZnO/CuO was then added to the solution and stirred with a magnetic stirrer for an hour to produce the homogeneous suspension. That suspension was then heated at 120°C for 3 hours, and the precipitate obtained by centrifugation was then dried at 70°C for 12 hours. For present purposes, the amount of NGP was varied at 5 wt.%, 10 wt.%, and 15 wt.%, respectively denoted as FZC, FZC-5 wt.% NGP, FZC-10 wt.% NGP, and FZC-15 wt.% NGP.

All the samples were characterized using X-ray diffraction (Rigaku Miniflex 600), Fourier transform infra-red spectroscopy, *Brunauer–Emmett–Teller (*BET) surface area analysis, UV-Vis Diffuse Reflectance Spectroscopy (DRS) Shimadzu, and a vibrating sample magnetometer Oxford Type 1.2 T.

Adsorption and sonocatalytic process were analyzed using methylene blue as a model of the organic pollutant. For present purposes, concentration of methylene blue was 20 mg/L, and catalyst dosage was 0.3 g/L. Typically, the methylene blue solution was poured into a 100 mL beaker; Fe_3O_4/ZnO/CuO/NGP nanocomposite was then added to the solution and stirred using a magnetic stirrer. Adsorption of methylene blue was performed in dark conditions for 4 hours. Meanwhile, the sonocatalytic process was analyzed using an ultrasonic bath operating at a frequency of 40 kHz and 150 W. At specified intervals, the methylene blue solution was taken and analyzed for color degradation by UV-Vis spectroscopy, using the following equation:

$$\text{Decolorization} = \frac{C_t}{C_0}$$

where C_t and C_o are the concentrations of methylene blue at time t and in the initial condition (mg/L). The adsorption capacity of the samples was analyzed using the equation

$$q_e = \frac{(C_0 - C_t)V}{m}$$

where q_e is the equilibrium adsorption capacity (mg/g), V is the volume of solution (mL), and m is the amount of catalyst used (g).

IOP Conf. Series: Materials Science and Engineering **188** (2017) 012005 doi:10.1088/1757-899X/188/1/012005

3. Results and discussion

Figure 1a shows the results of XRD measurement for Fe_3O_4/ZnO/CuO/NGP nanocomposite variations of 5, 10 and 15 wt.% (respectively denoted as FZC-5 wt.% NGP, FZC-10 wt.% NGP, and FZC-15 wt.% NGP). For comparison, the XRD spectra of Fe_3O_4/ZnO/CuO nanocomposite (FZC) were also plotted. These spectra show the cubic spinel crystal structure of Fe_3O_4, hexagonal wurtzite structure of ZnO, and monoclinic structure of CuO. The presence of NGP was confirmed at a value of $2\theta = 26°$, indicating the existence of a graphitic-like crystal structure. Lattice parameter values of Fe_3O_4, ZnO, CuO, and NGP were analyzed using the Rietveld refinement method. As seen in table 1, the lattice parameter values of Fe_3O_4, ZnO, CuO, and NGP did not change significantly, indicating that the formation of Fe_3O_4/ZnO/CuO/NGP nanocomposite does not alter the crystal structure of each material.

FT-IR measurement of FZC-10 wt.% NGP nanocomposite is shown in figure 1b and the results for FT-IR of NGP and FZC nanocomposite are also shown. NGP's vibration mode at wavenumbers 1050 and 1621 cm^{-1} indicates vibration mode C-C and C=O, while vibration mode of OH was detected at wavenumber 3400 cm^{-1} [20]. The FCZ nanocomposite material's vibration mode at wavenumbers 400 to 800 cm^{-1} indicates Cu-O, Fe-O and Zn-O vibration [21-23]. For the FZC-10 wt.% NGP nanocomposite, all vibration modes for FZC as well as NGP could be detected, indicating the presence of Fe_3O_4/ZnO/CuO nanocomposite and NGP in the sample.

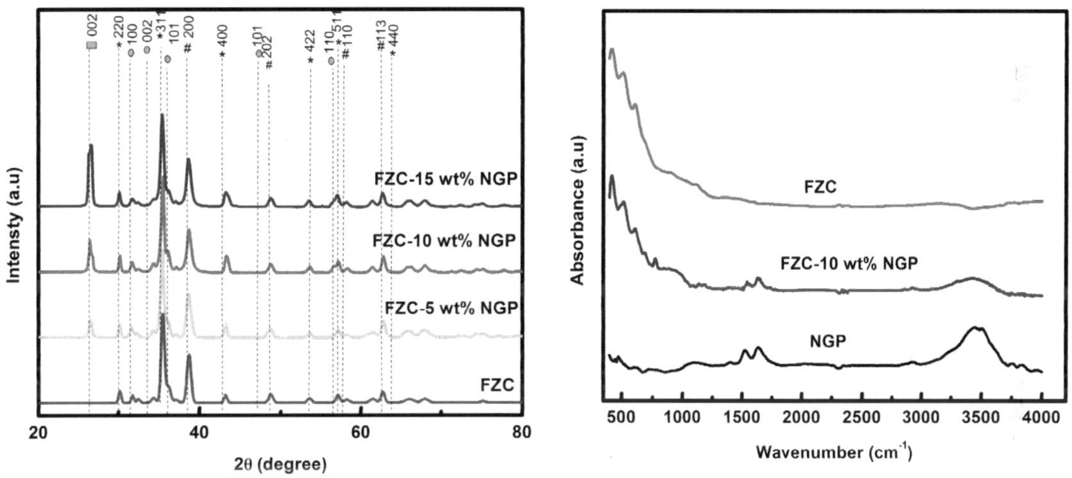

(a) (b)

Figure 1. (a) XRD and (b) FT-IR spectra of Fe_3O_4/ZnO/CuO/NGP.

Table 1. Lattice parameter and grain size of Fe_3O_4/ZnO/CuO/NGP with different NGP weight percent.

Sample	Lattice parameter (Å)						
	Fe_3O_4	ZnO		CuO			NGP
	a	a=b	c	a	b	c	a=b=c
FZC	8.373	3.249	5.217	4.689	3.436	5.138	
FZC-5 wt.% NGP	8.373	3.260	5.204	4.687	3.445	5.129	3.645
FZC-10 wt.% NGP	8.366	3.252	5.215	4.683	3.427	5.144	3.638
FZC-15 wt.% NGP	8.381	3.248	5.212	4.687	3.428	5.147	3.643

The surface areas of all samples are tabulated in table 2. The results confirm that addition of NGP increases sample surface area. The surface area of Fe_3O_4/ZnO/CuO nanocomposite modified with NGP increased from 14.42 m^2/g to 23.15 m^2/g. Surface area can influence the sample's adsorption capacity and sonocatalytic activity.

UV-Vis diffuse reflectance measurements for the samples (FZC, FZC-5 wt.% NGP, FZC-10 wt.% NGP and FZC-15 wt.% NGP) are shown in figure 2a, representing samples' reflectance ability. The results show that sample reflectance ability gradually decreased in the visible light range with incorporation of 5, 10, and 15 wt.% NGP. The value of the energy gap was analyzed using the Kubelka-Munk equation [24], extrapolating the $F(R)^2$ curve to the energy value in $F(R)^2 = 0$. The results confirm that addition of NGP reduced the energy gap value from 2.53 eV to 2.32 eV.

VSM measurements for FZC and FZC-10 wt.% NGP nanocomposites are shown in figure 2b, which also shows the VSM measurement of Fe_3O_4 nanoparticles for comparison. The results show that magnetization of FZC-10 wt.% NGP nanocomposites is lower than that of Fe_3O_4 nanoparticles but higher than the magnetic saturation value of FZC-10 nanocomposites. The high magnetization value is very potential for reusability of catalyst by using magnetic separation technique. The inset in figure 2b shows the separation of FZC-10 wt.% NGP nanocomposites material from degreded methylene blue solution using an external magnet, showing that the FZC-10 wt.% NGP nanocomposite can be easily attracted by an external magnetic.

The adsorption ability of FZC, FZC-5 wt.% NGP, FZC-10 wt.% NGP and FZC-15 wt.% NGP nanocomposite samples is shown in figure 3a. The results show that FZC-5 wt.% NGP, FZC-10 wt.% NGP and FZC-15 wt.% NGP nanocomposites have better adsorption ability than FZC nanocomposites; addition of 15 wt.% NGP shows the best decolorization results. Adsorption capacity values are presented in table 2, showing that FZC-15 wt.% NGP has the highest adsorption capacity (26.68 mg/g).

The sonocatalytic activity of all samples is shown in figure 3b. The results indicate that the addition of NGP to Fe_3O_4/ZnO/CuO nanocomposite can increase the degradation efficiency of methylene blue through sonocatalytic process, but only up to 10 wt.% NGP. Addition of 15 wt.% of NGP reduced the efficiency of methylene blue degradation through sonocatalytic process. It is probably due to the higher NGP loading reduce the active sites on the sonocatalytic process. The catalyst stability result detailed in figure 4a shows that repeated (four times) usage of Fe_3O_4/ZnO/CuO/NGP nanocomposite reduced degradation efficiency only until 92%, indicating the good stability of the Fe_3O_4/ZnO/CuO/NGP sample. On the other hand, separation using an external magnet (inset figure 2b) minimized the occurrence of catalyst weight loss.

To understand the sonocatalytic mechanism, various scavengers were added. There are three type of scavenger: (1) sodium sulfate as electron scavenger, (2) ammonium oxalate as hole scavenger, and (3) tert-butyl alcohol as hydroxyl radical scavenger. The results (figure 4b) indicate that the addition of electron, hole, and hydroxyl radical scavengers reduce sonocatalytic performance. The addition of a scavenger could trap active species involved in the sonocatalytic process and it cannot react in methylene blue degradation therefore the sonocatalytic performance decrease. The addition of hole and OH radical

Table 2. BET surface area, band gap and adsorption capacity of Fe_3O_4/ZnO/CuO/NGP with different NGP weight percent.

Sample	Surface Area (m^2/g)	Band gap (eV)	Adsorption Capacity (mg/g)
FZC	14.42	2.53	16,54
FZC-5 wt.% NGP	20.85	2.47	21.92
FZC-10 wt.% NGP	21.80	2.35	24.97
FZC-15 wt.% NGP	23.15	2.32	26.68

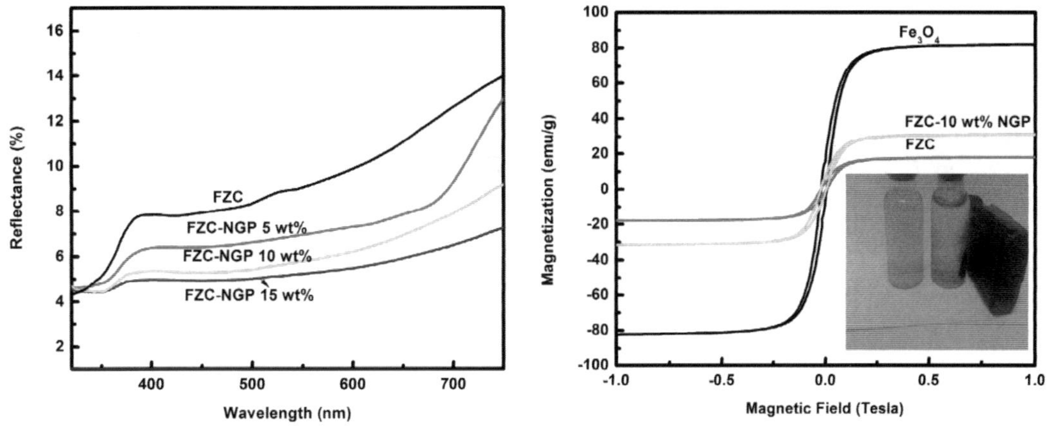

(a) (b)

Figure 2. (a) UV Vis DRS spectra and (b) magnetization of Fe_3O_4/ZnO/CuO/NGP

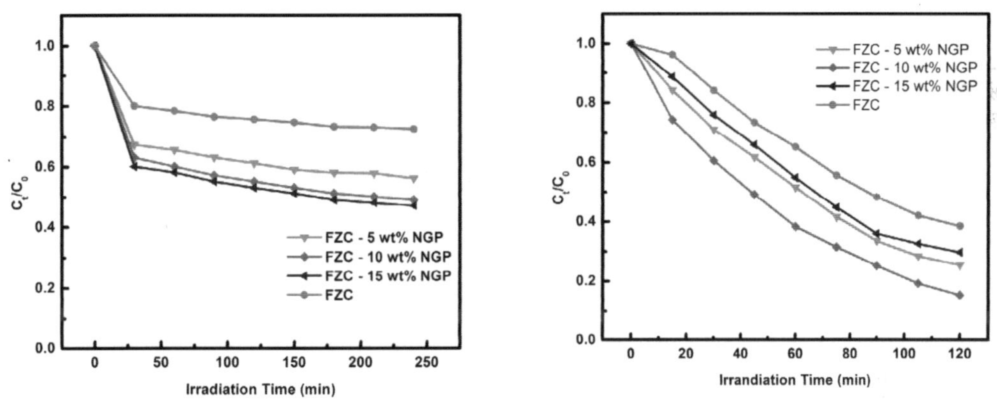

(a) (b)

Figure 3. (a) Adsorption and (b) sonocatalytic performance of Fe_3O_4/ZnO/CuO/NGP nanocomposites with different NGP weight percent.

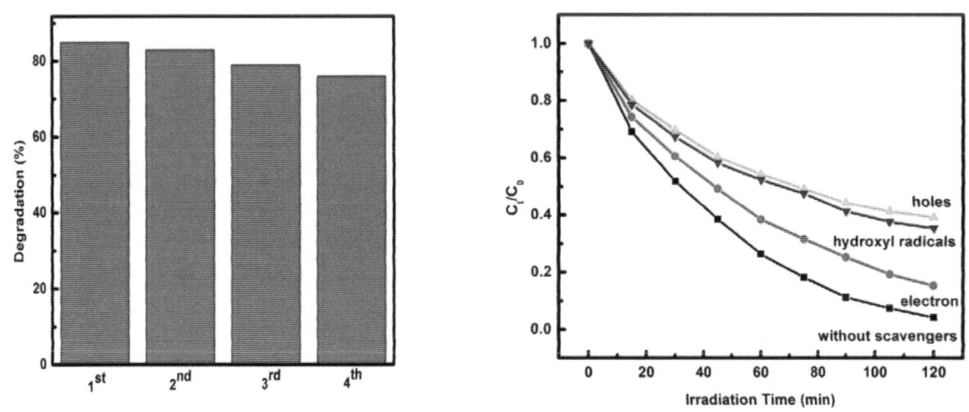

(a) (b)

Figure 4. (a) Reusability of catalyst and (b) the effect of scavenger addition on sonocatalytic performance of the samples.

scavengers exhibit the lowest sonocatalytic performance, indicating that holes and OH radicals are the main active species in the degradation process.

Under ultrasound irradiation, the Fe_3O_4/ZnO/CuO/NGP take energy from the sonoluminescence process that has wide range spectrum including visible and UV light. When Fe_3O_4/ZnO/CuO/NGP is irradiated by sonoluminescence, electron-hole pairs are created for both ZnO and CuO. The presence of Fe_3O_4 can improve both separation and degradation by acting as an electron acceptor that prevents recombination of electrons and holes. When electron was captured by Fe^{3+} holes are still has enough time to react with water molecules to produce hydroxyl radicals and attack directly the methylene blue solution therefore OH radicals and holes are the main active species on the degradation process. The addition of NGP materials could improve the charge separation process due to its could act as electron acceptor that could prevent recombination electron and holes [25].

4. Conclusions

The addition of NGP material can increase the adsorption capacity and sonocatalytic activity of Fe_3O_4/ZnO/CuO samples. Addition of NGP up to 15 wt.% shows the highest adsorption capacity; for the sonocatalytic process, addition of 10 wt.% NGP shows the best degradation capacity. After four repetitions of the sonocatalytic process the catalytic performance only reduced about 92% indicated the Fe_3O_4/ZnO/CuO/NGP nanocomposite have good stability and degradation efficiency.

References

[1] Guang L, Hui W and Xuejun Z 2016 *J. Solid State Chem.* **239** 259-64
[2] Thangavel S, Thangavel S, Raghavan N, Krishnamoorthy K and Venugopal G 2016 *J. Alloys Compd.* **665** 107-12
[3] Lin C C, Lin Y S and Ho J M 2016 *J. Alloys Compd.* **666** 153-8
[4] Crini G 2006 *Bioresour. Technol.* **97** 1061-85
[5] Chang J, Ma J, Ma Q, Zhang D, Qiao N, Hu M and Ma H 2016 *Appl. Clay Sci.* **119** 132-40
[6] Giri S K, Das N N and Pradhan G C 2011 *Colloids Surf., A* **389** 43-9
[7] Ikehata K, Naghashkar N J and Din M G E 2007 *Ozone Sci. Eng.* **28** 353-414
[8] Zhou M, Yang H, Xian T, Li R S, Zhang H M and Wang X X 2015 *J. Hazard. Mater.* **289** 149-57
[9] Zhang H, Wei C, Huang Y and Wang J 2016 *Ultrason. Sonochem.* **30** 61-9
[10] Flint E B and Suslick K S 1991 *Science.* **253** 1397-9
[11] Suslick K 1990 *Science.* **247** 1439-45
[12] Zhu L, Meng Z D, Park C Y, Gosh T and Oh W C 2013 *Ultrason. Sonochem.* **20** 478-84
[13] Bao Y, Wang C and Ma J Z 2016 *Mater. Des.* **101** 7-15
[14] Yin H H, Yu K, Song C Q, Huang R and Zhu Z Q 2014 *ACS Appl. Mater. Interfaces* **6** 14851-60
[15] Yang C, Cao X, Wang S, Zhang L, Xiao F, Su X and Wang J 2015 *Ceram. Int.* **41** 1749-56
[16] Xia J, Wang A, Liu X and Su Z 2011 *Appl. Surf. Sci.* **257** 9724-32
[17] Fei P, Wang Q, Zhong M and Su B 2016 *J. Alloys Compd.* **685** 411-7
[18] Hendry T, Taufik A and Saleh R 2015 *Proc. Int. Conf. on Advanced Materials Science and Technology (Semarang)* Vol **1725** ed Sutikno *et al.* 020027 (Melvile: New York American Institute of Physics)
[19] Taufik A, Susanto I K and Saleh R 2015 *Mater. Sci. Forum* **827** 37-42
[20] Shang Y, Li T, Li H, Dang A, Zhang L, Yin Y, Xiong C and Zhao T 2016 *Composites Part* B **99** 106-11
[21] Lamba R, Umar A, Mehta. Kumar S and Kansal S 2015 *J. Alloys Compd* **620** 67-73
[22] Sonia S, Jayram N D, Kumar P S, Mangalaraj D, Ponpandian N and Viswanathan C 2014 *Superlattices Microstruct.* **66** 1-9
[23] Lin C C and Ho J M 2014 *Ceram. Int.* **40** 10275-82
[24] Hapke B 1993 *Theory of Reflectance and Emittance Spectroscopy* (Cambridge: University Press)
[25] Zhu L, Chung J D and Oh W C 2015 *Ultrason. Sonochem* **27** 252-61

Preparation of hemoglobin-modified boron-doped diamond for acrylamide biosensors

K Umam, E Saepudin and T A Ivandini

Department of Chemistry, Faculty of Mathematics and Natural Sciences Universitas Indonesia, Kampus UI Depok, Depok 16424, Indonesia

Corresponding author's e-mail: ivandini.tri@sci.ui.ac.id

Abstract. Boron-doped diamond (BDD) electrode was modified with haemoglobin to develop electrochemical biosensors of acrylamide. Prior to modify with haemoglobin, the BDD was modified by gold nanoparticles to increase the affinity of BDD against haemoglobin. The electrochemical behaviour of the electrode in the presence of acrylamide was studied in comparison to haemoglobin-modified gold electrodes. Cyclic voltammetry indicated the optimum responses in 0.1 M sodium acetate buffer at pH 5. The responses were linear to the acrylamide concentration range of 5-50 μM with an estimated detection limit of 5.14 μM, suggesting that the electrode was promising for acrylamide biosensors.

Keywords: Acrylamide, Biosensor, Haemoglobin, Boron-Doped Diamond, Gold Nanoparticle

1. Introduction

In 2002, Swedish National Food Administration declared that heating process of foods which contains high carbohydrates such as potato chips, coffee, and bread leads to an acrylamide formation [1, 2]. The process is derived from the reaction between asparagine and reducing sugars, for example glucose and fructose, which is known as Maillard reaction [3-5]. Research over the years regarding the carcinogenic properties of acrylamide establishing the fact that acrylamide causes cancer in human [6, 7]. Therefore, International Agency For Research On Cancer [8] classifies acrylamide into a carcinogenic of Group 2A, while the European Commission [9] and the European Chemical Agency [10] classifies acrylamide as category 2 for carcinogenic and mutagenic.

The harmful effect of acrylamide leads to the requirement of a simple and accurate method for acrylamide detection. The conventional methods to detect acrylamide were generally based on chromatography techniques, including LC/MS, LC-MS/MS and GC/MS [2, 5, 11] which are considered as less efficient process. On the other hand, biosensors have been developed for a quick detection method of acrylamide [12].

Electrochemical biosensor of acrylamide was initially developed in 1997 [13]. Heavy metal ions or hemoglobin (Hb) was generally employed for the sensor elements. Comparison of these two methods showed that the performance of Hb-modified electrode was satisfactory. It was reported that the interaction formed at the double bond of acrylamide which are covalently bonded to the α-NH_2 at the N-terminal valine groups of Hb can be used as the responses) [2].

On the other hand, in comparison to other conventional electrodes, BDD is known to have some superior properties, such as wide potential window, low background currents, and high physical and chemical stability as well as good biocompatibility [14, 15]. In this study, Hb-modified BDD was prepared for the applications as acrylamide biosensors. Initially, BDD was modified with gold nanoparticles in order to increase the affinity of Hb on the surface of BDD. Then, gold nanoparticles-modified BDD (AuNPs-BDD) was modified with Hb to form Hb-BDD. The results showed that Hb-BDD is promising to be applied for a sensor of acrylamide.

2. Materials and methods

Allylamine, human hemoglobin, CH_3COOH, CH_3COONa, $HAuCl_4$, H_2SO_4, K_2HPO4, KH_2PO_4, 1-propanol, and other chemicals were of analytical grade and supplied by Merck.

Prior to use, the BDD film was pre-treated with ultrasonication in 1-propanol for 15 min and high purity water for 15 min. Then, BDD was oxidized at +3.0 V (vs. Ag/AgCl) in 0.1 M H_2SO_4. After cleaned and dried, gold particles were electrodeposited on BDD surface to form AuNPs-BDD. Amperometry technique was used by applying a potential of +0.5 V (vs. Ag/AgCl) for 100 s in solution of 0.2 M H_2SO_4 containing of 2 mM $HAuCl_4$. Then, Hb-BDD was prepared by immersing the prepared AuNPs-BDD in a solution of 0.1 acetate buffer solution (ABS) pH 5 containing of 1.5 mg/mL human hemoglobin.

The electrochemical cell was filled with 4 mL of 0.1 M ABS containing acrylamide at pH 5. Then, a linear calibration curve for acrylamide was constructed from various acrylamide concentrations (5-50 μM). Cyclic voltammetry technique was performed with a potential range of -200 mV - 800 mV.

3. Results and discussion

3.1. Characterization of Hb-BDD electrodes

Figure 1a shows the XPS spectra of Hb-AuNPs BDD. Two peaks observed at 283 eV and 534 eV were identified as the C-C and C-O bonds of the BDD electrode, while the peak at 400 eV was attributed to the property of the C-N bonds from haemoglobin. In addition, a small peak at 90 eV was observed, indicating the presence of gold at the electrode surface. SEM characterization in figure 1b shows the presence of fiber of haemoglobin that was immobilized on the electrode surface, indicating the Hb-AuNP BDD can be successfully prepared.

AuNPs deposited on the electrodes BDD is assured to increase the direct electron transfer between hemoglobin and bulk electrodes [16]. On the other hand, the presence of AuNP was expected to facilitate the adsorption of hemoglobin onto AuNPs through the interactions between cysteine and NH_4-lysine residues of hemoglobin [17].

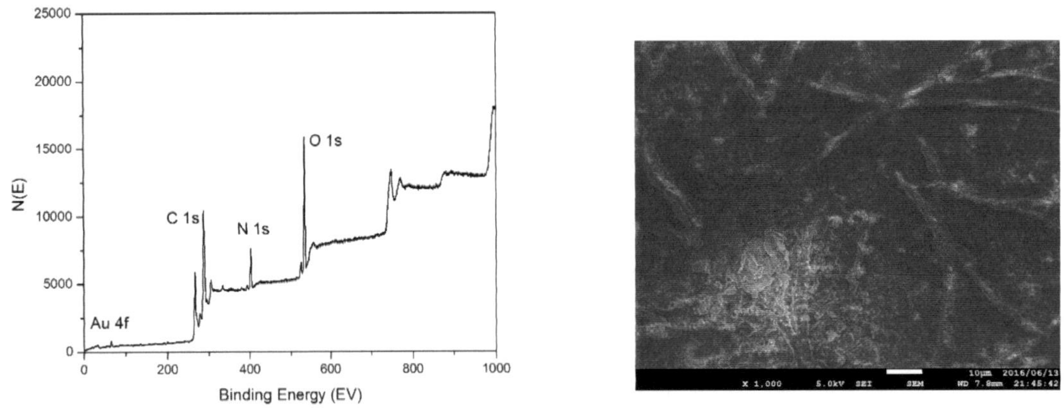

Figure 1. (a) XPS spectrum and (b) SEM image of Hb-AuNPs BDD.

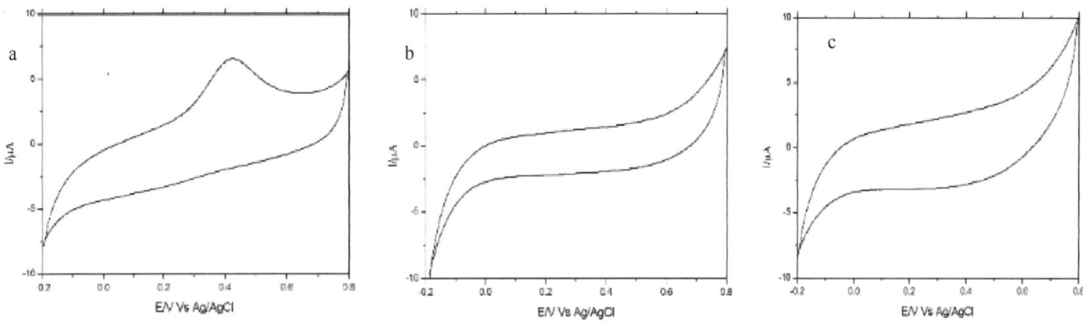

Figure 2. Cyclic voltammogram of haemoglobin modified gold electrode in (a) 0.1 M ABS pH 5 (b) 0.1 PBS pH 7, and (c) 0.1 M PBS pH 9.

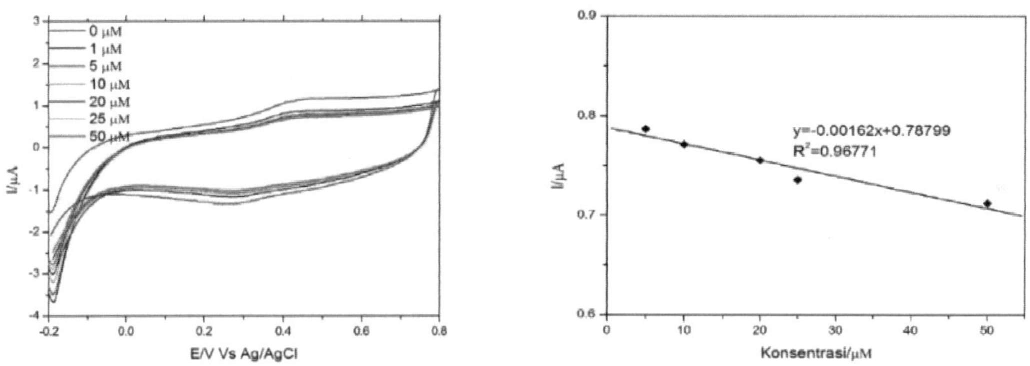

Figure 3. (a) Cyclic voltammetry response of Hb-AuNPs-BDD electrode toward acrylamide at various concentrations (0 – 50 μM) and (b) its linear dependence.

In order to study the electrochemical properties of the Hb-modified gold electrode, cyclic voltammetry was investigated at various pHs. While an oxidation peak at +0.45 V was observed at pH 5, no peak can be observed at pH 7 and pH 9 (figure 2). Therefore, the next experiments were conducted at pH 5.Conversion to the current density and power density is shown in figure 2.

3.2. The response of Hb-AuNPs-BDD electrode in the presence of acrylamide

Cyclic voltammetry of Hb-AuNPs-BDD electrode was studied in 0.1 M ABS pH 5 at a scan rate of 100 mV/s (figure 3a). Hemoglobin oxidation peak was observed at a potential of 0.43V (vs. Ag/AgCl). The oxidation peak decreased in the presence of acrylamide. Furthermore, investigation using various concentrations of acrylamide (0 – 50 μM) indicated that the electrode could be used for acrylamide sensors. A linear calibration curve ($r^2 = 0.97$) with an LOD of 8.5 μM can be achieved (figure 3b), indicating that hemoglobin-modified BDD is promising to be developed for the detection method of acrylamide.

4. Conclusions

BDD electrode was successfully modified with hemoglobin after the BDD modification with gold nanoparticles (Hb-AuNP-BDD). A linear calibration curve can be obtained at pH 5 at various acrylamide concentrations (0-50 μM) with an estimated LOD of 8.5 μM using this electrode.

Acknowledgements
This work was funded by Hibah Kompetensi Publikasi Internasional Terindeks Universitas Indonesia 2016, Contract No. 2016/UN2.R12/HKP.05.00/2016.

References

[1] Capuano E and Fogliano V 2016 *Encyclopedia of Food and Health* eds. Caballero B M, Finglas P and Toldra F (Amsterdam: Elsevier Ltd) 24-9
[2] Tareke E, Rydberg P, Karlsson P, Eriksson S and Törnqvist M 2002 *J. Agric. Food Chem.* **50** 4998-5006
[3] Stadler R H, Blank I, Varga N, Robert F, Hau J, Guy PA, Robert MC and Riediker S 2002 *Nature* **419** 449-50
[4] Mottram D S, Wedzicha B L and Dodson A T 2002 *Nature* **419** 448
[5] Zyzak DV, Sanders RA, Stojanovic M, Tallmadge DH, Eberhart BL, Ewald DK, Gruber DC, Morsch TR, Strothers MA, Rizzi GP, and Villagran MD 2003 *J. Agric. Food Chem.* **51**, 4782-7
[6] Fuhr U *et al.* 2006 *Am. Assoc. Cancer Res.* 266-71
[7] Chen M, Hsu H, Lin C and Ju W 2012 *Food Chem. Toxicol.* **50** 3867-76
[8] International Agency for Research on Cancer 1994 *Acrylamide* 389-433 (World Health Organization)
[9] European Commission 2002 *Opinion of the Scientific Committee on Food on New Findings Regarding the Presence of Acrylamide in Food* 1-16
[10] European Chemical Agency 2009 *Member State Committee Support Document for Identification of Acrylamide as A Subtance of Very High Concern Because of Its CMR Properties* 1-26
[11] Lingnert H *et al.* 2002 *Scand. J. Nutr.* **46** 159-72
[12] Hu Q, Xu X, Fu Y & Li Y 2015 *Food Control* **56** 135-46
[13] Ignatov O V, Rogatcheva S M, Kozulin S V and Khorkina N 1997 *Biosens. Bioelectron.* **12** 105-11
[14] Suzuki A, Ivandini T A, Kamiya A, Nomura S, Yamanuki M and Matsumoto K 2007 *Sensors Actuat. B: Chemical* **120** 500-7
[15] Ivandini T A, Wicaksono W P, Saepudin E, Rismetov B and Einaga Y 2015 *Talanta* **134** 136-43
[16] Garabagiu S and Micaelescu G 2011 *J. Electroanal. Chem.* **659**, 196-200
[17] Shao Q, Wu P, Gu PA, Xu XQ, Zhang H and Chai CX 2011 *J. Phys. Chem. B* **115**, 8627-37

Eu^{3+} complex of ligand4'-(4-carboxyphenyl)-2,2':6',2" -terpyridine as fluorosensor of heavy metals

A Zulys and N Rachmawati

Department of Chemistry, Faculty of Mathematics and Natural Sciences Universitas Indonesia, Kampus UI Depok, Depok 16424, Indonesia

Corresponding author's e-mail: zulys@ui.ac.id

Abstract. Ligand 4'-(4-carboxyphenyl)-2,2':6',2"-terpyridine (2-Hcptpy) has been synthesized by modification of Kröhnke method. The synthesize was performed using aldol condensation reaction. The white precipitate was collected and produced 62% yield (1.062 gr). Ligand 2-Hcptpy has been characterized by FTIR, Elemental analyzer, H-NMR, UV-vis, and UV-DRS spectrometer. It was then reacted with lanthanide group (Eu^{3+}) to form a complex by hydrothermal process. The result of Eu^{3+} complex was 0.352 gr. of white yellowish precipitate. The application of this research is for the fluorosensor of heavy metals (Pb^{2+}dan Cd^{2+}). The data of fluorescence showed two types of fluorescence, either turn on or turn off fluorosensor. Ligand 2-Hcptpy has an on-off type with the addition of Pb^{2+} and Cd^{2+}, while complex Eu^{3+} has two types of fluorosensor. The complex showed turn on-off and turn on-off by addition of Pb^{2+} and type of turn off by addition of Cd^{2+}. Either ligand or complex, showed fluorescence intensity by adding heavy metals up to concentration $5x10^{-8}$ M.

Keywords: Fluorosensor, Hcptpy, lanthanide, heavy metals.

1. Introduction

The lanthanides are unique among other elements. It has shielding effect by 4f orbitals resulting in the spectroscopic properties. The result of having a narrow-line emission is a wide range wavelength of the emission bands that do not overlap noticeably, and long life-time fluorescent to distinguish the background of fluorescence [1-3]. The complex compound between lanthanides and fluorophore organic ligand could form a fluorosensor. These properties attract some interests, for example to detect some metals or to apply as labeling agent, and also potential for heterogeneous catalysis [4-6].

Free lanthanide ions follow Laporte rule, thus it has a low extinction coefficient. The organic ligands which have fully π–π* transitions are able to pull through the lowlight absorbance of lanthanide [7-9]. Some organic ligands with functional group such carboxylic acid has the ability to work as sensitizer for lanthanide ions. Thus organic ligand with functional group could exhibit higher fluorescent stabilities than other sensitizers. The complex compound formed between sensitizer and lanthanide ions have been widely studied as fluorescence [4].

The organic ligand with functional group of carboxylic acid has been developed for several applications. Its fascinating structure lead some interests in *bifunctional function*, such as to act as bridge or chelating group to form building blocks by forming some linkage between carbonyl and metal ions, while the organic ligand of terpyridyl group can act as linker to extend network such as forming polymerization. To fulfill good connection of electron transport between the donor and acceptor is supported by the Hcptpy itself which has a large π-conjugated system [10].

2. Materials and methods

The 2-Hcptpy ligand and Eu^{3+} complex were prepared according to the literature [10] with some modification. The ligand was prepared using aldol condensation by Kröhnke method while the lanthanide complex by hydrothermal processes. All chemicals were obtained from commercial sources without further purification. The chemicals used were 2-acetylpyridine (Sigma-Aldrich), 4-carboxybenzaldehyde (Sigma-Aldrich), KOH (Merck), $NH_3.H_2O$ (Merck), Ethanol p.a (Merck), Methanol p.a (Merck), HCl (Merck), Eu_2O_3 (Merck), HNO_3 (Merck), $Cd(NO_3)_2.4H_2O$ (Merck), and $Pb(NO_3)_2$ (Merck). While the instrumentation used to determine the structure and molecular structure of ligand and complex was Elemental TruSpec LECO. ^1H NMR spectra were recorded at room temperature using JEOL 500 MHz apparatus with tetramethylsilane (TMS) as reference and DMSO as solvent. Infrared spectra were obtained from KBr pellets on Shimadzu 8400-S in the range of 400-4000 cm^{-1}, while the emission spectra fluorescence were recorded on a Hitachi F-2000 spectrophotometer.

2.1. Synthesis of ligand 2-Hcptpy

Ligand 2-Hcptpy was prepared according to the literature method with some modification by one pot synthesis [4]. The reaction process used aldol condensation reaction between 2-acetylpyridine (1.21 g) and 4-carboxybenzaldehyde (0.75 g) in basic condition by using KOH (0.8 g) and $NH_3.H_2O$ (20 mL) in 150 mL ethanol as solvent. The reaction mixture was stirred for 48 hours at room temperature for a while, then continued with the addition of HCl [aq., conc.] until slightly acidic (reached around pH 5). The result was light yellow precipitate then continued by washing it with mixture of ethanol and methanol by ratio of 1:1 until the precipitate changed into slightly white color. The precipitate was collected and dried in vacuum. The result was 1.09 g (62%). Based on the data, following are the result found for $C_{22}H_{15}N_3O_2$ (353.11 g/mol); C, 74.37; H, 5.39; N, 11.96. IR (KBr, cm^{-1}): 3061 (w), 1696 (s), 1579 (s), 1462 (m), 1363 (w), 1267 (s), 857 (w), 759 (w), 681 (w). 6.^1H-NMR (500 MHz, DMSO-d_6). δ: 8.76 (s, 1H), 8.68 (t, 3H), 8.17 (d, 2H), 7.76 (d, 2H), 7.56 (d, 2H).

2.2. Synthesis of Eu^{3+} complex

Complex of Eu^{3+} was prepared according to the literature method with some modification by hydrothermal process [5]. The reaction consist of $Ln(NO_3)_3.6H_2O$ (1 mmol), 2-Hcptpy (1 mmol), and NaOH (1 mmol) in 100 mL deionized water at 150^0C for 72 h. After cooling to room temperature at a rate 10°C/h, the separation process was performed by filtering the light brown precipitate and dried in vacuum until the humidity was lower than the initial condition. The result was 0.36 g. The data of elemental analyzer found for $C_{88}H_{57}N_{12}O_8$ (1561.07 g/mol) are C, 67.27; H, 6.73; N, 10.4. The IR spectra should be 3070 (m), 2560 (m), 2285 (m), 1960 (s), 1276 (s), 869 (s), 798 (s), 594 (m).

3. Results and discussion

3.1. Characterization of ligand and complex

The synthesis of 2-Hcptpy was used aldol condensation reaction by Kröhnke method with some modification. The reaction was done in one-pot synthesized with high yield percentage and simple preparation. Ligand of 2-hcptpy was characterized by infrared spectroscopy, H-NMR, elemental analyzer, and UV-DRS spectroscopy. The complexation process with Eu^{3+} was adapted by hydrothermal process with some modification[2]. The confirmation for the complexation reaction was performed by infrared spectra that showed the absence of the strong bands at around 1700 cm^{-1}. It indicated that the carbonyl group of ligand 2-Hcptpy were completely deprotonated. The other way to confirm the complexation successfully was by UV-DRS spectroscopy. The spectra showed band gap energy was decreased from ligand 2-Hcptpy to complex. Ligand 2-Hcptpy had a band gap of 2.89 eV, while Eu^{3+} complex had 2.67 eV.

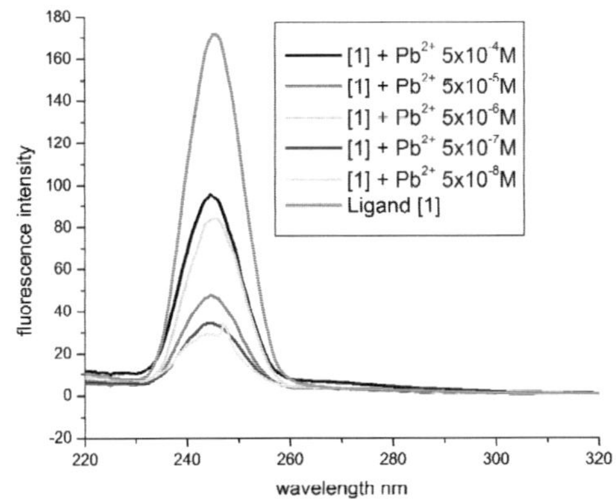

(a)

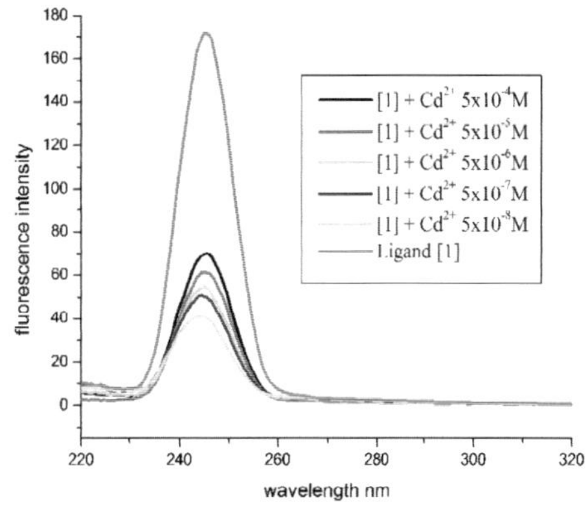

(b)

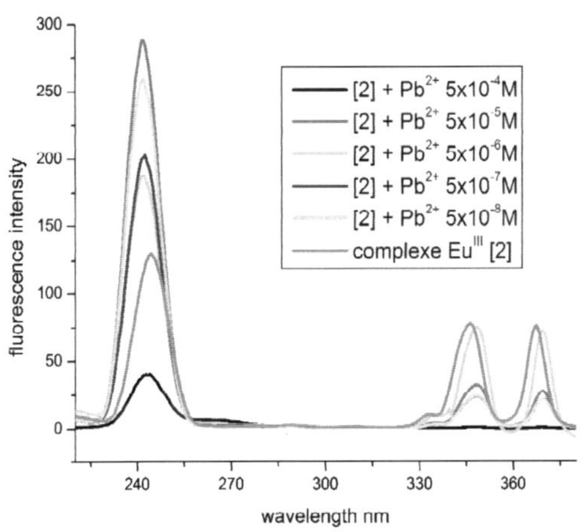

(c)

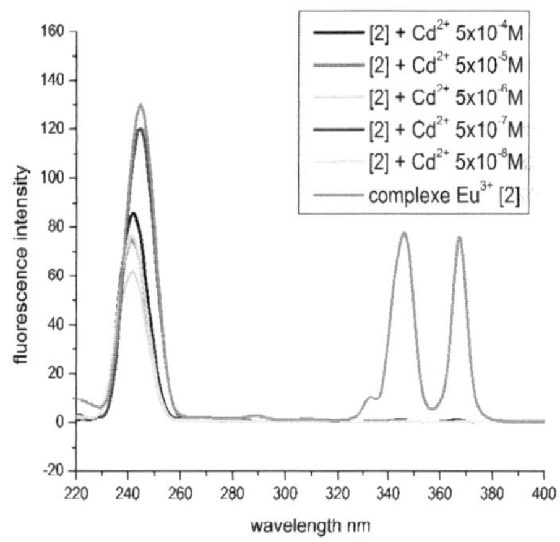

(d)

Figure 1. The data of fluorescence of ligand and complex by adding heavy metals in various concentrations.

3.2. Fluorescence properties

The ligand and complex were used as fluorosensor of heavy metals. Both compounds were added with heavy metals in various concentrations ($5x10^{-4} – 5x10^{-8}$) M. It showed that either ligand or complex was given fluorescence response by heavy metals addition up to $5x10^{-8}$ M. The data of fluorescence intensity of ligand and complex are shown in figure 1. The detection mechanism of heavy metals was conducted by PET (Photoinduced Electron Transfer) process. Fluorescence intensity of ligand was 170 a.u. while Eu^{3+} complex was 130 a.u. The band gap of complex was decreased from its ligand with fluorescence intensity. The addition of heavy metals to ligand or complex produced a quenching effect by decreasing the fluorescence intensity and an enhancement effect by showing increasing fluorescence intensity.

4. Conclusions

The synthesis of ligand 2-Hcptpy was done by Kröhnke method with some modification. The synthesis reaction was using aldol condensation with the yield of 62%. Eu^{3+} complex preparation was done by hydrothermal process with result of 0.35 gram. The confirmation of complexation was done by infrared spectra by showing the absence of carbonyl group at 1700 cm^{-1} and hydroxyl group at 3300 cm^{-1}. The other characterization to confirm the complexation was performed by UV-DRS spectroscopy. It showed that band gap energy from ligand and complex was decreased from 2.89 eV to 2.67 eV. The fluorescence properties of ligand and complex with the addition of heavy metals in various concentrations ($5x10^{-4}$ – $5x10^{-8}$) M showed a quenching effect of fluoresensor. The confirmation of quenching effect was showed by the decrease of fluorescence intensity before heavy metals was added into solution. The ligand showed a turn-off fluorosensor while Eu^{3+} complex showed a turn on and off fluorosensor.

Acknowledgements
This work was supported by university of Indonesia

References

[1] Mauro F, Vieri F, Luca G and Mauro M 2012 *Journal of Coordination Chemistry Reviews* **256** 170-92
[2] Bazzicalupi C *et al.* 2007 *Journal of Coordination Chemistry Reviews* **252** 1052-68
[3] Megan E A F, Sherryllene S P, Catherine N M and Wayne E J J 2012 *Journal of Inorganica Chimica Acta* **381** 78-84
[4] Wu Q R, Wang J J, Hu H M, Shangguan Y Q, Fu F, Yang M L, Dong F X and Xue G L 2011 *Inorganic Chemistry Communications* **14** 484-8
[5] Ali N K, Abolghasem B and Sara Z 2011 *Journal of Polyhedron* **30** 1134-42
[6] Zhu B H, Liu Y H, Jian X Y, Xu H Y, Han Y Y and Zhao Q 2014 *Polyhedron* **74** 67-71
[7] Yongsuk J and Juyoung Y 2012 *Journal of Inorganica Chimica Acta* **381** 2-14
[8] Lis S and Elbanowski M 2002 *Journal of Photochemistry and Photobiology A: Chemistry* **150** 233-47
[9] Valeur B and Berberan-Santos M N 2011 *J. Chem. Educ.* **88** 731-8
[10] Li N, Zhu Q E, Hu H M, Guo H L, Xie J, Wang F, Dong F X, Yang M L and Xue G L 2013 *Polyhedron* **49** 207-15

IOP Conf. Series: Materials Science and Engineering **188** (2017) 012008 doi:10.1088/1757-899X/188/1/012008

Citrate-capped superparamagnetic iron oxide (Fe_3O_4-CA) nanocatalyst for synthesis of pyrimidine derivative compound as antioxidative agent

A H Cahyana, D Pratiwi and B Ardiansah

Department of Chemistry, Faculty of Mathematics and Natural Sciences Universitas Indonesia, Kampus UI Depok, Depok 16424, Indonesia

Corresponding author's e-mail: herrykim@ui.ac.id

Abstract. The development of a recyclable catalyst based on magnetic nanoparticles has attracted an increasing interest as the emerging application in the heterogeneous catalyst field. Superparamagnetic iron oxide nanoparticle with citric acid as capping agent was successfully obtained from iron (III) chloride solution via two steps synthesis. The first step involving the formation of magnetite nanoparticle by bioreduction using *Sargassum* Sp, then its surface was modified by adding citric acid solution in the second step. The structural, surface morphology and magnetic properties of the nanocatalyst were investigated by various instrumentations such as scanning electron microscope with energy dispersive (SEM-EDS), and particle size analyser (PSA). Fe_3O_4-CA was then applied as reusable catalyst for Knoevenagel condensation of barbituric acid and cinnamaldehyde to produce (E)-5-(3-phenylallylidene)pyrimidine-2,4,6(1H,3H,5H)-trione. The optimum condition of this reaction was achieved by using 7.5% mole of catalyst at 50°C for 6 h to give 83% yield. Some spectroscopy techniques such as UV-Vis, FTIR, LC-MS and ^{1}H-NMR were used to confirm the product's structure. Furthermore, the synthesized compound has an attractive antioxidant activity based on the *in-vitro* analysis using DPPH method.

1. Introduction

In recent years, the development of Fe_3O_4 as magnetic nanoparticle became a trending topic due to their multiple functions, including information storage, electronic devices, rapid biologic separation, and drug delivery [1]. Fe_3O_4 was successfully synthesized by the reduction of ferric chloride solution using *Sargassum* Sp. brown seaweed extract, applying a green chemistry concept. Initially, the ferric chloride hydrolyzes to form ferric hydroxide and release H^+. The presence of aldehyde group in *Sargassum* Sp. extract reduced the ferric chloride to form Fe_3O_4 nanoparticle, and the other side of the aldehyde was oxidized to form the corresponding acid [2, 3]. However, the nanoparticle with no stabilizer will cause agglomeration. To solve this problem, the surface of Fe_3O_4 must be stabilized with citric acid through the coordination bond [4]. Fe_3O_4 capped by citric acid can be used as catalyst to form pyrimidine derivative compound. The properties of nano-sized Fe_3O_4-CA such as high surface area shows an important role in catalytic performance. The advantages of using this nanocatalyst are its low toxicity and cost [5], can be easily separated, and recycled from the product by an external magnet [6]. Bioactivity of pyrimidine derivative compound as an antioxidative agent was measured by UV-Vis spectrophotometer using DPPH method and interpreted as IC_{50} [7].

2. Materials and methods

2.1. General

Chemicals used were from pro-analytical grade. The *Sargassum* Sp. was obtained from Banten Province (Binuangeun Sea). The chemical used in this research were analytical grade from commercial suppliers.

2.2. Synthesis of Fe_3O_4-CA nanocatalyst

Fe_3O_4 nanoparticles were prepared based on earlier work with modification [3]. In short, $FeCl_3.6H_2O$ (4 mmol), $FeCl_2.4H_2O$ (1 mmol), *Sargassum* Sp. extract (5 mL) and deionized water (50 mL) were stirred in a 250 beaker glass at room temperature, and the mixture was adjusted to pH 10 by adding NaOH. After 2 h, the product will be capped by adding 1% of citric acid solution, the reaction temperature was raised up to 60°C and completed for 1 h with continuous stirring. The product was dried in the oven at 50°C and characterized using SEM-EDSand PSA [8].

2.3. Synthesis of pyrimidine derivative compound and its bioactivity

In a 250 Erlenmeyer flask, 1 mmol cinnamaldehyde, 2 mmol barbituric acid, 5%mol Fe_3O_4-CA were added to methanol as solvent and stirred for 6 h at 50°C. When the reaction was completed, the catalyst can be removed easily from the product using an external magnet. The product was recrystallized by adding hot ethanol and characterized with UV-Vis, FTIR, LC-MS, and NMR. The bioactivity was measured by UV-Vis spectrophotometer using DPPH method according to previous research [7, 9].

3. Results and discussion

The citrate-modified Fe_3O_4 was prepared based on our earlier work and re-characterized by SEM, EDS, and PSA [10]. Figure 1 is the morphology of Fe_3O_4 and Fe_3O_4-CA by SEM. Fe_3O_4 has granular grain. The other side of Fe_3O_4-CA has granular and fiber grain. The different surface of these compound show that the Fe_3O_4 was successfully capped by citrate from citric acid. Elemental analysis using EDS for Fe_3O_4 indicated that the sample contains Fe (65.33%), O (24.47%), C (8.31%) and the trace of Na element (1.89%). After capping process of Fe_3O_4 with citrate, there was an increase of carbon content in the sample. The result of EDS analysis of Fe_3O_4-CA is as follows: Fe (51.36%), O (38.49%), and C (10.14%). The re-synthesized catalyst has similar characteristic, and it is comparable with our previous work [10].

Figure 2 shows the particle size of Fe_3O_4 and Fe_3O_4-CA. From this measurement, the average particle size of Fe_3O_4 is 62.5 nm and 89.0 nm for Fe_3O_4-CA. The increasing of Fe_3O_4-CA particle size caused the capping of Fe_3O_4 surface by citrate and it increased the radius of particle. The Fe_3O_4-CA was then applied in synthesis of pyrimidine derivative compound. From FTIR analysis, the peak at 1567 cm^{-1} might be assigned to mono-substituted benzene. The peak at 1694 cm^{-1} signified the presence of secondary amide. The absorption at 1748 cm^{-1} is attributed to the C=O vibration. The peaks at around 2851 and 3188-3071 cm^{-1} are attributed to the C-H sp^2 and N-H functional group, respectively.

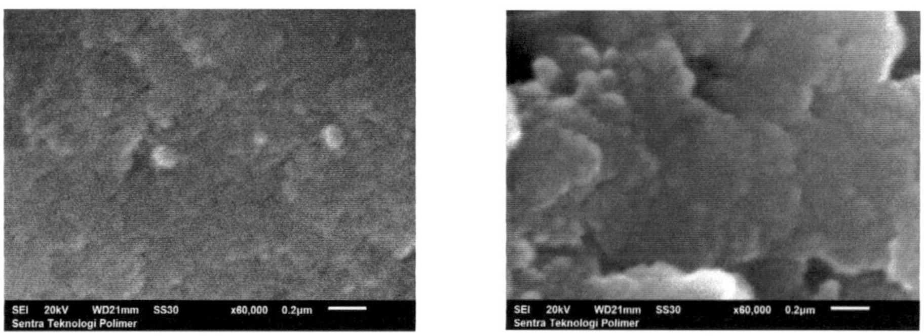

(a) (b)

Figure 1. SEM micrographs of (a) Fe_3O_4 and (b) Fe_3O_4 modified with citrate.

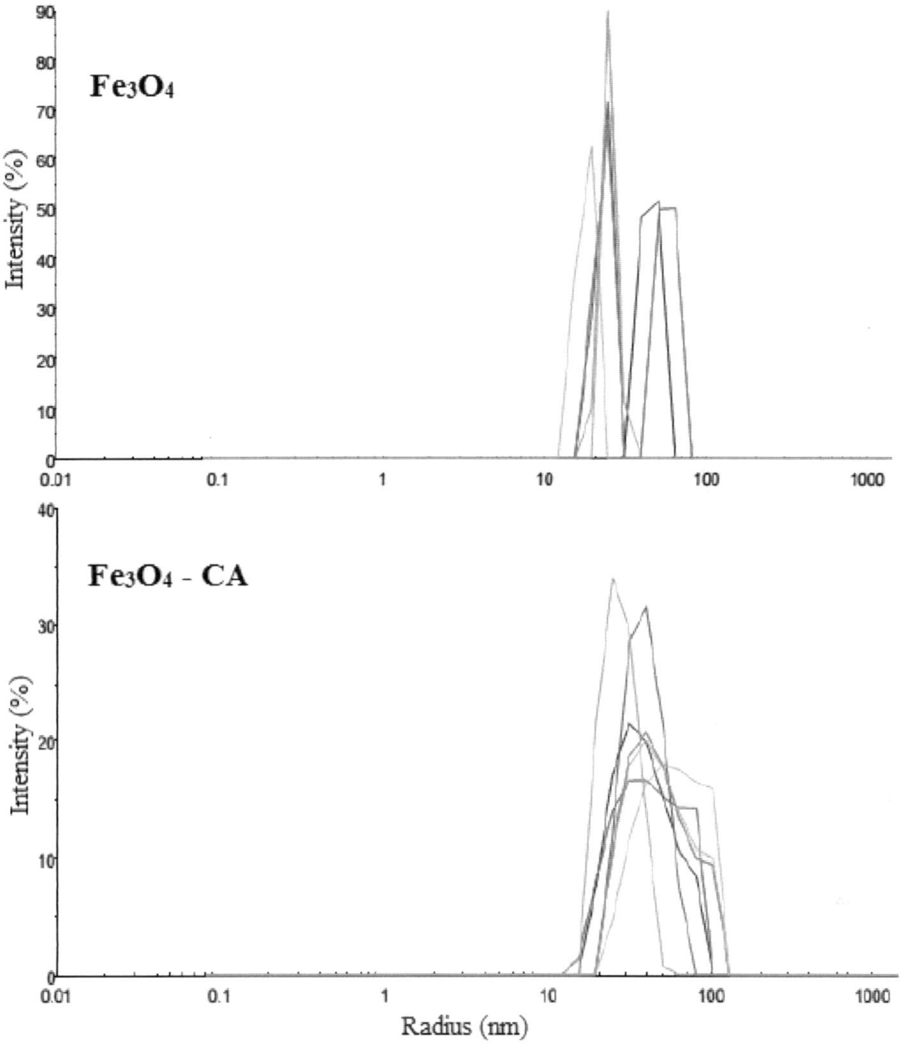

Figure 2. Size distribution of (a) Fe_3O_4 and (b) Fe_3O_4-CA.

The maximum wavelength of this pyrimidine derivative compound is 371 nm as shown in figure 3. Additionally, the LC-MS gave the M+1 base peak at 243 (Figure 3), means the molecular weight of this compound is 242 g mol^{-1}. Further, ^{1}H-NMR (500 MHz, DMSO-d6) was used to confirm the structure. The peaks were appeared at 7.48 ppm (*m*, 5H, phenyl); 7.69 ppm (*d*, 1H, CH, *J*= 15.6 Hz); 8.01 ppm (*d*, 1H, CH, *J* = 11.9 Hz); 8.44 ppm (*t*, 1H, CH); 11.25 ppm (*s*, 2H, 2NH). From the LC-MS and ^{1}H-NMR results, it was showed that the compound is (E)-5-(3-phenylallylidene)pyrimidine-2,4,6(1H,3H,5H)-trione ($C_{13}H_{10}N_2O_3$). The bioactivity of this pyrimidine derivative compound was measured using DPPH method. The result showed an antioxidative activity through the changes of violet colour from DPPH to yellow colour by adding 1 mL sample. The IC_{50} of this compound is 8.57 ppm.

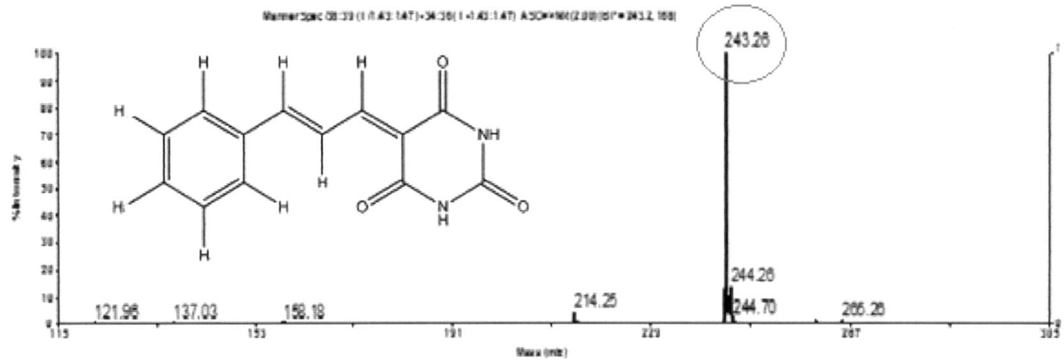

Figure 3. LC-MS of pyrimidine derivative compound.

4. Conclusions

Magnetic nanoparticle Fe_3O_4 was successfully synthesized using *Sargassum* Sp. as stabilizer and it was successfully capped by citrate to form Fe_3O_4-CA. The carboxyl group on its surface acted as catalyst for synthesis of pyrimidine derivative compound. This compound's structure was confirmed by some instruments such as FT-IR, LC-MS, and NMR. From DPPH test, we can conclude that the compound is very active as antioxidative agent.

Acknowledgements

The authors would like to thank Universitas Indonesia for funding this research through PITTA Grant Universitas Indonesia with contract No: 2035/UN2.R12/HKP.05.05/2016.

References

[1] Naeimi H and Nazifi Z S 2013 *J Nanopart. Res.* **15** 2026-37
[2] Kannan R R R, Stirk W A and Van Staden J 2013 *South African J. Botany* **86** 1-4
[3] Mahdavi M, Namvar F, Ahmad M B and Mohamad R 2013 *Molecules* **18** 5954-64
[4] Cheragipour E, Javadpour S and Mehdizadeh A R 2012 *J. Biomed. Sci. Eng.* **5** 715-9
[5] Safari J and Zarnegar Z 2013 *J. Mol. Catal. A: Chem.* **379** 269-76
[6] Kiasat A R and Davarpanah J *J. Mol. Catal. A: Chem.* **373** 46-54
[7] Vijayalakhsmi A, Kumar P R, Priyadarsini S S and Meenaxshi C 2013 *J. Chem.* **5** 1-9
[8] Nigam S, Barick KC and Bahadur D 2011 *J. Magn.Magn. Mater.* **323** 237-43
[9] Cahyana A H and Ardiansah B 2016 *AIP Conf. Proc.* **1729** 020057
[10] Cahyana A H, Pratiwi D and Ardiansah B 2016 *Rasayan J. Chem.* **9** 896-902

IOP Conf. Series: Materials Science and Engineering **188** (2017) 012009 doi:10.1088/1757-899X/188/1/012009

Theoretical study on the magnetic moments formation in Ta-doped anatase TiO_2

A Bupu[1], M A Majidi[1] and A Rusydi[2,3]

[1]Department of Physics, Faculty of Mathematics and Natural Sciences Universitas Indonesia, Kampus UI Depok, Depok 16424, Indonesia

[2]NUSNNI-NanoCore, Department of Physics, Faculty of Science, National University of Singapore, Singapore 117576, Singapore

[3]Singapore Synchrotron Light Source, National University of Singapore, Singapore 117603, Singapore

Corresponding author's email: aziz.majidi@sci.ui.ac.id

Abstract. We present a theoretical study on Ti-vacancy induced ferromagnetism in Ta-doped anatase TiO_2. Experimental study of $Ti_{1-x}Ta_xO_2$ thin film has shown that Ti-vacancies (assisted by Ta doping) induce the formation of localized magnetic moment around it, then, the observed ferromagnetism is caused by the alignment of localized magnetic moments through Ruderman-Kittel-Kasuya-Yosida (RKKY) interaction. In this study, we focus on the formation of the localized magnetic moments in this system. We hypothesize that on a unit cell, Ti-vacancy has caused four electrons from the surrounding oxygen atoms to become unpaired. These unpaired electrons then arrange themselves into a configuration with a non-zero net magnetic moment. To examine our hypothesis, we construct a Hamiltonian of the four unpaired electrons, incorporating the Coulomb intra- and inter-orbital interactions, in matrix form. Using a set of chosen parameter values, we diagonalize the Hamiltonian to get the eigenstates and eigenvalues, then, with the resulting eigenstates, we calculate the magnetic moment, μ, by obtaining the expectation value of the square of total spin operator. Our calculation results show that in the ground state, provided that the ratio of parameters satisfies some criterion, $\mu \approx 4\mu_B$, corresponding to the four electron spins being almost perfectly aligned, can be achieved. Further, as long as we keep the Coulomb intra-orbital interaction between 0.5 and 1 eV, we find that $\mu \approx 4\mu_B$ is robust up to far above room temperature. Our results demonstrate that Ti vacancies in anatase TiO_2 can form very stable localized magnetic moments.

1. Introduction

Research on room temperature ferromagnetism (RTFM) in dilute magnetic semiconductors (DMS) has attracted much attention because of its potential for spintronics applications. Many studies on oxide DMS, such as TiO_2 and ZnO doped with transition metals or other elements [1–4], have successfully shown RTFM. Despite those facts, however, studies on the mechanisms of how the magnetic moments formed and aligned to lead to ferromagnetism in oxide DMS are still scarce.

For doped TiO_2, either cationic vacancies (Ti vacancies) or oxygen vacancies can play an important role in inducing ferromagnetism [3, 5]. An experimental study on anatase $Ti_{1-x}Ta_xO_2$ [6] shows that Ti vacancies are responsible for the observed RTFM by inducing formation of localized magnetic moments around the vacant sites. Furthermore, another study by Barman et al. [7] also suggests that Ti vacancies in $Ti_{1-x}Ta_xO_2$ act as magnetic centers. Ta atoms in Ta-doped anatase TiO_2 substitute Ti atoms in some sites of Ti. These substitutions cause distortion in some unit cells and lead to a situation in which a Ti atom cannot occupy its site, resulting in formation of a Ti vacancy. This situation makes Ta-doped anatase TiO_2 contain three types of unit cells: pure TiO_2 unit cells, TiO_2 with Ta replacing the Ti's position and TiO_2 with the Ti site being vacant. It is this TiO_2 unit cell with the Ti site being vacant that is suspected to form the localozed magnetic moment. Assuming such Ti-vacancy-induced magnetic moments exist, a recent theoretical study demonstrates that the Ruderman-Kittel-Kasuya-Yosida (RKKY) exchange interaction is likely responsible for the alignments of the magnetic moments to form the ferromagnetic order [8].

In this study, we develop a theoretical model and perform calculations to explain how Ti vacancies induce the formation of localized magnetic moments in Ta-doped anatase TiO_2 system. In each unit cell that contains a Ti vacancy, we argue that as a result of the breaking of Ti-O bonds due to the missing of Ti ion, the Ti vacancy creates four unpaired electrons in the surrounding oxygen sites. The unpaired electrons then interact with one another through repulsive Coulomb interactions. These interactions combined with Pauli exclusion principle can result in Hund's coupings among spins of the four unpaired electrons, which eventually can lead to spin alignment among them. Previous studies [6, 7] indicate that there are three possible ways of spin alignment in this system, yielding the net magnetic moment of 0, 2 μ_B, and 4 μ_B. Since Ta-doped anatase TiO_2 has a pretty high T_C, a mean-field theory on feromagnetism suggests that the high value of T_C may be contributed by the high value of total magnitude of each consitituent spin. It means that the four unpaired electrons may maximally align to form a net magnetic moment of 4 μ_B. Here, we aim to find such an optimum condition when alignment of the unpaired electrons leads to a net magnetic moment of 4 μ_B. Further, we would like to analyze the stability of the formed magnetic moments against temperature, especially at a room temperature and above it.

2. Model

As we aim to explain the role of Ti vacancy in creating a net localized magnetic moment in its corresponding unit cell, we isolate such a unit cell from the rest of the material and consider it as our system to study. Starting with the actual unit cell of TiO_2, we select a region where four oxygen atoms (red spheres) surround a titanium atom (blue sphere) marked with the black-dashed oval and pointed by

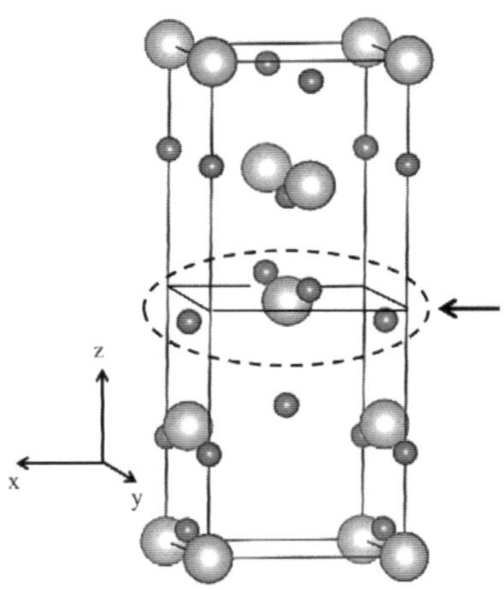

Figure 1. Unit cell of anatase TiO_2. The enclosed loop marked by the arrow shows the region we aim to explore.

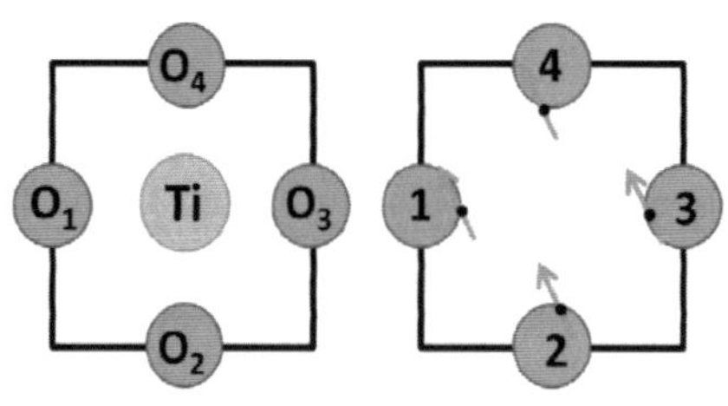

Figure 2. (a) The considered block around Ti site. (b) Model of the block aroud a Ti-vacancy site.

the black arrow in figure 1. Assuming that all atoms in that region lying on the same plane, we take this block shown in figure 2(a) to be the system we would like to model. We then construct the model with Ti vacancy by taking out the titanium atom from this block (figure 2(b)). The removal of Ti atom in that block causes the breaking of bonds between Ti and O atoms. As a result, four electrons from the surrounding oxygen atoms become unpaired. Each electron has a magnetic moment corresponding to it's spin (blue arrow in figure 2(b)). These electrons interact with one another through Coulomb repulsive interactions. We model the dynamics of the four unpaired electrons by constructing a Hamiltonian consisting of the kinetic (non-interacting) part and the interaction part as follows

$$H = H_{kin} + H_{int}. \tag{1}$$

The kinetic part of the Hamiltonian reads

$$H_{kin} = \sum_{m,m',\sigma} t_{m,m'} a^{\dagger}_{m,\sigma} a_{m',\sigma}. \tag{2}$$

Here, we assume that only one orbital from each oxygen atom containing the unpaired electrons is relevant to be considered in this model. The four electrons are assumed to be able to hop only within these four such orbitals. Within this restriction we define the size of the Hilbert space of our model. In equation (2), $t_{m,m'}$ is the hopping parameter characterizing the energy associated with a hopping of an electron from site m to site m', $a^{\dagger}_{m,\sigma}$ ($a_{m',\sigma}$) is creation (annihilation) operator each of which is associated with its basis orbital and spin orientation. $t_{m,m'}$ is formally derived from

$$t_{m,m'} = \int \phi^{*}_{m}(\mathbf{r}) \left(-\frac{\hbar^2}{2m_{el}} \nabla^2 + V(\mathbf{r}) \right) \phi_{m'} d^3\mathbf{r}, \tag{3}$$

with m_{el} being the electron mass, $V(\mathbf{r})$ the potential energy experienced by one electron due to the nearby atomic nuclei, and $\phi(\mathbf{r})$ and $\phi^{*}(\mathbf{r})$ the single electron wave function and its conjugate corresponding to each oxygen's atomic orbital at site m and m', respectively. The interaction part represents the Coulomb repulsive interaction between two electrons that, in second quantized form, reads as

$$H_{int} = \sum_{\substack{m,m' \\ m'',m'''}} \frac{1}{2} U_{m,m',m'',m'''} \sum_{\sigma,\sigma'} a^{\dagger}_{m,\sigma} a^{\dagger}_{m',\sigma'} a_{m'',\sigma'} a_{m''',\sigma}, \tag{4}$$

with

$$U_{m,m',m'',m'''} = \int \phi^{*}_{m}(\mathbf{r}_1) \phi^{*}_{m'}(\mathbf{r}_2) \frac{e^2}{|\mathbf{r}_1 - \mathbf{r}_2|} \phi_{m''}(\mathbf{r}_2) \phi_{m'''}(\mathbf{r}_1) d^3\mathbf{r}_1 d^3\mathbf{r}_2, \tag{5}$$

being the Coulomb interaction strength associated with a scattering of a pair of electrons initially at sites m'' and m''' scattered into sites m and m'.

We further simplify the interaction part by restricting the scatterings of two electrons only to those with final occupied sites being the same as the initial ones, as we consider these to be the most dominant contribution. This restriction implies that there are only three possible conditions to be considered: (1) $m = m' = m'' = m'''$; (2) $m = m'''$ and $m' = m''$, but $m \neq m'$; (3) $m = m''$ and $m' = m'''$, but $m \neq m'$. With the above restriction, equation (4) is split into three terms as

$$H_{int} = \sum_m \frac{1}{2} U \sum_{\substack{\sigma,\sigma' \\ \sigma\neq\sigma'}} a^\dagger_{m,\sigma} a^\dagger_{m,\sigma'} a_{m,\sigma'} a_{m,\sigma} + \sum_{\substack{m,m' \\ m\neq m'}} \frac{1}{2} V_{m,m'} \sum_{\sigma,\sigma'} a^\dagger_{m,\sigma} a^\dagger_{m',\sigma'} a_{m',\sigma'} a_{m,\sigma} + \sum_{\substack{m,m' \\ m\neq m'}} \frac{1}{2} J_{m,m'} \sum_{\sigma,\sigma'} a^\dagger_{m,\sigma} a^\dagger_{m',\sigma'} a_{m,\sigma'} a_{m',\sigma}. \tag{6}$$

The first term of the above equation is the Coulomb intra-orbital interaction (with $m = m' = m'' = m'''$) with U being the Coulomb intra-orbital parameter, while the last two terms are the Coulomb inter-orbital interactions with $V_{m,m'}$ and $J_{m,m'}$ being two different Coulomb inter-orbital parameters. Here, $V_{m,m'}$ corresponds to scattering between two electrons initially occupying different orbitals with no exchange of orbitals after the scattering ($m = m''', m' = m''$), while $J_{m,m'}$ corresponds to scattering with an exchange of orbitals after the scattering ($m = m'', m' = m'''$). Further rearrangement yields

$$H_{int} = \sum_m \frac{1}{2} U \sum_{\substack{\sigma,\sigma' \\ \sigma\neq\sigma'}} n_{m,\sigma} n_{m,\sigma'} + \sum_{\substack{m,m' \\ m\neq m'}} \frac{1}{2} V_{m,m'} \sum_{\sigma,\sigma'} n_{m,\sigma} n_{m',\sigma'} - \sum_{\substack{m,m' \\ m\neq m'}} \frac{1}{2} J_{m,m'} \left(\frac{1}{2} + 2\mathbf{s}_m \cdot \mathbf{s}_{m'} \right), \tag{7}$$

with $n_{m,\sigma} = a^\dagger_{m,\sigma} a_{m,\sigma}$ being the occupation number operator of electron at site m with spin component σ, and $\mathbf{s}_m = s^x_m \hat{\mathbf{x}} + s^y_m \hat{\mathbf{y}} + s^z_m \hat{\mathbf{z}}$ being the spin operator of electron at site m, with $s^i_m = (1/2) \sum_{\alpha,\beta} a^\dagger_{m\alpha} \sigma^i_{\alpha\beta} a_{m\beta}$, and $\sigma^i_{\alpha\beta}$ elements of the i-th component of the spin Pauli matrices. Note that from the above equation, $J_{m,m'}$ is nothing but the Hund's coupling parameter expected to lower the energy of the electrons occupying different sites with the same spin orientations. Hence, our full Hamiltonian

$$H = \sum_{m,m',\sigma} t_{m,m'} a^\dagger_{m,\sigma} a_{m',\sigma} + \sum_m \frac{1}{2} U \sum_{\substack{\sigma,\sigma' \\ \sigma\neq\sigma'}} n_{m,\sigma} n_{m,\sigma'} + \sum_{\substack{m,m' \\ m\neq m'}} \frac{1}{2} V_{m,m'} \sum_{\sigma,\sigma'} n_{m,\sigma} n_{m',\sigma'} - \sum_{\substack{m,m' \\ m\neq m'}} \frac{1}{2} J_{m,m'} \left(\frac{1}{2} + 2\mathbf{s}_m \cdot \mathbf{s}_{m'} \right), \tag{8}$$

comes up to be the well known extended Hubbard model with magnetic exchange interaction.

3. Method

With only four atomic orbitals, two possible spin orientations, and four electrons, the size of our Hilbert space is quite manageable to handle using exact diagonalization method. To proceed with this method, we first need to bring the Hamiltonian in equation (8) into matrix form by operating it on the 4-particle basis states, $|\phi_i\rangle$. Here, the 4-particle basis states are all the possible configurations of the four electrons among the four p orbitals of the oxygen atoms surrounding the Ti vacancy. The total number of such states is $8!/(4!4!) = 70$. The Hamiltonian matrix is constructed out of 70×70 elements $\langle \phi_i | H | \phi_j \rangle$. Once the matrix is set up, we diagonalize it to obtain eigenstates and eigenvalues of the Hamiltonian. The diagonalization is done numerically using DSYEV routine of LAPACK (Linear Algebra Package). As our goal is to calculate the net magnetic moment of our system, we then use the eigenstates to calculate the ground-state expectation value of the square of total spin operator, $\langle S^2 \rangle$, with $S^2 = \mathbf{S} \cdot \mathbf{S} = (\sum_m \mathbf{s}_m) \cdot (\sum_{m'} \mathbf{s}_{m'})$, and its corresponding thermodynamic averages for various temperatures.

Having obtained $\langle S^2 \rangle$, we can extract the total spin (S) of the system using relation $\langle S^2 \rangle = S(S+1)$, yielding

$$S = \frac{-1+\sqrt{1+4\langle S^2\rangle}}{2}. \tag{9}$$

The magnetic moment can then be calculated as

$$\mu = gS\mu_B, \tag{10}$$

with g being the Lande factor, which may be set to the free-electron value, $g \approx 2$.

Further, we also need to calculate the temperature depenpence of the magnetic moment. For this purpose, we first take the thermodynamic averages of $\langle S^2\rangle$ as function of temperature as follows

$$\langle S^2\rangle(T) = \frac{1}{Z}\sum_i \langle S^2\rangle_i e^{-\beta E_i}, \tag{11}$$

where

$$Z = \sum_i e^{-\beta E_i}, \tag{12}$$

is the partition function, with E_i being the energy eigenvalue of state i, and $\beta = \frac{1}{k_B T}$, where k_B is the Boltzmann constant. Having this in hand, we then calculate the corresponding total spin as a function of temperature, $S(T)$, by putting $\langle S^2\rangle(T)$ into equation (9). Then, the magnetic moment as a function of temperature is obtained by replacing S in equation (10) with $S(T)$.

4. Results and discussion

To perform the computations, we need to specify the parameter values, $t_{m,m'}$, U, $V_{m,m'}$ and $J_{m,m'}$. There are some constraints in determining these parameters. Firstly, U, $V_{m,m'}$ and $J_{m,m'}$ may have to be small values as they describe the Coulomb interactions occuring between electrons occupying p orbitals which, unlike d orbitals, are not very localized. Secondly, according to equation (5), the intra-orbital Coulomb interaction coupling U has to be larger than the inter-orbital couplings $V_{m,m'}$ and $J_{m,m'}$. Furthermore, $t_{m,m'}$, $V_{m,m'}$ and $J_{m,m'}$ may depend on the distance between sites m and m'. Here, we set $t_{m,m'} = t$, $V_{m,m'} = V$ and $J_{m,m'} = J$ for m and m' located next to each other, while $t_{m,m'} = t/2$, $V_{m,m'} = V/2$ and $J_{m,m'} = J/2$ for m and m' located opposite to each other (see figure 2(b)).

Figures 3 and 4 demonstrate how we vary the parameters to explore how the magnitude of magnetic moment at the ground state depends on the parameters based on the considerations explained above. In figure 3, we present the magnetic moment (μ) as a function of t/J while setting $J = V$. We find that increasing t/J at fixed U/J decreases μ. However, the decrease in μ with increasing V/J is insignificant. While, increasing U/J at fixed t/J or fixed V/J increases μ. The results presented in figure 3 show that to obtain maximum value of μ, we need to minimize t/J and maximize U/J. The maximum value of μ is reached when the ratio of t/J is small enough (≤ 0.2). We obtain $\mu = 3.995\mu_B$, which is close to $4\mu_B$, when $t/J = 0.2$ and $U/J = 10$ with $U = 1$ eV and $V = J$.

Figure 4 shows the calculated μ as a function of V/J. We use a constant ratio of $t/J = 0.1$ and $U = 1$ eV with $U \geq J$. Our calculation shows that, at small values of U/J, increasing V/J makes μ decrease slightly on the order of 10^{-3}, whereas increasing V/J at higher value of U/J (U/J = 10) does not change the magnitude of magnetic moment. Here, at U/J = 10, we obtain $\mu = 3.998\mu_B$. Thus, our overall result suggests that the maximum value of μ can be achieved by requiring that $U > J > t$. Our results are consistent with previous calculation by Peng et al. [9] and Qi et al. [10] which show that a Ti vacancy could generate a $4\mu_B$ magnetic moment.

From figure 3 we have found that the highest value of μ requires t/J be as small as possible while U/J should be large. It may be also interesting to explore the other limit, that is for very high values of

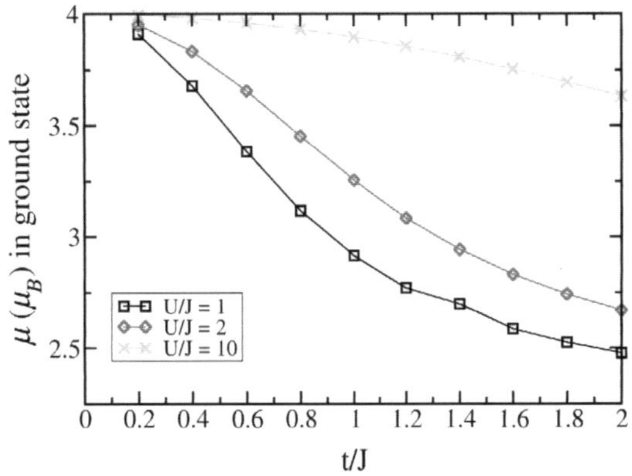

Figure 3. Magnetic moment, μ, as a function of t/J for various U/J, with $U = 1$ eV and $J = V$, at $T = 0$.

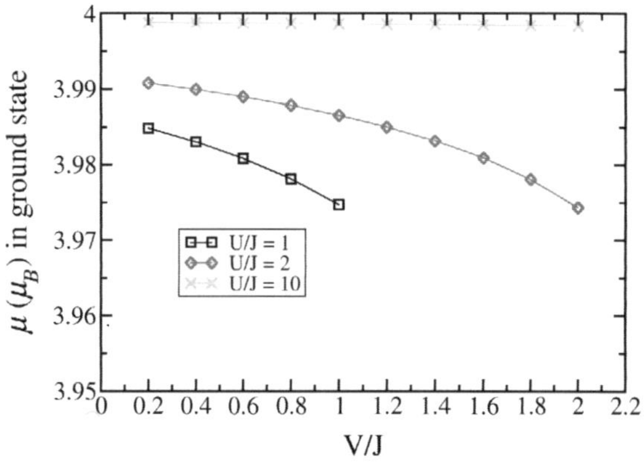

Figure 4. Magnetic moment, μ, as a function of V/J for various U/J, with $U = 1$ eV and $t/J = 0.1$, at $T = 0$.

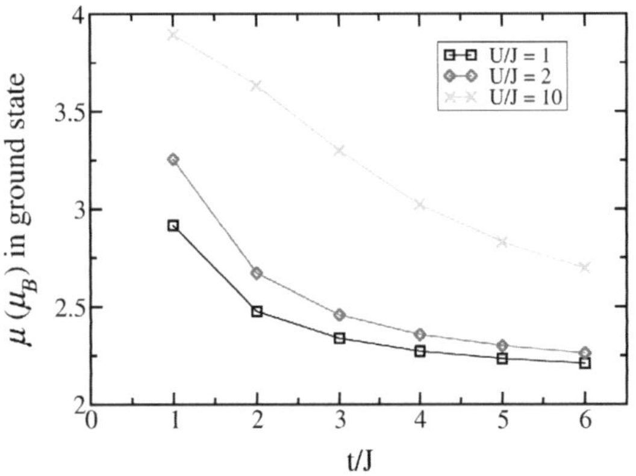

Figure 5. Magnetic moment, μ, as a function of t/J at various high values of t/J.

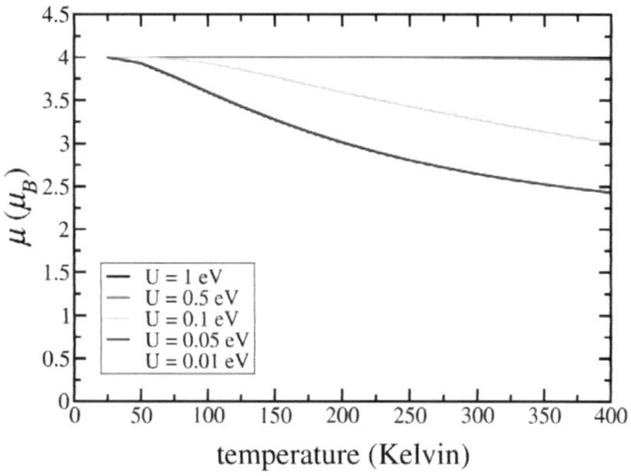

Figure 6. Magnetic moment, μ, as a function of temperature with fixed ratio of parameters.

t/J. Figure 5 shows how μ evolves whent/J is taken from 1 up to 6. At very high value of t/J, it appears that μ reduces exponentially to some finite values between2 and 2.5 μ_B. We argue that the non-vanishing value of μ is the consequence of keeping J finite no matter how big the value of t/J is.

Finally, we investigate the temperature dependence of μ for various U with a fixed ratio among all the parameters. We use a constant ratio of $U : J : V : t = 1 : 0.1 : 0.1 : 0.01$ and vary the temperature from 25 to 400 K. As expected, increasing temperature would excite the energy of the electron spin at each oxygen site, that in turn, would destroy the spin alignment and decrease the total magnetic moment in the unit cell. Figure 6 shows our results for various values of U. When $U = 1$ eV and $U = 0.5$ eV we observe that μ does not change as temperature increases. μ only starts to fall off as a function of temperature if we reduce U to ~ 0.1 eV or smaller. Furthermore, at the smallest chosen value of U, $U = 0.01$ eV, a full spin alignment corresponding to $\mu = 4\mu_B$ is not achieved even at an extremely low temperature. Another important observation is that, at room temperature, we find that for all values of U, provided $J > 0$, the system always reveals a magnetic moment of the value $> 2\mu_B$.

5. Conclusions

We have developed a model and performed calculations to demonstrate that Ti vacancy, induced by Ta, in Ta-doped anatase TiO_2 leads to the formation of magnetic moment. The Ti vacancy causes the breaking of Ti-O bonds and generates four unpaired electrons hopping around the oxygen sites. The hopping integral $t_{m,m'}$ along with the Coulomb inteactions among the electrons manifested in the parameters U, V and V, at the proper circumstances generate a maximum magnetic moment, $\mu \approx 4\mu_B$, at the ground state. In our simulation this condition is achieved when $U : J : V : t = 1 : 0.1 : 0.1 : 0.01$. As expected, temperature increase causes the magnetic moment to decrease, but this effect is only significant when U is very small. At room temperature, the system always has a non-zero magnetic moment, provided $J > 0$. As we keep U between 0.5 and 1 eV, the maximum magnetic moment of $\sim 4\mu_B$ is robust upto 400 K. Our results confirm our hypothesis that Ti vacancies in Ta-doped anatase TiO_2 induce the formation of magnetic moments, presumming the above proper circumstances of the parameters are met in the system.

Acknowledgement

We are very grateful to Universitas Indonesia for providing us a full funding for this project through PITTA Research Grant No. 2020/UN2.R12/HKP.05.00/2016.

References

[1] Pan H, Yi J B, Shen L, Wu R Q, Yang J H, Lin J Y, Feng Y P, Ding J, Van L H and Ying J H 2007 *Phys. Rev. Lett.* **99** 127201
[2] Hu Y M, Li S S, Kuang C H, Han T C and Yu C C 2015 *J. Appl. Phys.* **117** 17B901
[3] Zhang S, Ogale S B, Yu W, Cao X, Liu T, Ghosh S, Das G P, Wee A T, Greene R L and Venkatesan T 2009 *Adv. Mater.* **21** 2282-71
[4] Choudhury B, Choudhury A and Borah D 2015 *J. Alloys Compd.* **646** 692-8
[5] Wang X *et al.* 2014 *Appl. Phys. Lett.* **105** 262402
[6] Rusydi A *et al.* 2012 *Phil. Trans. R. Soc. A* **370** 4927-43
[7] Barman A R, Annadi A, Gopinadhan K, Lu W M, Ariando, Dhar S and Venkatesan T 2013 *AIP Adv.* **2** 012148
[8] Majidi M A, Bupu A and Fauzi A D 2016 *Physica B* http://dx.doi.org/10.1016/j.physb.2016.08.031
[9] Peng H, Li J, Li S and Xia J 2009 *Phys. Rev. B* **79** 092411
[10] Qi D C *et al.* 2013 *Phys. Rev. B* **87** 245201

The study of DNA adduct 8-hydroxy-2'deoxyguanosine (8-OHdG) formation of butylated hydroxyanisole (BHA) and its metabolite ter-butyl hydroquinone (TBHQ) through in vitro reaction with Calf Thymus DNA and 2'deoxyguanosine

Budiawan, S S Purwaningsih and D I Cahaya

Department of Chemistry, Faculty of Mathematics and Natural Sciences Universitas Indonesia, Kampus UI Depok, Depok 16424, Indonesia

Corresponding author's e-mail: dr.budiawan@gmail.com

Abstract. Butylated Hydroxyanisole (BHA) and its metabolite Tert-Butyl Hydroquinone (TBHQ) are synthetic antioxidants, commonly used as food and beverage preservatives. Although WHO declared their safety, the use of these preservatives are still controversial because some studies showed that BHA induced proliferative effects in animal testing and TBHQ is considered as carcinogenic and causes DNA cleavage. This study is aimed to analyze the interaction between Calf Thymus DNA with BHA and TBHQ which are mediated with Copper (II) Chloride. The result of the study in spectrophotometric showed there was bathochromic shift as much as 2-3 nm in DNA treated with TBHQ. The next analysis used HPLC method in stationary phase of ODS, mobile phase of 10mM Natrium Hydrogen Phosphate Buffer and Methanol (85 : 15) for DNA adduct formation, 8-Hydroxy-2-Deoxyguanosine (8-OHDG) as biomarker of risk cancer. The resultof the study showed the formation of DNA adduct 8-OHDG in the interaction between DNA and 20-500 ppm of TBHQ. The 8-OHdG formation was greatly increased by the higher concentration of TBHQ. The relative amount of 8 OHDG which formed was reached $946/10^5$ deoxyguanosine in DNA bases. Confirmation test by LCMS/MS was characterized with the detection of mother ion peak (m/z 284); fragment ion peaks at m/z 167.9, and 139.9; at retention time 3.52 min. Meanwhile the interaction between DNA and 50-250 ppm BHA did not induce 8-OHDG.

Keyword: Butylated hydroxyanisole (BHA), tert-butyl hydroquinone (TBHQ), 8-hydroxy-2-deoxyguanosine (8-OHDG), DNA adduct, spectrophotometric, HPLC, LCMS/MS

1. Introduction

Butylated Hydroxyianisole (BHA) and Tert - Butyl Hydroquionone (TBHQ) are synthetic antioxidants in phenolic groups which commonly used as a preservative in food and beverage industry. They have a highly solubility in animals fat and stable in various conditions during food processing. BHA is expected to prevent free radicals formation and prevent lipid oxidation [1]. This substance is able to significantly obstruct or inhibit the oxidation of a substance that is easily oxidized even in low concentrations, not more than 0.02% of oil or fat contained in food [2]. TBHQ is formed by BHA through an O-methylated process [3]. TBHQ is a preservative in products of fats and oils; animal fat; margarine and similar products; snacks made from potatoes, cereals, flour and starch; and nut products [3].

Although the use of preservatives such as BHA and TBHQ is permitted by the WHO with BHA acceptable daily intake (ADI) of 0.5 mg/kg body weight [3], and TBHQ at 0.70 mg/kg body weight [4-6] until now, BHA and TBHQ acceptable consumption still makes controversy because several studies reported that BHA and TBHQ could damage DNA, leading to initiate and promote tumors in several animal tissues. TBHQ can cause the breakdown of DNA in vitro [7] and form 8-hydroxy-2'-deoxyguanosine on Calf Thymus DNA [8].

Study on xenobiotics (chemical compounds) interaction with DNA became a hot topic in this recent years. The covalent interaction between the DNA bases with chemical compounds can produce a form of adducts known as DNAadduct and expected as one of the important stages in the process of cancer initiation [9]. DNA adducts can cause either the coding (miscoding) during transcription or sometimes cause mutations and cancer [10]. DNA adduct can be useful as a molecular biomarker that provides the best approach to early detection of cancer risk [11]. DNA adduct formation by BHA and TBHQ interaction with DNA can be proved through in vitro and in vivo study.This research is aimed to confirm if the interaction between DNA, BHA and TBHQ could lead to DNA adducts formation as biomarkers of cancer risk.

2. Materials and methods

In this research, the study conducted was DNA binding formation in spectrophotometry and study of DNA adducts formation by using HPLC and LC MS/MS. The analytical method has been validated in the previous research.

Study of DNA binding interaction was observed by incubating 100 µL Calf Thymus DNA 10 mg/mL and various concentrations (20, 25, 30, 35 dan 40 ppm) of BHA and TBHQ compounds in 10 mM Tris HCl solvent, pH 7.4; the various incubation times were held in 1, 2 and 3 hours at room temperature 25°C. The interaction between DNA with BHA and TBHQ was analyzed using spectrophotometry.

Study of DNA Adduct formation (8-OHdG) from BHA and TBHQ interaction with Calf Thymus DNA was observed by incubating 40 µL Calf Thymus DNA 10 mg/mL with each compound of 100ppm BHA and 80ppm TBHQ, in 50 mM Tris HCl solvent, pH 7.4; the reaction mixture was incubated for 1 hour at temperature 37°C. After incubation, the reaction mixture was hydrolyzed by adding 20 µL volume of natrium succinate mixture 0.1 M, and calcium chloride 0.05 M in ratio 1:1, and 30 µL volume of micrococcal nuclease mixture (2 mU/µg DNA) and phosphodiesterase II (0.04 mU/µg DNA). The mixture was incubated at temperature 37°C for 4 hours. The hydrolysis products were analyzed by using HPLC and LC MS/MS with analysis condition shown in table 1 and 2.

Study of DNA adduct formation was performed using DNA base 2' deoxyguanosine with BHA and TBHQ. Around 100 ppm of 2' deoxyguanosine was incubated with BHA in various concentrations, 50, 100, 150, 200, and 250 ppm in 50 mM Tris HCl solvent, pH 7.4; the reaction mixture was incubated for 1 hour at temperature 37°C. Meanwhile, incubation with TBHQ was conducted in various concentrations, 20, 40, 60, 80, 100, 120, 160, 200, 300, 400, and 500 ppm in 50 mM Tris HCl solvent,

Table 1. Analysis condition using HPLC

Column	X Bridge C 18 Waters (4.6 X 250 mm, particle size 5 µm)
Column temperature	31°C
Eluent	Sodium hydrogen phosphate buffer 10 mM and Methanol (85 : 15)
Flow rate	1.0 mL/minute
Injection Volume	20 µL
Detector	UV wavelength 254 nm

Table 2. Analysis Condition using LCMS/MS LC Condition

Column	Atlantis T3 diameter 5 μm, size 4.6 x 150 mm (Waters)
Column temperature	35°C
Eluent	Eluent A : Eluent B = Formic acid 0.1 % : Formic acid 0.1 % in Methanol
Flow rate	0.5 mL/minute
Injection Volume	10 μL
Detector	MS/MS, Triple Quadrupole
Analysis Time	10 min.
Analysis Technique	Binary Gradient Programme

Table 3. Binary Gradient Programme (with modification) [12]

Time (minute)	Eluent A (%)	Eluent B (%)
1.0	50	50
3	5	95
5	5	95
5.5	50	50
10	50	50
10	Stop	

MS/MS Condition

Ionization	: positive
Mode	: ESI
Parent ion	: 284
Product ion	: 167.9 ; 139.9

pH 7.4; the reaction mixture was incubated for 1 hour at temperature 37°C. The formation of DNA adduct was analyzed by using HPLC, and 8 – OHDG was measured towards 2' deoxyguanosine (table 3 [12]).

3. Results and discussion

3.1. DNA binding formation of BHA and TBHQ using spectrophotometry

Binding or interaction of the compounds such as BHA and TBHQ causes spectra changes. Calf thymus DNA was incubated with BHA in various concentrations 20, 25, 30, 35 and 40 ppm in 0.01 M Tris HCl solvent, pH 7.4 at temperature 25°C in three various incubation times (1 hour, 2 hours, and 3 hours). The spectra profile shows that in each incubation time, DNA control has a maximum wavelength absorbance at around 258 nm, meanwhile BHA control has a maximum wavelength absorbance at 286.6 (figure 1). Interaction between DNA and BHA in various concentrations provides maximum wavelength absorbance at around 282-283.4 nm. There is no specific difference in spectra of DNA and BHA in each incubation time. There is only an increasing of absorbance in the higher BHA concentrations. The result stated that there is no binding interaction between DNA and BHA. Calf thymus DNAincubation with TBHQ in various concentrations, 0, 20, 25, 30, 35, and 40 ppm, in 0.01 M Tris HCl pH 7.4, at temperature 37°C for 30 minutes showed that the formed spectra have a maximum wavelength absorbance at around 282.4-284.8 nm. The increasing amount of spectra absorbance was equal with the increasing of TBHQ concentration. Meanwhile, the DNA control has a maximum wavelength absorbance at around 258-258.4nm after incubation for 1, 2, and 3 hours and the TBHQ control has a maximum wavelength absorbance at around 288 nm. The spectra showed maximum wavelength absorbance at around 259.6-261 nm indicated the bathochromic shift, the shift of peak absorbance to the higher wavelength around 2-3 nm from maximum wavelength absorbance of DNA. In addition, there was a hyperchromic shift, the increasing of maximum wavelength absorbance intensity in the interaction of DNA and TBHQ which was in line with the increasing of TBHQ (figure 2). The spectra produced in the treatment of DNA with TBHQ in various concentrations show that there is binding interaction between DNA and TBHQ. According to the previous research, this bonding interaction is

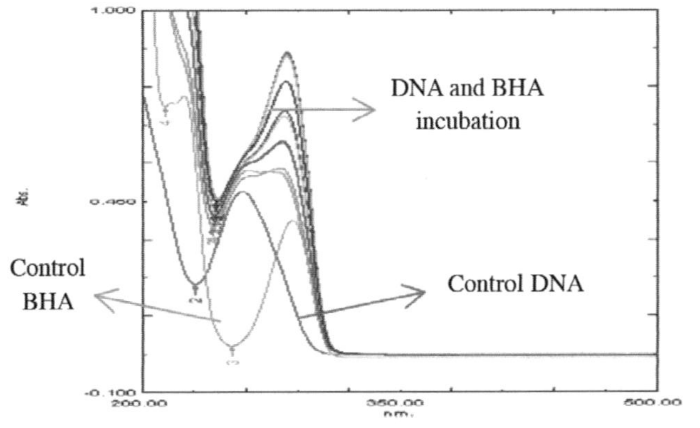

Figure 1. Spectra profile of DNA Binding with BHA in 0.01 M Tris HCl pH 7.4; DNA concentration was 20 ppm; BHA in various concentrations 0, 20, 25, 30, 35 and 40 ppm with 3 hours incubation time

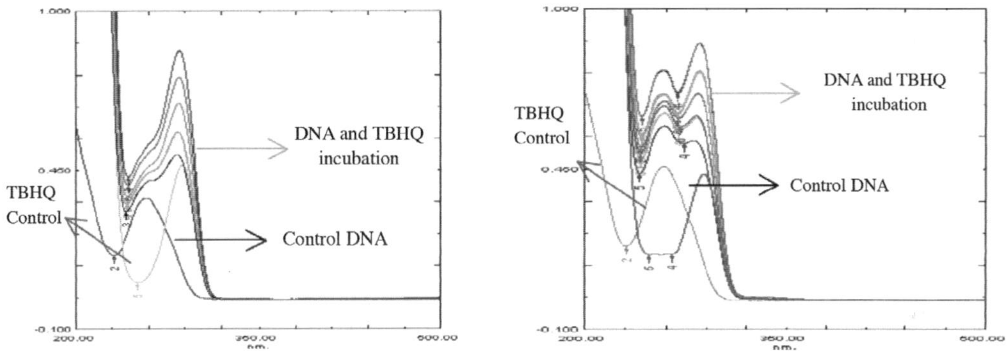

Figure 2. Spectra profile of DNA Binding with TBHQ in 0.01 M Tris HCl pH 7.4; DNA concentration was 20 ppm; TBHQ in various concentrations 0, 20, 25, 30, 35 and 40 ppm with incubation time of (a) 30 minutes and (b) 3 hours

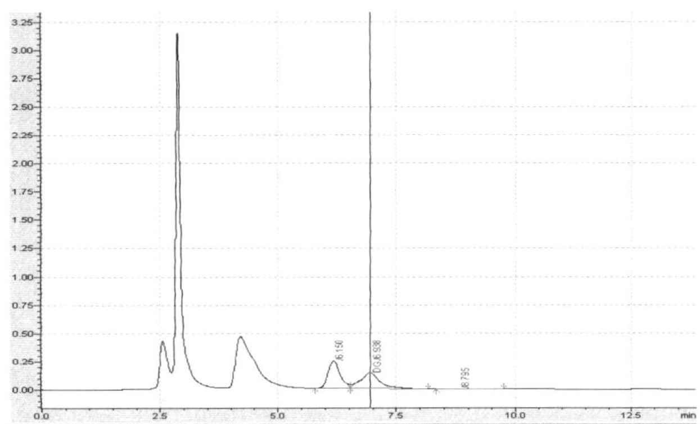

Figure 3. Chromatogram profile of enzimatically hydrolyzed DNA using HPLC

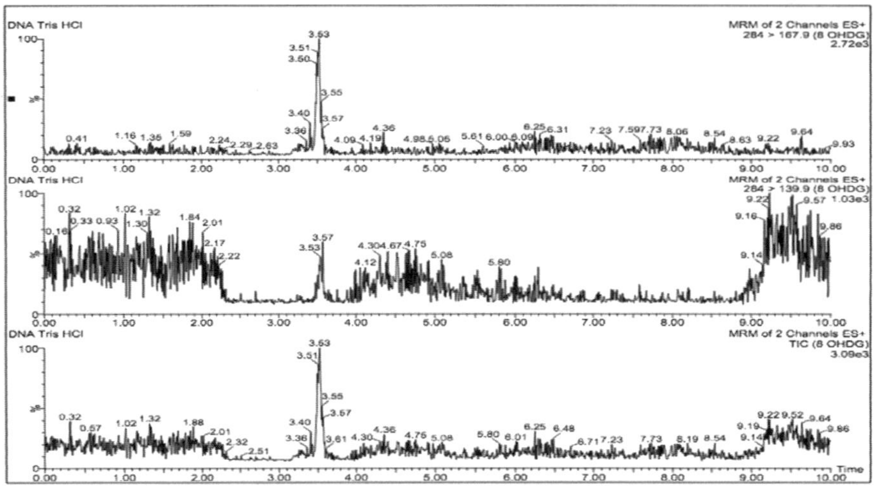

Figure 4. Chromatogram profile of DNA hydrolysis using LCMS/MS

called as intercalation, in which the whole or the half of molecules enter the gap between two base-pairs in double helix DNA which can affect the conformation structure and the structure of DNA - chain. The responses observed in the spectrum of the intercalation are 'hyperchromic' (the increasing of maximum wavelength absorbance) or 'hypochromic' (the decreasing of maximum wavelength absorbance).

3.2. DNA hydrolysis

Hydrolysis of DNA was conducted enzimatically by using Micrococcal Nuclease enzymes from Staphylococcus aureus and Phospodiesterase II from Spleen bovine. Both enzymes are functionate to hydrolyze DNA into single nucleoside. The result of hydrolysis of the mixture compounds was then analyzed using HPLC. The chromatogram results seemed to form 4 chromatogram peaks, one of them is 2'Deoxyguanosine at retention time 6.842 (figure 3). Neverthless, the DNA adduct 8-OHdG can not be detected using HPLC. Further analysis was performed using LCMS/MS, and the result showed that 8OHdG was formed in very low concentration 7.356 ppb with area of 169.8 (figure 4). The concentrations of 8OHdG are below limit of detection (8 ppb).

3.3. Incubation of 2' Deoxyguanosine with BHA and TBHQ

Incubation of 2'Deoxyguanosine with BHA in various concentrations 50, 100, 150, 200 and 250 ppm at pH 7.4 mediated with Cu^{2+} did not trigger the formation of *adduct* 8-OHDG (figure 5). The formation of 8-OHDG from incubation with TBHQ was analyzed by incubating 2'Deoxyguanosine with TBHQ in various concentrations 20, 40, 60, 80, 100, 120, 140, 200, 300, 400 and 500 ppm, pH 7.4, mediated with Cu^{2+} at temperature 37°C for 1 hour. TBHQ triggered the formation of 8-OHdG in the increasing of TBHQ concentrations. In TBHQ concentration of 500 ppm, 8-OHdG has formed approximately at $946/10^5$ DG (figure 6).

3.4. DNA incubation with BHA and TBHQ

Interaction between TBHQ and DNA was studied by incubating 80 ppm TBHQ and DNA in the presence of Cu^{2+} at temperature 37°C, pH 7.4 for 1 hour, then hydrolyzed enzimatically for 3 hours. The reaction mixture was analyzed using LCMS/MS. DNA Adduct 8-OHdG formed was about 10 ppb (figure 7). Neverthless, the interaction between DNA and 100 ppm BHA by using LCMS/MS was not detected.

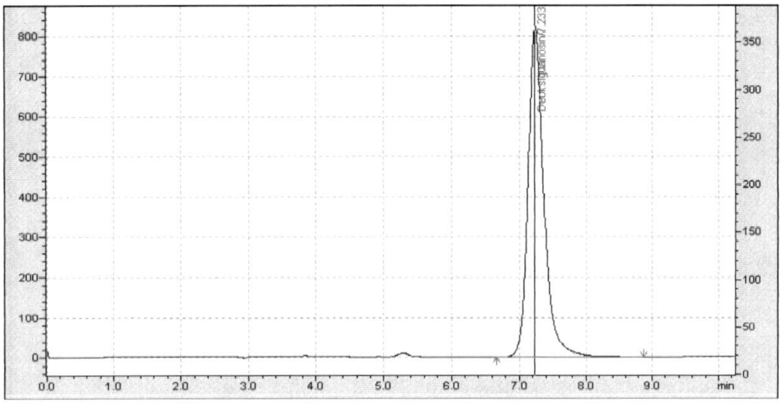

Figure 5. Chromatoram profile of HPLC after 2'deoxyguanosine incubation with 250 ppm BHA at pH 7.4 for 1 hour

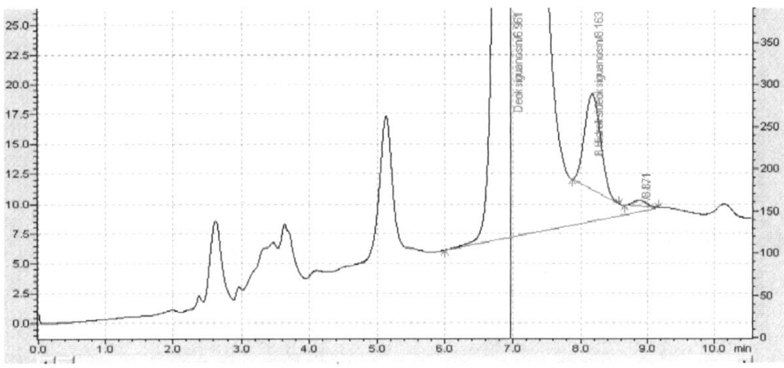

Figure 6. Chromatogram profile of 2'deoxyguanosine incubation with 500 ppm TBHQ at pH 7.4 for 1 hour

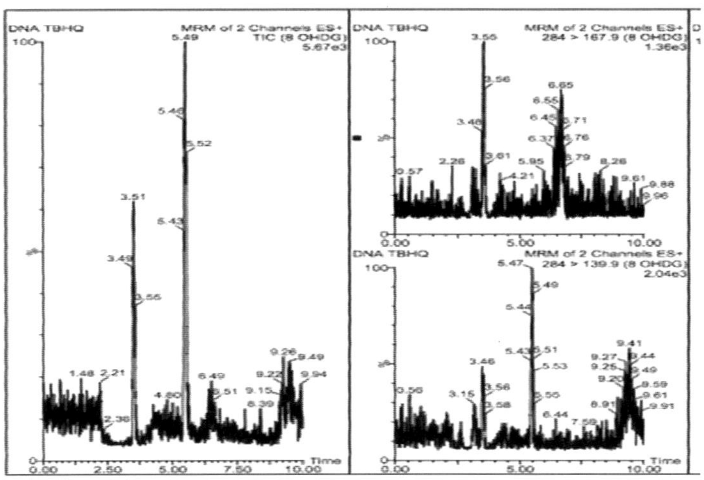

Figure 7. Chromatogram profile of DNA hydrolisis after incubated with TBHQ using LCMS/MS.

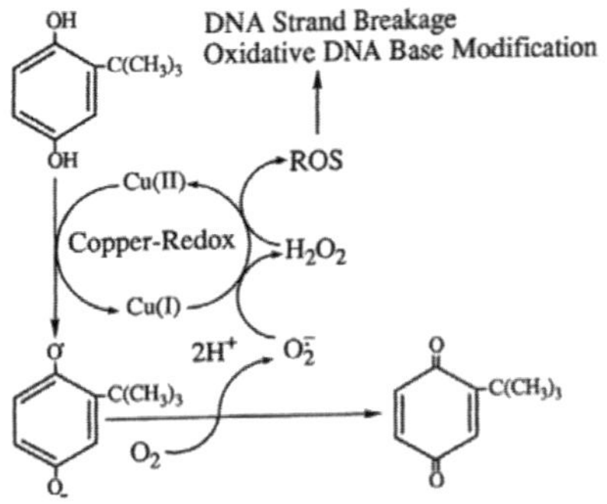

Figure 8. The reaction mechanism of TBHQ oxidation stimulated with Cu^{2+}into TBQ) and ROS which trigger the DNA damage [7].

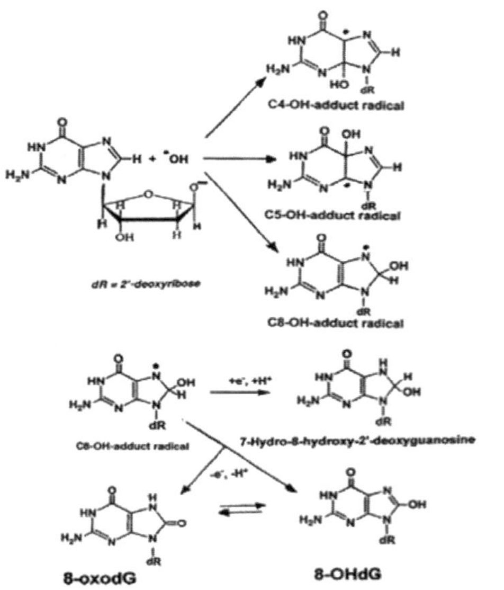

Figure 9. The reaction mechanism of 2' deoxyguanosine with hydroxyl radicals generating 8-OHDG [13]

The formation of 8OHdG induced by TBHQ was stimulated by Cu^{2+} ion which induced the formation of reactive oxygen species. The reaction showed the role of redox cycle of Cu(II)/Cu(I). The ion of Cu^{2+} can promote oxidation process of TBHQ and generate ROS in the form of semiquinone anion radicals; Tert Butyl Quinone (TBQ), H_2O_2, superoxidant, and hydroxyl radicals (figure 8). The oxidized form of TBHQ generated intermediates of TBQ anion radical, and released one electron and two H^+. protons The missing electron will alter oxygen into superoxide. In the presence of H^+, superoxide will generate H_2O_2.

Copper (Cu^{2+}) as free ions are present in the nucleus at a concentration of approximately 16 x 10^{-6} M. In the presence of H_2O_2, Cu^{2+} can produce hydroxyl radicals continuously through the Fenton reaction as follows

$$Cu^{2+} + O_2\bullet^{-} Cu^{+} + O_2$$
$$Cu^{+} + H_2O_2 \longrightarrow Cu^{2+} + HO^{-} + HO\bullet$$

$$O_2\bullet^{-} + H_2O_2 \longrightarrow O_2 + HO^{-} + HO\bullet$$

The last reaction ($O_2\bullet^{-} + H_2O_2 \longrightarrow O_2 + HO^{-} + HO\bullet$) is known as the Haber Weiss reaction.

The interaction between hydroxyl radicals with DNA nucleoside caused the DNA damage, as it can oxidize single base of DNA, 2'deoxyguanosine generating 8-OHDG. The formation of 8-OHDG was occurred through two stages of reactions, the hydroxyl radical attack on carbon C-8 of 2'deoxyguanosine generating C8-OH- radical*adduct*, followed by the release of electrons and H^+, generating 8-OHDG (figure 9).

The conformation test of 8-OHdG formation using LCMS/MS with electro spray ionization (ESI) is aimed to identify molecular mass from DNA adduct 8-OHDG formation and its fragmentations which generated from molecule collided with high-energy electrons. LCMS/MS chromatogram profile and mass spectra in figure 10 show the chromatogram of 8-OHdG at retention time 3.5 min and base peak (M^++1) at m/z = 284. This is equal with 8-OHdG molecular mass (C10H13N5O5) = 283.24. Pieta *et al.* [14] reported base peak of 8-OHdG (M^++1) = 284. The fragmentation of 8-OHDG with positive ion generated two main fragments at m/z = 167.9 and m/z = 139.9. The first fragment ion m/z

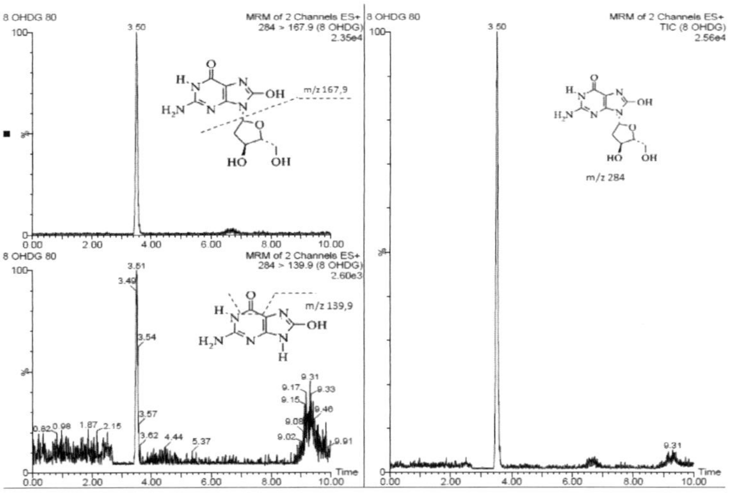

Figure 10. Base peak profile of 8-OHDG (M^++1 = 284) and its fragmentations generating fragment ions m/z =167.9 and m/z = 139.9

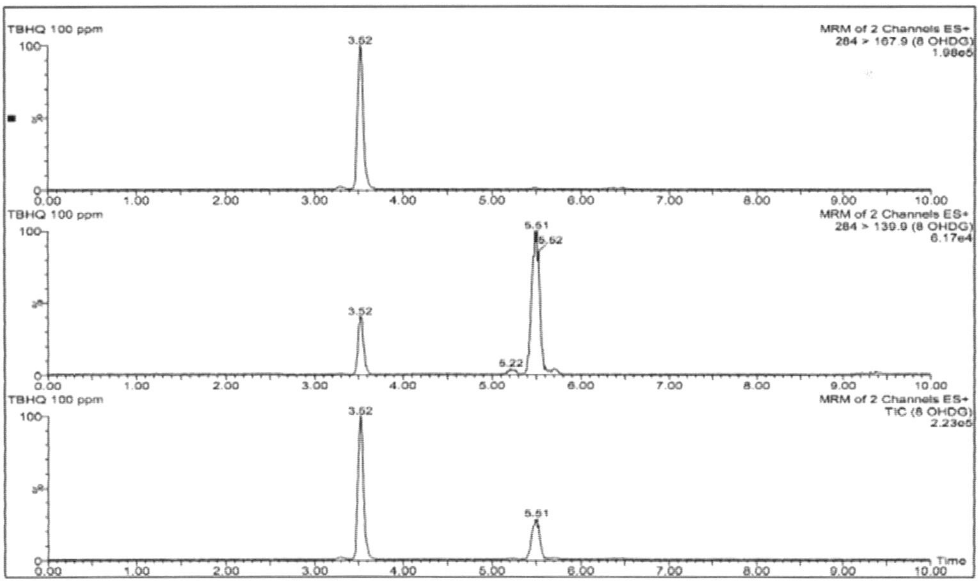

Figure 11. Chromatogram profile of 2'Deoxyguanosine and 50 ppm TBHQ mediated with Cu^{2+} incubation at pH 7.4 for 1 hour using LCMS/MS generating base peak (M^++1) = 284 and two main fragment ions at m/z = 167.9 and m/z = 139.9

= 169.9 was a result of N-glicoside bond breaking, thus the glycosides (C5H8O3, M^+ 116) were separated. Meanwhile, the fragment ion m/z = 139.9 was formed from the losing of carbon monoxide (CO, M^+ 28). The same result was occurred in conformation study using 2'deoxyguanosine treated with 50 ppm TBHQ (figure 11).

4. Conclusions

From the result of this study, it was concluded that there was no binding interaction between DNA and BHA in various concentrations 20, 25, 30, 35 and 40 ppm in 10 mM Tris HCl solvent for 1-3 hours incubation time, at temperature 25°C. Based on the analysis by spectrophotometry, it can be concluded that there was DNA binding interactions between DNA and TBHQ in various concentrations 20, 25, 30, 35 and 40 ppm in 10 mM Tris HCl solvent for 1-3 hours incubation time, at temperature 25 ° C. The interaction has been confirmed by the occurance of bathochromic shift as much as 2 – 3 nm while spectrum absorption increased with the increase of TBHQ concentrations. Calf thymus DNA treated with 100 ppm BHA mediated with Copper (II) Chloride and incubated for 4 hours at temperature 37°C did not trigger the DNA Adduct formation 8-OHdG. Calf thymus DNA and 80 ppm TBHQ incubation at 37°C for 4 hours, mediated with Copper (II) Chloride triggered the DNA Adduct 8-OHDG formation. BHA and 2'Deoxyguanosine incubation at 37°C for 1 hour, mediated with Copper (II) Chloride did not trigger the DNA Adduct 8-OHDG formation. Incubation of TBHQ in various concentrations 20-500ppm with 2'Deoxyguanosine at 37°C, for 1 hour mediated with Copper (II) Chloride, triggered the DNA Adduct 8-OHDG formation. There was an increase of DNA Adduct 8-OHDG formation as TBHQ concentrations increased. According to HPLC analysis, relative amount of 8-OHDG formation was $946/10^5$ Deoxyguanosine (DG) from DNA. The conformation test of 8-OHdG formation using LCMS/MS showed the peak of 8-OHdG at retention time 3.52 min and base peak of 8-OHdG (M^++1) = 284, generating two main fragment ions: m/z = 167.9 and m/z = 139.9.

References

[1] Aksu M I and Kaya M 2005 *Meat Sci.* **71** 277-83
[2] Omaye S T 2004 *Food and nutritional toxicology* (New York: CRC Press)
[3] Okubo T, Yokoyama Y, Kano K and Kano I 2003 *Food Chem. Toxicol.* **41** 679-88
[4] Badan POM RI 2013 *Peraturan Kepala Badan POM RI No. 38 Tahun 2013 tentang Batas Maksimum Penggunaan Bahan Tambahan Pangan Antioksidan* (Jakarta: Badan POM RI, Indonesia)
[5] WHO 1994 *JECFA Monographs* http://www.fao.org/ag/agn/jecfa-additives/specs/Monograph1/Additive-459.pdf
[6] EFSA Panel on ANS 2012 Scientific Opinion on the Re-evaluation of Butylated Hidroksitoluene BHT (E 321) as A Food Additive *EFSA Journal* **10** 2588
[7] Okubo T, Nagai F, Ushiyama K and Kano I 1997 *Toxicol. Lett.* **90** 11-8
[8] Nagai F, Okubo T, Ushiyama K, Satoh K and Kano I 1996 *Toxicol. Lett.* **89** 163-7
[9] Neumann J R, Morency C A and Russian K O 1987 *BioTechniques* 5 444-8
[10] Koc H and Swenberg J A 2002 *J. Chromatogr.* B **778** 323-43
[11] Vineis P and Husgafvel-Pursiainen K 2005 *Carcinogenesis* **26** 1846-55
[12] Arnett S D, Osbourn D M, Moore K D, Vandaveer S S and Lunthe C E 2005 *J. Chromatogr. B Analyt. Technol. Biomed. Life Sci.* **827** 16-25
[13] Valavanidis A, Vlahogianni T, Dassenakis M and Scoullos M 2009 *Journal of Environmental Science and Health Part* C **27** 120-39
[14] Pietta P G, Simonetti P, Gardana C, Christoni S, Bramati L and Mauri P L 2002 *Journal of Pharmaceutical and Biomedical Analysis* **32** 657-61

IOP Conf. Series: Materials Science and Engineering **188** (2017) 012011 doi:10.1088/1757-899X/188/1/012011

Preparation of copper oxide modified boron-doped diamond electrodes and its preliminary study for CO_2 reduction

N Y Yetri, T A Ivandini and J Gunlazuardi

Department of Chemistry, Faculty of Mathematics and Natural Sciences Universitas Indonesia, Kampus UI Depok, Depok 16424, Indonesia

Corresponding author's e-mail: ivandini.tri@sci.ui.ac.id

Abstract. Preparation of boron-doped diamond (BDD) modified with Cu_2O (Cu_2O-BDD) electrodes was conducted to study the electrochemical reduction of CO_2. The electrodes were prepared by electrochemical reduction using a solution containing 1mM $Cu(CH_3COO)_2$ and 0.1 M CH_3COONa (1: 1) at pH 5.7 for 60 s. The electrodeposition of Cu_2O at BDD surface was performed by chronoamperometry technique at -0.4 V (vs Ag/AgCl). SEM-EDS and XPS were utilized to characterize the electrodes. At Cu_2O-BDD electrodes, cyclic voltammetry of dissolved CO_2 in 0.1 M NaCl solution exhibited a reduction peak at around -1.3 V (vs Ag/AgCl), indicated the possibility for application in electrochemical reduction of CO_2.

Keywords: Copper oxide, boron-doped diamond, carbon dioxide, electrochemical reduction

1. Introduction

Carbon dioxide is one of main components contained in the atmosphere and is known as one of the greenhouse gas that causes global warming. Since the 19th century, emission of carbon dioxide gas increases dramatically with the rate of around 2 ppm every year. Pletcher [1] has stated that the current content of CO_2 gas has reached around 400 ppm. This phenomenon has attracted the interest of researchers to reduce the emission of CO_2 converting it to more useful compounds. In order to achieve this purpose, some electrochemists have applied certain types of electrodes, including Sn, Pb, Cu, and their related oxides [2, 3] to study the electrochemical reduction of CO_2. It was reported that electrode surface contained copper oxide and can increase the formation of hydrocarbons [4]. The use of copper as a catalyst in CO_2 electroreduction by coating the electrode with Cu_2O was also reported to efficiently produce hydrocarbons in ambient temperature [5].

On the other hand, boron-doped diamond (BDD) electrodes are reported to have various advantages, such as high stability, wide working potential, as well as very low background current [6, 7]. Nakata *et al.* [8] reported that the electrochemical reduction of CO_2 by using BDD electrodes can be performed in NaCl or sea water as the electrolyte. Further, Panglipur *et al.* [9] have employed copper-modified BDD electrodes for the electroreduction of CO_2. In this work, we modify BDD with copper oxide. It is expected that by applying this electrode the percent yield of useful products can be increased.

2. Materials and methods

The materials used were CO_2, N_2, $Cu(CH_3COO)_2$, CH_3COONa, CH_3COOH, H_2SO_4, NaCl, 2-propanol, $HClO_4$, phosphate buffer, and acetonitrile.

Prior to copper deposition, cyclic voltammetry of a solution containing 1 mM Cu $(CH_3COO)_2$ and 0.1 M CH_3COONa with a ratio of 1: 1 at pH 5.7 was performed using BDD as the working electrode in the potential range of -0.2 V to -1.0 V (vs. Ag/AgCl). Chronoamperometry technique with a deposition time of 60 s was then applied at the peak potential attributed to the Cu(I) formation. The characterization of the modified was performed using SEM-EDS and XPS.

Electrochemical reduction of CO_2 was performed in a cell with three-electrode system with 0.1 M NaCl as the electrolyte. Cu_2O-BDD was employed as the working electrode, while Ag/AgCl and Pt foil were used as the reference and the counter electrodes, respectively. In order to remove other gas in the solution, N_2 was bubbled to the solution for 30 min. Further, CO_2 was bubbled followed by performing the cyclic voltammetry of the solution at the potential range of 0.0 to -2.0 V. The solution produced was characterized using HPLC.

3. Results and discussion

Cyclic voltammograms of the solution containing 1 mM $Cu(CH_3COO)_2$ and 0.1 M CH_3COONa (1:1) pH 5.7 (figure 1a) shows that two reduction peaks of copper were observed. The first peak at -0.40 V was attributed to the formation Cu(I), while the second peak at -0.60 V indicated the formation of Cu(0). Accordingly, Cu_2O-BDD was prepared by electrochemical reduction of the same solution at the potential of -0.40 V. Calculation based on the amperogram (figure 1b) indicated that 1.72 μg Cu_2O could be deposited at the surface of BDD.

SEM image of the electrode surface (figure 2 a) showed that copper particles could be deposited at the surface of the electrode with the average size of ~1 μm. EDS result showed the surface contained copper with the percent yields of 25.71% as well as oxygen with the percent yields of 10.67%. Meanwhile, the XPS spectrum shows that besides the peaks of C 1s at 285 eV and of O 1s at 530.5 eV, which characterized the surface of BDD [10], a couple peaks of Cu $2p_{3/2}$ and $2p_{1/2}$ at 952.8 eV and 933 eV, respectively (figure 2b) were also observed. The results confirmed the existence of bonds associated with Cu^+ in Cu_2O [11].

The cyclic voltammogram (figure 3) of the solution showed that the longer the time of CO_2 bubbling, the higher the current which starting at about -1.2 V, suggested that onset of electrochemical reduction of CO_2 was around the potential. Chronoamperometry technique at this potential was then required to investigate the products of electrochemical reduction of CO_2 in NaCl solution.

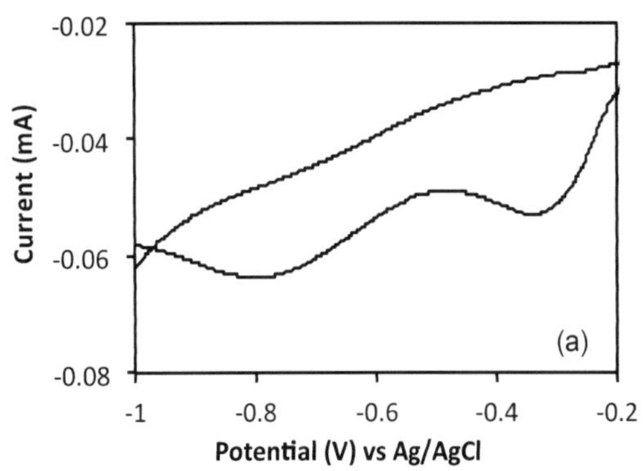

(a)

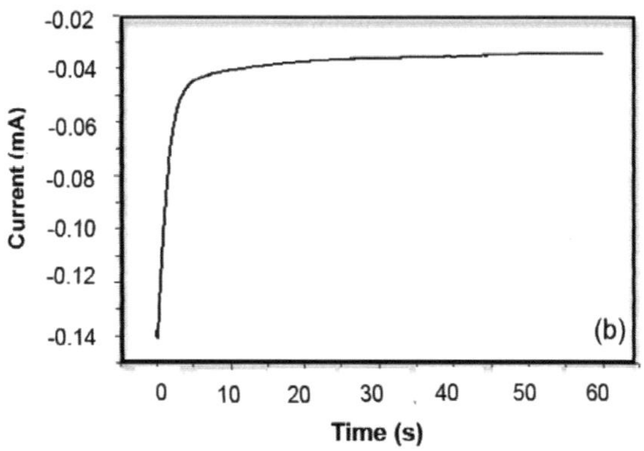

(b)

Figure 1. (a) Cyclic voltammogram and (b) chronoamperogram of the solution containing 1 mM $Cu(CH_3COO)_2$ and 0.1 M CH_3COONa (1:1) pH 5.7. Scan rate of 100 mV/s was applied with BDD electrode as the working electrode.

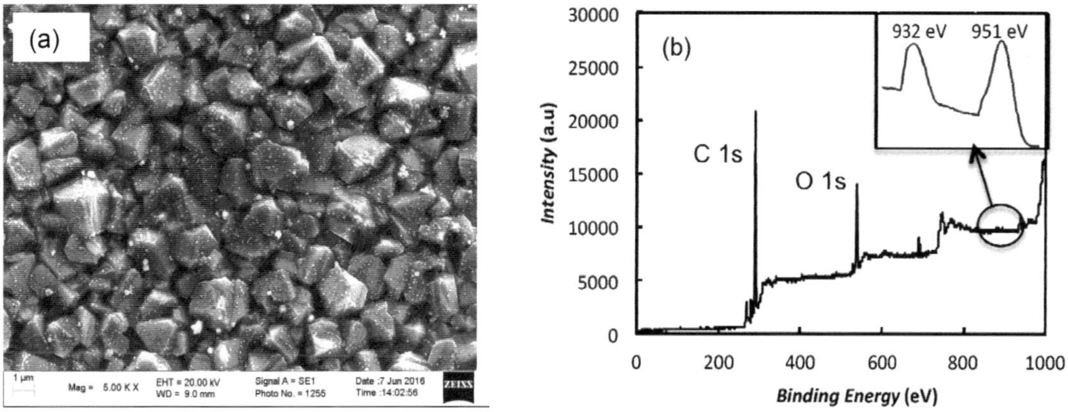

(a) (b)

Figure 2. The characterization of the prepared Cu_2O-BDD electrodes shown by (a) SEM image and (b) XPS spectrum with the magnification of Cu 2p peaks shown in the inset.

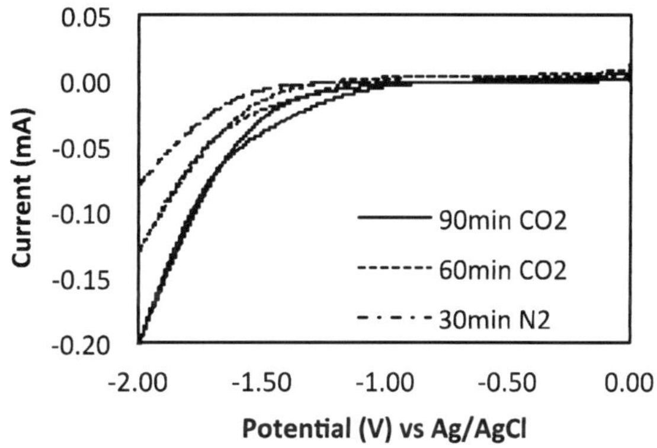

Figure 3. Cyclic voltammograms of 0.1 M NaCl solution containing N_2 (solid line) and CO_2 (dashed lines) at Cu_2O- BDD

4. Conclusions

Cu_2O-modified boron-doped diamond electrodes have been successfully prepared. The electrodes showed the onset of CO_2 reduction in 0.1 M NaCl solution at the potential around -1.3 V (vs. Ag/AgCl), indicated that the electrode is promising for application in CO_2 reduction. More experiments will be conducted to investigate this possibility.

Acknowledgements

This work was funded by Hibah Desentralisasi PUPT Universitas Indonesia, Contract No. 1127/UN2.R12/HKP.05.00/2016.

References

[1] Pletcher D 2015 *Electrochemistry Communications* **61** 97-101
[2] Chang T Y, Liang R M, Wu P W, Chen J Y and Hsieh Y C 2009 *Materials Letters* **63** 1001-3
[3] Albo J, Alvarez-Guerra M, Castano P and Irabien A 2015 *Green Chemistry* **17** 2304-24
[4] Ohya S, Kaneco S, Katsumata H, Suzuki T and Ohta K 2009 *Catalysis Today* **148** 329-34

[5] Kas R, Kortlever R, Yilmaz H, Koper M T M and Mul G 2015 *Chem. Electro. Chem.* **2** 354-8
[6] Fujishima A, Einaga Y, Rao T N and Tryk D A 2005 *Diamond Electrochemistry* (Tokyo: BKC; Amsterdam: Elsevier)
[7] Pino F, Ivandini T A, Nakata K, Fujishima A, Merkoçi A and Einaga Y 2015 *Anal. Sci.* **31** 1061-8
[8] Nakata K, Ozaki T, Terashima C, Fujishima A and Einaga Y 2014 *Angew. Chem. Int. Ed.* **53** 871-4
[9] Panglipur H S, Ivandini T A, Wibowo R and Einaga Y 2016 *AIP Conference Proceedings* **1729** 020047
[10] Ivandini T A, Saepudin E, Wardah H, Dewangga N and Einaga Y 2012 *Anal. Chem.* **84** 9825-32
[11] Zhao W *et al.* 2011 *Crystal Eng. Comm.* **13** 2871-7

Enhanced stability and photocatalytic performance of transition metal-doped ZnO with magnetite nanoparticle and zeolite

M I Pratiwi[1,2], **N Afifah**[1,2] **and R Saleh**[1,2]

[1]Department of Physics, Faculty of Mathematics and Natural Sciences Universitas Indonesia, Kampus UI Depok, Depok 16424, Indonesia
[2]Integrated Laboratory of Energy and Environment, Faculty of Mathematics and Natural Sciences Universitas Indonesia, Kampus UI Depok, Depok 16424, Indonesia

Corresponding author's e-mail: rosari.saleh@gmail.com

Abstract. The combination of zeolite and transition metal-doped ZnO nanoparticles for improved electron and hole photogeneration and inhibited electron-hole recombination, due to the trapping states, has been studied in our previous work. However, the photocatalyst has not been separated and reused after successfully degrading the organic dye. Therefore, in this study, we incorporated four different variations of magnetite nanoparticles into zeolite-supported Fe-doped ZnO, using the co-precipitation method. The samples were characterized with the aid of various measurements, such as x-ray diffraction, infrared absorption, diffuse reflectance spectroscopy, vibrating sample magnetometer (VSM), and Burneur-Emment-Teller (BET). The photocatalytic activity of nanocomposites was examined by photodegradation of methylene blue under UV light irradiation. The results show that the presence of a certain amount of magnetite nanoparticles in a zeolite-supported Fe-doped ZnO nanocomposite improved its efficiency in degrading methylene blue. The role of charged carriers and the active radical involved in the photocatalytic activity is discussed.

Keywords: ZnO, photocatalytic, natural zeolite, Fe-doped

1. Introduction

Photocatalysis is a process of organic pollutant degradation using light irradiation and a catalyst, which has attracted researchers' attention over the past few years [1-4]. Research into the issue has reached a high level of development, because it could be deployed to manage the water pollution issue wrought by the dye wastes produced by various chemical factories. In the photocatalytic process, the catalyst that could be used to degrade organic pollutant in water is a semiconductor. This is a promising approach, since it could degrade the organic pollutant without causing secondary pollution [1-2].

One of semiconductor that is commonly used as a photocatalytic is ZnO, mainly because its low-cost and band gap within the UV light absorption spectrum, it is non-toxic, a strong oxidator, and able to degrade organic pollutants within photocatalytic activity [3-6]. To increase the photocatalytic activity of ZnO, modifications using the doping transition metal ferrite (Fe) [7], as well as the addition of zeolite, were performed in our previous work [2]. Zeolite, with its large surface area and high absorption capacity, could also increase the photocatalytic activity of Fe-doped ZnO (Fe:ZnO) [8-10].

Besides increasing photocatalytic activity, a catalyst with good stability also attracts the interest of researchers. It is known that the use of the magnetic material Fe_3O_4 could increase the magnetic saturation of the sample and enhance the catalyst's stability. Therefore, in this research, the Fe:ZnO/zeolite/Fe_3O_4 nanocomposite was synthesized by using co-precipitation method. Its structural properties, optical properties, magnetic properties, and surface area were also studied. The photocatalytic activity of the nanocomposite, in degrading the methylene blue (MB) as the organic pollutant, was evaluated under UV light radiation. The catalyst's stability was checked after four cycle processes. The main active species in photocatalytic activity was also studied, by adding various scavengers.

2. Experimental details

2.1. Materials

The chemical materials used in this research were purchased from Merck and Sukabumi, Indonesia. Zinc (II) sulfate hepta hydrate ($ZnSO_4.7H_2O$), iron (II) sulfate hepta hydrate ($FeSO_4.7H_2O$), sodium hydroxide (NaOH), ethanol, natural zeolite, methylene blue (MB), ethylene glycol, and acetic acid (20%) were analytical grade and used without further purification. Distilled water was used throughout the experiment.

2.2. Synthesis of the Fe:ZnO/Zeolite/Fe_3O_4 nanocomposites

The Fe:ZnO/zeolite nanocomposite and Fe_3O_4 nanoparticles were each synthesized by using co-precipitation [2] and sol-gel [11] method, as described in our previous works. The Fe-doped ZnO/zeolite/Fe_3O_4 (the zeolite amount was held constant by 30 wt.%) nanocomposite was obtained through the following steps: first, the Fe-doped ZnO/zeolite nanocomposite was dissolved into distilled water, while Fe_3O_4 was stirred into ethanol. Then, the two solutions were mixed and ultrasonicated for two hours. A centrifugation process was conducted to separate the precipitations and solutions. Thereafter, the precipitations obtained were allowed to stand at room temperature for 18 hours, before being dried in the oven at 100°C for one hour in vacuum conditions. The Fe-doped ZnO/zeolite/Fe_3O_4 nanocomposite was produced four times, with weight ratios of Fe-doped ZnO/zeolite to Fe_3O_4 of 1:0.05, 1:0.1, 1:0.5, and 1:1, respectively.

2.3. Characterization

The structure, optical properties, magnetic properties, and surface area of the samples were characterized using spectroscopy X-ray diffraction (Rigaku Miniflex 600), fourier transform infrared (FEI Spectrophotometer Shimazu IR Prestige-21), UV-vis diffuse reflectance (Shimadzu UV-Vis spectrophotometer), vibrating sample magnetometer (Oxford type 1.2 T), and Burneur-Emment-Teller (BET).

2.4. Catalytic activity

The photocatalytic activity of the Fe:ZnO/zeolite/Fe_3O_4 nanocomposite was studied using methylene blue (MB) as the model of organic pollutant. A 40 W UV lamp was used as the light irradiation source for the photocatalytic activity. Then, the MB degradation rate was observed for two hours, at intervals of 15 minutes, using a spectrophotometer from UV-Vis Hitachi UH5300. The catalyst's stability was checked after four times cycle processes. The main active species in the degradation process was also observed by adding some materials, such as $Na_2S_2O_8$, di-ammonium oxalate, and tert-butyl-alcohol, as electrons, holes, and hydroxyl radical scavengers.

3. Results and discussion

XRD spectroscopy was used to study the crystal structure of the synthesized samples. Figure 1a shows the diffraction patterns of the Fe:ZnO/zeolite/Fe_3O_4 nanocomposite sample, with weight ratios of 1:0.05,

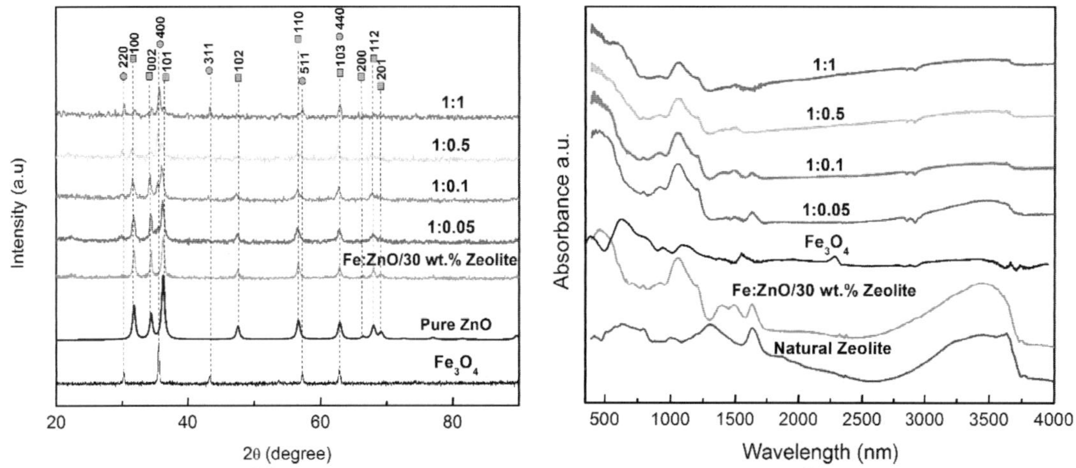

(a) (b)

Figure 1. (a) XRD pattern of Fe_3O_4 nanoparticles, ZnO nanoparticles, Fe:ZnO/Zeolite nanoparticles, and Fe:ZnO/Zeolite/Fe_3O_4 nanocomposites; and (b) FTIR spectra of natural zeolite, Fe:ZnO/Zeolite nanoparticles, Fe_3O_4 nanoparticles and Fe:ZnO/Zeolite/Fe_3O_4 nanocomposites.

Table 1. Lattice parameters, unit cell volume, and crystallite size of nanoparticles.

Sample	Weight ratio	Lattice Parameter (Å) ZnO a=b	Lattice Parameter (Å) ZnO c	Lattice Parameter (Å) Fe_3O_4 a=b=c	Unit cell volume (Å³) ZnO	Unit cell volume (Å³) Fe_3O_4	Crystallite Size (nm) ZnO	Crystallite Size (nm) Fe_3O_4
Fe_3O_4	-	-	-	8.369	-	507.607	-	41
Fe:ZnO/Zeolite	-	3.260	5.191	-	47.775	-	20	-
Fe:ZnO/Zeolite/ Fe_3O_4	1:0.05	3.246	5.183	10.118	47.297	897.212	18	18
	1:0.1	3.247	5.194	12.338	47.419	1626.851	17	21
	1:0.5	3.247	5.195	15.226	47.417	3057.257	13	41
	1:1	3.245	5.191	17.025	47.343	4273.494	11	48

1:0.1, 1:0.5, and 1:1. As comparison, diffraction patterns of the Fe_3O_4 nanoparticle, ZnO nanoparticle, and Fe:ZnO/30 wt.% zeolite nanocomposite can also be seen in the figure. The diffraction pattern of the Fe:ZnO/zeolite/Fe_3O_4 nanocomposite shows the existence of diffraction peaks at 2θ = 30.36°, 35.76°, 43.47°, 53.94°, 57.51°, and 63.74°, which represent the areas (220), (311), (400), (442), (511), and (440) that were from the cubic spinel structure of Fe_3O_4 [12], and diffraction peaks at 2θ = 31.82°, 34.55°, 36.29°, 47.57°, 56.65°, 62.85°, 66.49°, 67.99°, and 69.13°, which represent the areas (100), (002), (101), (102), (110), (103), (200), (112), and (201) that are from the hexagonal wurtzite ZnO structure [13]. No other phase or impurity was found on the obtained diffraction patterns. The lattice parameter value, obtained by the Rietveld refinement technique using the program MAUD, and the crystal size was calculated through the Debye-Scherer formula [14], are summarized in table 1.

The presence of zeolite in the sample could be confirmed using FTIR spectroscopy. Figure 1b shows the absorbent spectra of each sample when exposed to infrared at wave numbers of 400-4000 cm^{-1}. The existence of the stretching vibration of Fe-O at wave number 594 cm^{-1} [15] confirmed the existence of the dopant ion Fe and Fe_3O_4 material in the sample. Meanwhile, the absorption peaks at

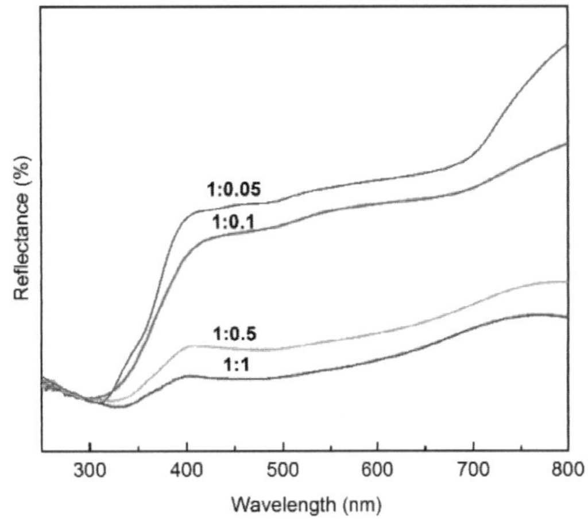

(a)

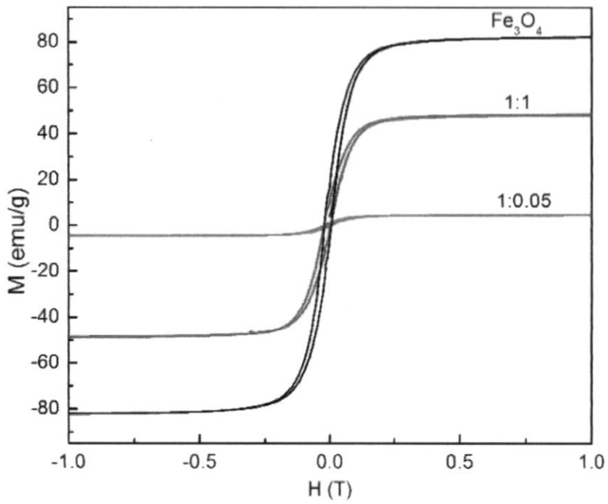

(b)

Figure 2. (a) UV-Vis reflectance spectra of Fe:ZnO/Zeolite/Fe_3O_4 nanocomposites; and (b) VSM spectra of Fe_3O_4 nanoparticles and Fe:ZnO/Zeolite/Fe_3O_4 nanocomposites.

Table 2. Band gap and surface area of Fe:ZnO/Zeolite/Fe_3O_4 nanocomposites.

Sample Weight Ratio	Band Gap (eV)	Surface Area (m^2/g)
1:0.05	3.18	82.11
1:0.1	3.15	79.46
1:0.5	3.08	73.99
1:1	3.06	54.01

wave numbers 1049, 1384, and 1723 cm^{-1}, which stemmed from the stretching vibrations of C-O, O-H, and C = O [16]. A vibration mode of the O-H bond at wave numbers 3000-3500 cm^{-1} [17] was also obtained. The Si-O-Al bending vibration and Si-O-Si stretching vibration were found at wave number 987 cm^{-1} and 1090 cm^{-1}, respectively, confirm the existence of zeolite in the sample [18, 19].

The reflectance of the UV-Vis measurement result of the Fe:ZnO/zeolite/Fe_3O_4 nanocomposite can also be observed in figure 2a. The band gap value of the samples was obtained by extrapolating the UV-Vis reflectance spectra, using the Kubelka-Munk equation [20]. Table 2 shows the band gap value of the samples. The increasing content of Fe_3O_4 in Fe:ZnO/zeolite decreases the band gap energy of ZnO to the smaller energy (red shift).

The magnetic properties of Fe_3O_4 nanoparticles and Fe:ZnO/zeolite/Fe_3O_4 nanocomposite were measured using VSM spectroscopy. The hysteresis magnetic curve of the samples can be seen in figure 2b. The incorporation of Fe_3O_4 into Fe:ZnO/zeolite, with the ratio 1:1, may have increased the magnetic saturation of the sample to 48.355 emu/g, compared to the magnetic saturation of Fe:ZnO/zeolite/Fe_3O_4 with a ratio of 1:0.05 at 5.97 emu/g. All samples showed ferromagnetic behavior at room temperature, so that, in this research, the separation process of the catalyst from the aqueous organic dye became easier and the stability of the former could also be checked.

Degradation of the organic pollutant model MB, using the Fe:ZnO/zeolite/Fe_3O_4 nancomposite, was evaluated under UV light irradiation. Figure 3a shows the photocatalytic activity of the nanocomposite, with a different weight ratio of Fe_3O_4. The result obtained shows that the nanocomposite, with a ratio of 1:0.05, displayed maximum degradation of MB. The increase in photocatalytic activity could have been influenced by several factors, including the surface area of the catalyst. The surface area of the Fe:ZnO/Zeolite/Fe_3O_4 nanocomposite is summarized in table 2. The increasing content of Fe_3O_4 in

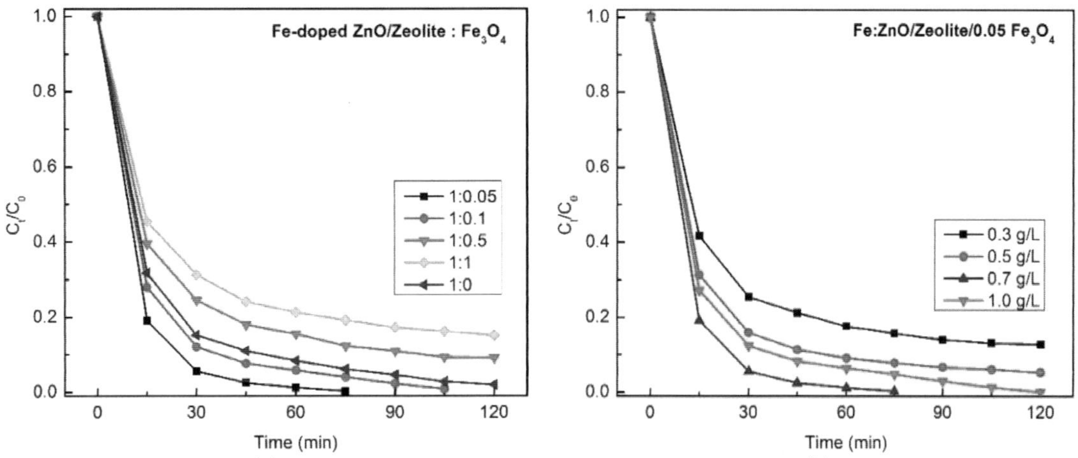

(a) (b)

Figure 3. Photocatalytic activity of Fe:ZnO/Zeolite/Fe_3O_4 nanocomposites with various (a) weight ratio of Fe_3O_4; and (b) dosage concentration.

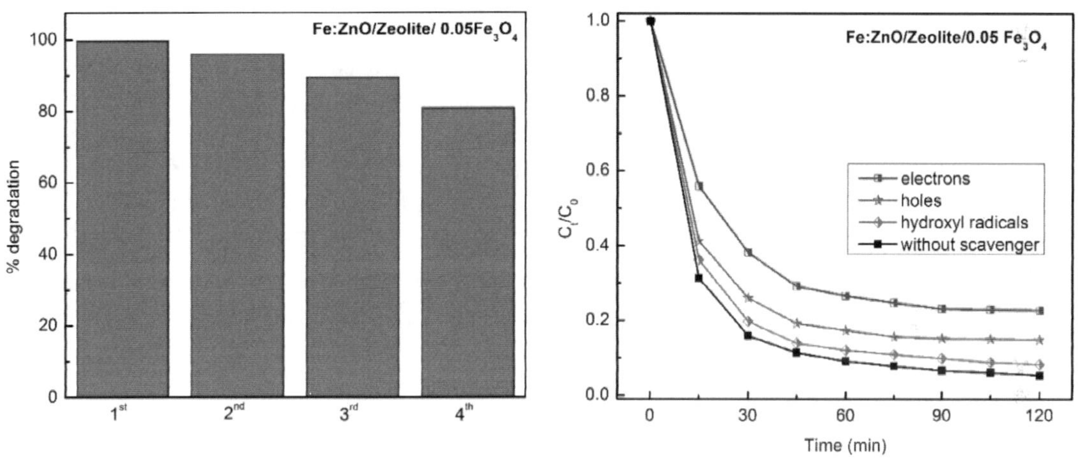

(a) (b)

Figure 4. Photocatalytic activity of Fe:ZnO/Zeolite/Fe_3O_4 nanocomposites (a) after 4 cycles processes; and (b) with the effect of scavenger.

Fe:ZnO/Zeolite decreases the surface area value of the catalyst. This result corresponded with the photocatalytic activity obtained. The reduction of the surface area of the catalyst may decrease the contact probability between the catalyst and the MB while irradiated. Therefore, the photocatalytic performance obtained could also decrease accordingly. Figure 3a also shows that the addition of Fe_3O_4 to the Fe:ZnO/zeolite, with ratio of 1:0.05, could enhance the photocatalytic activity of the nanocomposite if compared to Fe:ZnO/zeolite. It is known that the incorporation of Fe_3O_4 in the nanocomposite could also inhibit the recombination rate of the electron and hole, so that the photocatalytic activity obtained also increases more markedly. To ensure optimum conditions in the photodegradation process, the effect of the dosage catalyst has also been evaluated. Figure 3b shows the photocatalytic degradation of MB with various dosage concentrations of the catalyst (Fe:ZnO/zeolite/0.05 Fe_3O_4), from 0.3 g/L to 1.0 g/L. The results show that a 0.7 g/L catalyst is the maximum dosage concentration for degrading MB.

Besides increasing the photocatalytic activity, the advantage of Fe_3O_4 incorporation in the nanocomposite is that it makes the separation process of the catalyst and solutions easier, by using external magnetism. Then, the separated catalyst was redissolved into a new aqueous MB solution and the photocatalytic activity was rechecked in the same way. The percentage of catalyst degradation after reusing can be seen in figure 4a. The result obtained shows that after four cycles of photocatalytic process, the degradation obtained was decreased, although not significantly. This result indicates that the Fe:ZnO/zeolite/Fe_3O_4 nanocomposite has good stability as a catalyst.

The main active species in the photocatalytic activity was also checked by varying some scavengers. As seen in figure 4b, the addition of tert-butil alcohol, di-amonium oxalate, sodium sulfate (Na_2SO_4) as a hydroxyl radical (•OH), holes, and an electron scavenger in a row was able to inhibit the degradation rate of MB. This demonstrates that the main active species in descending order, in this research, was electron > holes > hydroxyl radical (•OH).

4. Conclusions

The Fe:ZnO/zeolite/Fe_3O_4 nanocomposite was successfully synthesized using the co-precipitation method. The photocatalytic activity of Fe:ZnO/zeolite/ Fe_3O_4, with a ratio of 1:0.05, showed higher degradation than Fe:ZnO/zeolite. The incorporation of Fe_3O_4 in the nanocomposite could also increase the magnetic saturation of the sample, so that its stability as a catalyst could be checked in the photocatalytic activity. After four cycles processes, the Fe:ZnO/zeolite/ Fe_3O_4 nanocomposite showed good stability in degrading methylene blue. The main active species in the photocatalytic activity, in descending order, was electron > holes > hydroxyl radical (•OH).

References

[1] Ruijun C, Kai S, Mingxiang X, Dinghan X and Qingyu Z 2016 *Mater. Sci. Semi. Proc.* **43** 155-62
[2] Afifah N, Adriani S, Djaja N F and Saleh R 2015 *Adv. Mater. Res.* **1123** 295-302
[3] Dong S, Xu K, Liu J and Cui H 2011 *Phys.* B **406** 3609-12
[4] Kumar K, Chitkara M, Sandhu I S, Mehta D and Kumar S 2014 *J. Alloy. Compd.* **588** 681-9
[5] Hassan M S, Touseef A, Yang O B, Kim H C and Khil M S 2012 *Ceram. Int.* **38** 5925-30
[6] Anandan S, Vinu A, Mori T, Gokulakrishnan N, Srinivasu P, Murugesan V and Ariga K 2007 *Catal. Commun.* **8** 1377-82
[7] Saleh R and Djaja N F 2014 *Superlattice. Microst.* **74** 217-33
[8] Fang L M, Zu X T, Li Z J, Zhu S, Liu C M, Zhou W L and Wang L M 2008 *J. Alloy. Compd.* **454** 261-7
[9] Corma A and Garcia H 2004 *Chem. Commun.* **13** 1443-59
[10] Guo Y N, Zu B Y and Dou X C 2013 *J. Thermodyn. Catal.* **4:2** 1000e120
[11] Taufik I, Susanto K and Saleh R 2015 *Mater. Sci. Forum* **827** 37-42
[12] Chidambaram S, Pari B, Kasi N and Muthusamy S 2016 *J. Alloy. Compd.* **665** 404-10
[13] Estévez-Hernández O, Santiago-Jacinto P and Reguera E 2013 *Mater. Focus* **2** 438-42
[14] Khan M E, Alam F, Parveen A and Naqvi A H 2013 *J. Adv. Microscopy Res.* **8** 45-52
[15] Chong S, Zhang G, Tian H, Zhao H 2016 *J. Environ. Sci.* **44** 148-57
[16] Stankovich S, Piner R D, Nguyen S T and Ruoff R S 2006 *Carbon* **44** 3342-7
[17] Xing S, Zhou Z, Ma Z and Wu Y 2011 *Appl. Catal.* B **107** 386-92
[18] Du E, Yu S, Zuo L, Zhang J, Huang X and Wang Y 2011 *Appl. Clay Sci.* **51** 94-101
[19] Chen D, Honglei Z and Wang X 2014 *Appl. Surf. Sci.* **319** 158-66
[20] Lavand A B and Malghe Y S 2015 *J. Saudi Chem. Soc.* **19** 471-8

IOP Conf. Series: Materials Science and Engineering **188** (2017) 012013 doi:10.1088/1757-899X/188/1/012013

The role of aqueous leaf extract of *Tinospora crispa* as reducing and capping agents for synthesis of gold nanoparticles

D O B Apriandanu and Y Yulizar

Department of Chemistry, Faculty of Mathematics and Natural Sciences Universitas Indonesia, Kampus UI Depok, Depok 16424, Indonesia

Corresponding author's e-mail: yokiy@ui.ac.id

Abstract. Environmentally friendly method for green synthesis of Au nanoparticles (AuNP) using aqueous leaf extract of *Tinospora crispa* (TLE) was reported. TLE has the ability for reducing and capping AuNP. Identification of active compounds in aqueous leaf extract was obtained by phytochemical analysis and Fourier transform infrared spectroscopy (FTIR). The AuNP-TLE growth was characterized using UV-Vis spectrophotometer. The particle size and the distribution of AuNP were confirmed by particle size analyzer (PSA). AuNP-TLE formation was optimized by varying the extract concentration and time of the synthesis process. UV-Vis absorption spectrum of optimum AuNP formation displayed by the surface plasmon resonance at maximum wavelength of λ_{max} 536 nm. The PSA result showed that AuNP has size distribution of 80.60 nm and stable up to 21 days. TEM images showed that the size of the AuNP is ± 25 nm.

Keywords: Au nanoparticle, *Tinospora crispa,* green synthesis, capping agent

1. Introduction

Au nanoparticle (AuNP) is (one) of the most stable metal nanoparticles which used in various fields such as medicine, cosmetics, water treatment and catalysis [1-4]. AuNP is known to be selective in various catalytic reactions [5-6].

Synthesis of AuNP has been successfully carried out by the chemical reduction method. It was obtained by using reducing agents such as sodium borohydride, sodium citrate, hydrazine hydrate, dimethylformamide, tollens reagents, and polyethylene glycol [7]. However, these methods used toxic chemical compounds, so the compound can be adsorbed on the surface of AuNP, and caused a toxic side effect [8]. Along with the development of science and technology, AuNP can be obtained by green synthesis methods using environmentally friendly raw materials. This method also has advantages such as lower cost, easily produced in large scale, lower energy, temperature, and pressure requirement, when compared with the physical and chemical methods. Many raw materials are used for reducing and capping agent in AuNP synthesis, such as microorganisms, either bacteria or fungi [9], biopolymer like alginate [10-11], or plant extract [12]. Green synthesis of AuNP has been developed by using a various of plants, namely Guava, Pomelo, Moringa, noni, *Nepenthes khasiana*, hibiscus flower, Ginger, *Solanum nigrum*, *Gymnema sylvestre*, and *Plumeria alba* [13-19].

Indonesian plants can be used as reducing agents in AuNP synthesis. One of them is brotowali leaf (*Tinospora crispa*) as flavonoids source of *glucosylflavones* type as *a-glucosidase* inhibitor [20].

Traditionally, brotowali leaf has been used for the treatment of fever, diabetes, skin infections and antimalarial [21].

Tinospora crispa contains metabolites compounds such as phenols, polyols, amines, flavonoids, tannins, saponins, gallic acid, protein, and reducing sugar which have the ability as a reducing agent [21]. It also has an antioxidant activity [22] that acts as a substitute for the chemical reducing agent in synthesis of AuNP. There is no report of AuNP synthesis from *Tinospora crispa* leaf. The present study describes the synthesis of AuNP by using the aqueous extract of *Tinospora crispa.*

2. Materials and methods

2.1. Material

Gold metal 99.99% was obtained from PT Antam, Indonesia, and dissolved into aqua regia. Stock solution of $HAuCl_4$ was prepared in 100 mL deionized water. Precursor solutions of the desired molar concentrations were prepared by a proper dilution of the stock solution. *Tinospora crispa* leaves were collected from Balitro in Bogor, Indonesia.

2.2. Preparation of Tinospora crispa leaves extract

Tinospora crispa leaves were thoroughly washed with distilled water. They were dried by placing them outdoor for a week. The dried *Tinospora crispa* leaves were grinded to get the fine powder. The 50 g leaf powder was soaked in 250 mL methanol for a week while regularly stirred every day. The result of soaking was separated by filtration. The filtered *Tinospora crispa* leaves was extracted by hexane, ethyl acetate, and water gradually to get the aqueous leaf extract of *Tinospora crispa* (TLE).

2.3. Preliminary phytochemical screening

Each extract was phytochemically tested for the presence of flavonoids, tannins, phenolic, saponosides, terpenoids, steroids, and alkaloids.

2.4. Synthesis of gold nanoparticles

Synthesis of AuNP was done at room temperature by adding 1ml of aqueous leaf extract of *Tinospora crispa* in 9 mL of 0.1 mM $HAuCl_4$ solution. AuNP formation was detected by observing colloids color changed from yellow to light red.

2.5. Characterization of AuNP

Shimadzu 2600 UV-Visible spectrophotometer was used to get absorption spectra of AuNP. Average particles diameter and particle size distribution were measured with Particle Size Analyzer MalvernZEN 1600. Identification of functional groups of biocompounds in *Tinospora crispa* leaves extract before and after the reaction were done using FTIR spectrophotometer (Perkin-Elmer). The Transmission Electron Microscopy JEM 1400 was used to determine the morphology and particle size of the nanoparticles.

2.6. Concentration effect on aqueous AuNP synthesis and its stability

AuNP stability was observed at various concentration of aqueous leaf extract of *Tinospora crispa* (0.20-0.40%). The Effect of concentration on AuNP synthesis was investigated by monitoring the UV-Vis spectrophotometer.

3. Results and discussion

The preliminary phytochemical was perfomed to investigate the secondary metabolite presence of flavonoids, tannins, phenolic, steroids, saponosides, terpenoids, alkaloid in the *Tinospora crispa* leaves extracts. The methanol fraction of *Tinospora crispa* leaves extract showed a positive result of all secondary metabolite. Meanwhile, the aqueous fraction showed the presence of flavonoids and phenolic compounds.

UV-Vis absorption spectra of AuNP-TLE observed at various different times are displayed below in figure 1a. There was no significant reduction of $HAuCl_4$ occurred at 25 min of the reaction which was proved from SPR peak formation. A color change was observed from yellow to pink after 90min reaction at room temperature, confirmed the AuNP formation.

Figure 1b shows the various concentrations of AuNP formation and observation of UV-Vis spectra with high absorbance value of 0.329, the smallest λ_{max} and the sharpest peak shape at 0.25% concentration indicated the optimum concentration of AuNP synthesis.

AuNP has a tendency to agglomerate at the certain time and conditions. Figure 1c shows the stability of AuNP for 21 days, observed from the UV-Vis absorption. At the first day of the synthesis, AuNP-TLE 0.25% has λ_{max} of 536 nm and absorbance value of 0.329. After 21 days, λ_{max} of AuNP shifted to 532 nm. This shifting shows that AuNP is stable for 21 days. AuNP stability is affected by the active compounds in TLE which act as a reducing and capping agent of AuNP. The absorbance values continued to increase up to the 21st days. It decreased after 21 days due to the increasing of AuNP formation. However, after day 21 the decrease was due to the AuNP agglomeration.

TEM characterization was performed to determine the morphology, shape and size of AuNP. Figure 2a and 2b shows the TEM images with a magnification of 40000 and 150000 times. The results showed that the morphology of AuNP-TLE has a sphere shape with a diameter ± 25 nm. To confirm the AuNP, the Miller indices were adjusted with data Join Committee on Powder Diffraction Standards of

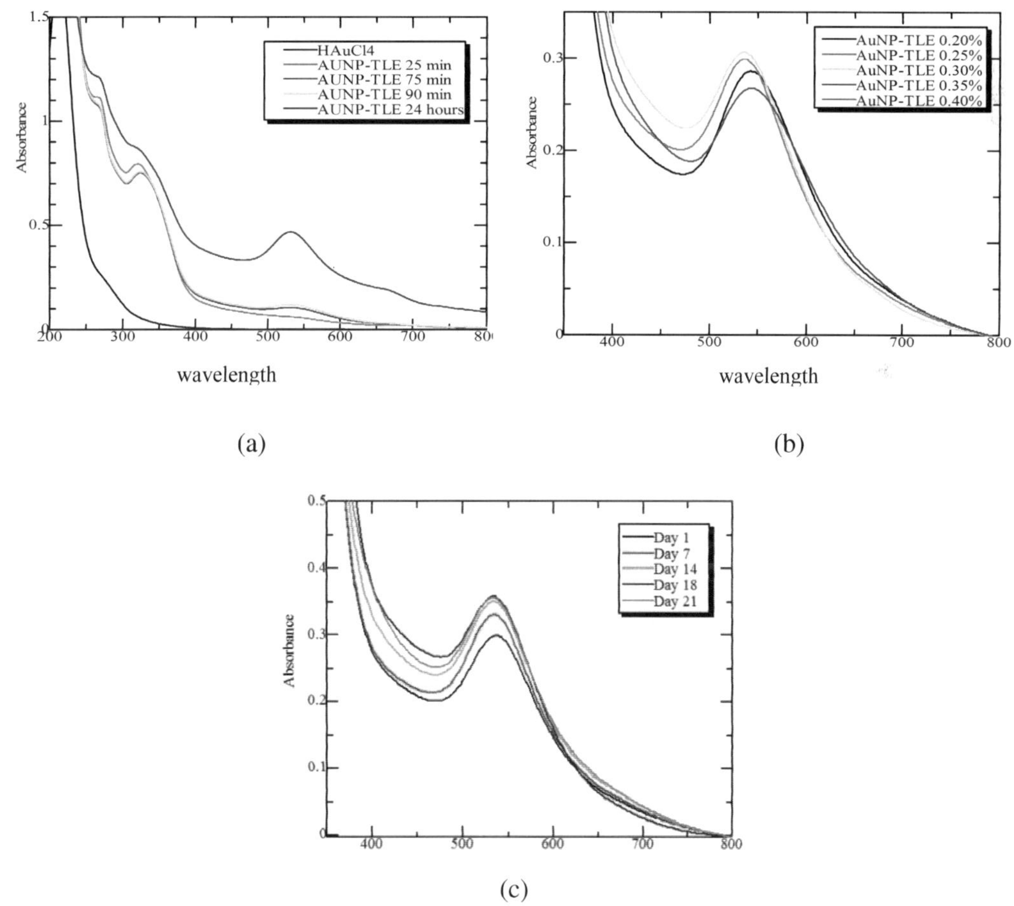

Figure 1. UV–Vis absorption spectra of AuNP-TLE (a) against time reaction (b) at various concentration (c) stability observation in 21 days.

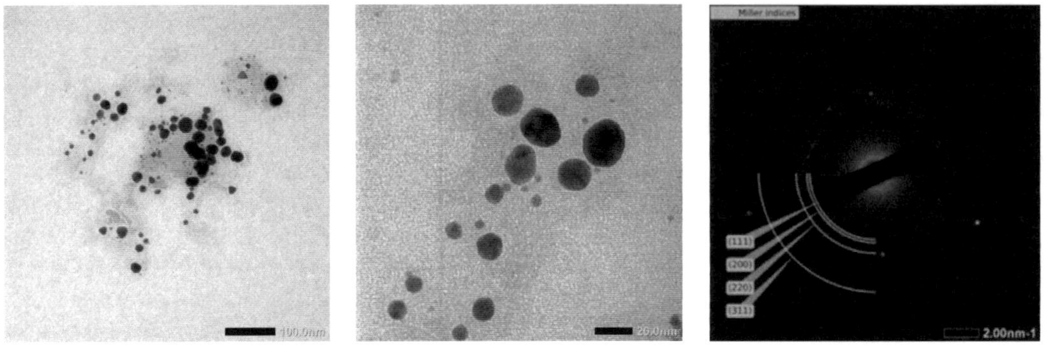

(a) (b) (c)

Figure 2. (a) and (b) TEM images of AuNP-TLE, (c) SAED images of AuNP-TLE diffraction pattern.

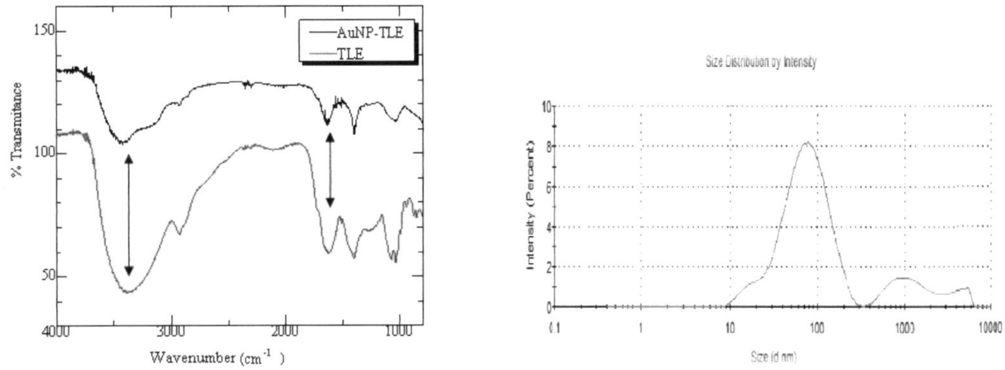

(a) (b)

Figure 3. (a) FTIR spectra for AuNP-TLE and TLE, (b) PSA profile: size distribution of AuNP-TLE.

Au (JCPDSNo. 04-0748) by SAED analysis. The results are shown in figure 2c. The miller indices are (111), (200), (220) and (311) indicated the synthesized AuNP.

FTIR characterization was conducted to determine the interaction between the functional groups of TLE in the AuNP synthesis as shown in figure 3a. FTIR spectra show the wavenumber shifting of TLE before and after AuNP formation. The stretch of -OH group shifted from 3086 to 3138 cm^{-1}. C = O group shifted from 1766 to 1706 cm^{-1}. This shift was due to the interaction of functional group (-OH and -C=O) in TLE compounds with AuNP.

The significant shifts of 3086 and 1766 cm^{-1} was supposed to be the role of –OH group (phenol, alcohol) and -C=O (carboxylic acid and its derivative) as the reducing and capping agent of AuNP. The phenolic, carboxylic, nitrogen compounds, vitamins, and reducing sugars have been reported to play the role in reduction of $HAuCl_4$ and capping of AuNP [22].

PSA Characterization was conducted to determine the particle size and size distribution of AuNP as shown in figure 3b. PSA spectrum peak of AuNP is 80.60 nm. The controlled size of AuNP proves the TLE act as a good reducing agent in the AuNP synthesis.

4. Conclusions

AuNP was obtained using aqueous *Tinospora crispa* leaves extract by green synthesis method. Effect of TLE concentration as $HAuCl_4$ reducing agent in the AuNP formation was investigated. The variations of TLE concentration affected the wavelength, absorbance, and the spectrum shape of AuNP. The AuNP was observed to be stable up to 21 days and the AuNP formation was optimum at λ_{max} 536 nm. The size distribution of AuNP-TLE was confirmed at 80.60 nm using PSA. Based on FTIR characterization, the phenolic and carboxylic compounds acted as reducing and capping agents in the AuNP synthesis. TEM images showed the AuNP size of ± 25 nm.

Acknowledgements
The authors would like to thank Universitas Indonesia for funding this research through PITTA Grant Universitas Indonesia with contract No: 2039/UN2.R12/HKP. 05.00/2016.

References

[1] Meng D L, Shang L, Feng X H, Huang X F and Che X 2016 *Int. Pharmaceutics* **506** 184-90
[2] Lam A T N, Ganbold E O, Cho K H, Kang D S and Joo S W 2014 *Vibrational Spectroscopy* **73** 15-8
[3] Monerris M J, D´Eramo F, Arévalo F J, Fernández H, Zon M A and Molina P G 2016 *Microchemical Journal* **129** 71-7
[4] Mata R, Bhaskaran A and Shadras S R 2016 *Journal of Particuology* **24** 78-86
[5] Corma A and Garcia H 2008 *Chemical Society Reviews* **37** 2096-126
[6] Mitsudomz T and Kaneda K 2013 *Green Chemistry* **15** 2636-54
[7] Ahmed S, Ahmad M, Swami B L and Ikram S 2016 *J. Adv. Research* **7** 17-28
[8] Asmathunisha N and Kathiresan K 2013 *Colloids and Surfaces B, Biointerfaces* **103** 283-7
[9] Shabani M G, Ghahfarokhi M S, Shabani Z G, Akbarzadeh A, Riazi G, Ajdari S, Aman A and Abyaneh M R 2015 *Process Biochemistry* **50** 1076-85
[10] Yulizar Y, Foliatini and Ha□zah M A E 2016 *Arabian Journal of Chemistry* http://dx.doi.org/10.1016/j.arabjc.2016.05.013
[11] Foliatini, Yulizar Y and Hafizah M A E 2015 *Journal of Mathematical and Fundamental Sciences* **47** 31-50
[12] Iravani S 2011 *International Journal Green Chemistry* **13** 2638-50
[13] Yu J *et al.* 2016 *Journal of Materials Letters* **166** 110-2
[14] Anand K, Gengan R M, Phulukdaree A and Chuturgoon A 2015 *Journal of Industrial and Engineering Chemistry* **21** 1105-11
[15] Suman T Y, Radhika Rajasree S R, Ramkumar R, Rajthilak C and Perumal P 2014 *Journal of Spectrochimica Acta Part A: Molecular and Biomolecular Spectroscopy* **118** 11-6
[16] Dhamecha D, Jalalpure S and Jadhav K 2016 *Journal of Photochemistry and Photobiology B: Biology* **154** 108-17
[17] Philip D 2010 *Physica* E **42** 1417-24
[18] Kumar K P, Paul W and Sharma C P 2011 *Journal of Process Biochemistry* **46** 2007-13
[19] Adil S F, Assal M E, Khan M, Al-Warthan A, Siddiqui M R H and Liz-Marzan L M 2015 *Dalton Trans.* **44** 9709-17
[20] Chang C C, Ho S L and Lee S S 2015 *Journal of Bioorganic and Medicinal Chemistry* **23** 3388-96
[21] Abdillah S, Tambunan R M, Farida Y, Sandhiutami N M D and Dewi R M 2015 *Journal of Asian Pac J Trop Dis* **5(6)** 454-7
[22] SaeYoon A, Sukkarn B, Nosoongnoen W, Jantarat C, Hiransai P, Sakdiset P. Chingunpitak J, Makchuchit S and Itharat A 2015 *Asian Journal of Pharmaceutical Sciences* http://dx.doi.org/10.1016/j.ajps.2015.11.001

IOP Conf. Series: Materials Science and Engineering **188** (2017) 012014 doi:10.1088/1757-899X/188/1/012014

Role of temperature-dependent O-p–Fe-d hybridization parameter in the metal-insulator transition of Fe_3O_4: a theoretical study

A D Fauzi[1], M A Majidi[1] and A Rusydi[2,3]
[1]Department of Physics, Faculty of Mathematics and Natural Sciences Universitas Indonesia, Kampus UI Depok, Depok 16424, Indonesia
[2]NUSNNI-NanoCore, Department of Physics, Faculty of Science, National University of Singapore, Singapore 117576, Singapore
[3]Singapore Synchrotron Light Source, National University of Singapore, Singapore 117603, Singapore

Corresponding author's email: aziz.majidi@sci.ui.ac.id

Abstract. We propose a simple tight-binding based model for Fe_3O_4 that captures the preference of ferrimagnetic over ferromagnetic spin configuration of the system. Our model is consistent with previous theoretical and experimental studies suggesting that the system is half metallic, in which spin polarized electrons hop only among the Fe B sites. To address the metal-insulator transition (MIT) we propose that the strong correlation among electrons, which may also be influenced by the electron-phonon interactions, manifest as the temperature-dependence of the O-p–Fe-d hybridization parameter, particularly Fe-d belonging to one of the Fe B sites (denoted as $t^{(2)}_{\mathrm{FeB-O}}$). By proposing that this parameter increases as the temperature decreases, our density-of-states calculation successfully captures a gap opening at the Fermi level, transforming the system from half metal to insulator. Within this model along with the corresponding choice of parameters and a certain profile of the temperature dependence of $t^{(2)}_{\mathrm{FeB-O}}$, we calculate the resistivity of the system as a function of temperature. Our calculation result reveals the drastic uprising trend of the resistivity profile as the temperature decreases, with the MIT transition temperature located around 100 K, which is in agreement with experimental data.

1. Introduction

Transition-metal oxides have attracted a lot of attention of researchers in both fundamental condensed-matter and applied physics area and are intensely studied in last decades. This is due to their interesting properties that could lead to promising technological applications such as superconductivity [1], giant magnetoresistance [2, 3, 4] and metal-insulator transition [5, 6]. Among them, Fe_3O_4, a ferrimagnet with $T_C \simeq 860$ K with a saturation magnetization of 0.6 Tesla ($\sim$471 emu/cm^3) [7], is considered as an important material owing to its technological applications to be a new generation soft magnet. This material has been predicted theoretically and observed experimentally to be a half-metal [8, 9, 10], making it be prospective for spintronic devices [11, 12, 13].

Fe_3O_4, belonging to the $Fd3m$ symmetry group, crystallizes in cubic spinel inverse structure with lattice constant of $a = 8.396$ Å where oxygen anions form a close-packed face-centered-cubic (fcc) sublattice with iron cations located in interstitial sites [14]. One-third of Fe atoms are tetrahedrally coordinated with four oxygen ions as Fe^{3+} at tetrahedral sites (A sites) and the remaining two-third of the Fe atoms consist of a 1:1 mixture of Fe^{2+} and Fe^{3+} that are octahedrally coordinated with six oxygen ions at octahedral sites (B sites). At $T_V \simeq 120$ K, the system experiences metal-insulator transition (MIT) known as Verwey transition [15]. At this transition, the structure distorts from cubic symmetry [16]. In addition, charge ordering (CO) occurs at B sites [17, 18], thus increasing the resistivity by two orders of magnitude.

In this work, using an effective model that explores only the dynamics of the electrons, we aim to show how the MIT occurs in Fe_3O_4, and explain the mechanism underlying it. For this purpose, we construct a tight-binding based Hamiltonian [19] and then calculate the density of states, the electronic energy, and the resistivity of the system. We adjust our model parameters so as to capture the preference of ferrimagnetic over ferromagnetic ordering. Further, within the model, we search for reasonable temperature dependence of a certain parameter that could lead to the occurence of MIT. We find that temperature-dependent parameter of a particular hybridization between O $2p$ and Fe $3d$, which we denote as $t^{(2)}_{\text{FeB-O}}$, plays the key role in the MIT of Fe_3O_4. The increase of $t^{(2)}_{\text{FeB-O}}$ as temperature decreases remarkably also leads to a CO transition that occurs slightly below the MIT temperature. We argue that this temperature dependence of $t^{(2)}_{\text{FeB-O}}$ results from strong correlations due to electron-electron as well as electron-phonon interactions involved in the Verwey structural transition. Our results show that the mechanisms of MIT and CO transition in Fe_3O_4 have the same origin, and it may be connected to the Verwey structural transition.

2. Model

Hinted by the indication of half-metallic behavior [8, 9, 10] and charge ordering [17, 18] occuring only at Fe ions at B sites or B layers, and to simplify our calculations, we define the unit cell of our model by selecting the presumably most important layer in the actual unit cell of Fe_3O_4, as marked with the orange rectangle in figure 1. With this effective two-dimensional unit cell, we propose a tight-binding based Hamiltonian as follows:

$$H = \sum_{\mathbf{k}} \eta^{\dagger}_{\mathbf{k}} [H_0(\mathbf{k})] \eta_{\mathbf{k}} - J_H \sum_{i,j} \mathbf{S}^{(i)}_j \cdot \mathbf{s}^{(i)}_j + U \sum_{i,j} n^{(i)}_{j\uparrow} n^{(i)}_{j\downarrow}, \tag{1}$$

where $\mathbf{k}$ denotes $\mathbf{k}$-point in the Brillouin zone, i is index of Fe sites in a unit cell, while j is index of unit cells in the entire system. The first term of the Hamiltonian represents the kinetic part, whereof $\eta^{\dagger}_{\mathbf{k}}$ is a row vector whose elements are the creation operators associated with the basis orbitals used in the calculation, and $\eta_{\mathbf{k}}$ is its Hermitian conjugate containing the corresponding annihilation operators. $[H_0(\mathbf{k})]$ is a matrix in momentum space. The second term represents the magnetic interactions between spins of the Fe ions and spins of the itinerant Fe electrons in the e_g orbital. The third term describes the local Coulomb repulsive interactions between two Fe electrons occupying the same e_g orbitals.

The $[H_0(\mathbf{k})]$ matrix shown in equation (1) is constructed through tight-binding approximation out of twenty two atomic orbitals forming our Hilbert space. Here, each of four oxygen atoms contributes two p orbitals, and each of three iron atoms contributes one upper e_g orbitals. We order these basis states as follows: $|\mathrm{O_A}p_{x\uparrow}\rangle$, $|\mathrm{O_A}p_{x\downarrow}\rangle$, $|\mathrm{O_A}p_{y\uparrow}\rangle$, $|\mathrm{O_A}p_{y\downarrow}\rangle$, $|\mathrm{O_B}p_{x\uparrow}\rangle$, $|\mathrm{O_B}p_{x\downarrow}\rangle$, $|\mathrm{O_B}p_{y\uparrow}\rangle$, $|\mathrm{O_B}p_{y\downarrow}\rangle$, $|\mathrm{O_C}p_{x\uparrow}\rangle$, $|\mathrm{O_C}p_{x\downarrow}\rangle$, $|\mathrm{O_C}p_{y\uparrow}\rangle$, $|\mathrm{O_C}p_{y\downarrow}\rangle$, $|\mathrm{O_D}p_{x\uparrow}\rangle$, $|\mathrm{O_D}p_{x\downarrow}\rangle$, $|\mathrm{O_D}p_{y\uparrow}\rangle$, $|\mathrm{O_D}p_{y\downarrow}\rangle$, $|\mathrm{FeA}e_{g\uparrow}\rangle$, $|\mathrm{FeA}e_{g\downarrow}\rangle$, $|\mathrm{FeB_1}e_{g\uparrow}\rangle$, $|\mathrm{FeB_1}e_{g\downarrow}\rangle$, $|\mathrm{FeB_2}e_{g\uparrow}\rangle$, and $|\mathrm{FeB_2}e_{g\downarrow}\rangle$. These basis states form a 22×22 matrix in momentum space whose structure is arranged in four 11×11 blocks corresponding to their spin directions as

$$[H_0(\mathbf{k})] = \begin{bmatrix} H_0(\mathbf{k})_{\uparrow} & \mathbf{O} \\ \mathbf{O} & H_0(\mathbf{k})_{\downarrow} \end{bmatrix}, \tag{2}$$

where $\mathbf{O}$ is a 11×11-size zero matrix, and

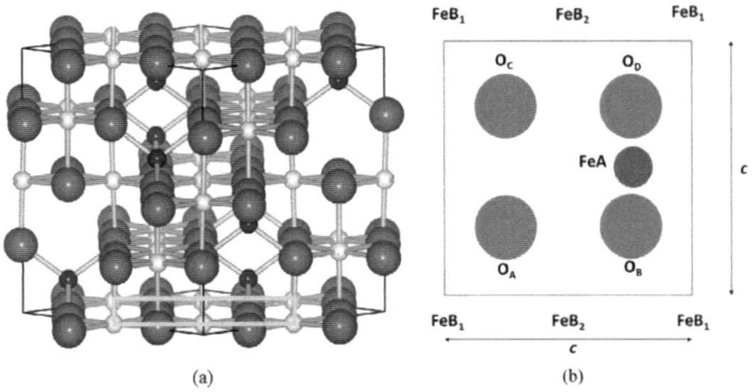

Figure 1. (a) Actual crystal structure of Fe_3O_4 of which the two-dimensional block marked with the orange rectangle is taken to be the effective unit cell of the model. (b) Detail of the model unit cell.

$$H_0(\mathbf{k})_{\uparrow(\downarrow)} =$$

$$\begin{bmatrix}
\varepsilon_{O_A} & 0 & 0 & -2t^{(1)}_{O-O}\cos\alpha & 0 & -2t^{(1)}_{O-O}\cos\beta & -t^{(2)}_{O-O}e^{-i\delta} & 0 & 0 & -t_{FeB-O}e^{i\delta/2} & 0 \\
0 & \varepsilon_{O_A} & -2t^{(1)}_{O-O}\cos\alpha & 0 & -2t^{(1)}_{O-O}\cos\beta & 0 & 0 & -t^{(2)}_{O-O}e^{i\lambda} & 0 & 0 & -t_{FeB-O}e^{-i\lambda/2} \\
0 & -2t^{(1)}_{O-O}\cos\alpha & \varepsilon_{O_B} & 0 & -t^{(2)}_{O-O}e^{-i\delta} & 0 & 0 & -2t^{(1)}_{O-O}\cos\beta & -t_{FeA-O}e^{-i\beta/2} & 0 & -t_{FeB-O}e^{i\delta/2} \\
-2t^{(1)}_{O-O}\cos\alpha & 0 & 0 & \varepsilon_{O_B} & 0 & -t^{(2)}_{O-O}e^{i\lambda} & -2t^{(1)}_{O-O}\cos\beta & 0 & -t_{FeA-O}e^{-i\beta/2} & -t_{FeB-O}e^{-i\lambda/2} & 0 \\
0 & -2t^{(1)}_{O-O}\cos\beta & -t^{(2)}_{O-O}e^{i\delta} & 0 & \varepsilon_{O_C} & 0 & 0 & -2t^{(1)}_{O-O}\cos\alpha & 0 & 0 & -t_{FeB-O}e^{-i\delta/2} \\
-2t^{(1)}_{O-O}\cos\beta & 0 & 0 & -t^{(2)}_{O-O}e^{-i\lambda} & 0 & \varepsilon_{O_C} & -2t^{(1)}_{O-O}\cos\alpha & 0 & 0 & -t_{FeB-O}e^{i\lambda/2} & 0 \\
-t^{(2)}_{O-O}e^{i\delta} & 0 & 0 & -2t^{(1)}_{O-O}\cos\beta & 0 & -2t^{(1)}_{O-O}\cos\alpha & \varepsilon_{O_D} & 0 & -t_{FeA-O}e^{i\beta/2} & -t_{FeB-O}e^{-i\delta/2} & 0 \\
0 & -t^{(2)}_{O-O}e^{-i\lambda} & -2t^{(1)}_{O-O}\cos\beta & 0 & -2t^{(1)}_{O-O}\cos\alpha & 0 & 0 & \varepsilon_{O_D} & -t_{FeA-O}e^{i\beta/2} & 0 & -t_{FeB-O}e^{i\lambda/2} \\
0 & 0 & -t_{FeA-O}e^{i\beta/2} & -t_{FeA-O}e^{i\beta/2} & 0 & 0 & -t_{FeA-O}e^{-i\beta/2} & -t_{FeA-O}e^{-i\beta/2} & \varepsilon_{FeA} & 0 & 0 \\
-t_{FeB-O}e^{-i\delta/2} & 0 & 0 & -t_{FeB-O}e^{i\lambda/2} & 0 & -t_{FeB-O}e^{-i\lambda/2} & -t_{FeB-O}e^{i\delta/2} & 0 & 0 & \varepsilon_{FeB} & -t_{FeB-FeB}\cos\alpha \\
0 & -t_{FeB-O}e^{i\lambda/2} & -t_{FeB-O}e^{-i\delta/2} & 0 & -t_{FeB-O}e^{i\delta/2} & 0 & 0 & -t_{FeB-O}e^{-i\lambda/2} & 0 & -2t_{FeB-FeB}\cos\alpha & \varepsilon_{FeB}
\end{bmatrix}, \quad (3)$$

where $\alpha = k_x c/2$, $\beta = k_y c/2$, $\delta = (k_x + k_y)c/2$ and $\lambda = (k_x - k_y)c/2$. Here, ε, t and $c = a\sqrt{2}/2$, with $a = 8.396$ Å [14], indicate the on-site energy, hybridization parameter and lattice parameter of the model, respectively, while the subscripts indicate the corresponding atoms.

We treat the second and the third terms of the Hamiltonian in equation (1) as a self-consistent local self-energy matrix $[\Sigma_{loc}(z)]$ that depends on the angles representing the relative direction between the electron spin and the spin of the Fe ion at the corresponding site. The elements of this 22×22 matrix are all zero, except for the blocks

$$\begin{bmatrix} \Sigma_{9,9} & \Sigma_{9,20} \\ \Sigma_{20,9} & \Sigma_{20,20} \end{bmatrix} = \begin{bmatrix} -\frac{J_H S\hbar}{2}\cos\theta_{FeA} + U\langle n_{FeA\downarrow}\rangle & -\frac{J_H S\hbar}{2}\sin\theta_{FeA}e^{-i\phi_{FeA}} \\ -\frac{J_H S\hbar}{2}\sin\theta_{FeA}e^{i\phi_{FeA}} & -\frac{J_H S\hbar}{2}\cos\theta_{FeA} + U\langle n_{FeA\uparrow}\rangle \end{bmatrix},$$

$$\begin{bmatrix} \Sigma_{10,10} & \Sigma_{10,21} \\ \Sigma_{21,10} & \Sigma_{21,21} \end{bmatrix} = \begin{bmatrix} -\frac{J_H S\hbar}{2}\cos\theta_{FeB_1} + U\langle n_{FeB_1\downarrow}\rangle & -\frac{J_H S\hbar}{2}\sin\theta_{FeB_1}e^{-i\phi_{FeB_1}} \\ -\frac{J_H S\hbar}{2}\sin\theta_{FeB_1}e^{i\phi_{FeB_1}} & -\frac{J_H S\hbar}{2}\cos\theta_{FeB_1} + U\langle n_{FeB_1\uparrow}\rangle \end{bmatrix},$$

$$\begin{bmatrix} \Sigma_{11,11} & \Sigma_{11,22} \\ \Sigma_{22,11} & \Sigma_{22,22} \end{bmatrix} = \begin{bmatrix} -\frac{J_H S\hbar}{2}\cos\theta_{FeB_2} + U\langle n_{FeB_2\downarrow}\rangle & -\frac{J_H S\hbar}{2}\sin\theta_{FeB_2}e^{-i\phi_{FeB_2}} \\ -\frac{J_H S\hbar}{2}\sin\theta_{FeB_2}e^{i\phi_{FeB_2}} & -\frac{J_H S\hbar}{2}\cos\theta_{FeB_2} + U\langle n_{FeB_2\uparrow}\rangle \end{bmatrix}. \quad (4)$$

Here, the second term, i.e. the spin-spin interaction term, is treated following Furukawa's implementation [20] of the dynamical mean-field theory (DMFT) [21]. Whereas, the third term of the Hamiltonian, i.e. the Hubbard U term, is treated within the mean-field approximation, where $\langle n_{FeA\uparrow}\rangle$, $\langle n_{FeB_1\uparrow}\rangle$, $\langle n_{FeB_2\uparrow}\rangle$, $\langle n_{FeA\downarrow}\rangle$, $\langle n_{FeB_1\downarrow}\rangle$, and $\langle n_{FeB_2\downarrow}\rangle$ are the averaged occupation values of the electrons in Fe ions with spin up and down. In order to make sure that our choice of parameters leads to the preference of ferrimagnetic (up-up-down) over ferromagnetic (up-up-up) configurations of the spins of the Fe ions in each unit cell, we perform our calculations under each different spin configuration. For ferrimagnetic configuration, we set $\theta_{FeA} = \pi$ and $\theta_{FeB_1} = \theta_{FeB_2} = \phi_{FeA} = \phi_{FeB_1} = \phi_{FeB_2} = 0$, while for ferromagnetic

IOP Conf. Series: Materials Science and Engineering **188** (2017) 012014 doi:10.1088/1757-899X/188/1/012014

configuration, $\theta_{\mathrm{FeA}} = \theta_{\mathrm{FeB}_1} = \theta_{\mathrm{FeB}_2} = \phi_{\mathrm{FeA}} = \phi_{\mathrm{FeB}_1} = \phi_{\mathrm{FeB}_2} = 0$.

3. Method

We solve the model by applying the DMFT algorithm as used in reference [22] to calculate the Green function and the corresponding physical quantities of interest. The total electronic energy of the system can be calculated as

$$E = \int \omega\, \mathrm{DOS}(\omega)\, f(\mu,T,\omega)\, d\omega \;-\; U(\langle n_{\mathrm{FeA}\uparrow}\rangle\langle n_{\mathrm{FeA}\downarrow}\rangle + \langle n_{\mathrm{FeB}_1\uparrow}\rangle\langle n_{\mathrm{FeB}_1\downarrow}\rangle + \langle n_{\mathrm{FeB}_2\uparrow}\rangle\langle n_{\mathrm{FeB}_2\downarrow}\rangle). \quad (5)$$

Here, $f(\mu,T,\omega)$ is Fermi-Dirac distribution function with μ indicating the chemical potential of the system. We also calculate the conductivity tensor as

$$\sigma_{\alpha\beta}(T) = \frac{\pi e^2}{\hbar c}\int d\nu \frac{(f(\mu,T,\nu))(1-f(\mu,T,\nu))}{T}\frac{1}{N}\sum_{\mathbf{k}} \mathrm{Tr}[v_\alpha(\mathbf{k})][A(\mathbf{k},\nu)][v_\beta(\mathbf{k})][A(\mathbf{k},\nu)]. \quad (6)$$

Here, $[v_\lambda(\mathbf{k})] = \partial[H_0(\mathbf{k})]/\partial k_\lambda$ denotes the Cartesian component of the velocity matrix and $[A(\mathbf{k},\nu)] = ([G(\mathbf{k},\nu+i0^+)] - [G(\mathbf{k},\nu-i0^+)])/(2\pi i)$ denotes the spectral function matrix. In order to restore the proper physical unit, the dimensional prefactor $\pi e^2/(\hbar c)$ is introduced since the rest of the expression was derived by setting $e = \hbar = 1$. As we are interested to explore the temperature dependence of resistivity, we define the resistivity as inverse of the corresponding diagonal element of the conductivity tensor, that is $\rho_{\alpha\alpha} = 1/\sigma_{\alpha\alpha}$. Here, as x and y directions are not completely equivalent in our model, we expect that ρ_{xx} and ρ_{yy} to be slightly different quantitatively, but pretty much the same qualitatively.

4. Results and discussion

Based on some physical considerations, we estimate and set the following parameter values for our calculations: $\varepsilon_{\mathrm{O_A}} = \varepsilon_{\mathrm{O_C}} = -1.5$ eV, $\varepsilon_{\mathrm{O_B}} = \varepsilon_{\mathrm{O_D}} = -1.8$ eV, $\varepsilon_{\mathrm{FeA}} = 0.5$ eV, $\varepsilon_{\mathrm{FeB}} = 0.4$ eV, $t_{\mathrm{FeA-O}} = 1.8$ eV, $t^{(1)}_{\mathrm{FeB-O}} = 0.9$ eV, $t^{(1)}_{\mathrm{O-O}} = 0.4$ eV, $t^{(2)}_{\mathrm{O-O}} = 0.3$ eV, $t_{\mathrm{FeB-FeB}} = 0.825$ eV, $t_{\mathrm{FeA-FeB}} = 1$ eV, $U = 4$ eV and $J_H = 1$ eV. As for $t^{(2)}_{\mathrm{FeB-O}}$, we vary this parameter as a function of temperature for the reason we elaborate later. Our calculated densities of states for various temperatures are shown in figure 2 for which the energy value is defined relative to the chemical potential. The inset of figure 2 clearly shows that at low temperatures ($T < 100$ K) the system forms a gap around the chemical potential, indicating that it is in an insulating phase. The bands on the left side of the chemical potential arises from O $2p$ orbitals, while Fe $3d$ orbitals contribute to the bands on the right side of the chemical potential. By setting $t^{(2)}_{\mathrm{FeB-O}}$ to decrease as temperature increases following a certain functional form (to be discussed further later), we can see that the gap gradually disappears turning the system into a metal at high temperatures ($T > 100$ K). This demonstrates that the MIT is driven by the temperature dependence of $t^{(2)}_{\mathrm{FeB-O}}$.

To analyze the origin of the gap and investigate the mechanism of MIT in Fe_3O_4, we show in figure 3 the projected densities of states contributing to states around the chemical potential at high temperatures, and show how they evolve as temperature decreases. These states come from both Fe at B sites (i.e. Fe B_1 and Fe B_2 in figure 1). It is clearly seen in the figure that the gap forms due to splitting of Fe B_1 and Fe B_2 bands. At high temperatures Fe B_1 and Fe B_2 bands are roughly equally filled, but at low temperatures the Fe B_1 states are filled with more electrons than the Fe B_2 states are, suggesting a charge ordering occurs. Hence, these two bands are responsible for the closing and opening of the gap around the chemical potential, and thus the mechanism of MIT. The redistribution of electrons among the Fe B_1 and Fe B_2 sites occuring around the MIT transition signifies that the MIT arises concomitantly with the formation of CO phase among electrons in Fe B sites.

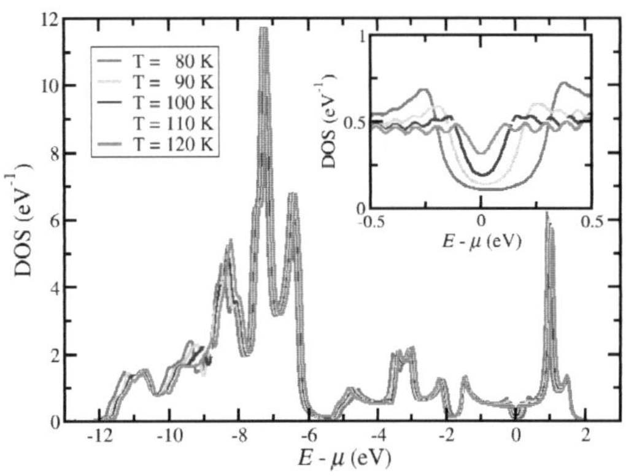

Figure 2. Calculated density of states of Fe_3O_4. The gap is closed at high temperature, indicating the metal-insulator transition occurs in this system.

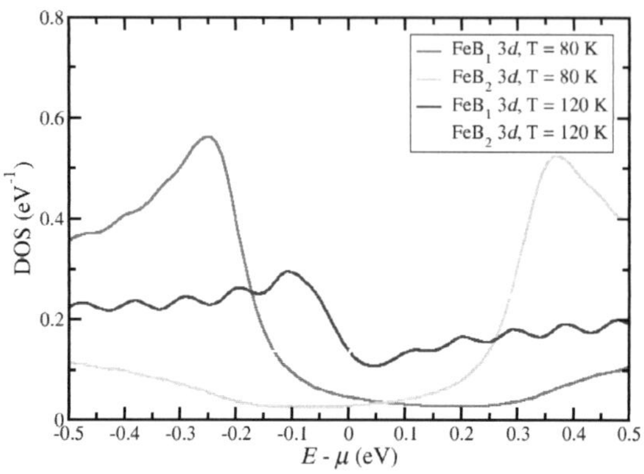

Figure 3. Projected density of states of Fe *d* bands at low (80 K) and high (120 K) temperatures. The dynamics of these two bands control the formation of the gap.

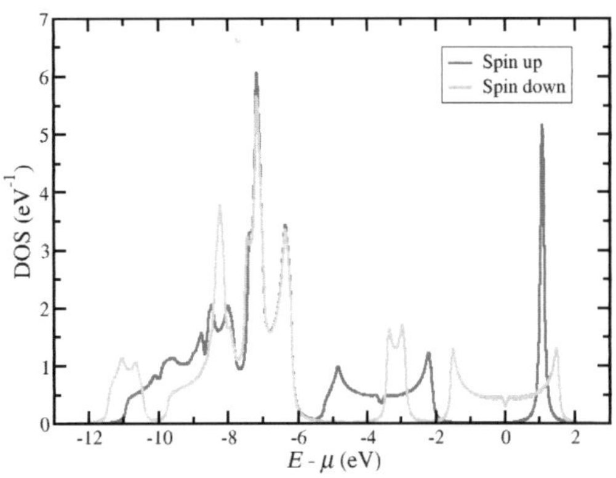

Figure 4. Spin-polarized density of states at $T = 120$ K. The system is half-metal with insulating majority spin channel and metallic minority spin channel.

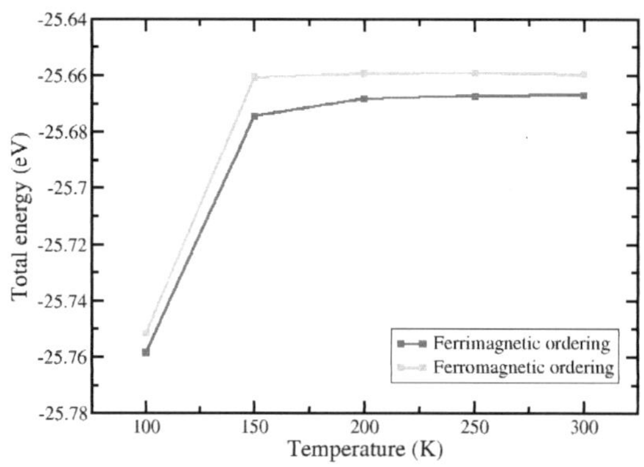

Figure 5. Plot of total electronic energy versus temperature. The system energetically prefers to be in ferrimagnetic spin configuration.

Further, our model also captures the half-metallicity of Fe_3O_4 at high temperatures. As shown in figure 4, the majority (up) spin channel is in insulating state, whereas the minority (down) spin channel has a metallic characteristic. This result confirms the half-metallic nature of Fe_3O_4 which has been observed experimentally [10].

Figure 5 shows the calculated total electronic energy versus temperature for both ferrimagnetic and ferromagnetic spin configurations. It can be seen that the system has lower energy when the Fe spins are in ferrimagnetic configuration compared to that in ferromagnetic configuration. In the ferrimagnetic configuration, for every unit cell in our model the Fe spins at B sites are antiparallel to the spin of Fe at A site. Hence, the two magnetic moments of 5 μ_B from Fe^{3+} at A and B sites are cancelled, leaving the net magnetic moment of 4 μ_B from the remaining Fe^{2+} at B site. This magnetic moment corresponds to the saturation magnetization value of 0.6 Tesla observed in experiment [7].

In order to show the MIT process as observed in experiment, we calculate temperature-dependent resistivity of our Fe_3O_4 model and show the results in figure 6. Within our model, the resistivity calculated along the x direction, ρ_{xx}, has slightly different values from that calculated along the y

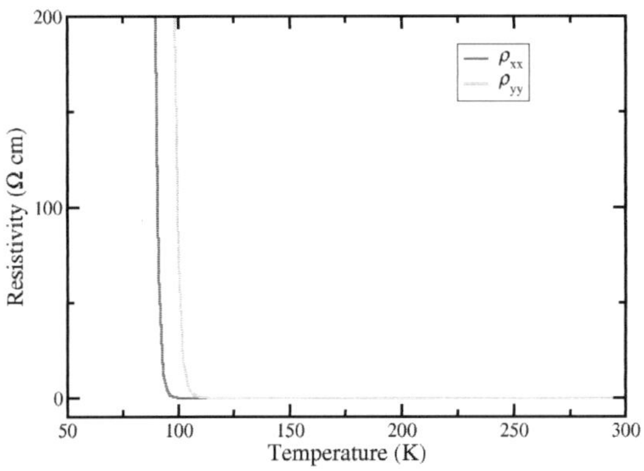

Figure 6. Resistivity as a function of temperature calculated along x and y directions described in the model unit cell. The Verwey transition is located at $\sim$100 K.

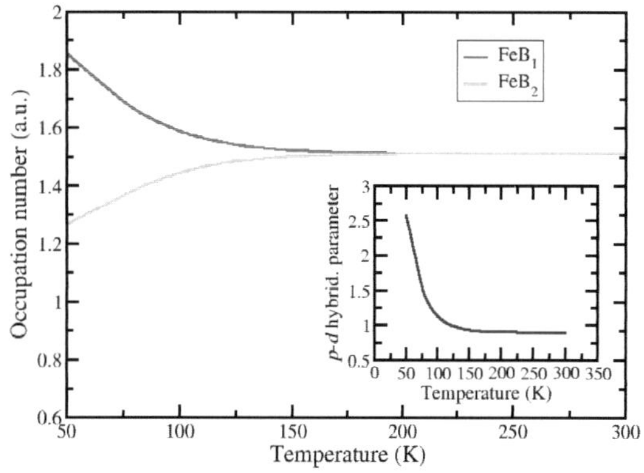

Figure 7. Occupation numbers of electrons at Fe B_1 and Fe B_2 sites as functions of temperature. Inset: the temperature-dependence of $t^{(2)}_{\mathrm{FeB-O}}$.

direction, ρ_{yy}. However, both have exactly the same profile in which an abrupt change indicating the MIT is observed at $\sim$100 K, which was initially referred to as Verwey transition [15]. The temperatures about which the abrupt resistivity change occurs is close to the transition temperature T_V obtained from experiment ($\sim$120 K). As already mentioned before, the transition from metal at high temperatures to insulator at low temperatures is due to charge localization at one of the two Fe B sites (i.e. the B1 sites in this case). This charge localization can be illustrated by the imbalanced occupation numbers of electrons at of Fe B_1 and B_2 sites as the system is brought to low temperatures, as shown in figure 7. This result is consistent with the experimental observation of charge ordering occurring in Fe_3O_4 at low temperatures [17, 18], which according to our model, is responsible for the insulating behavior of the system.

Now we elaborate why and how we propose the temperature dependence of $t^{(2)}_{\mathrm{FeB-O}}$. Within our model that does not incorporate electron-phonon interactions explicitly, the possible abrupt change in conductivity with temperature can only be accommodated if we allow for some parameter(s) to depend on temperature in some manner, presumably due to correlation effects arising from electron-phonon interactions. To find out which parameter(s) having such a temperature dependence, we analyze the projected DOS profiles around the chemical potential as shown in figure 3. From there, we see that the gap can be opened and closed by adjusting the parameter that controls the distribution of states below and above the chemical potential between Fe B_1 and Fe B_2. Hence, we infer that the parameter playing such a role is the hybridization between O $2p$ to Fe $3d$ at B_1 sites, i.e. the $t^{(2)}_{\mathrm{FeB-O}}$. Therefore, to properly capture the temperature-dependent physics phenomenologically within our model, we propose that $t^{(2)}_{\mathrm{FeB-O}}$ depends on temperature in the following way:

$$t^{(2)}_{\mathrm{FeB-O}}(T) = Ae^{-\frac{T}{T^*}} + B, \tag{7}$$

where $A \approx 12$ eV, $T^* \approx 25.34$ K and $B \approx 0.9$ eV. The plot of $t^{(2)}_{\mathrm{FeB-O}}(T)$ as a function temperature is shown in the inset of figure 7. Such a temperature-dependent behavior of a hybridization parameter between O p and d orbital of a transition metal element is not a completely new phenomenon as it has been observed experimentally in some other systems [23]. For our model of Fe_3O_4, as discussed above, this temperature-dependence of $t^{(2)}_{\mathrm{FeB-O}}$ leads to MIT as well as CO transition in the system.

5. Concluding remarks

In conclusion, we have developed a theoretical model to investigate the role of temperature-dependent O-p−Fe-d hybridization parameter in the metal-insulator transition of Fe_3O_4. Our model has successfully explained the transition, capturing the Verwey transition temperature at $\sim$100 K, the ferrimagnetic configuration and the half-metallicity of the system. Our calculation results suggest that the MIT occurs concomitantly with charge ordering among electrons of Fe at B sites. We argue that the temperature dependence of the O-p−Fe-d hybridization parameter arises as a manifestation of Verwey structural transition felt by the electrons through electron-phonon interactions.

Acknowledgements

We thank Universitas Indonesia, Indonesia, for the funding support for this project through PITTA Research Grant No. 2022/UN2.R12/HKP.05.00/2016.

References

[1] Bednorz J G and Müller K A 1986 *Phys. B: Condensed Matter* **64** 189; Bednorz J G and Müller K A 1998 *Rev. Mod. Phys.* **60** 585
[2] Von Helmolt R, Wecker J, Holzapfel B, Schultz L and Samwer K 1993 *Phys. Rev. Lett.* **71** 2331
[3] Ramirez A P 1997 *J. Phys.: Condens. Matter* **9** 8171
[4] Coey J M D, Viret M and von Molnár S 1999 *Adv. Phys.* **48** 167-293
[5] Tsuda N, Nasu K, Yanase A and Siratori K 1991 *Electronic Conduction in Oxides* (Berlin: Springer)
[6] Fujimori A and Tokura Y 1995 *Spectroscopy of Mott Insulators and Correlated Metals* (Berlin: Springer)
[7] Smit J and Wijn H P J 1959 *Ferrites* (Eindhoven: Philips' Technical Library)
[8] Yanase A and Siratori K 1984 *J. Phys. Soc. Jpn.* **53** 312-7
[9] Zhang Z and Satpathy S 1991 *Phys. Rev. B* **44** 13319-31
[10] Wang W *et al.* 2013 *Phys. Rev. B* **87** 085118
[11] Seneor P, Fert A, Maurice J L, Montaigne F, Petroff F and Vaurès A 1999 *Appl. Phys. Lett.* **74** 4017
[12] Matsuda H, Takeuchi M, Adachi H, Hiramoto M, Matsukawa N, Odagawa A, Setsune K and Sakakima H 2002 *Jpn. J. Appl. Phys.* **41** L387
[13] Hu G and Suzuki Y 2002 *Phys. Rev. Lett.* **89** 276601
[14] Fonin M, Pentcheva R, Dedkov Y S, Sperlich M, Vyalikh D V, Scheffler M, Rüdiger U and Güntherodt G 2005 *Phys. Rev. B* **72** 104436
[15] Verwey E J W 1939 *Nature* **144** 327-8
[16] Verwey E J, Haayman P W and Romeijn F C 1947 *J. Chem. Phys.* **15** 181
[17] Toyoda T, Sasaki S and Tanaka M 1999 *Am. Mineral.* **84** 294-8
[18] Wright J P, Attfield J P and Radaelli P G 2001 *Phys. Rev. Lett.* **87** 266401
[19] Slater J C and Koster G F 1954 *Phys. Rev.* **94** 1498-524
[20] Furukawa N 1994 *J. Phys. Soc. Jpn.* **63** 3214
[21] Georges A, Kotliar G, Krauth W and Rozenberg M J 1996 *Rev. Mod. Phys.* **68** 13
[22] Majidi M A, Su H, Feng Y P, Rübhausen M and Rusydi A 2011 *Phys. Rev. B* **84** 075136
[23] Yin X *et al.* 2015 *NPG Asia Mater.* **7** e196

Adsorption of heavy metals on amine-functionalized MCM-48

P Taba, P Budi and A Y Puspitasari

Physical Chemistry Laboratory, Universitas Hasanuddin, Jl. Perintis Kemerdekaan KM. 10, Tamalanrea Indah, Kota Makassar, Sulawesi Selatan 90245, Indonesia

Corresponding author's e-mail: paulinataba@unhas.ac.id

Abstract. The ordered mesoporous silica with cubic structure, MCM-48 was synthesized by post-synthesis under basic media using colloidal silica, cetyltrimethylammonium bromide, and Triton X-100. The modified material, NH_2-MCM-48 was prepared using 3-aminopropyl trimetoxysilane (3-APTMS). X-ray diffraction and FT-IR were used to characterize the samples. The modified material was utilized for adsorption of $Cu^{2+}$and Mn^{2+} from aqueous solution. Parameters used for studying the adsorption process were pH, time of contact, and the initial concentrations of Cu^{2+} and Mn^{2+} ions. Desorption of ions from the adsorbent was also studied using several desorbing agents. The pseudo-second order was found to be the kinetic order for the metals adsorption. The adsorption of Cu^{2+} and Mn^{2+} on NH_2-MCM-48 was fixed by the Langmuir model better than the Freundlich model with the capacity of 0.52 and 0.80 mmol g^{-1} for Cu^{2+} and Mn^{2+}, respectively. The best desorbing agents for removing the adsorbed Cu^{2+} and $Mn^{2+}$from the adsorbent were 1 M HNO_3 and 1 M HCl, respectively.

1. Introduction

The presence of heavy metals in the water and soil environments is a worldwide concern because the metals are toxic to living organism, non-degradable, and can accumulate in the food chain. Manganese (Mn) and copper (Cu) are essential metals for human life. However, in high concentration, they can cause health problems. If Cu accumulates inside the body of humans, it will create health problems in some parts, such as brain, skin, pancreas and heart [1, 2], whereas the high levels of Mn exposure can result in a typical Parkinsonism [3]. Therefore, they have to be removed from aqueous solutions the water source. Several ways have been developed to solve the pollution problem caused by metal ions. To name a few are membrane separation [4, 5] bioremediation [6], liquid-liquid and solid phase extractions [7-9], cationic exchange [10, 11], precipitation [12, 13], and adsorption [14-16]. Among them, adsorption is the most popular method in removing heavy metals because it is efficient, simple, and can be applied at low concentration, so it is easy to be conducted [17]. It has been approved that adsorption is effective to reduce the concentration of heavy metal ions from solution as reported by some studies, including the use of zeolite [18-20]. Zeolites were widely used as adsorbents because they have three dimension frameworks with opening pores and high surface area. However, their pores are small (micropores). Therefore, zeolites have limited applications. Many efforts have been performed to find adsorbents with high adsorption capacity. In this case, mesoporous materials, such as mesoporous silica becomes an alternative adsorbent.

Since the recovery of mesoporous silica (M41S family) by Mobile corporation [21], many studies have been conducted using this material as adsorbent because it has pores (higher than zeolites have) that can be adjusted to a wide range sizes using various templates, high surface area, pore volume, thermal stability, and can be made under various pHs, compositions, precursors, temperatures, as well as processing times [22-26]. The materials have been utilized to adsorb organic substances [28-30]) and heavy metals [28]. One member of the family, MCM-48, has a three-dimensional channels [31]. The material has a greater potency for many applications than other M41S materials (MCM-41 and MCM-50) because it is easier for different host molecules to access the pore [32]. This is due to faster diffusion of adsorbates and resistance of pore blocking [33]. Functionalized MCM-48 has attracted many researchers to use it as adsorbent for heavy metals because the selectivity to remove heavy metals from aqueous solution will be improved as reviewed by Benhamou *et al.* [34].

The objective of this work is to investigate the capability of functionalized MCM-48 with triaminopropyltrimetoxysilane (3-APTMS) toward adsorption of Cu^{2+} and Mn^{2+} ions.

2. Materials and methods

2.1. Materials

Reagents utilized in this study were produced by Merck or Sigma-Aldrich and used directly without any treatment. Cetyltrimethylammonium bromide (CTAB), Ludox HS-40 (35% b/b SiO_2, 0.4% b/b Na_2O, and 60.1% b/b H_2O), Triton X-100, 3-aminopropyl-trimethoxisilane, 3-APTMS (Sigma-Aldrich), acetic acid, 30%, sodium hydroxide, and hydrochloric acid were used in this research. The solutions of heavy metals (Cu^{2+} and Mn^{2+}) were made in stock solutions up to 1000 ppm of metal from the corresponding nitrate salts (Algara). Solutions of 1 M nitric acid, and 1 M hydrochloric acid were prepared from analytical grade of corresponding materials.

2.2. Synthesis of MCM-48

A hydrothermal method was used to synthesize MCM-48 following the procedure of Ryoo *et al.* [26] with modification as described elsewhere [35] without addition of sodium chloride as mentioned in the procedure. The removal of surfactant was conducted by washing with a mixture of HCl and ethanol as described in the procedure [35]. However, calcination in a muffle furnace was excluded. After filtering and washing with double distilled water, the remaining water was removed from the product by heating at a temperature of 378 K in an oven. The X-ray powder diffraction (XRD) method was used to characterize MCM-48 before and after the surfactant removal at room temperature with the use of a Panalytical X'Pert Powder diffractometer using a Cu K-alpha as the source of X-ray. The FTIR spectra of the two materials were taken using a Shimadzu: IR Prestige-21 FTIR Spectrometer scanned at wavenumbers ranging from 340-4500 cm^{-1}, a resolution of 4, and the scan number of 300.

2.3. Modification of MCM-48 with 3-APTMS

Modification was conducted following the method of Pirouzmand *et al.* [36] with some adjustment. In this procedure, 1.3 g of $(CH_3O)_3Si(CH_2)_3NH_2$ was used instead of 1.6 g of $(C_2H_5O)_3Si(CH_2)_3NH_2$ as mentioned in the method. The functionalized material produced was NH_2-MCM-48. The FTIR spectrum of the material was taken at the condition mentioned above.

2.4. Adsorption kinetic experiments

To determine the metals removal rate by NH_2-MCM-48, a kinetic study was separately carried out at series of copper nitrate and manganese nitrate solutions with the same initial concentrations of 50 mg/L (Cu^{2+} and Mn^{2+}). The solutions were stirred in the presence of 100 mg adsorbent at room temperature for various selected times ranging from 100 to 220 min for Cu^{2+} ion and 600 to 1200 min for Mn^{2+} on a magnetic stirrer at the certain stirring rate. The solutions were filtered to remove the solids and analyzed by atomic absorption spectrophotometer (AAS). The amount of ion adsorbed by NH_2-MCM-48 was calculated by subtracting the initial concentration against the final one at any contact time, *t* [37]. The optimum contact times for Cu^{2+} and Mn^{2+} ions were also obtained from these experiments. A blank experiment was conducted without any adsorbent.

The adsorption process was studied using either pseudo-first order or pseudo-second order kinetics. The Lagergren rate equation is generally utilized [38] for pseudo-first order process. The equation is given in equation (1),

$$\ln (q_e - q_t) = \ln q_e - k_1 t \quad (1)$$

where q_e is the amount of ions adsorbed at equilibrium, q_t is the one adsorbed at time t and k_1 is the rate constant of pseudo-first order process.
The pseudo-second order process is shown in equation (2)

$$\frac{t}{q_t} = \frac{1}{k_2 x q_e^2} + \frac{t}{q_e} \quad (2)$$

For the equation, q_e is the amount of ions adsorbed at equilibrium, q_t is the one adsorbed at time t and k_2 is the rate constant of the pseudo-second order process [38, 39].

2.5. Effect of pH on adsorption of Cu^{2+} and Mn^{2+} ions

The effect of pH on metal adsorption was studied by equilibrating the mixtures at different initial pH values (ranging from 2 to 7) of 50 mg/L of ion solutions before the addition of 100 mg of NH_2-MCM-48. The initial pH was adjusted using HCL and NaOH solutions before the addition of the adsorbent. The solution and the adsorbent were stirred at the optimum contact time. Blank experiments were carried out at each pH value without the addition of adsorbent.

2.6. Effect of initial concentration of metal ions on adsorption

The solutions used in these experiments were made by dilution of the stock solution in aquabidest. The experiments were conducted by putting 100 mg of NH_2-MCM-48 into beakers containing 50 mL of single metal ion solutions and mixed carefully at the optimum contact time and pH. Concentrations used were in the range of 50-600 mg/L to obtain the data to study the adsorption isotherms.

2.7. Desorption studies

Modified mesoporous silica, NH_2-MCM-48, previously in contact with 50 ppm of Cu^{2+} and Mn^{2+} ions, was washed with aquabidest to remove the excess ions from the surface. Furthermore, 0.1 g of adsorbent was inserted into three beaker glasses; each of them contained 50 mL of aquabidest, 1 M HNO_3, 1 M HCl, and 1 M. The mixture was stirred at the optimum time of the ion adsorption. The filtrate was then analyzed to determine the amount of ions removed from the adsorbent. The amount of ions desorbed from the adsorbent was calculated using equation (3).

$$\%\, desorption = \frac{amount\ of\ ions\ desorbed\ from\ the\ desorbent}{amount\ of\ ions\ desorbed\ from\ the\ adsorbent}\ x\ 100\% \quad (3)$$

3. Results and discussion

3.1. Characterization of unmodified and modified MCM-48s

X-Ray diffractograms of unmodified and modified MCM-48s are given in figure 1, which are the characteristic of well-ordered materials as described in previous work [21, 26]. The patterns were indexed in the cubic space group Ia3d. After calcination, the intensities of peaks increased due to the absence of scattering from the surfactant. The peaks also shifted to lower d spacing because of condensation and constriction of the pores.

FTIR spectra of MCM-48 before and after extraction of surfactant are shown in figure 2. Two typical regions for organic molecules (surfactants) are observed in as-synthesized MCM-48. C-H stretching modes consist of CH_2 symmetric ($\nu_s CH_2$ at ca. 2851 cm^{-1}), CH_2 antisymmetric ($\nu_{as} CH_2$ at ca. 2920 cm^{-1}), and terminal CH_3 asymmetric ($\nu_{as} CH_3$ at ca. 3015 cm^{-1}). C-H bending modes are observed at 1512 and 1481 cm^{-1}. After the removal of surfactants C-H stretching and bending are almost disappear. The Si-O stretching vibrations of silicate lattice appear with strong bands at 1225 and 1065 cm^{-1} and the weaker ones at 990 and 793 cm^{-1}. After the surfactant removal, the peak at 1065 cm^{-1} shifts to a higher wave number by approximately 25 cm^{-1}. This indicates that the lattice is contracted during the removal of template as mentioned in the earlier study [27]. The spectra of MCM-48 before and after modification can be seen in figure 2(c). It is clear that after modification of MCM-48 with 3-APTMS, two peaks appear at 4243 and 3368 cm^{-1} indicating the presence of $-NH_2$ functional group. The stretching vibrations at 2932 and 2885 cm^{-1} and bending vibrations at 1643, 1560, and 1491 cm^{-1} belong to –C-H group from 3-APTMS. The shift of peaks at 962 and 800 cm^{-1} to 793 and 694 cm^{-1} indicate an interaction between silanol group and 3-AMTMS. The vibration at 1560 cm^{-1} shows the presence of bending vibration of N-H from the primary amine [40]. The stretching vibration of C-N appears at 1389 cm^{-1}. The peak at the wavenumber of 694 cm^{-1} belongs to the vibration of Si-CH_2-R.

3.2. Optimum time of adsorption and kinetic study

Figure 3 shows the amount of ion adsorbed as a function of the contact time and pH. It is obvious that the adsorption of Cu^{2+} and Mn^{2+} ions increases with the increase of the adsorption time until the equilibrium is reached. Figure 3a shows that the optimum time to adsorb $Cu^{2+}$and Mn^{2+} ions are 200 and 1200 min, respectively. Therefore, these contact times were used for further experiments.

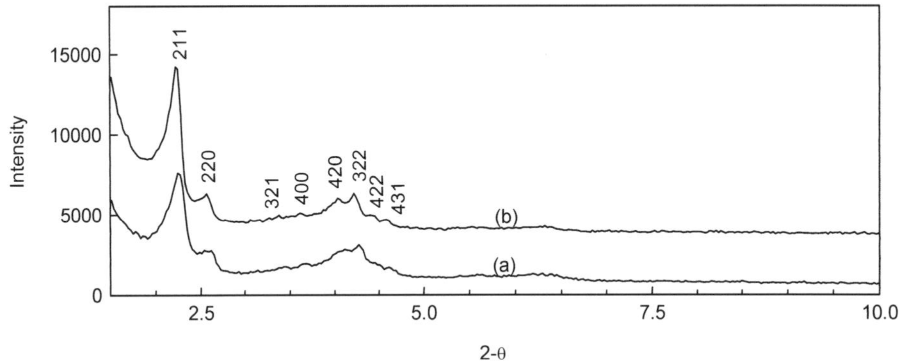

Figure 1. XRD patterns of (a) MCM-48 and (b) NH_2-MCM-48

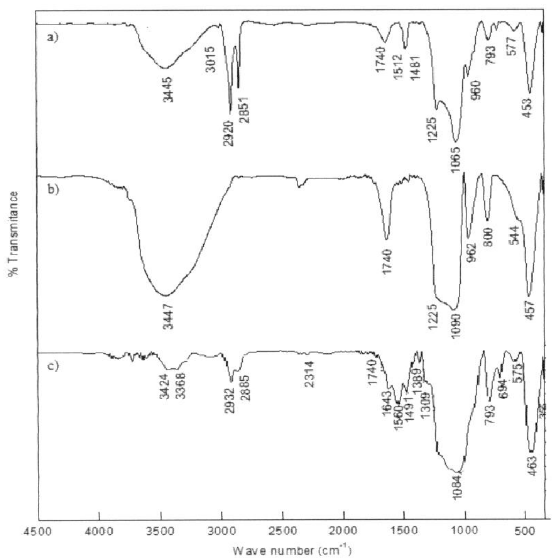

Figure 2. FTIR spectra of materials: (a) as-synthesized MCM-48, (b) MCM-48 after removal of templateand (c) NH_2-MCM-48

IOP Conf. Series: Materials Science and Engineering **188** (2017) 012015 doi:10.1088/1757-899X/188/1/012015

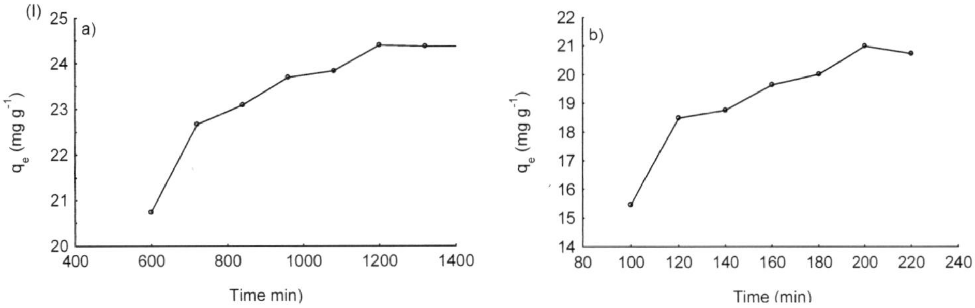

(a)

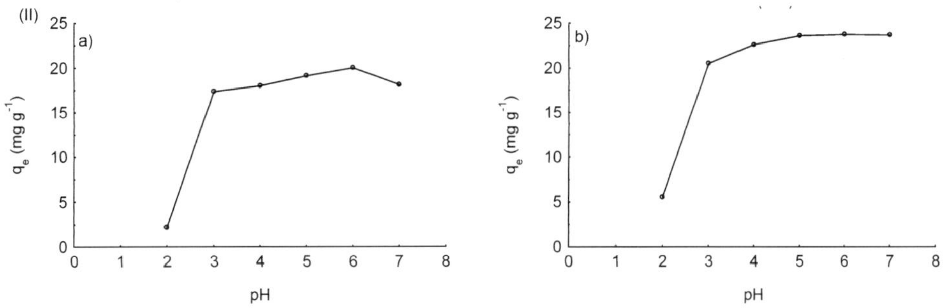

(b)

Figure 3. The influence of (a) contact time and (b) pH on the amount of a) Cu^{2+} and b) Mn^{2+} ions adsorbed on NH_2-MCM-48

Table 1. Kinetic data of metals adsorption on NH_2-MCM-48

Data	Cu^{2+}	Mn^{2+}
Pseudo first order		
k_1(min^{-1})	0.02	3.68 x 10^{-3}
q_e(mg g^{-1})	36.38	30.76
R^2	0.95	0.95
Linear equation	y = -0.0088x + 1.5609	y = -0.0016x + 1.4880
Pseudo second order		
k_2(g mg min^{-1})	3.98 x 10^{-4}	1.62 x 10^{-4}
q_e(mg g^{-1})	29.85	28.82
R^2	0.98	0.99
Linear equation	y = 0.0355x +2.8176	y = 0.0347x + 7.4414
Experimental q_e	20.99	24.40

Using the data of amount adsorbed with contact times, the kinetic study was performed and the kinetic data is given in table 1. All correlation coefficient uses pseudo and second order equations close to 1. However, the amount of ions adsorbed at the equilibrium obtained from the second order equation is closer to the one obtained from experiments. The results indicated that the adsorption of metal ions follow the pseudo-second order with the rate constant (k_2) of 3.98 x 10^{-4} and 1.62 x 10^{-4}g mg min^{-1} for Cu^{2+} and Mn^{2+} adsorption, respectively. The rate constant of Cu^{2+} adsorption is higher than that of Mn^{2+} adsorption. The results were in agreement with the results of the optimum contact times where the optimum contact time of Cu^{2+} adsorption is lower than that of Mn^{2+} adsorption.

3.3. The optimum pH of adsorption

The amount of Cu^{2+} and Mn^{2+} ions adsorbed on MCM-48-NH_2 with various pH solutions is given in figure 3b. The amount of Cu^{2+} and Mn^{2+} ions adsorbed on MCM-48-NH_2 was studied at the pH range of 2-7. It is clear that the amount of ions adsorbed increases sharply from pH of 2 to 3 for both ions. The amount of ions adsorbed increases gradually from pH of 3 to 6 and achieves the optimum at the pH of 6. The amount of ions adsorbed at the pH of 2 is low because the ions compete with the H^+ ion in the solution. In addition, at the lower pH the surface of MCM-48-NH_2 will be protonated to form MCM-48-NH_3^+ that inhibit the ions to be interacted with the adsorbent. At higher pH, the protonated surface is lesser and as a result, the amount of ions adsorbed is higher. At the pH higher than 6, precipitates will be formed so that the amount of ions in the solution will be reduced. From figure 3b, the optimum pH of adsorption for both ions is 6. The pH was used for studying the effect of initial concentrations on the adsorption in order to find the adsorption capacity.

3.4. Adsorption isotherm

The adsorption isotherms of Cu^{2+} and Mn^{2+} ions on MCM-48-NH_2 are given in figure 4. The adsorption of ions increases with the increase of initial concentration of ions. Figure 4 shows that the ions adsorption continues to increase at the measured concentration. Therefore, to find the capacity of adsorption, Langmuir and Freundlich isotherms were utilized. In the former isotherm, monolayer adsorption is assumed to be occurred on the adsorbent's surface with a fixedamount of identical sites [41-42]. The isotherm is shown in equation (4) as the form of linier equation:

$$\frac{C_e}{q_e} = \frac{1}{Q_0 b} + \frac{C_e}{Q_0} \quad (4)$$

where C_e is the metal ion concentration at equilibrium (mg L^{-1}), q_e is the amount of ions adsorbed (mg g^{-1}), Q_o is the capacity of adsorption (mg g^{-1}), and b is the Langmuir coefficient (L mg^{-1}). The latter isotherm (empirical equation) assumes that multilayer adsorption occurs on a heterogeneous surface [41-42]. The equation can be seen in equation (5):

$$log\ q_e = log\ k + \frac{1}{n}\ log\ C_e \quad (5)$$

where q_e is the amount of metal ions adsorbed (mg g^{-1}), C_e is the concentration of metal ions at equilibrium (mg L^{-1}), k is the Freundlich constant associated with the capacity of adsorption (mg g^{-1}), and n is the adsorption intensity. Table 2 summarizes the Langmuir and Freundlich constants of the metal ion adsorption on NH_2-MCM-48.

The adsorption of Cu^{2+} and Mn^{2+} ions on MCM-48-NH_2 fits both Langmuir and Freundlich models. Table 2 shows that the adsorption capacity of Mn^{2+} is higher than that of Cu^{2+}. This is consistent with the Hard Soft Acid Base (HSAB) theory where $-NH_2$ is a hard base that can well interact with Mn^{2+} (a hard acid) compared to Cu^{2+} (a borderline acid).

3.5. Desorption studies

Figure 5 shows the percentage of ions desorbed by several desorbing agent. The amount of both ions desorbed by aquabidest is very small, namely 0.02 and 0.01% for Cu^{2+} and Mn^{2+} ions. The amount desorbed by 1 M HNO_3 and 1 M HCl is considerably higher than that by aquabidest. This indicated that the adsorption of ions was mainly occurred by chemical adsorption. The best desorbing agents for Cu^{2+} and Mn^{2+} ions were 1 M HCl and 1 M HNO_3, respectively. The desorbing agents that behave as an acid can protonate the surface. As a consequence, the ions will be removed from the surface of the adsorbent.

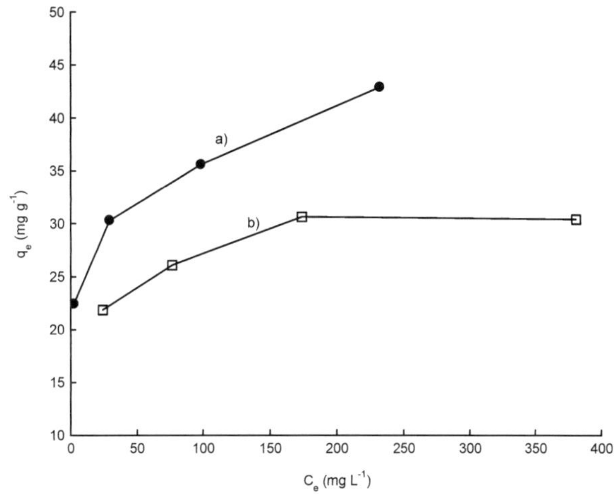

Figure 4. Adsorption isotherms of Cu^{2+}dan Mn^{2+} ions

Table 2. The constants of Langmuir and Freundlich models for the metal ions adsorption on NH_2-MCM-48

Ion	Langmuir model Q_o mg/g	Q_o mmol/g	b (L.mg^{-1})	R^2	Freundlich model k mg/g	k mmol/g	n (g.L^{-1})	R^2
Cu^{2+}	33.00	0.52	0.05	0.99	13.26	0.21	6.65	0.98
Mn^{2+}	44.05	0.80	0.09	0.99	19.35	0.35	7.17	0.98

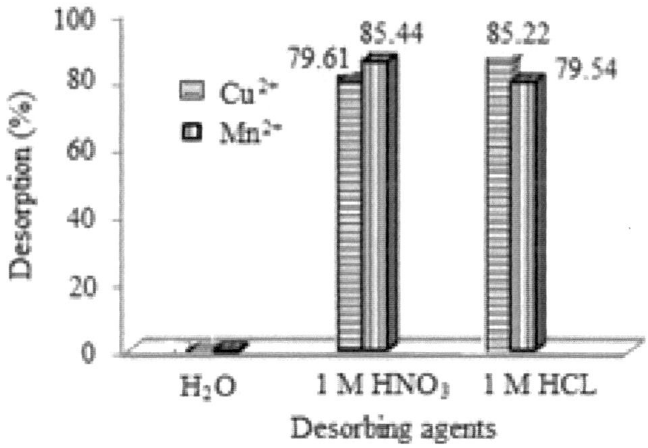

Figure 5. Percentage of ion desorbed from MCM-48 after the adsorption

4. Conclusions

The study concluded that the amount of Cu^{2+} and Mn^{2+} ions adsorbed on MCM-48-NH_2 achieved the optimum at the contact time of 200 and 1200 min, respectively and at the pH of 6 for both ions. Both ion adsorptions fitted the pseudo-second order, in which the rate constant of Cu^{2+}adsorption was higher than that of Mn^{2+} adsorption. In addition, both ion adsorptions followed the Langmuir and Freundlich isotherms. The adsorption capacity for Mn^{2+} ion (0.80 mmol g^{-1}) was higher than that for Cu^{2+} ion (0.52

mmol g^{-1}). Furthermore, the best desorbing agent for Cu^{2+} ion was 1 M HCl solution, whereas for Mn^{2+} ion was 1 M HNO_3

Acknowledgements
Professor Russell F. Howe and Bayu Adi Samodro from the University of Aberdeen are greatly acknowledged for their assistance in XRD measurements.

References

[1] Veli S and Alyuz B 2007 *J. Hazard. Mater.* **149** 226-33
[2] Wang N, Han Y, Liu Y, Bai T, Gao H, Zhang P, Wang W and Liu W 2012 *J. Hazard. Mater.* **213-214** 258-64
[3] Guilarte T R and Gonzales K K 2015 *Toxicol. Sci.* **146** 204-12
[4] Fatin-Rouge N, Dupont A, Vidonne A, Dejeu J, Fievet P and Foissy A 2006 *Water Res.* **40** 1303-9
[5] Gao J, Sun S P, Zhu W P and Chung T S 2014 *Water Res.* **63** 252-61
[6] Kang S Y, Lee J U and Kim K W 2007 *Biochem. Eng. J.* **36** 54-8
[7] Mane C P, Mahamuni S V, Kolekar S S, Han S H and Anuse M A 2012 *Arabian J. Chem.* **9** S1420-7
[8] Su B L, Ma X C, Xu F, Chen L H, Fu Z Y, Moniotte N, Maamar S B, Lamartine R and Vocanson F 2011 *J. Colloid Interface Sci.* **360** 86-92
[9] Al-bishri H M, Abdel-Fattah T M and Mahmoud M E 2012 *J. Ind. Eng. Chem.* **18** 1252-7
[10] Abdel-Aziz M H, Amin N K and El-Ashtoukhy E S Z 2013 *Hydrometallurgy* **137** 126-32
[11] Zewail T M and Yousef N S 2015 *Alexandria Eng. J.* **54** 83-90
[12] Sakai H, Matsuoka S, Zinckenko A A and Murata S 2009 *Colloids Surf. A: Physicochem. Eng. Aspects* **347** 210-4
[13] Fu F, Xie L, Tang B, Wang Q and Jiang S 2012 *Chem. Eng. J.* **189-190** 283-7
[14] Wu S, Li F, Xu R, Wei S and Li G 2010 *J. Nanoparticle Res.* **12** 2111-24
[15] Hu X N, Han Y, Li J Y, Wu J Y, Chen J R and Tang M J 2012 *Adv. Mater. Res.* **413** 148-53
[16] Tresintsi S, Mitrakas M, Simeonidis K and Kostoglou M 2015 *J. Colloid Interface Sci.* **460** 1-7
[17] Cao J, Wu Y, Jin Y, Yilihan P and Huang W 2014 *Taiwan Inst. Chem. Eng.* **45** 860-8
[18] Oliviera L C A, Petkowicz D I, Smaniotto A and Pergher S B C 2004 *Water Res.* **38** 3699-704
[19] Jamil T S, Ibrahim H S, Abd El-Maksoud I H and El-Wakeel S T 2010 *Desalination* **258** 34-40
[20] Motsi T, Rowson N A and Simmons M J H 2011 *Int. J. Miner. Process.* **101** 42-9
[21] Beck J S *et al.* 1992 *J. Am. Chem. Soc.* **114** 10834-43
[22] Kresge C T, Leonowicz M E, Roth W J, Vartuli J C and Beck J S 1992 *Nature* **359** 710-2
[23] Vartuli J C, Kresge C T, Leonowicz M E, Chu A S, McCullen S B, Johnson I D and Sheppard E W 1994 *Chem. Mater.* **6** 2070-7
[24] Huo Q, Leon R, Petroff P M and Stucky G D 1995 *Science* **268** 1324-7
[25] Ryoo R, Kim J M, Ko C H and Shin C H 1996 *J. Phys. Chem.* **100** 17718-21
[26] Ryoo R, Joo S H and Kim J M 1999 *J. Phys. Chem. B* **103** 7435-40
[27] Taba P 2001 *Mesoporous Solids as Adsorbents* PhD Thesis (Sydney: The University of New South Wales, Australia)
[28] Sayari A, Hamoudi S and Yang Y 2005 *Chem. Mater.* **17** 212-6
[29] Taba P, Howe R F and Moran G 2008 *Indo. J. Chem.* **8** 1-5
[30] Taba P 2009 *Indo. J. Chem.* **9** 184-8
[31] Schumacher K, Grün M and Unger K K 1999 *Microporous Mesoporous Materials* **27** 201-6
[32] Olkhovyk O, Antochshuk V and Jaroniec M 2004 *Colloids Surf. A: Physicochem. Eng. Aspects* **236** 69-72
[33] Shim W G, Lee J W and Moon H 2006 *Microporous and Mesoporous Materials* **88** 112-25
[34] Benhamou A, Baudu M, Derriche Z and Basly J P 2009 *J. Hazard. Mater.* **171** 1001-8
[35] Taba P 2008 Makara Sains **12** 120-5

[36] Pirouzmand M, Amini M M and Safari N 2008 *J. Colloid Interface Sci.* **319** 199-205
[37] Deng S and Ting Y P 2005 *Environ. Sci. Technol.* **39** 8490-6
[38] Hossain K Z, Monreal C M and Sayari A 2008 *Colloids Surf. B: Biointerfaces* **62** 42-50
[39] Yurdakoç M, Seki Y, Karahan S and Yurdakoç K 2005 *J. Colloid Interface Sci.* **286** 440-6
[40] Yokoi T, Yoshitake H and Tatsumi T 2004 *J. Mater. Chem.* **14** 951-7
[41] Temoçin Z and Yigitoglu M 2010 *Water, Air, and Soil Pollut.* **210** 463-72
[42] Namasivayam C, Radhika R and Suba S 2001 *Waste Manag.* **21** 381-7

IOP Conf. Series: Materials Science and Engineering **188** (2017) 012016 doi:10.1088/1757-899X/188/1/012016

A numerical study of the sensitivity of surface plasmon resonance bimetallic silver-gold alloys using boundary element method

M H Putra, D Djuhana, V Fauzia, A Harmoko and C Imawan

Department of Physics, Faculty of Mathematics and Natural Sciences Universitas Indonesia, Kampus UI Depok, Depok 16424, Indonesia

Corresponding author's e-mail: dede.djuhana@sci.ui.ac.id

Abstract. In this study, we have systematically investigated the sensitivity of surface plasmon resonance bimetallic silver-gold alloys by means of metallic nanoparticle simulation, MNPBEM with respect to the diameter, mole fraction, and refractive index medium variation. The sensitivity of surface plasmon resonance was obtained from a fitting linear between the value of the peak position LSPR spectra and the refractive index medium. Interestingly, it was found the sensitivity of surface plasmon resonance increased as the diameter increased for a diameter less than 40 nm, whereas the sensitivity of surface plasmon resonance exhibited the fluctuation values for diameter above 40 nm. The highest value of the sensitivity occurred at the silver mole fraction at $x = 0.6$ rather than at $x = 1.0$ (pure silver particle). This result also showed the sensitivity of surface plasmon resonance pure silver particle ($x = 1$) was greater than gold particle ($x = 0$). The sensitivity of surface plasmon resonance was not only affected by the diameter but also influenced by the mole fraction.

1. Introduction

Metallic and bimetallic noble nanoparticle have been attracted much attention due to their potential application for sensing such as biosensor, biochemical sensing, biomedical application, etc. [1, 2]. They produce unique optical properties from the interaction between free electron conduction in noble metallic with the electromagnetic field that known as localized surface plasmon resonance (LSPR). Furthermore, LSPR spectra depends on the size, shape, medium, and configuration [3, 4]. Understanding LSPR spectra includes shifting of the peak position and dielectric sensitivity is very important for designing sensor device based on surface plasmon behavior [5]. Several observations have been reported for the peak LSPR shifting and the sensitivity of surface plasmon resonance, both experimental [6, 7] and simulation [8-10]. However, few observations reported about LSPR shifting such as amplitude and peak position spectra and the sensitivity of surface plasmon resonance bimetallic silver-gold alloy as the diameter, mole fraction, and medium variation.

In this study, we calculated the sensitivity of surface plasmon resonance bimetallic silver-gold alloys with the variation of the silver mole fraction, diameter, and the refractive index medium. We produced LSPR spectra bimetallic silver-gold alloy using a public metallic nanoparticle simulation based on boundary element method. We have observed two behaviors of the LSPR spectra of the silver-gold alloys such as amplitude and the peak position. The LSPR spectra showed shifting to a higher wavelength (red-shift) as the diameter of alloys increased. Interestingly, as the mole fraction of silver increased, the amplitude increased and the peak position shifted to a smaller wavelength (blue-shift).

Then, we calculated the sensitivity of surface plasmon resonance based on the peak position of the LSPR spectra with respect to the refractive index medium and diameter variation.

2. Simulation procedure

We have systematically performed the sensitivity of surface plasmon resonance bimetallic silver-gold alloys by means of a public metallic nanoparticle simulation, MNPBEM based on boundary element method [11]. Bimetallica silver-gold alloys (Ag_xAu_{1-x}) was modeled as a spherical particle with the diameter variation from 10 nm to 100 nm. The mole fraction x was varied from 0 to 1 with 0.2 increment. We have also prepared the variation of the refractive index medium to observe the sensitivity of surface plasmon resonance. In this study, we varied the refractive index medium by mixture of water and glycerol concentration based on Lorentz-Lorentz approximation [12, 13], n = 1.3334 (100% water pure), n = 1.3605 (80% water and 20% glycerol), n = 1.3881 (60% water and 40% glycerol), n = 1.4164 (40% water and 60% glycerol), and n = 1.4452 (20% water and 80% glycerol). The dielectric bimetallic silver-gold alloys were calculated based on the dielectric silver and gold combination with respect to the mole fraction variation [9, 14]. This simulation produced the LSPR spectra such as extinction, absorption, and scattering curve with respect to the wavelength. We only focused on the extinction curve as representative of LSPR behavior in this study.

3. Results and discussion

Figure 1 shows the extinction curve of the silver-gold alloys for the diameter and the mole fraction variation under the refractive index medium water n = 1.3334. For understanding, we showed the extinction curve for small (D = 20 nm), intermediate (D = 60 nm), and large diameter (D = 100 nm). Interestingly, the peak position of the extinction curve shifted to a higher wavelength (red-shift) as the diameter of alloys increased and also occurred in the visible light range. The red-shift phenomenon originated from the bimetallic particle interacted with the electromagnetic field that depended on the material, size, shape, and local dielectric properties. The scattering exhibited to be more dominant than the absorption as the particle size increased and the peak position of the LSPR spectra will shift to a higher wavelength [15, 4]. Further, we found two behaviors of the extinction curve such as amplitude and the peak position. As the silver mole fraction increased, we observed the amplitude also increased and the peak position shifted to a lower wavelength (blue-shift). Similar observations have been reported [16, 8].

Interestingly, the extinction curve bimetallic silver-gold alloys exhibited different behavior compared to the extinction curve from pure silver and pure gold particle. Introducing a hybrid surface plasmon resonance from bimetallic particle that originated from d-band and conduction band. We observed, for small and intermediate diameter, the amplitude of the extinction curve slowly increased until the silver mole fraction at x = 0.6, whereas greater than x = 0.6, the amplitude clearly changed. For the case the mole fraction x = 0 and x = 1 related to LSPR spectra pure gold and silver particles, respectively.

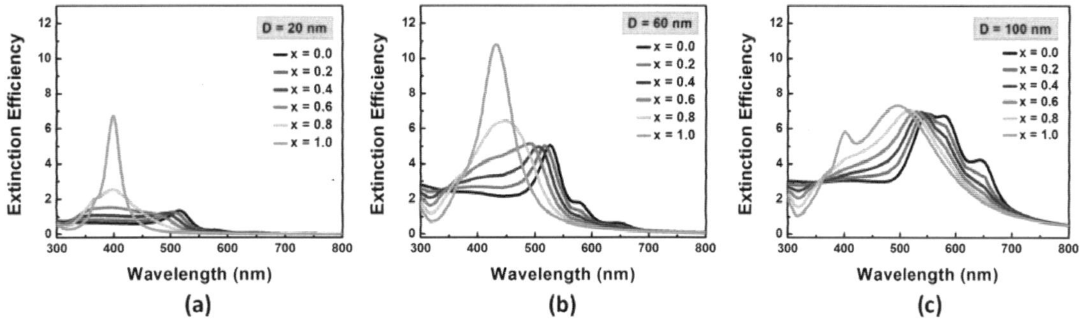

Figure 1. The extinction curve bimetallic silver-gold alloys for the mole fraction variation from x = 0 to x = 1 (a) D = 20 nm, (b) D = 60 nm, and (c) D = 100 nm.

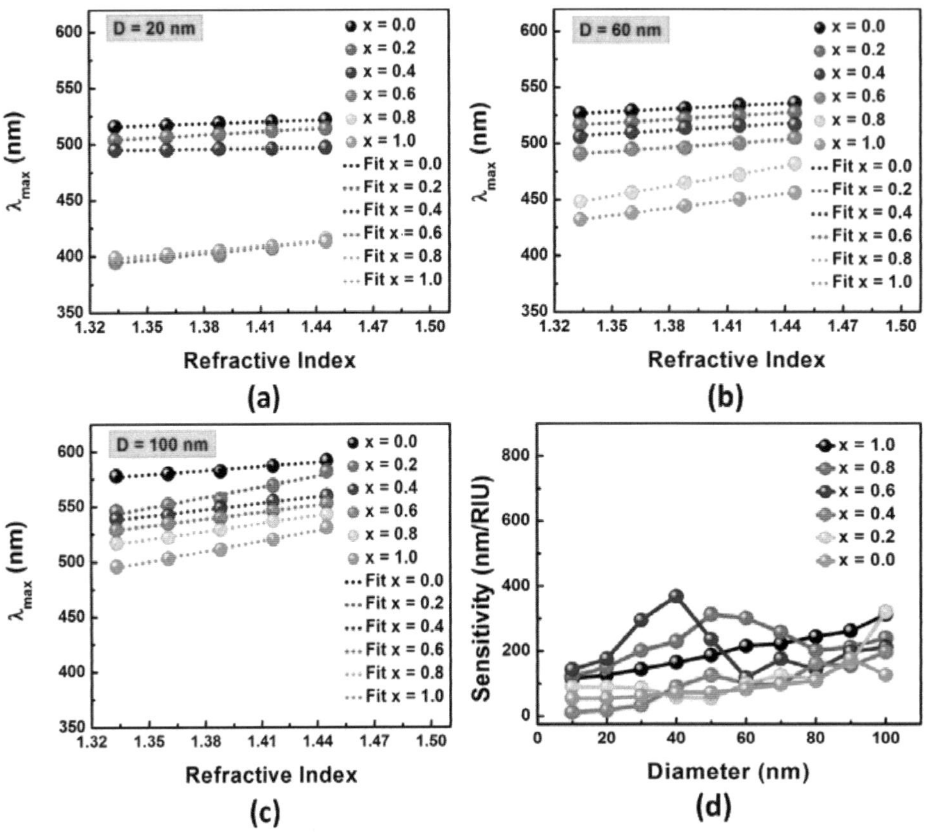

Figure 2. The peak position with respect to the refractive index medium variation and the mole fraction (a) D = 20 nm, (b) D = 60 nm, and (c) 100 nm as represented in dot symbol and the dot-line represents the linear fitting. (d) The sensitivity of surface plasmon resonance relates to the diameter bimetallic silver-gold alloys and the mole fraction

Next, we have also produced the extinction curve of the silver-gold alloys corresponded to the refractive index medium variation. Figure 2 exhibits the peak position of the extinction curve for small (D = 20 nm), intermediate (D = 60 nm), and large diameter (D = 100 nm) as the mole fraction and the refractive index medium variation. According the figure, we have found that the peak position still occurred in the visible light range as changing the refractive index medium. The peak position also shifted to red-shift as the refractive index medium increased and blue-shift as the silver mole fraction increased. The values of the peak position from the extinction curve exhibited a linear-like peak position. For small diameter (D = 20 nm), the peak position slowly decreased until the silver mole fraction at x = 0.4 and relatively same value for x greater than 0.4. We have observed there was a large difference value between x = 0.4 and 0.6, such as a transition. In contrast to intermediate and large diameter, the value of peak position decreased as the silver mole fraction increased. Based on Kuzma's work, shifting of the peak position in bimetallic the silver-gold alloys originated from the dipole term rather than interparticle coupling [8].

Furthermore, we calculated the sensitivity of surface plasmon resonance bimetallic silver-gold alloys using a linear fitting based on the peak position and the refractive index medium relation. The fitting results are shown in figure 2a for D = 20 nm, figure 2b for D = 60 nm, and figure 2c for D = 100 nm (denoted in dotted lines). The slope value of the linear fitting was a value of the sensitivity of surface plasmon resonance bimetallic silver-gold alloys [5, 17]. Figure 2d shows the sensitivity of surface plasmon resonance corresponded to the diameter and the mole fraction variation. We observed, for diameter less than D = 40 nm the sensitivity tended to increase as the diameter increased. Above the diameter 40 nm, the sensitivity exhibited the fluctuation value that depended on the changing of the mole fraction. Interestingly, the highest sensitivity of bimetallic silver-gold alloys occurred in the silver mole fraction at x = 0.6 rather than x = 1.0 at the diameter D = 40 nm. This figure also shows the sensitivity of surface plasmon resonance the silver particle (x = 1.0) was higher than gold particle (x = 0) for all range the diameter. In other words, the sensitivity of surface plasmon resonance the bimetallic silver-gold alloys exhibited to be more stable at the mole fraction x = 0.6 at small diameter. From this result, we suggested to consider the mole fraction when preparing the bimetallic nanoparticle.

4. Conclusions

In conclusion, we have calculated the sensitivity of surface plasmon resonance bimetallic silver-gold alloy with respect to the diameter, mole fraction, and refractive index medium variation. The value of the sensitivity was obtained by a linear fitting based on the peak position and the refractive index medium relation. The sensitivity tended to increase as the diameter increase to 40 nm. Above the diameter 40 nm the sensitivity showed the fluctuation values depended on the changing of the mole fraction. The highest sensitivity occurred in the mole fraction of silver at x = 0.6 at the diameter 40 nm. Beside the diameter, the mole fraction should be considered for observing the LSPR spectra in bimetallic nanoparticle.

Acknowledgements

The authors would like to thank Universitas Indonesia for funding this research through PITTA Grant Universitas Indonesia with contract No:1987/UN2.R12/HKP.05.00/2016.

References

[1] Hutter E and Fendler J H 2004 *Adv. Mater.* **16** 1685-706
[2] Guo L, Jackman J A, Yang H H, Chen P, Cho N J and Kim D H 2015 Nano Today **10** 213-39
[3] Yeshchenko O A, Dmitruk I M, Alexeenko A A, Kotko A V, Verdal J and Pinchuk A O 2012 *Plasmonic* **7** 685-94
[4] Willet K A and Van Duyne RP 2007 *Annu. Rev. Phys. Chem.* **58** 267-97
[5] Sherry L J, Chang S H, Schatz G C, Van Duyne R P, Wiley B J, and Xia Y 2005 *Nano Lett.* **5** 2034-8
[6] Pal A, Shah S, Devi S 2007 *Colloids Surf. A: Physcochem. Eng. Asp.* **302** 51-7
[7] Jia K, Khaywah M Y, Li Y, Bijeon J L, Adam P M, Déturche R, Guelorget B, François M, Louarn G, and Ionescu R E 2014 *ACS Appl. Mater. Interfaces* **6** 219-27
[8] Kuzma A, Weis M, Daricek M, Uhrik J, Horinek F, Donoval M, Uherek F, and Donoval D 2014 *J. Appl. Phys.* **115** 053517
[9] Bansal A and Verma S S 2014 *AIP Advances* **4** 057104
[10] Bansal A, Sekhon J S, and Verma S S 2014 *Plasmonics* **9** 143-50
[11] Hohenester U and Trügler A 2012 *Comp. Phys. Commun.* **183** 370-81
[12] Bohren C F and Huffman D R 1983 *Absorption and Scattering of Light by Small Particles* (New York: Wiley)
[13] Chen H, Kou X, Yang Z, Ni W and Wang J 2008 *Langmuir* **24** 5233-7
[14] Palik E D 1997 *Handbook of Optical Constant of Solids I* (Cambridge: Academic Press) 350-7
[15] Link S and El-Sayed M A 1999 *J. Phys. Chem.* B **103** 8410-26
[16] Link S, Wang Z L and El-Sayed M A 1999 *J. Phys. Chem.* B **103** 3529-33
[17] Lee K S and El-Sayed M A 2006 *J. Phys. Chem.* B **110** 19220-5

IOP Conf. Series: Materials Science and Engineering **188** (2017) 012017 doi:10.1088/1757-899X/188/1/012017

Virtual screening of commercial cyclic peptides as NS2B-NS3 protease inhibitor of dengue virus serotype 2 through molecular docking simulation

M A F Nasution, R N Aini and U S F Tambunan

Bioinformatics Research Group, Department of Chemistry, Faculty of Mathematics and Natural Sciences Universitas Indonesia, Kampus UI Depok, Depok 16424, Indonesia

Corresponding author's e-mail: usman@ui.ac.id

Abstract. A disease caused by dengue virus infection has become one of the major health problems in the world, particularly in Asia, Africa, and South America. This disease has become endemic in more than 100 countries, and approximately 100 million cases occur each year with 2.5 billion people or 40% of the world population at risk of having this virus infection. Therefore, we need an antiviral drug that can inhibit the activity of the enzymes that involved in the virus replication in the body. Lately, the peptide-based drug design has been developed and proved to have interesting pharmacological properties. This study uses commercially cyclic peptides that have already marketed. The purpose of this study is to screen the commercial cyclic peptides that can be used as an inhibitor of the NS2B-NS3 protease of dengue virus serotype 2 (DENV-2) through molecular docking simulations. Inhibition of NS3 protease enzyme can lead to enzymatic inhibition activity so the formed polyprotein from the translation of RNA cannot be cut into pieces and remain in the long strand form. Consequently, proteins that are vital for the sustainability of dengue virus replication cannot be formed. This research resulted in [alpha]-ANF (1-28), rat, Brain Natriuretic Peptide, porcine, Atrial Natriuretic Factor (3-28) (human) and Atrial Natriuretic Peptide (126-150) (rat) as the best drug candidate for inhibiting the NS2B-NS3 protease of DENV-2.

1. Introduction

Dengue virus infection has become one of the major health problems worldwide, especially in subtropical and tropical regions such as Asia, Africa, and South America. According to the latest data from World Health Organization (WHO), this disease has put approximately 3.9 billion people in 128 countries or roughly 40% of the world population at risk of infection, and the number of dengue incidence has been raised exponentially in latest decades [1-3]. Moreover, this disease infected more than 96 million people each year, resulting 20.000 deaths [4, 5]. Dengue virus infection may cause Dengue Hemorrhagic Fever (DHF), which leads to high body temperature (up to 41°C) and several symptoms such as a headache and muscle pain. In severe cases, the DHF can lead into Dengue Shock Syndrome (DSS), which significantly increase the mortality rate from 5% to 40% [6, 7]. Dengue virus infection is transmitted through female mosquito bites (*A. Aegypti* and *A. Albopictus*) [1]. Unfortunately, although several vaccines go through into clinical trial phase, there is no available commercial antiviral drug, nor vaccine to treat dengue fever efficiently. The development of dengue virus drugs and vaccines are difficult to conduct due to the pathogenicity of the virus itself [8]. Therefore, we need a new antiviral drug and vaccines that can combat the dengue fever by inhibiting its replication activity inside our body.

To date, there are five serotypes of dengue virus (DENV-1, DENV-2, DENV-3, DENV-4, and DENV-5), although only the former four serotype proved to be lethal to human [9]. These serotypes share almost identical genomes and morphology but show a different type of antigens. This unique property makes the body won't protect itself from another serotype's infection while increasing its pathogenicity [10]. Dengue virus is a positive-strand RNA virus that belongs to the *Flavivirus* genus of the *Flaviviridae* family [11]. It has a genome that encodes ten proteins; three structural protein (the capsid, envelope and membrane protein) and seven non-structural protein (NS1, NS2A, NS2B, NS3, NS4A, NS4B, and NS5), all seven non-structural protein have an important role in the replication cycle of the virus [12, 13].

NS3 protease in DENV belongs to the serine protease family that consists of 184 amino acid residues at the N-terminal sequences. The NS3 protease is the second-largest viral protein (at 69 kDa) that has a crucial role in the life cycle of virus [14]. Recent studies indicated that the NS3 protease has the active site which consists of three functional amino acids (His51, Asp75, and Ser135), known as the catalytic triad [15]. The catalytic activity of NS3 protease is assisted by NS2B cofactors. They have a crucial role in cutting the polyprotein precursor to NS2A/NS2B, NS2B/NS3, NS3/NS4A, and NS4B/NS5, as well as the internal cutting of capsid protein, NS2A, NS3, and NS4A [16]. Therefore, targeting NS2B-NS3B protein for treating the dengue fever is viable and crucial for the development of a new antiviral drug for DENV.

Cyclic peptides are the polypeptide chains build from two or more amino acids that form a cyclic ring structure [17]. In general, the peptides have unique pharmacological properties, due to its selectivity and low toxicity. Moreover, the peptide-based drug has been widely distributed in the market and approved by FDA, such as Buserelin and Leuprorelin. However, because peptides consist of the amino acid, they are very vulnerable to hydrolysis and oxidation reaction. Additionally, the peptides also tend to have low oral bioavailability due to enormous molecular weight and huge Total Polar Surface Area (TPSA) value [18, 19]. Thus, reduce their potency to become the leading drug. To enhance the pharmacological properties of peptides, the cyclic peptides can be made to improve the chemical and physical stability of the peptides. In this research, we selected the commercial cyclic peptide from the vendors, to be simulated with the NS2B-NS3 protease of DENV via molecular docking simulation, in hope to get a leading compound to inhibit the protease, and later can be helpful in the dengue fever therapeutics.

2. Methods

This research was done by using the general research pipeline that has been utilized by this research group [20, 21]. Several kinds of software were used in this research, such as Molecular Operating Environment (MOE) 2008.10 [22], Vega ZZ 3.0.5.12, and PerkinElmer ChemBioDraw 13.0.

2.1. 3D structure preparation of the DENV NS2B-NS3 protease

Three-dimensional structure of DENV NS2B-NS3 protease was obtained from the Research Collaboratory for Structural Bioinformatics – Protein Data Bank (RCSB-PDB) database. The selected 3D structure was downloaded in PDB format file. After that, the visualization on the active site of NS2B-NS3 protease was observed by using MOE 2008.10 software. Afterward, the geometry optimization and energy minimization of DENV NS2B-NS3 protease was conducted by removing the water molecules. Then, the protonation of the DENV NS2B-NS3 protease was done through 'Protonate 3D' feature, followed by the optimization of the partial charge and energy minimization. The energy minimization was performed with an AMBER99 force field, gas phase solvation mode and RMS gradient of 0.05, while the other parameters were set at default setting. The optimized, minimized 3D structure of DENV NS2B-NS3 protease then was saved in the .moe format.

2.2. Commercial cyclic peptides preparation

The molecular structure of commercial cyclic peptides were collected from the chemical company's website, such as BaChem, Mimotopes, and American Peptide. First, the structures were drawn by using PerkinElmer ChemBioDraw 13.0 software. Then, the structure were optimized by using Vega ZZ 3.0.5.12 software. Afterward, the optimization of ligand geometry process was conducted by using MOE

IOP Conf. Series: Materials Science and Engineering **188** (2017) 012017 doi:10.1088/1757-899X/188/1/012017

2008.10 software, and it started by importing all ligands into MOE database viewer, followed by 'Wash' process. Furthermore, the optimization of peptide ligands were done by using MMFF94x force field and gas phase solvation. The RMS gradient of 0.001 kcal/Å was selected as well. The rest parameters were set at default setting.

2.3. Molecular docking simulation of DENV NS2B-NS3 protease and commercial cyclic peptides
The molecular docking simulation of DENV NS2B-NS3 protease and commercial cyclic peptides was done by using MOE 2008.10 software. At first, the docking simulation process began by selecting the 'Dock' feature from the 'Compute' panel. Then, we selected 'Triangle Matcher' placement method, with LondondG rescoring (1) function and retain value by 30. Additionally, the 'Forcefield' refine method was selected, followed by Generalized Born solvation model (GBVI/WSA) rescoring (2) function and retain value by 1. The rest parameters were set at default setting. The docking result can be saved in .mdb file format. The determination of the best ligand from docking simulation was decided on their molecular interaction (by observing the protein-ligand complex from the 'LigX' feature) and the Gibbs free binding energy that generated from the simulation.

3. Results and discussion

3.1. The preparation of commercial cyclic peptides and DENV NS2B-NS3 protease 3D structure
The three-dimensional protein structure of DENV NS2B-NS3 protease was taken from RCSB PDB (http://www.rscb.org/pdb/). This database contains the information about the structure of various proteins and nucleic acids. In this research, we downloaded the 3D structure of DENV NS2B-NS3B protease under the name of PDB ID: 2FOM [23]. This 3D structure itself was taken from the crystallized enzyme of DENV2 NS2B-NS3 protease, and it contains 247 amino acid residues, 62 amino acids from the NS2B, while the rest 185 amino acids from the NS3 protease. In this 3D structure, the NS2B is responsible as the cofactor of the NS3 protease. NS3 protease has three functional amino acids that act as the active site of the enzyme, namely His51, Asp75 and Ser135. These amino acids are commonly known as the catalytic triad of NS3 protease, and they have the polar and hydrophilic properties in their binding site area [24]. The visualization of theNS2B-NS3 protease was conducted with MOE 2008.10 software by using 'Gaussian Contact' feature from 'Surface and Maps' menu. This visualization was done to observe the important residues in the enzyme that will be used later in the docking process, as we can see the figure 1.

Before the geometry optimization and energy minimization process was conducted, first we removed the chlorine ions, water molecules and glycerol that exist in the 3D structure. Then, the process was started by protonated the NS2B-NS3 protease structure and added the hydrogen atom through 'Protonate 3D' protocol, since the 3D structure generated from the X-ray crystallography doesn't include the hydrogen atom. After the enzyme had been protonated, we added the partial charge via 'Partial Charge' protocol with the current method forcefield parameters. Finally, we minimized the structure, through 'Energy Minimization' menu. The gas phase solvation mode and AMBER99 forcefield was selected for this simulation, and the selected forcefield was used due to its reliability in optimizing the protein geometry than other types of forcefield, such as MMFF94x. The energy minimization process was carried out at RMS value of 0.05 kcal/mol.Å. In general, this process aims to create the most stable protein structure, by eliminating the bad contact or undesirable interactions and generate the lowest energy conformation of the protein structure. Hence, the protein structure will be ready later for the molecular docking simulation process.

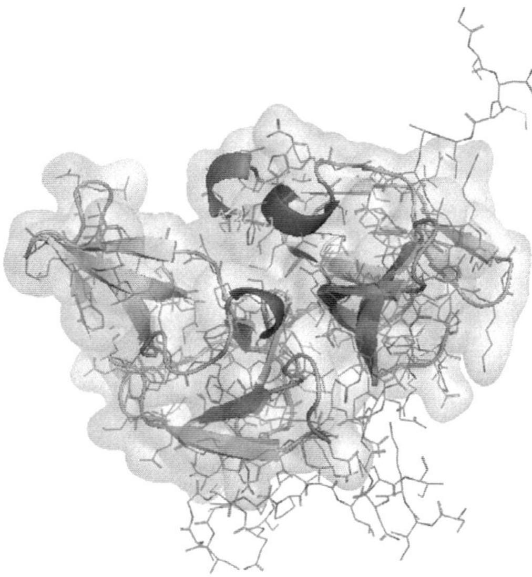

Figure 1. The three-dimensional protein structure of DENV NS2B-NS3 protease (PDB ID: 2FOM).

The next step of this research is the preparation of commercial cyclic peptides. We drawn all the cyclic peptides based on their structure in several online databases and vendors, such as American Peptide, Mimotopes, and BaChem. The molecular structure of cyclic peptides was drawn by using PerkinElmer ChemBioDraw 13 software, and the structure was later optimized by using Vega ZZ 3.0.5.12 software, resulting 308 commercial cyclic peptide structures in the process. As the comparison of these ligands, the linear peptide Bz-Nle-Lys-Arg-H was chosen as the standard ligand in this experiment, as this ligand activity and anti-dengue potency have been studied before by Yin *et al.* in 2005 [25]. The ligands that have been designed were saved in Molfile MDL format (.mol). Afterward, the geometry optimization and energy minimization of commercial cyclic peptides was done by using MOE 2008.10 software through "Database Viewer" menu. First, the 'Wash' protocol was conducted to improve the position of the hydrogen atom in the ligand. Then, the 'Partial Charge' protocol was done, prior to the 'Energy Minimization' process. The latter step was done by using MMFF94x forcefield, Gas Phase solvation mode and RMS gradient of 0.001 kcal/Å. The output file was stored in a.mdb format, and it will be ready for the molecular docking simulation.

3.2. Molecular docking simulation of DENV NS2B-NS3 protease and commercial cyclic peptides

The docking process between 308 commercial cyclic peptides as ligands and NS2B-NS3 protease was done by using MOE 2008.10 software. The docking scoring was done to determine the binding affinity between the ligand and receptor, as well as the refinement that was done to adjust the ligand-receptor complex with their respective forcefield [24]. After the ligand interacts with the receptor, the binding affinity of the ligand was evaluated by estimating the Gibbs free binding energy [26]. In this research, we used Triangle Matcher as the placement method and London dG as the scoring function, which are the default parameters of molecular docking simulation on MOE 2008.10. The docking process was done twice, each with 'Retain' value of 30 and 100, respectively. The aim of this treatment is to regulate the amount of 'Retain' pose interactions between the ligands and the receptor in molecular docking process. Out of 308 ligands that screened during the initial screening, 100 ligands, with the lowest Gibbs free binding energy value, were selected and taken into the secondary screening. The screening process was conducted twice to obtain the best conformation and interaction between the each ligand with NS2B-NS3 protease. In the end, we acquire 18 best ligands that have the lowest Gibbs free binding energy among all. The interaction between the ligand and the NS2B-NS3 protease can be seen in figure

2 below, while the results of 18 best ligands during the molecular docking simulation process is shown in table 1.

The best 18 ligands were selected not only because they have the lower Gibbs free binding energy ($\Delta G_{binding}$) value, but also good interaction with the catalytic triad of NS3 protease. The calculation of the $\Delta G_{binding}$ value is based on the laws of thermodynamics, which involves the equilibrium constant (K) value. The lower (negative) $\Delta G_{binding}$ value means the ligand-enzyme complex interaction is stronger and tends to be more stable. This is due to the stability and strength of non-covalent interactions (e.g. hydrogen bonds) on ligand-enzyme complex, which affects the molecular interaction of the complex

Table 1. The result of molecular docking simulation phase two (retain value at 100)

No.	Ligand Name	Gibbs free binding energy ($\Delta G_{binding}$) (kcal/mol)	Inhibition Constant (pKi)
1	[alpha]-ANF (1-28), rat	-53,469	38,952
2	[Tyr8]-Atrial Natriuretic Peptide (5-27), rat; [Tyr8]-Atriopeptin II, rat	-46,883	34,153
3	gp38	-46,696	34,017
4	[Tyr0] Calcitonin Gene Related Peptide II, human	-45,868	33,414
5	Vasonatrin Peptide (VNP)	-44,931	32,732
6	[CysCys21] Atrial Natriuretic Factor (3-28), Rat	-40,597	29,574
7	[beta]-Calcitonin Gene Related Peptide, rat	-39,878	29,051
8	[Tyr1] - Somatostatin 14	-39,812	29,002
9	Urodilatin (95-126)	-39,567	28,824
10	Calcitonin, eel	-52.613	38,327
11	Calcitonin Gene Related Peptide, human	-50,846	37,041
12	Calcitonin	-49,347	35,948
13	Brain Natriuretic Peptide, porcine	-47,732	34,772
14	Atrial Natriuretic Factor (3-28) (human)	-44,615	32,501
15	Atrial Natriuretic Peptide (126-150) (rat)	-44,206	32,203
16	Neuron Specific Peptide	-43,288	31,534
17	Amylin,human	-40,467	29,479
18	ANP (11-30), frog	-39,599	28,847
19	Bz-Nle-Lys-Arg-H (standard ligand)	-10,467	7,940

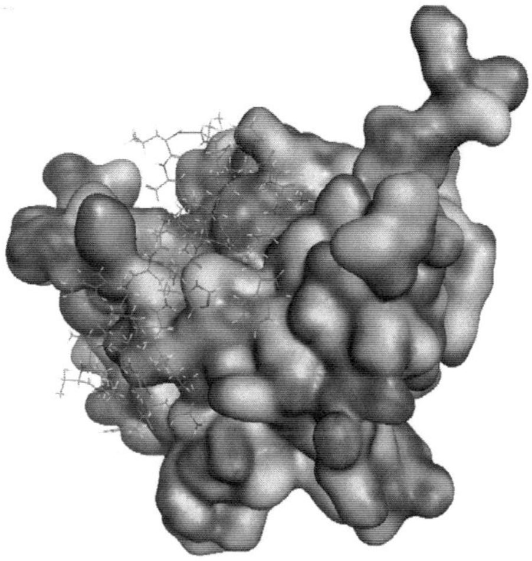

Figure 2. The visualization of molecular docking simulation between the cyclic peptide ligand and the NS2-NS3B protease.

and lowering the free binding energy value. Furthermore, the lower $\Delta G_{binding}$value also inversely proportional to inhibitory constant (Ki) value, the smaller Ki value, the greater the concentration of the ligand-enzyme complex, which means the complex is more stable and the ligand also has a higher binding affinity for the enzyme [27]. Therefore, the smaller the Ki value or the greater the pKi value of a ligand-enzyme complex means the ligand has a higher potential to interact with the enzyme. Thus, the dissociation reaction between the ligand-enzyme complex tends to be more difficult to happen. From the table 1, we can see that the [alpha]-ANF (1-28), rat ligand has the best results by having the lowest Gibbs free binding energy at -53,469 kcal/mol, far below the second-most low Gibbs free binding energy, [Tyr8]-Atrial Natriuretic Peptide (5-27), rat; [Tyr8]-Atriopeptin II, rat, at -46,883 kcal/mol. Moreover, the standard ligand, Bz-Nle-Lys-Arg-H, surprisingly has -10,467 kcal/mol, higher than the highest Gibbs free binding energy ligand, ANP (11-30), frog, at -39,599 kcal/mol. Therefore, based on the current Gibbs free binding energy results, we can consider that the [alpha]-ANF (1-28), rat ligand can be selected as the potential inhibitor of DENV NS2B-NS3 protease.

After we observed the $\Delta G_{binding}$ and the inhibition constant value from the previous result, we continue to view the visualization between the selected cyclic peptides and the NS2B-NS3 protease after the docking simulation was conducted. This interaction can be seen via 'LigX' feature in MOE 2008.10. The interaction between the cyclic peptides and the NS2B-NS3 protease can happen not only in the hydrogen bond form, but also with other non-covalent interaction, such as Van der Waals interaction, which can affect the binding affinity between them. The hydrogen bond itself can be defined as the intermolecular force that occurs between high electronegative atoms with hydrogen atoms that covalently bonded to an electronegative atom [28]. In general, the interactions that involved in the ligand-enzyme complex are weak, compared to covalent bond [27]. The visualization between selected cyclic peptides and the NS2B-NS3 protease can be seen in figure 3, while the list of the binding interaction between them is shown in table 2.

From the figure 3, we notice that the Calcitonin, Brain Natriuretic Peptide, porcine, and Atrial Natriuretic Factor (3-28) (human) ligand may form at least eight, eleven, and fourteen hydrogen bonds with the NS2-NS3 protein, respectively. The hydrogen bonds itself between the ligand and the receptor are essential in inhibiting the receptor's activity, although the other non-covalent bonds between them are important as well.

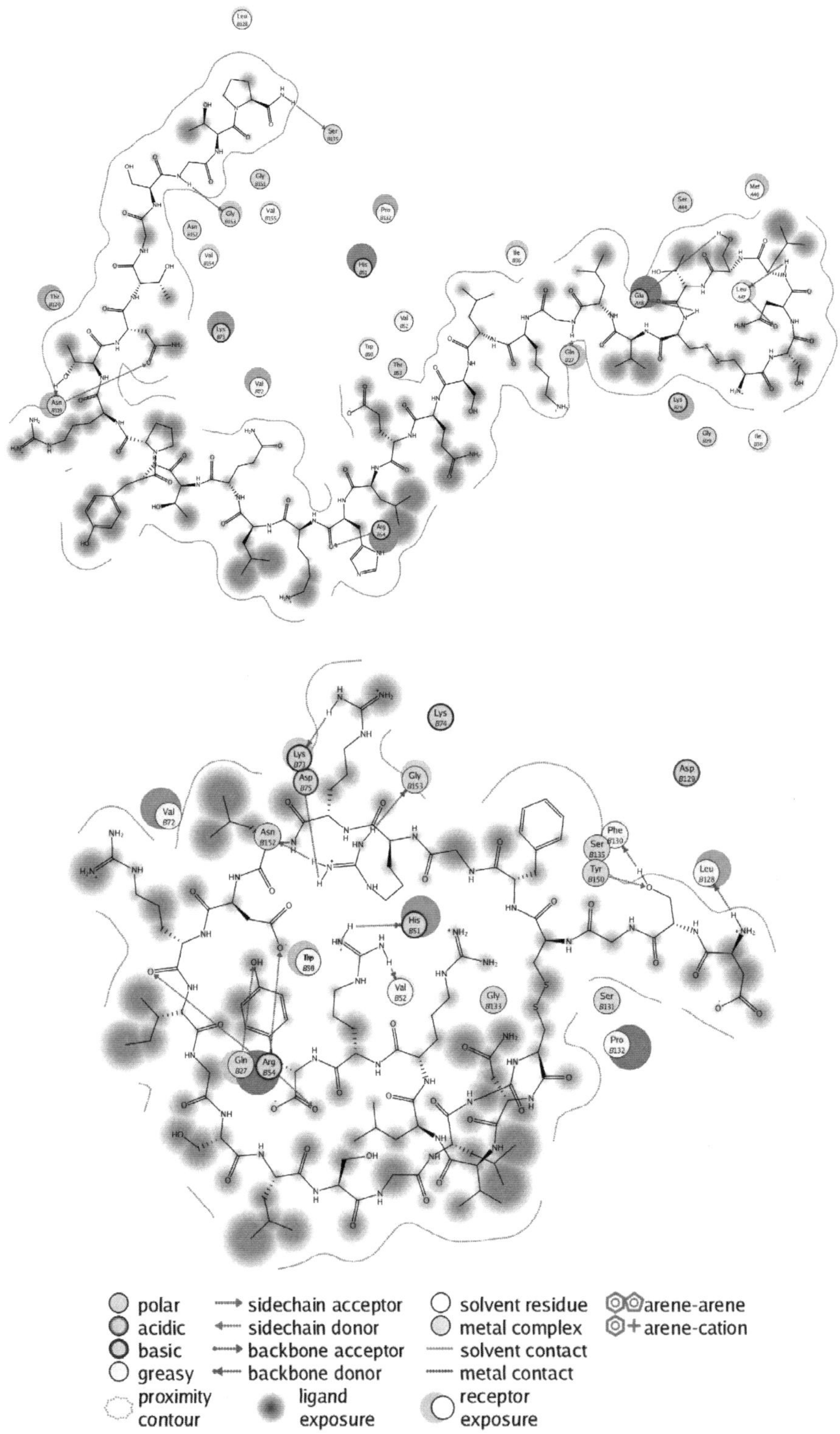

Figure 3. The 2D visualization of molecular docking simulation between the NS2-NS3B protease and calcitonin (top) and brain natriuretic peptide, porcine (bottom)

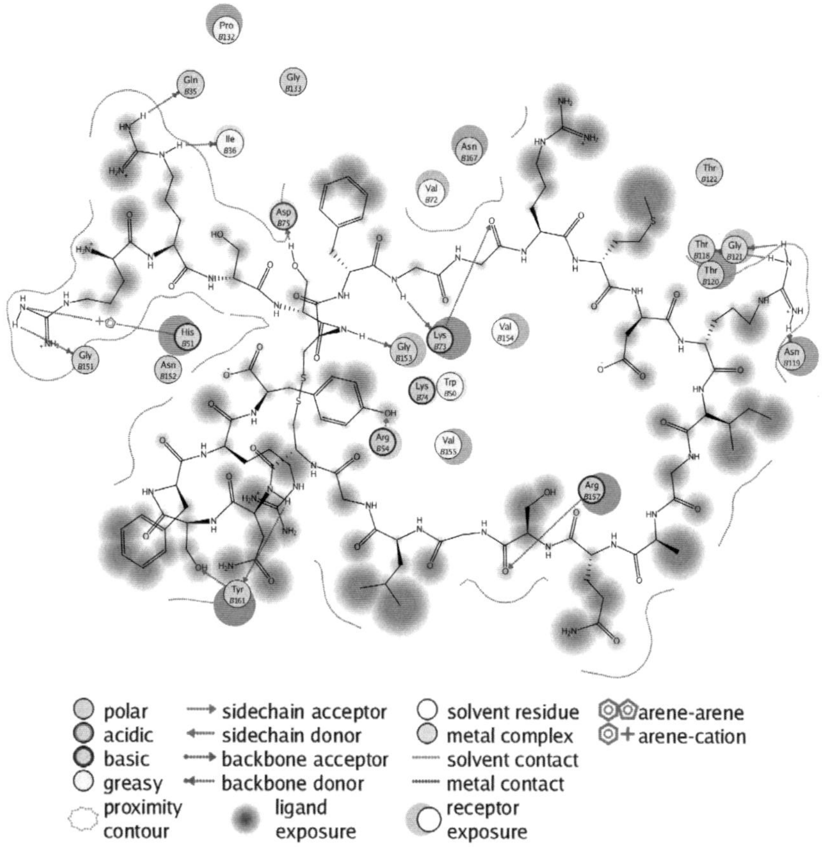

Figure 3 (cont.). The 2D visualization of molecular docking simulation between the NS2-NS3B protease and atrial natriuretic factor (3-28) (human)

As we can see in table 2, the catalytic triad of NS2B-NS3 protease plays an important role in the binding interaction between the cyclic peptide and the NS2B-NS3 protease, with at least ten, eleven and three interactions between the best 18 ligands that involve His51, Asp75, and Ser135, respectively. Moreover, Brain Natriuretic Peptide, porcine, Atrial Natriuretic Factor (3-28) (human), and Atrial Natriuretic Peptide (126-150) (rat) ligand can interact with two catalytic sites of NS2B-NS3 protease, means that they are more likely to bind well with the NS2B-NS3 protease, thus interrupting its function and may disturb the life cycle of the DENV. After we observed the molecular interaction among all of the best 18 cyclic peptides ligands, we concluded that the [alpha]-ANF (1-28), rat-, Brain Natriuretic Peptide, porcine, Atrial Natriuretic Factor (3-28) (human) and Atrial Natriuretic Peptide (126-150) (rat) ligand are the best ligands from this research to inhibit the NS2B-NS3 protease.

4. Conclusions

Cyclic peptides are one of the most fascinating molecules to become a leading compound due to their unique properties and exhibit various bioactivities, such as antiviral. In this research, we screened about 308 commercial cyclic peptides with the NS2B-NS3 protease to obtain the best ligands for inhibiting that protein, as well as treating the dengue fever. The screening was done by using molecular docking simulation, which resulted about 18 ligands that have lower Gibbs free binding energy than the standards. These ligands tend to interact with the NS2B-NS3 protease active site as well. In the end, we concluded that the [alpha]-ANF (1-28), rat-, Brain Natriuretic Peptide, porcine, Atrial Natriuretic Factor (3-28) (human) and Atrial Natriuretic Peptide (126-150) (rat) ligand are the best ligands to interact with

Table 2. Contact residue of the best 18 ligands with NS2B-NS3 protease after docking, red highlight shows the catalytic site of the NS3 protease

No.	Ligand Name	Binding Residue
1	[alpha]-ANF (1-28), rat	Glu 92, Lys 104, Asp 129, Ser 131, Pro 132, Thr 53, Val 52, Ile 36, Arg 54, His B51, Asp 75, Gly 153, Gly 151, Asn 152
2	[Tyr8]-Atrial Natriuretic Peptide (5-27), rat; [Tyr8]-Atriopeptin II, rat	Pro 132, Arg 54, Ile 36, Thr 53, Gln 27, Val 52, Gly 153, Arg 54, Glu 48, Asp 75
3	gp38	Leu 128, Glu 92, Ser 131, His 51, Arg 54, Lys 73, Gly 153, Asn 152
4	[Tyr0] Calcitonin Gene Related Peptide II, human	Glu 92, Gln 27, Gly 32, Ser 131, Val 155, Asn 152, Lys 73, Arg 54, His 51, Val 52
5	Vasonatrin Peptide (VNP)	Glu 48, Gln 27, Asn 152, His 51, Arg 54, Gly 153, Gly 151, Leu 128, Trp 50
6	[CysCys21] Atrial Natriuretic Factor (3-28), Rat	Arg 157, His 51, Arg 54, Thr 120, Val 155, Asn 119, Lys 73, Gly 153
7	[beta]-Calcitonin Gene Related Peptide, rat	Glu 88, Gly 87, Val 147, Ile 165, Lys 74, Trp 83, Asp 71, Lys 73, Asn 152, Asp 75, Gly 153, Tyr 161, Arg 54, Glu 92
8	[Tyr1] - Somatostatin 14	Arg 54, Gln 27, Asp 75, Val 72, Gly 153, Asn 152
9	Urodilatin (95-126)	Glu 92, Ile 36, Gln 27, Val 52, Lys 73, Asp 75, Gly 153, Asn 152
10	Calcitonin, eel	Lys 74, Asp 71, Lys 73, Asp 75, Asn 152, Gly 153, Ile 36,
11	Calcitonin Gene Related Peptide, human	Gln 35, His 51, Ile 36, Asn 167, Lys 75, Thr 118, Ala 70, Arg 54
12	Calcitonin	Ser 135, Gly 153, Asn 119, Arg 54, Glu 48, Gln 27, Leu 47,
13	Brain Natriuretic Peptide, porcine	Lys 73, Asp 75, Gly 153, Asn 152, His 51, Val 52, Gln 27, Arg 54, Tyr 150, Phe 130, Leu 128
14	Atrial Natriuretic Factor (3-28) (human)	Gln 35, Ile 36, Asp 75, His 51, Gly 151, Tyr 161, Arg 54, Gly 153, Lys B73, Thr 118, Gly 121, Asn 119, Arg B157
15	Atrial Natriuretic Peptide (126-150) (rat)	Arg 54, Lys 73, Asp 75, Thr 12, Ser 135, His 151, Val 155, Asn 119, Arg 54
16	Neuron Specific Peptide	Thr 120, Lys 73, His 51, Ser 34
17	Amylin,human	Gln 35, Glu 92, His 51, Arg 54, Gly 153, Asn 152, Asp 71, Val 72
18	ANP (11-30), frog	Lys 73, Val 22, Asp 71, Asn 152, Asp 75, Asn 119, Arg 54, Val 52, Gln 27, Glu 48, Val 57
19	Bz-Nle-Lys-Arg-H (standard ligand)	Arg 54, Asp 75, Val 52, Ile 36,His 51, Gly 133, Gln 27, Ser 135, Pro 132, Gly 153, Asn 152

the NS2B-NS3 protease. Therefore we suggest to continue this study by observing the ligand-protease stability through molecular dynamics simulation, or go into the in vitro test, to validate these results and determine their potency to become the leading compound for treating dengue fever.

Acknowledgements
The authors would like to thank Universitas Indonesia for funding this research through PITTA Grant Universitas Indonesia with contract No: 2049/UN2.R12/HKP.05.05/2016. We also like to thank Arli Aditya Parikesit for his outstanding work to improve the reliability of this research.

References
[1] World Health Organization. Fact Sheet 117: Dengue and severe dengue [Internet]. Fact Sheet. World Health Organization; updated July 2016. p. 1–4. Available from: http://www.who.int/mediacentre/factsheets/fs117/en/
[2] Brady O J, Gething P W, Bhatt S, Messina J P, Brownstein J S, Hoen A G, Moyes C L, Farlow A W, Scott T W and Hay S I 2012 *PLoS Negl. Trop. Dis.* **6** e1760
[3] Monath T P 1994 *Proc. Natl. Acad. Sci. U S A* **91** 2395-400
[4] Bhatt S *et al.* 2013 *Nature* **496** 504-7
[5] Noble C G and Shi P Y 2012 *Antiviral Res.* **96** 115-26
[6] Stevens A J, Gahan M E, Mahalingam S and Keller P A 2009 *J. Med. Chem.* **52** 7911-26
[7] Gubler D J *Clin. Microbiol. Rev.* 1998 **11** 480-96
[8] Geiss B J, Stahla H, Hannah A M, Gari A M and Keenan S M 2009 *Future Med. Chem.* **1** 327-44
[9] Mustafa M S, Rasotgi V, Jain S and Gupta V 2015 *Med. J. Armed Forces India* **71** 67-70
[10] Rajamanonmani R, Nkenfou C, Clancy P, Yau Y H, Shochat S G, Sukupolvi-Petty S, Schul W, Diamond M S, Vasudevan S G and Lescar J 2009 *J. Gen. Virol.* **90** 799-809
[11] Podvinec M, Lim S P, Schmidt T, Scarsi M, Wen D, Sonntag L S, Sanschagrin P, Shenkin P S and Schwede T 2010 *J. Med. Chem.* **53** 1483-95
[12] Guzman M G *et al.* 2010 *Nat. Rev. Microbiol.* **8** S7-16
[13] Henchal E A and Putnak J R 1990 *Clin. Microbiol. Rev.* **3** 376-96
[14] Luo D, Vasudevan S G and Lescar J 2015 *Antiviral Res.* **118** 148-58
[15] Shafee N and AbuBakar S 2003 *J. Gen. Virol.* **84** 2191-5
[16] Brinkworth R I, Fairlie D P, Leung D and Young P R 1999 *J. Gen. Virol.* **80** 1167-77
[17] Joo S H 2012 *Biomol. Ther. (Seoul)* **20** 19-26
[18] Fosgerau K and Hoffmann T 2015 *Drug Discov. Today* **20** 122-8
[19] Kaspar A A and Reichert J M *Drug Discov. Today* **18** 807-17
[20] Tambunan U S F, Parikesit A A, Ghifari A S and Satriyanto C P 2015 *Arabian J. Chem.* (doi:10.1016/j.arabjc.2015.07.010)
[21] Tambunan U S F, Zahroh H, Parikesit A A, Idrus S and Kerami D 2015 *Drug Target Insights* **9** 33-49
[22] Molecular Operating Environment (MOE) ver. 2008.10. 2008 (Montreal: Chemical Computing Group Inc.)
[23] Erbel P, Schiering N, D'Arcy A, Renatus M, Kroemer M, Lim S P, Yin Z, Keller T H, Vasudevan S G and Hommel U 2006 *Nat. Struct. Mol. Biol.* **13** 372-3
[24] Tambunan U S F and Alamudi S 2010 *Bioinformation* **5** 250-4
[25] Yin Z *et al.* 2006 *Bioorganic Med. Chem. Lett.* **16** 40-3
[26] Kapetanovic I M 2008 *Chem. Biol. Interact.* **171** 165-76
[27] Silverman R B and Holladay M W 2014 *The Organic Chemistry of Drug Design and Drug Action* 3rd ed. (New York: Academic Press)
[28] Arunan E *et al.* 2011 *Pure Appl. Chem.* **83** 1637-41

Potency of *Parkia speciosa* Hassk seed extract for green synthesis of CdO nanoparticles and its characterization

Y N Permana and Y Yulizar

Department of Chemistry, Faculty of Mathematics and Natural Sciences Universitas Indonesia, Kampus UI Depok, Depok 16424, Indonesia

Corresponding author's e-mail: yokiy@ui.ac.id

Abstract. CdO Nanoparticles (CdO NPs) have been green synthesized via sol-gel method using aqueous extract of *Parkia speciosa* Hassk (PSH) seed as the base source and stabilizing agent. The structural and optical properties of the formed CdO were investigated by X-ray diffraction (XRD), Fourier transform infrared (FTIR) spectroscopy, and UV-visible diffuse reflectance spectroscopy (UV-Vis DRS). The optical band gap of CdO NPs was estimated by UV-Vis DRS spectroscopy and calculated from reflectance spectrum ~2.00 eV. FTIR spectrum showed the functional groups of active compound of PSH seed extract which plays a role in the CdO NPs formation.

Keywords: *Parkia speciosa* Hassk, CdO nanoparticles, green synthesis, sol-gel synthesis method

1. Introduction

Nanoparticles (NPs) show the unique chemical, physical, optical, electrical and transport properties, which are very different from those of bulk materials and single atoms [1, 2]. The NPs have high surface to volume ratio and high surface energy. The formation of stable NPs to avoid agglomeration can be achieved by minimizing the surface energy [3, 4].

Semiconductor nanoparticles have been the subject of great interest in the past years due to their optical and electrical properties. Among the group II–VI semiconductor nanoparticles, CdO is considered to be one of the promising metal oxides with n-type semiconductor and band gap of ~2.5 eV [5]. It has low resistivity owed to oxygen vacancies and cadmium interstitials. Due to the high electrical conductivity and optical transmittance in the visible region of solar spectrum, CdO has potential applications in flat panel display, organic light emitting diodes, gas sensors, solar cells etc. [6].

CdO NPs are known to be acutely and chronically toxic to cells and organisms among the majority components of nanomaterials [7], and the commonly used capping molecules for the synthesis of nanoparticles such as mercaptoacetic acid, ethanolamine etc. are also toxic in nature and their large scale production can induce a potential threat to the environment [8, 9]. Therefore, it is necessary to design and use methods which are based on green chemistry to synthesize nanomaterials in order to improve and to protect the environment [10].

Recently, green chemistry has attracted researchers in nanosciences field [11]. Different green sources such as bacteria [12], fungus [13], plants [14], and algae [15] are used for nanoparticles synthesis. In the present study, we have prepared the aqueous extract of *Parkia speciosa* Hassk (PSH) seeds as the base source and stabilizing agent for the synthesis of CdO NPs through a sol-gel method.

2. Materials and methods

2.1. Preparation of PSH seed extract
Parkia speciosa Hassk (PSH) plant seeds were taken from University of Indonesia area. The PSH fresh seed was dried at room temperature and 5 g of the seed powder was stirred with 100 mL double distilled water for 1 h at 70°C until the colour of colloid changes to light yellow, and filtered using a Whatman No. 1 filter paper.

2.2. Green synthesis of CdO NPs
25 mL of PSH seeds extract was used for CdO NPs synthesis through sol-gel method. 1.5 g of $Cd(NO_3)_2.4H_2O$ was added into the extract and kept under continuous stirring at 85°C for 6 h. This paste was collected and heated in a furnace at 400°C for 2 h.

2.3. Characterization of CdO NPs
The UV-Vis absorption spectrum was recorded using SHIMADZU UV-Vis 2600 spectrophotometer. UV-Vis diffuse reflectance spectrum (UV-Vis DRS) was carried out using SHIMADZU UV-Vis 2450 spectrophotometer in reflectance mode in the wavelength range of 200-850 nm. The functional group of the extract compound was examined using Fourier transform infrared spectroscopy (PRESTGE 21 SHIMADZU). The crystal phase information was characterized from 2θ = 20-80° by XRD with Cu K (λ = 0.1546 nm) radiation (D8-Advanced, Bruker). The surface morphology of NPs was examined by JEOL JSM 6390 Scanning electron microscope (SEM) instrument operated at an accelerating voltage at 10 kV. The chemical composition of the product was examined by energy dispersive X-ray spectroscopy (EDS).

3. Results and discussion

3.1. UV-Vis absorption spectrum and DRS of CdO NPs
Figure 1 shows the UV-Vis absorption spectrum of the synthesized CdO NPs in colloidal system at maximum wavelength, λ_{max} 294 nm using UV-Vis spectrophotometry.

Figure 2a shows the DRS of CdO NPs prepared using 25 mL of PSH seeds extract. It shows an absorption inflection point at around 480 nm, which could be assigned to the excitonic absorption feature of CdO. The calculated direct band gap (E_g) from diffused reflectance spectrum was 2.0 eV as shown in

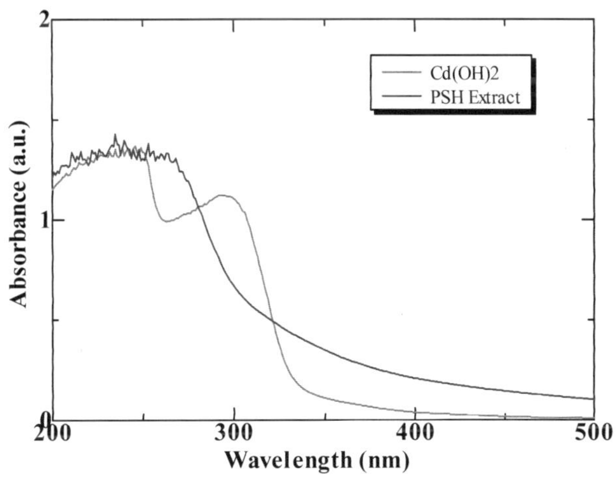

Figure 1. UV-Vis absorption spectra of the synthesized CdO NPs colloid

figure 2b, which was in good agreement with the reported values [16]. However, a small difference in E_g value was noticed due to various parameters, such as method of synthesis, size of particle, crystallinity, morphology, surface area, etc [17].

3.2. FTIR spectroscopy of CdO NPs

FTIR was used to study the purity and functional group composition of CdO NPs. The existence of strong FTIR spectrum of interactions between CdO and capping molecules were confirmed in the wavenumber region of 4000-400 cm^{-1} for the synthesized CdO NPs as shown in figure 3. The broad FTIR absorption band at 3457 cm^{-1} is assigned to stretching vibrations of hydroxyl group (O–H) of water molecules. The FTIR bands at 3254 and 3248 cm^{-1} are assigned to N–H asymmetric and symmetric stretching vibrations, respectively. The FTIR bands at 2970 and 2876 cm^{-1} are assigned to asymmetric and symmetric stretching vibrations of C–H group. A strong absorption band at 1366 cm^{-1} is due to wagging vibrations of C–H group [18]. The above peaks confirmed the presence of stabilizing molecules from PSH seeds extract. In the synthesized product, the metallic bonds identified near the 862 cm^{-1} due to metal-hydroxide (M-OH), and the band observed at 705 cm^{-1} is due to metal–oxygen (M–O) bond. It indicates the formation of CdO NPs.

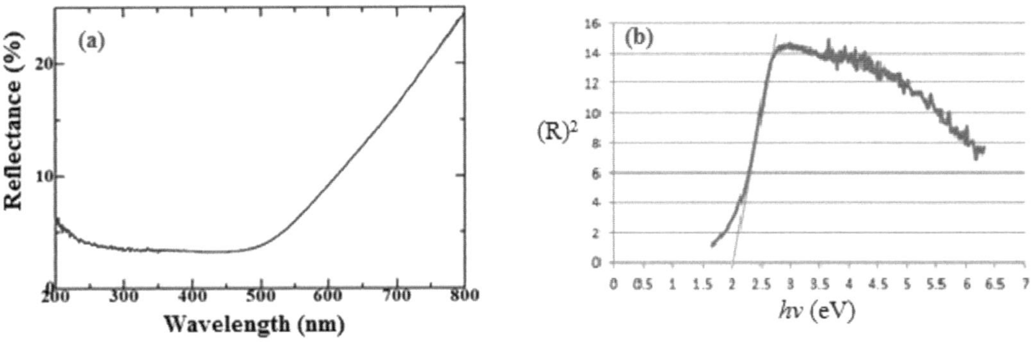

(a) (b)

Figure 2. (a) Diffuse reflectance spectrum (b) direct band gap energy of CdO NPs.

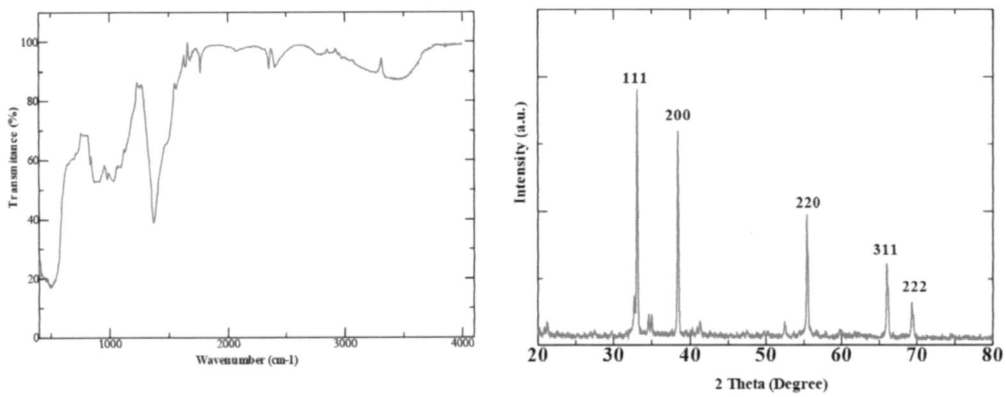

(a) (b)

Figure 3. (a) FTIR spectrum and (b) XRD spectrum of synthesized CdO NPs.

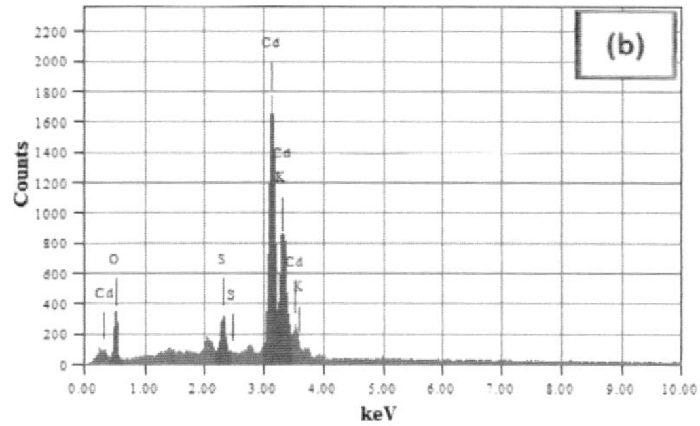

(a) (b)

Figure 4. (a). SEM images (b). EDS spectrum of CdO NPs.

3.3. XRD of CdO NPs

Figure 3b displays a typical XRD profile of the highly crystallized sample, i.e. the powder annealed at 400°C. It exhibits five diffraction peaks at 33.07°; 38.37°; 55.37°; 66.01°; 69.35°. More accurately, the diffraction peaks can be indexed as (111), (200), (220), (311) and (222) reticular orientations of the single phase cubic Monteponite CdO structure after adjusted with JCPDS card no. 05-0640 with an average lattice parameter $\langle a_{exp} \rangle = d_{hkl}\,[(h^2 + k^2 + l^2)]^{1/2} = 2.23617$ Å. From the broadening of Bragg peaks and using the Debye-Scherrer approximation, the average size of nanocrystals was ranged from 28.92 to 44.95 nm.

3.4 SEM and EDX of CdO NPs

SEM image of calcined CdO (figure 4a) exhibits a clear alteration in morphology. The diameter of CdO NPs becomes heterogeneous with particle size between 20 and 50 nm. These CdO NPs have scales around the stick to form nanoscale stick-like morphology. In figure 4b, the observed peaks of EDS spectrum at the different energy channels can be attributed to Oxygen (O), Cadmium (Cd), Potassium (K) and Sulfur (S). Consequentially, there are the other elements observed in CdO NPs as impurities.

4. Conclusions

The CdO NPs were successfully green synthesized using *Parkia speciosa* Hassk seed extract as the base source and stabilizing agents. UV-Vis DRS data confirmed the direct band gap at 2.0 eV. FTIR spectrum revealed the CdO bond was found in the range of 400-705 cm^{-1}. The XRD data showed the well crystallized in cubic phase of CdO Monteponite structure with the average size of the nanocrystals ranges within 28.92.9 and 44.95 nm.

Acknowledgements

The authors would like to thank Universitas Indonesia for funding this research through PITTA Grant Universitas Indonesia with contract No: 2039/UN2.R12/HKP.05.00/2016.

References

[1] Dong W and Zhu C 2003 *Opt. Mater.* **22** 227-33
[2] Yoffe A D 1993 *Adv. Phys.* **42** 173-262
[3] Rajeshwar K, de Tacconi N R and Chenthamarakshan C R 2001 *Chem. Mater.* **13** 2765-82
[4] Singh V and Chauhan P 2009 *Chalcogenide Lett.* **6(8)** 421-6
[5] Ortega M, Santana G and Morales-Acevedo A 2000 *Solid State Electron.* **44** 1765-9
[6] Chang Q, Chang C, Zhang X, Ye H, Shi G, Zhang W, Wang Y, Xin X and Song Y 2007 *Opt. Commun.* **274** 201-5
[7] Sreenkanth T V M, Pandurangan M, Dillip G R, Kim D H and Lee Y R 2016 *J. Photochem. and Photobiology B: Biology* **164** 174-81
[8] Ravindran T R, Arora A K, Balamurugan B and Mehta B R 1999 *Nanostruct. Mater.* **11** 603-9

[9] Pattabi M and Uchil J 2000 *Sol. Energy Mater. Sol. Cells* **63** 309-14
[10] Ariyanta H A and Yulizar Y 2016 *IOP Conf. Series: Materials Science and Engineering* **107** 012002-8
[11] Bansal P, Jaggi P and Rohilla S K 2012 *Res. J. Chem. Sci.* **2** 69-71
[12] Srivastava N and Mukhopadhyay M 2013 *Powder Technol.* **244** 26-9
[13] Fathima B S and Balakrishnan R M 2014 *Materials Lett.* **132** 428-31
[14] Thovhogi N, Park E, Manikandan E, Maaza E and Gurib-Fakim A 2016 *J. Alloy and Compounds* **655** 314-20
[15] Salari Z, Danafar F, Dabaghi S and Ataei S A 2016 *J. Saudi Chem. Soc.* **20** 459-64
[16] Niehl N 1981 *J. Lumin.* **24-25** 335-42
[17] Nagabhushana H, Basavaraj R B, Daruka-Prasad B, Sharma S C, Premkumar H B, Udayabhanu and Vijayakumar G R 2016 *J. Alloy Compd.* **669** 232-9
[18] Prabu K M, Anbarasan P M, Janarthanan S and Sivakumar G 2015 *Int. J. Sci. Res. Dev.* **2** 368-9

Characterization and properties of sodium alginate from brown algae used as an ecofriendly superabsorbent

Helmiyati and M Aprilliza

Department of Chemistry, Faculty of Mathematics and Natural Sciences Universitas Indonesia, Kampus UI Depok, Depok 16424, Indonesia

Corresponding author's e-mail: helmi-yt@ui.ac.id

Abstract. Sodium alginate obtained from the extraction of brown algae is used as the backbone for the synthesis of superabsorbent nanocomposite copolymerization. The first stage of extraction is the demineralization process using 0.1 M HCl solution and then 2% Na_2CO_3 solution for 2 hours at 60°C. The rendement of sodium alginate obtained was 44.32% with molecular weight of 40680 g/mol with measurement of the intrinsic viscosity. FTIR spectra of sodium alginate showed mannuronic acid functional group at wavenumber 884 cm^{-1} and the uronic acid at wavenumber 939 cm^{-1}, OH functional group at wavenumber 3200-3400 cm^{-1}, and CH_2 stretching at wavenumber 2928 cm^{-1}. The diffraction pattern of isolated sodium alginate has specific 2θ at 13.068 and 21.096, amorphous intensity found specific 2θ at 18.058, and the obtained crystallinity degree of the sodium alginate is equal to 29.292% from the XRD analysis. The morphological analysis by SEM shows fibrils of isolated sodium alginate. The success isolation of sodium alginate from brown algae is supported by DSC which shows the decomposition temperature of pure sodium alginate and isolated alginate have close values, namely 251.12°C for pure sodium alginate and 229.90°C for isolated sodium alginate.

Keywords: sodium alginate, nanocomposite, brown algae, copolymerization

1. Introduction

Alginates naturally present in brown algae cell walls [1, 2]. Alginates are polysaccharides found in brown algae, composed mainly by linear polymers of β-(1-4)-D-mannuronic (M), and α-L-gluronic (G) acids, which are different in terms of their proportions and linear arrangements [3, 4]. The sequence of monomers (M, G) changes greatly depending on the source [5]. In brown algae, alginates are the most abundant polysaccharides reaching up to 40% of the dry weight [6]. Amongst the recent applications of alginates, we can highlight their use as a matrix to encapsulate and release fertilizer and also for the synthesis of superabsorbent [7-9]. Alginates are typically described by their M/G ratio and average molecular weight, since these parameters are closely related to the functionality of the alginates as the encapsulation material [10]. Generally, the extraction and purification processes of alginates are based on the conversion from the insoluble form in the algae cell walls to the soluble one, normally by the sodium salt, followed by successive dissolutions and precipitations to eliminate impurities [11].

In this research, sodium alginate has been isolated from brown algae with acetone-methanol and chloroform, and can be separated from pigments, fat and other extractive substances. Then, the success isolation of the sodium alginate was observed by the molecular weight, and characterization by SEM, FTIR, XRD and DSC.

2. Materials and methods

2.1. Materials
Brown algae that used in this work, was obtained from Banten, Indonesia. Methanol (Merck), Acetone (Merck), and Chloroform (Merck) were used to remove pigment and fat from brown algae. Chloride Acid (Merck) was used to demineralize brown algae, Sodium Carbonate (Merck) to dissolve the alginate in the form of sodium salt, potassium hydroxide (Merck) as bleaching and impurity removal from brown algae. Ethanol (Merck) was used to precipitate the sodium alginate.

2.2. Sodium Alginate Isolation from Brown Algae
Isolation of the sodium alginate from brown algae referred to Sellimi *et al.* [12] with slight modifications. Brown Algae (50 g) was depigmented and defatted sequentially with acetone-methanol (7 : 3) and chloroform for 24 h at 30°C under constant stirring. The rendement was treated at pH around 2.0 with HCl solution for 2 h at 60°C under constant stirring for demineralization process. The supernatant was eliminated by centrifugation then the residue was washed with distilled water and treated with 2% Na_2CO_3 at 60°C for 2 h under constant stirring to form sodium alginate, followed by the solution bleaching with H_2O_2. The rendement was purified with methanol and acetone and then dried at room temperature. Sodium Alginate was analyzed by FTIR, XRD, SEM, and DSC and compared with the standard sodium alginate.

3. Results and discussion

3.1. Rendement of sodium alginate from brown algae
After the pretreatment process, the obtained average yield of sodium alginate can be seen in table 1. It can be seen the rendement of sodium alginate obtained after isolation is 44.32%.

Table 1. The Average of Sodium Alginate

Mass of Brown Algae (g)	Mass of Sodium Alginate (g)	Yield (%)
25.00	11.13	44.52
25.00	11.03	44.12
Average		44.32

3.2. Molecular weight determination
The average molecular weight (M) was calculated from the Mark-Houwink equation:

$$[\eta] = k\,M^a \qquad (1)$$

where $[\eta]$ is the instrinsic viscosity, and the constants a and k are empirical parameters which are depended on the system (polymer, solvent, and temperature). The obtained intrinsic viscosity in 0.1 M NaCl at 25°C for the isolated sodium alginate was 81.932 cm^3/g, so the molecular weight of sodium alginate was 40680 g/mol.

3.3. Analysis of functional groups of sodium alginate by FTIR
The FTIR spectra of the isolated sodium alginate showed in curve a and pure sodium alginate in curve b (figure 1). Both spectra have the similarities in which FTIR spectra of isolated sodium alginate and standard showed mannuronic acid functional group at wavenumber 884 cm^{-1} and the uronic acid at wavenumber 939 cm^{-1}, OH functional group at wavenumber 3200-3400 cm^{-1}, and CH_2 stretching at wavenumber 2928 cm^{-1}.

IOP Conf. Series: Materials Science and Engineering **188** (2017) 012019 doi:10.1088/1757-899X/188/1/012019

3.4. Analysis of crystallinity index of sodium alginate by XRD

The XRD patterns of brown algae, isolated sodium alginate and pure sodium alginate are shown in figure 2. Figure 2a shows the diffraction pattern of brown algae. It can be seen that brown algae has an amorphous structure with a peak $2\theta = 22.89°$. Figure 2b is diffraction pattern of pure sodium alginate and Figure 2c is the diffraction pattern of the isolated sodium alginate. It can be seen that both the diffraction patterns have similarities with the crystallinity index 35.62% and 29.292%, respectively. Crystallinity index of pure alginate is higher than the isolated alginate, because the purity of the isolated alginate is less than pure alginate. This shows that the pure sodium alginate and isolated alginate have semi-crystalline structure.

3.5. Analysis of surface morphology of sodium alginate by SEM

The surface morphology of brown algae and sodium alginate can be seen in figure 3. Figure 3a shows that the surface morphology of brown algae is not clearly indicating the fibrils sodium alginate and figure 3b shows the surface morphology of isolated sodium alginate indicating that the fibrils of sodium alginate fibers can be clearly observed because the impurities have been removed in the isolation process.

3.6. DSC analysis of sodium alginate

The success isolation of sodium alginate was supported by DSC and can be seen in figure 4. Figure 4a is thermogram from pure sodium alginate and figure 4b is thermogram from isolated sodium alginate It can be seen that the decomposition temperature of pure sodium alginate and isolated alginate have value of 251.12°C for pure sodium alginate and 229.90°C for the isolated sodium alginate. The decomposition temperature value of pure sodium alginate is higher than isolated sodium alginate because the crystallinity index of pure alginate is higher than the isolated alginate, as obtained from the XRD diffraction patterns that have been discussed.

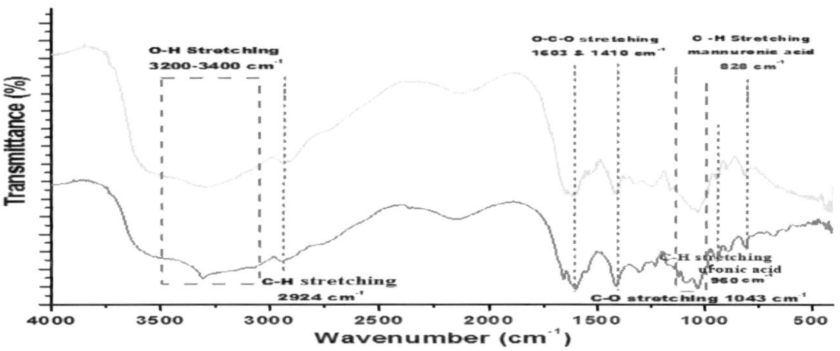

Figure 1. FTIR spectra of (a) sodium alginate isolation and (b) pure sodium alginate

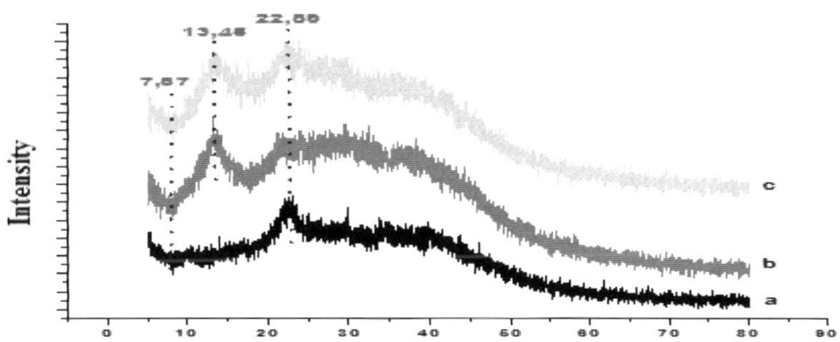

Figure 2. XRD pattern of (a) brown algae, (b) pure sodium alginate, and (c) isolated sodium alginate

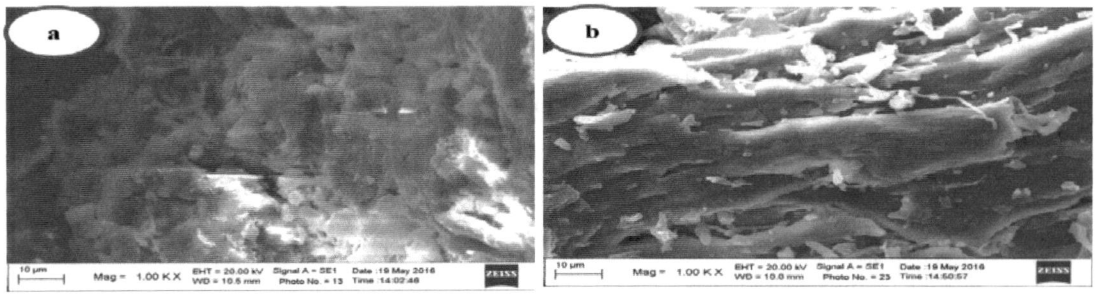

(a) (b)

Figure 3. Micrograph of (a) brown algae and (b) sodium alginate isolation

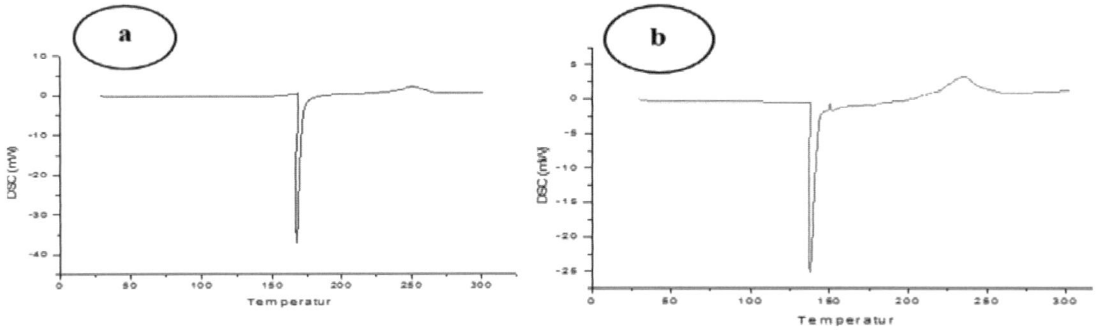

(a) (b)

Figure 4. DSC thermogram of (a) pure sodium alginate and (b) isolated sodium alginate

4. Conclusions

Isolation of sodium alginate from brown algae has been successfully performed. The obtained rendement of sodium alginate was 44.32% with molecular weight of 40680 g/mol. This successful process was supported by FTIR spectra which showed mannuronic acid functional group at wavenumber 884 cm^{-1} and uronic acid at wavenumber 939 cm^{-1}. The diffraction patterns of the isolated and pure sodium alginate have similarities with the crystallinity index of 35.62% and 29.292%, respectively. It showed that the pure and isolated sodium alginate have semi-crystalline structure. The surface morphology of the isolated sodium alginate analyzed by SEM showed that the fibrils of sodium alginate fibers can be clearly seen because the impurities have been removed in the isolation process. By the support of DSC the decomposition temperature of pure sodium alginate and isolated alginate showed close values, namely 251.12°C for pure sodium alginate, and 229.90°C for isolated sodium alginate.

Acknowledgements

The authors would like to thank Universitas Indonesia for funding this research through PITTA Grant Universitas Indonesia with contract No: 2045/UN2.R12/HKP.05.05/2016.

References

[1] Fertah M, Belfkira A, Dahmane E M, Taourirte M and Brouillette F 2014 Extraction and characterization of sodium alginate from Moroccan *Laminaria digitata* brown seaweed *Arabian Journal of Chemistry* http://dx.doi.org/10.1016/j.arabjc.2014.05.003

[2] Larsen B, Salem D M, Sallam M A, Mishrikey M M and Beltagy A 2003 *Carbohydr. Res.* **338** 2325-36

[3] Torres M R, Sousa A P A, Filho E A T S, Melo D F, Feitosa J P A, de Paula R C M and Lima M G S 2007 *Carbohydr. Res.* **342** 2067-74

[4] Sharma A and Gupta M N 2002 *Carbohydr. Polym.* **48** 391-5

[5] Donati I, Vetere A, Gamini A, Skajak-Braek G, Coslovi A, Campa C and Paoletti S 2003 *Biomacromolecules* **4** 624-31

[6] Moe S T, Draget K I, Skajak-Braek G and Smidsrod O 1995 *Food Polysaccharides and Their Applications* ed A M Stephen (New York: Marcel Dekker) pp 245-86
[7] Zhang Y, Liang X, Yang X, Liu H and Yao J 2014 *ACS Sustainable Chem. Eng.* **2** 1871-8
[8] Rashidzadeh A and Olad A 2014 *Carbohydr. Polym.* **114** 269-78
[9] Rashidzadeh A, Olad A and Reyhanitabar A 2015 Hydrogel/clinoptilolite nanocomposite-coated fertilizer: swelling, water-retention and slow-release fertilizer properties *Polym. Bull.* **72** DOI 10.1007/s00289-015-1428-y
[10] Finotelli P V, Da Silva D, Sola-Penna M, Rossi A M, Farina M, AndradeL R, Takeuchi A Y and Rocha-Leao M H 2010 *Colloids Surf.* B **81** 206-11
[11] Rinaudo M 2008 *Polym. Int.* **57** 397-430
[12] Sellimi S, Kadri N, Barragan-Montero V, Laouer H, Hajji M and Nasri M 2014 *Int. J. Biol. Macromol.* **66** 281-8

Anodic stripping voltammetry of nickel ions and nickel hydroxide nanoparticles at boron-doped diamond electrodes

N R R Musyarofah[1], J Gunlazuardi[1], Y Einaga[2,3] and T A Ivandini[1]

[1]Department of Chemistry, Faculty of Mathematics and Natural Sciences Universitas Indonesia, Kampus UI Depok, Depok 16424, Indonesia
[2]Department of Chemistry, Keio University 3-14-1 Hiyoshi, Yokohama 223-8522, Japan
[3]JST Accel 3-14-1 Hiyoshi, Yokohama 223-8522, Japan

Corresponding author's e-mail: ivandini.tri@sci.ui.ac.id

Abstract. Anodic stripping voltammetry (ASV) of nickel ions in phosphate buffer solution (PBS) have been investigated at boron-doped diamond (BDD) electrodes. The deposition potential at 0.1 V (vs. Ag/AgCl) for 300 s in 0.1 M PBS pH 3 was found as the optimum condition. The condition was applied for the determination of nickel contained in nickel hydroxide nanoparticles. A linear calibration curve can be achieved of $Ni(OH)_2$-NPs in the concentration range of x to x mM with an estimated limit of detection (LOD) of 5.73 x 10^{-6} mol/L.

Keywords: Boron-doped diamond, melamine, $Ni(OH)_2$ nanoparticles, strip test immunokromatografi

1. Introduction

Anodic stripping voltammetry (ASV) using boron-doped diamond (BDD) electrodes have been reported as a superior technique for trace metals detection, including metal ions and metal nanoparticles. The stability of BDD with its sp^3 carbon contributes to the high reproducibility of the measurements [1-5]. On the other hand, the development of sensor and biosensor employing metal nanoparticles as the probe is growing fast because of the excellence of this material, including the possibility to miniaturize the devices as well as the high sensitivity of optical and electrochemical signals, which induces to the high precision and accuracy of measurements [6-8].

In this study, ASV of nickel ions and nickel nanoparticles has been studied. $Ni(OH)_2$ is widely applied in electrochemistry due to its earth-abundant nature and its catalytic activity [9]. However, its application for a probe in sensors and biosensors are very limited. Meanwhile, synthesis of $Ni(OH)_2$ nanoparticles was reported [10]. ASV of nickel ions and nanoparticles showed that selective detection can be performed, indicated the possibility to apply the nanoparticles for a probe in sensors and biosensors.

2. Materials and methods

2.1. Anodic stripping voltammetry (ASV) of $Ni(NO_3)_2$ using BDD electrode

ASV of $Ni(NO_3)_2$ was performed in 0.1 M phosphate buffer solution (PBS) with a BDD films as the working electrode. Pt wire and Ag/AgCl system were used as the counter and the reference electrodes, respectively. Parameters of pH, potential deposition and deposition times were investigated. Linear calibration curves were performed in a concentration range from 5 to 200 mM.

2.2. Preparation of Ni(OH)2 nanoparticles

$Ni(OH)_2$ nanoparticles were synthesized by complexation-precipitation method on hydrothermal condition. Mechanism reaction was estimated to occur through the formation of nickel-citrate complex, followed by disassociation of the complex in the presence of NaOH solution to form $Ni(OH)_2$-NPs in hydrothermal condition [10].

2.3. Synthesis Ni(OH)2 nanoparticles (Ni(OH)2-NPs)

$Ni(OH)_2$-NPs was synthesized based on the previous study [10]. Briefly, a volume of 10 mL $Ni(NO_3)_2$ solution was mixed with 10 mL of 2.86 x 10^{-2} M $NaBH_4$ solution, then added by 5.71 x 10^{-1} $MC_6H_5Na_3O_7$. The mixture was stirred at 35°C for 10 h. Then, 5 mL of NaOH solution was added and continuously stirred for 5 min. Subsequently, the mixture was put into an autoclave and heated at a temperature of 120°C for 24 h. The synthesized $Ni(OH)_2$-NPs were characterized using UV-Vis spectrophotometer, FTIR, and TEM.

3. Results and discussion

3.1. Anodic stripping voltammetry of Ni(NO3)2

ASV is basically performed in two steps, including the pre-concentration and the stripping steps of the metal analytes. At the pre-concentration step, Ni^{2+} was expected to be Ni^0, while Ni^0 can be oxidized to be Ni^+ or Ni^{2+}at the stripping step. Therefore, ASV of Ni^{2+} at BDD electrodes was investigated at pH 3, 4, and 5 because Ni^0 can be immediately oxidized and produce sufficient oxidation current responses in acidic condition. Two oxidation peaks at around +0.4 V and +1.1 V (vs. Ag/AgCl), attributed to the oxidation of Ni^0 to Ni^+ and Ni^{2+}, respectively, were observed at all pHs of 3, 4, and 5. The peak at +1.1 V and pH 3 was selected since it generates the highest current. Furthermore, potential depositions were varied from 0 to -700 mV for 60 s with a scan rate of 100 mV/s, suggested the optimum condition at -100 mV. The deposition times were varied from 0 to 300 s at deposition potential of -100 mV, suggested the optimum time at 300 s. In addition, scan rate dependence was also investigated, which revealed that 100 mV/s was the optimum scan rate. Based on the result, a linear calibration curve can be achieved in the concentration range of 5-200 mM in the optimum condition (figure 1).

UV-Vis spectrum of $Ni(OH)_2$-NPs in figure 2a reflects a peak at 220-230 nm, which conforms the criteria of $Ni(OH)_2$-NPs characteristic [10]. This absorption peak shifted from those of $Ni(NO_3)_2$

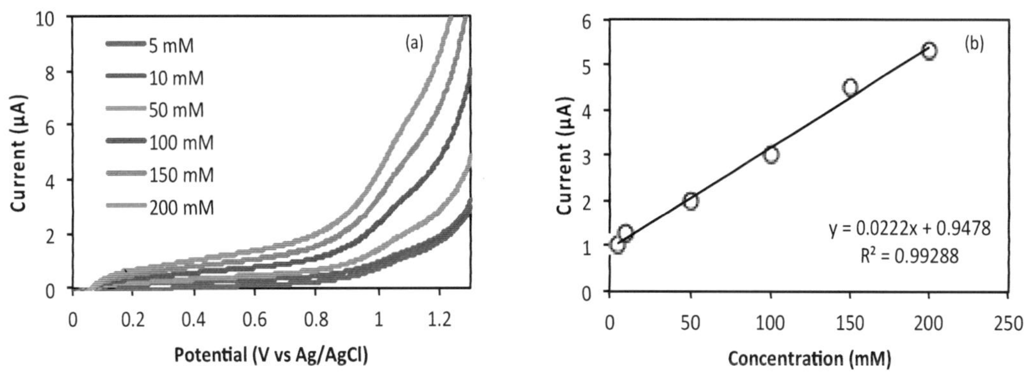

(a) (b)

Figure 1. (a) Anodic stripping voltammograms of $Ni((NO_3)_2$ in PBS 0.1 M pH 3 and (b) its linear calibrations curve

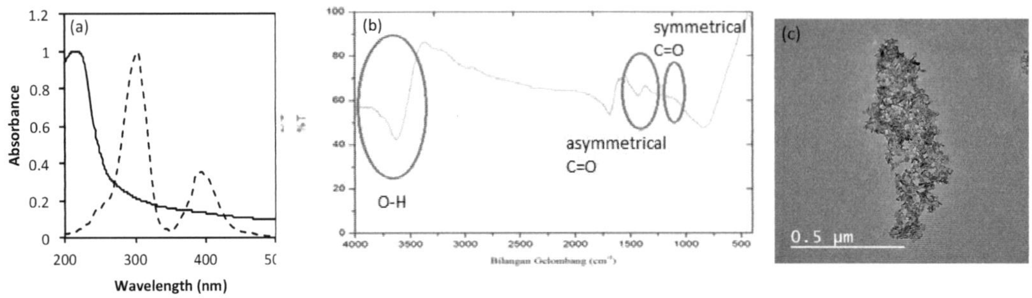

(a) (b) (c)

Figure 2. (a) UV-Vis and (b) spectra as well as (c) TEM image of $Ni(OH)_2$-nanoparticles

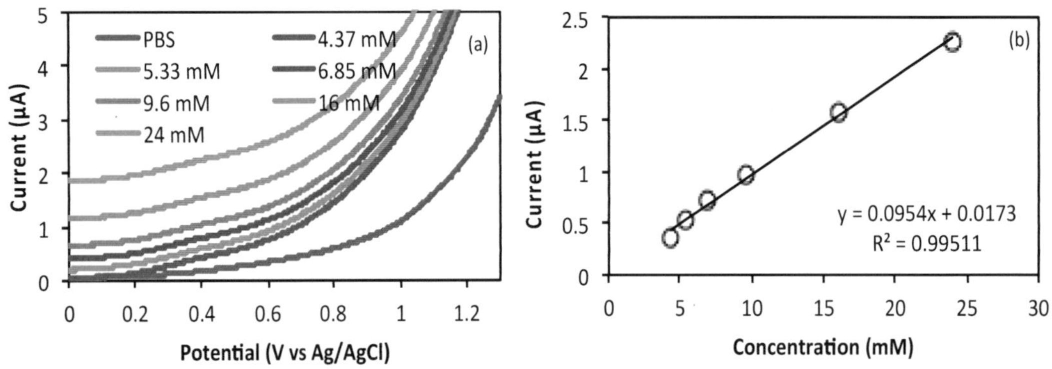

(a) (b)

Figure 3. (a) Anodic stripping voltammograms of $Ni(OH)_2$-NPs in 0.1 M PBS pH 3 and (b) its linear calibrations curve

solution, indicated that Ni^{2+} ions has changed. Investigation of FTIR spectra (figure 2b) shows a broad absorption at the wavenumber around 3600 $cm^{-1}$and a strong peak at 1650 cm^{-1}, indicated the vibration stretching of O-H and asymmetrical stretching of carbonyl functional group (C= O), respectively. In addition, symmetrical stretching C= O was also observed as a results of the trisodium citrate excess in the sample. Furthermore, TEM image shows that the $Ni(OH)_2$-NPs were hexagonal crystals with an average size of 14 nm (figure 2c)).

3.2. Anodic stripping voltammetry of $Ni(OH)_2$ nanoparticles

Determination of the nanoparticles was performed using the optimum condition described in section 3.1. Solution of 0.1 M PBS pH 3 was used as the electrolyte. A deposition potential of -100 mV was applied for 300 s with the scan rate of 100 mV/s. Figure 3a shows the voltammograms of various concentrations of $Ni(OH)_2$ nanoparticles with the linear calibration curve in figure 3b. High linearity (R^2 = 0.99) with an estimated LOD of 0.057 mM could be achieved, indicating that the method can be applied for the detection method employing $Ni(OH)_2$ nanoparticles as a probe.

4. Conclusions

Anodic stripping voltammetry of nickel at boron-doped diamond electrodes have been investigated. A deposition potential of -100 mV for 300 s with a scan rate of 100 mV/s applied in a solution of 0.1 M PBS pH 3 was found to be optimum condition. The method was successfully applied for the determination of $Ni(OH)_2$ nanoparticles, indicating that application of this nanoparticles is promising for a probe in sensors and biosensors.

Acknowledgements
This work was funded by Hibah Kompetensi Publikasi Internasional Terindeks Universitas Indonesia, Contract No. 2016/UN2.R12/HKP.05.00/2016.

References

[1] Yamada D, Ivandini T A, Komatsu M, Fujishima A and Einaga Y 2008 *J. Electroanal. Chem.* **615** 145-53
[2] Fierro S, Watanabea T, Akai K and Einaga Y 2012 *Electrochem. Acta* **82** 9-11
[3] Ivandini T A and Einaga Y 2013 *Electrocatal.* **4** 367-74
[4] Ivandini T A, Yamada D, Watanabe T, Matsuura H, Nakano N, Fujishima A and Einaga Y 2010 *J. Electroanal. Chem.* **645** 58-63
[5] Hayat M, Ivandini T A, Saepudin E and Einaga Y 2016 *AIP Conference Proceedings* **1729** 020052
[6] Thobhani S, Attree S, Boyd R, Kumarswami N, Noble J, Szymanski M and Porter R A 2010 *J. Immunol. Methods* **356** 60–9
[7] Ivandini T A, Wicaksono W P, Saepudin E, Rismetov B and Einaga Y 2015 *Talanta* **134** 136-43
[8] Gong Y and Fan Z 2014 *Sensors and Actuators B Chemical* **202** 638-44
[9] Martínez-Periñán E, Gennari M, Revenga-Parra M, Abad J M, Mateo-Martí E, Pariente F, Castillo O, Mas-Ballesté R, Zamora F and Lorenzo E 2015 *J. Catal.* **329** 22-31
[10] Wang R, Lang J, Liu Y, Lin Z and Yan X 2015 *NPG Asia Mater.* **7** e183-7

IOP Conf. Series: Materials Science and Engineering **188** (2017) 012021 doi:10.1088/1757-899X/188/1/012021

One-step microwave-assisted colloidal synthesis of hybrid silver oxide/silver nanoparticles: characterization and catalytic study

S P Prakoso[1], **A Taufik**[1,2] **and R Saleh**[1,2]

[1]Integrated Laboratory of Energy and Environment, Faculty of Mathematics and Natural Sciences Universitas Indonesia, Kampus UI Depok, Depok 16424, Indonesia
[2]Department of Physics, Faculty of Mathematics and Natural Sciences Universitas Indonesia, Kampus UI Depok, Depok 16424, Indonesia

Corresponding author's e-mail: rosari.saleh@gmail.com

Abstract. This study reports the characterization and catalytic activities of silver-oxide/silver nanoparticles (Ag_2O/Ag NPs) synthesized by microwave-assisted colloidal method in the presence of anionic sodium dodecyl sulfate (SDS) surfactant. To promote different contents of silver in silver oxide, the volume ratio (VR) of ethylene glycol (EG) was varied (VR: 10% to 14%) in relation to the total volume of distilled water solvent. The plasmonic resonance of Ag_2O/Ag NPs could be detected around a wavelength of 350 nm, and it is suggested that Ag_2O/Ag NPs were successfully formed in the colloid solution following exposure to microwaves. Additionally, the growth rate for each crystal phase within Ag_2O and Ag was influenced by an increase of EG as revealed by x-ray diffraction patterns. The morphology, average diameter, and uniformity of Ag_2O/Ag NPs were studied simultaneously by transmission electron microscopy. Infrared absorption measurement of Ag_2O/Ag NPs confirmed the existence of SDS surfactant as a protective agent. Based on the characterization data, Ag_2O/Ag NPs synthesized using this technique exhibited good properties, with high-yield production of NPs. The photocatalytic experiments demonstrate the key role of the crystal phase of Ag_2O/Ag NPs in photocatalytic efficiency.

1. Introduction

Quantum effects in nanoparticles indicate novel properties for future potential applications as distinct from bulk state. Dependency on size, shape, and elements or compounds in nanoscale particles governs alteration of these properties. For instance, the effects of adsorption-desorption on the catalytic process greatly influence the specific surface area of particles even for nanoparticles, as the surface is usually first to interact directly with the "outside world". In addition, the properties of nanoparticles depend on the method used in their preparation. Fazlali *et al.* [1] reported the photocatalytic activity of nickel oxide nanoparticles obtained by various means (thermal-decomposition, sol-gel, hydrothermal, and emulsion nanoreactors), showing differences of performance regardless of particle size. The use of different alkalis (e.g., sodium hydroxide, ammonia solution, urea, and triethanolamine) to obtain nanoparticles through precipitation gives rise to similar phenomena in relation to efficiency of photocatalytic activity [2]. Synthesizing nanoparticles using microwaves has become more popular by virtue of their higher heating efficiency and uniformity as compared to conventional methods of heating such as hot-plate, heating elements, or heat conduction [3]. Rapid and easy production of nanoparticles using microwaves can be achieved with any suitable method of synthesis.

To date, diverse nanoparticles have been considered in the search for an effective photocatalyst in degrading pollutant organic dyes, which are residual effluents of the textile industry. Of these, oxide semiconductors have attracted interest because of their optical absorption properties, which can be controlled on the basis of gap energies. Silver oxide (Ag_2O) is one of the most suitable oxides for photocatalysis in the visible region, naturally exhibiting p-type conductivity with gap energy (Eg) of around 1.4 eV at room temperature [4]. Despite its good photoabsorption in the visible regions, spontaneous recombination of electron-hole pairs is an issue that hinders the photocatalytic process. To overcome this issue, one technique for inhibiting charge carriers from recombination is to incorporate highly conductive metals to generate localized surface plasmonic resonance (LSPR) effects [5-9]. Along with gold (Au), silver (Ag) is a good metal for generating these LSPR effects [10], and the results of Hu *et al.* [5] demonstrated the extraordinary photocatalytic activities of hybrid Ag_2O/Ag nanoparticles.

Attempts to obtain nanohybrid Ag_2O/Ag particles have included such approaches as co-precipitation and thermal decomposition [5, 8, 9], photochemical reaction [6], and impulse electrochemical deposition [7]. However, to our knowledge, there are few reported studies of the rapid and easy synthesis of controllable amounts of silver in silver oxide nanohybrid particles. The present study reports the characterization and catalytic performance of hybrid Ag_2O/Ag nanoparticles with controlled Ag content as produced in a one-step microwave-assisted colloidal process. To investigate the photodegradation rate of nanohybrid Ag_2O/Ag particles with various Ag contents, methylene blue was employed as a model organic pollutant.

2. Materials and methods

2.1. Synthesis of hybrid Ag_2O/Ag nanoparticles

Production of hybrid Ag_2O/Ag nanoparticles used a household microwave and colloidal synthesis. All precursors were analytical grade reagents, used directly as received and without any further treatment; materials included silver nitrate ($AgNO_3$; Merck), sodium hydroxide (NaOH; Merck), sodium dodecyl sulfate (SDS/$C_{12}H_{25}NaO_4S$; Merck), ethylene glycol (EG/$C_2H_6O_2$; Merck), ethanol, and distilled water. Two solutions (named solution A and solution B) were made prior to starting the reaction. For solution A, a specified amount of $AgNO_3$ was diluted into the 50 mL mixed-solvent, which varied in terms of total volume ratio (VR): 10%, 12%, and 14% EG of 100 mL total volume, making a concentration of 0.09 M. For solution B, 50 mL of distilled water was incorporated by SDS with 1:1 mole ratio to silver ion concentration and an additional 2.6 mL of 2 M NaOH. Solution A was then reacted by solution B under constant stirring for 10 min at room temperature, and this mixed-solution was then irradiated inside the microwave cavity (Electrolux EMM2007X) for 30 sec at an irradiation level of 800 W. The solution was allowed to cool at room temperature while being stirred for 30 min. The precipitate of Ag_2O/Ag was collected by centrifugal force and washed several times with distilled water and ethanol. The final product of Ag_2O/Ag powders was obtained after drying the precipitate at ~120°C for 5 hours in ambient air.

2.2. Characterization methods

The crystal phase of Ag_2O and Ag was examined by x-ray diffraction (XRD) (MiniFlex 600, Rigaku) with an x-ray source operating voltage of 30 kV and a current of 15 mA. Diffraction pattern was recorded within 2θ of 10° to 90°, with a scan step size of 0.02°. Fourier transform infrared (FTIR) measurements were performed using a Shimadzu IR Prestige-21 to confirm the presence of SDS and other possible molecular vibrations in the IR-active region. Potassium bromide (KBr) was employed as a background in recording IR absorption spectra in the 400-4000 cm^{-1} range. Diffuse reflectance spectra (DRS) of UV-Visible were captured by a Shimadzu UV-2450 system, using barium sulfate (Ba_2SO_4) as the baseline correction. The photograph of particle shape was captured by tunneling electron microscopy (TEM) (FEI Tecnai G2 SuperTwin).

2.3. Photocatalytic experiments

Nano-hybrid Ag_2O/Ag catalysts were implemented to degrade methylene blue (MB), which was chosen as a model of organic pollutant in water. A 0.03 g sample of Ag_2O/Ag with varying silver content was dispersed in 100 mL MB solution (20 mg-L^{-1}) at a constant pH of 13. Prior to illumination, the suspensions of Ag_2O/Ag photocatalysts in the MB solution were magnetically stirred in the dark chamber for 30 min to ensure steady state adsorption/desorption of MB on the surface of catalyst particles. The mixtures were then illuminated by 40 W visible light for 120 min. Sampling was repeated every 15 min, and the degraded MB solutions were centrifuged to separate the catalyst. Quantitative analysis of MB in the degraded MB solutions was conducted by UV-Vis spectrophotometer (Hitachi UH5300), which recorded the absorption spectrum in the 450-750 nm range.

3. Results and discussion

The optical absorption spectra of Ag_2O/Ag NPs with various Ag contents are displayed in figure 1. A typical surface plasmonic resonance (SPR) from metallic Ag could be observed at ca. 350 nm [11], suggesting the formation of Ag in Ag_2O for whole samples. While SPR signals became narrower and more intense, the peak position remained unchanged, with increased EG as described in the inset (normalized optical absorption spectra) of figure 1.

The growth of hybrid Ag_2O/Ag NPs with different EG concentrations promoted a variety of Ag loadings, and more EG molecules were drawn to the surface of nanoparticles as revealed by the infrared absorption spectra (see figure 2). Several bands of organic compounds from EG could be identified at 705.9, 883.4, and 1060.9 cm^{-1}, corresponding to C-H bending in plane, C-H stretching mode, and stretching mode of –C=O bonds, respectively [12, 13]. Infrared absorption peaks at 1380.1 cm^{-1} and 1448.6 cm^{-1} were probably caused by the vibration of hydrocarbon groups bonding to Ag clusters [14]. On the other hand, the absorption band at 1663.7 cm^{-1} can be attributed to the –CHO group of acetaldehyde that formed as an outcome of dehydration of EG in the reduction reaction of Ag^+ ions [15]. The appearance of a broad shoulder at ca. 2500-3750 cm^{-1} is associated with the stretching modes of hydrocarbon and hydroxyl groups commonly encountered in oxide semiconductors [13, 16]. Aside from that, the presence of SDS surfactant is barely distinguished. As reported by Ramimoghadam *et al.* [13] the –S=O stretching modes of sulphonic acid groups on the infrared spectrum should occur at around 1027 and 1224 cm^{-1} from pure SDS. It follows that the protective and stabilizing agent from SDS may be diminished during the cleaning process, in contrast to the EG that remains on the nanoparticles.

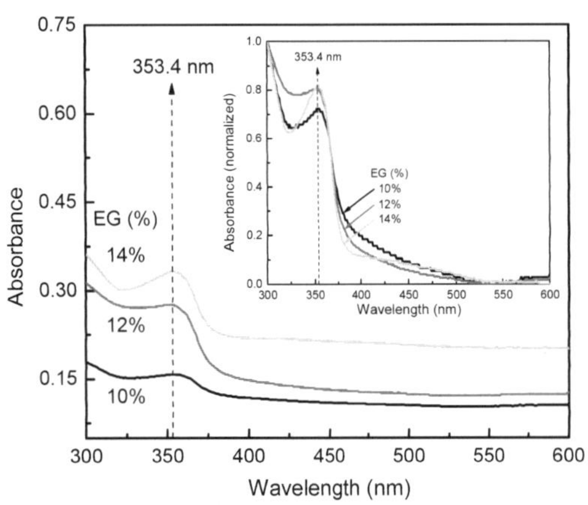

Figure 1. Exhibition of plasmonic resonance from Ag_2O/Ag UV-Vis absorption spectra.

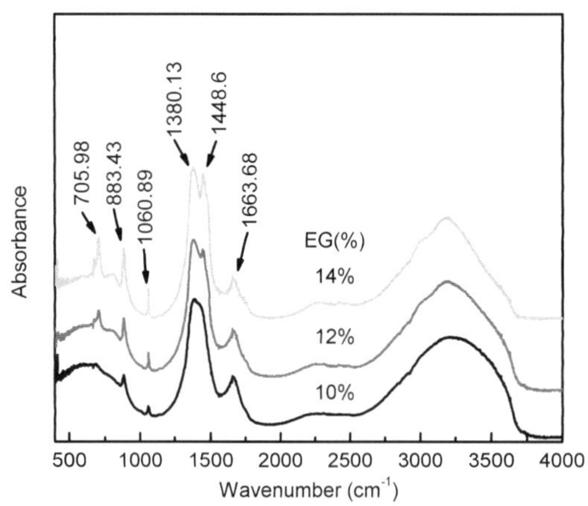

Figure 2. Infrared absorption spectra of hybrid Ag_2O/Ag NPs with various Ag.

Figure 4 shows the diffraction spectra from Ag_2O/Ag NPs with variation of EG volume ratio. Obvious diffraction peaks from cubic-Ag_2O [5] with Pn-3m space group could be indexed at 2θ of 32.5°, 37.8°, 54.6°, and 64.9°, correlated to the (111), (200), (220), and (311) crystallographic planes, respectively. On the other hand, intense growth of cubic-Ag phase (Fm-3m) is clear following the addition of EG from 10% to 14%. The crystal planes of (111) at 37.8° and (200) at 44° can be seen to have grown immensely as compared to the crystal plane of (111) of cubic-Ag_2O phase. Phases other

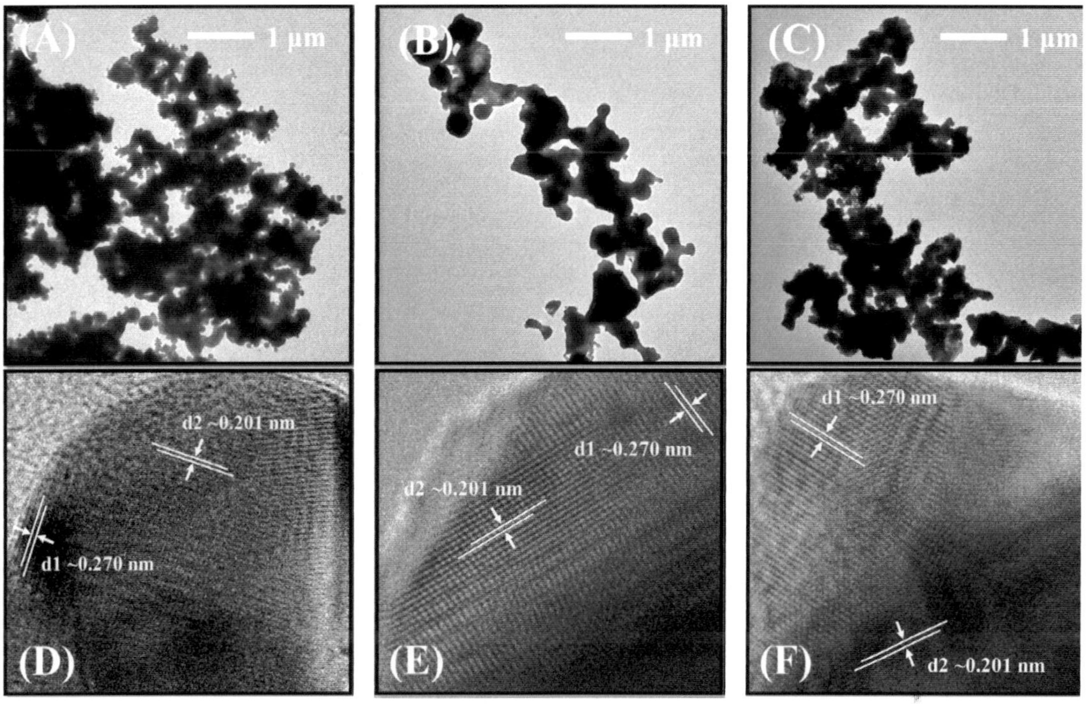

Figure 3. TEM imaging of Ag_2O/Ag NPs grown with 10% (A and D), 12% (B and E) and 14% (C and F) EG.

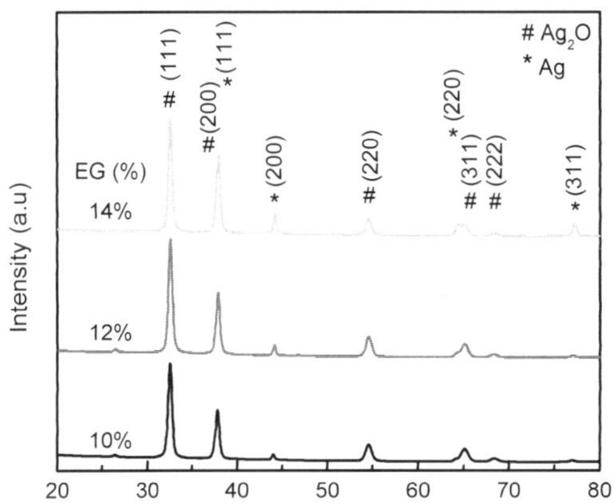

Figure 4. X-ray diffraction spectra of hybrid Ag_2O/Ag NPs grown by various EG concentrations.

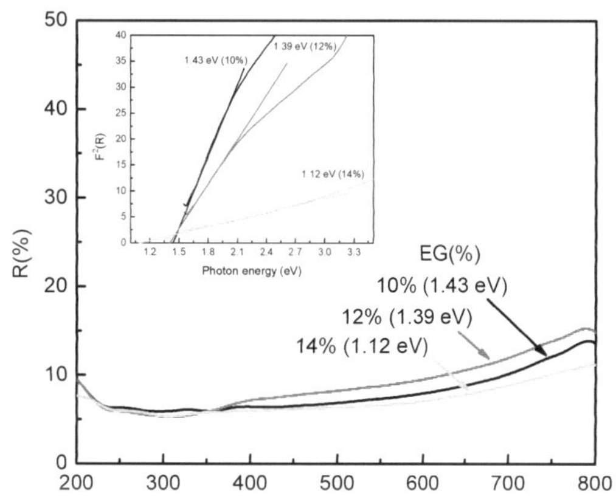

Figure 5. Optical reflectance spectra of various Ag contents in hybrid Ag_2O/Ag NPs.

Table 1. Quantitative analysis of x-ray diffraction patterns from hybrid Ag_2O/Ag NPs.

Crystal Properties	10% EG			12% EG			14% EG		
	Ag_2O (111)	Ag (200)	Ag_2O/Ag (200/111)	Ag_2O (111)	Ag (200)	Ag_2O/Ag (200/111)	Ag_2O (111)	Ag (200)	Ag_2O/Ag (200/111)
<D> (nm)	18.9	30.6	19	19.1	31.1	19.6	21	28.9	18.6
a=b=c (Å)	4.756	4.105	-	4.749	4.101	-	4.744	4.096	-
d-space (Å)	2.75	2.06	2.38	2.75	2.06	2.37	2.75	2.05	2.37
Rel. Intens. (%)	100	6.1	52.02	100	7.76	50.85	100	17.27	75.5
Center (°)	32.56	44.04	37.82	32.59	44.12	37.88	32.59	44.21	37.93
FWHM (°)	0.559	0.4005	0.563	0.553	0.396	0.548	0.515	0.417	0.573
Int. Area (%)	64.39	2.63	32.98	62.67	3.95	33.38	52.48	7.64	39.88

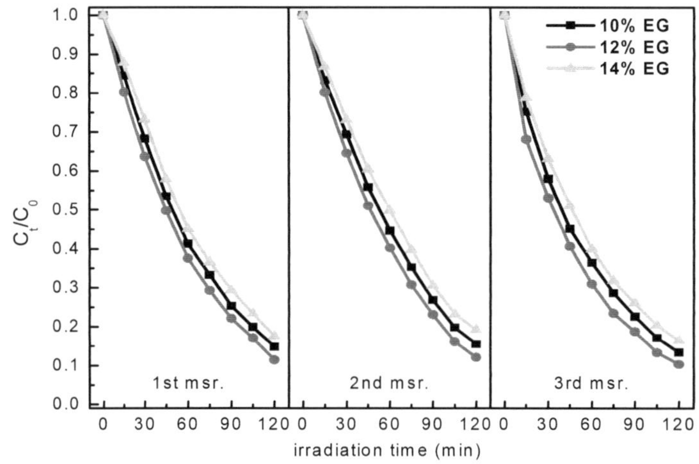

Figure 6. Photodegradation performance of photocatalysts hybrid Ag_2O/Ag NPs with various Ag contents.

than those two were undetectable, which means that the nanoparticles built up solely from Ag_2O and Ag material. Once again, this indicates that the number of Ag clusters can be adjusted by using an appropriate amount of EG in the chemical reaction. Briefly, the calculated average crystallite sizes, lattice parameters, d-spacing, relative intensities, peak centers, full width at half maximum (FWHM), and integrated areas are set out in table 1. According to table 1, several crystal properties are affected by the presence of EG, and the integrated area shows the elevation of Ag phase against Ag_2O phase. However, the crystallite size of Ag phase increases a little (from 10% to 12% EG) and then drops for 14% EG. The computed d-spacing ascribed to (111) plane of Ag_2O phase is about 0.275 nm and 0.206 nm for the (200) plane of Ag phase, as indicated by the TEM images.

To ascertain the optical gap energies of hybrid Ag_2O/Ag NPs with varying Ag content, optical reflectance spectra were recorded (figure 5). Ag_2O/Ag NPs with 14% EG show the lowest reflectance (about 8.84% at 700 nm), insignificantly altered to 11.9% and 10.2% for 12% and 10% EG. These trivial changes in optical reflectance can be attributed to the presence of metallic Ag, as metal tends to strongly absorb light because of SPR generation [10]. As demonstrated in previous studies [16], band gap energies for various Ag_2O/Ag NPs were derived from the reflectance spectra using the Kubelka-Munk function (inset, figure 5). The obtained gap energies were found at 1.43, 1.39, and 1.12 eV, with increased Ag concentration in Ag_2O/Ag NPs. The narrower optical window gap seen in the computed data owes to the electronic structure alteration of Ag_2O following the onset of Ag phase [17]. Similar results have been reported by Huijuan *et al.* [17], indicating that the redshift occurrence caused by different Ag contents can be attributed to change of particle size. The growth of Ag can interfere with the growth rate of Ag_2O, leading to an increase in crystal defects.

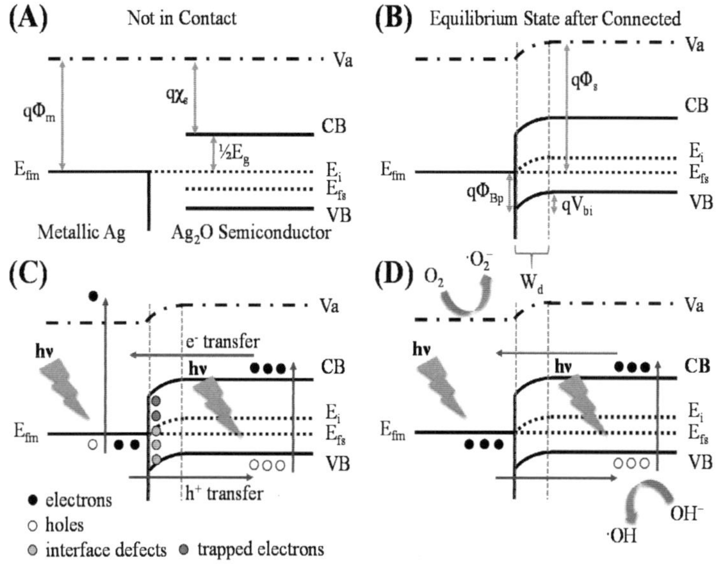

Figure 7. (A and B) Simple schematic of energy band diagrams dwelling on photocatalyst Ag_2O/Ag, (C) possibly charge transport mechanism, and (D) radical groups production process.

Figure 6 displays the photodegradation ability of hybrid Ag_2O/Ag NPs prepared with different EG percentages. Measurements were repeated three times to preserve the reproducibility of photocatalytic activity data. The highest removal of MB under visible light exposure was achieved up to 90% within 120 min with Ag_2O/Ag photocatalyst synthesized with 12% EG. Ag_2O/Ag photocatalyst with 10% and 14% EG reduced MB only up to 87% and 83%, respectively. Reaction rate of photocatalytic performance was estimated from pseudo first-order reaction kinetic studies of linear fitting from the plot of ln (C_0/C_t) vs kt (not shown), where C_0 is initial concentration, C_t is a concentration at time t, and k is a constant rate (min^{-1}) [9]. The observed photodegradation rate for 12% EG was 0.0184 min^{-1}, which is faster than the 10% and 14% EG photocatalyst rates of 0.0164 min^{-1} and 0.015 min^{-1}, respectively. It follows that the photocatalytic activity enhancement occurring in Ag_2O/Ag at 12% EG is probably due to the superior and highly crystalline interfacial contact revealed by x-ray diffraction and TEM imaging analyses [5].

Further corroboration of photodegradation improvement by Ag_2O/Ag is provided by the rough schematic of energy band diagrams in figure 7. To begin, we assume that Ag_2O/Ag NPs consist of metallic Ag and p-type Ag_2O semiconductor where they make Schottky metal-semiconductor contact [18]. Energy bands of isolated metallic Ag and isolated Ag_2O semiconductor are constructed in figure 7a. Upon connection, under thermal equilibrium, these will match each other's Fermi level energy to maintain charge balance. Consequently, a potential barrier (Φ_{Bp}) is formed, together with a depletion layer (W_d) and an upward band bending on the semiconductor side with built-in potential (V_{bi}) (see figure 7b). During visible light exposure, photon energy (hυ) excites electrons in the valence band (VB), generating electron-hole pairs in the semiconductor region (see figure 7c), and an SPR effect occurs in the metal region. The photogenerated electrons are further transferred to the metal region owing to the built-in electric field of the Schottky junction and the SPR effect [8]. The photogenerated holes remain on the semiconductor side, as it can interact directly with the dye pollutant and oxidizing water molecules to generate radical groups [9]. Eventually, charge separation is effectively improved where the photogenerated electrons gathered in metal side while the photogenerated holes remain in the semiconductor VB (see figure 7d).

The downgraded photocatalytic activity of Ag_2O/Ag (14% EG) reported here was also reported by Hu *et al.* [5], who noted the influence of a better interfacial contact layer and appropriate size of Ag in

Ag_2O NPs on charge migration, recombination, and charge storage in enhancing photocatalytic activities. In their study, a grain size of Ag phase of about 7-22 nm was the maximum size that could enhance photocatalytic performance, as compared to maximum grain size of Ag phase up to 31.1 nm (Ag_2O/Ag with 12% EG) in the present study. Despite having the highest Ag content in Ag_2O/Ag 14% EG as shown by the integrated area of Ag phase, the calculated grain size is lowest (28.9 nm) as compared to the other samples. Random growth rate and orientation of crystal phase may lead to imperfections and higher defect density. Defects normally increase recombination activities, so preventing charge migration. Linking to the metal-semiconductor contact model, interface defects may account for this issue. Figure 7c demonstrates the scheme case if the interface defect states presence. Such defects may also relate to grain boundary defects [19, 20]. If, for instance, there are several acceptor interface defect centers above Fermi level energy, these centers will be negatively ionized upon capturing the photogenerated electron migration from CB Ag_2O to metallic Ag site (see figure 7c); in consequence, there was a lower photodegradation rate for Ag_2O/Ag produced with 14% EG. Moreover, the TEM images in figure 3d-f offer additional evidence of interfacial/grain boundary defects. Perpendicular crystal orientation among Ag_2O and Ag phase is clearly visible in figures 3d and 3e. For photocatalyst Ag_2O/Ag with 14%, between those phases are inclined rather than perpendicular, probably inducing a larger interfacial defect.

4. Conclusions

In conclusion, various Ag contents of hybrid Ag_2O/Ag nanoparticles were successfully obtained in a one-step microwave-assisted colloidal process. By employing EG as a promoter of reducing agents, different contents of Ag phase within Ag_2O could be controlled, as revealed by diffraction patterns and optical spectra. Infrared absorption showed an increase of EG molecules attached to Ag_2O/Ag particles, along with additional EG concentrations. A specific crystal orientation for Ag_2O and Ag phases was clearly observed in high resolution TEM images. Among Ag_2O/Ag photocatalysts, the one prepared with 12% EG exhibited prominent photodegradation performance towards MB (with a photodegradation rate of 0.0184 min^{-1}). It was posited that the noble metallic Ag improved charge separation through involvement of Schottky junction and SPR effects in the photocatalytic activities of Ag_2O/Ag.

Acknowledgements

Thank you to Hendry Tju who had been assisting in photocatalytic measurements.

References

[1] Fazlali F, Mahjoub A R and Abazari R 2015 *Solid State Sci.* **48** 263-9
[2] Wan X, Yuan M, Tie S and Lan S 2013 *Appl. Surf. Sci.* **277** 40-6
[3] Zhu Y J and Chen F 2014 *Chem. Rev.* **114** 12 6462-555
[4] Ida Y, Watase S, Shinagawa T, Watanabe M, Chigane M, Inaba M, Tasaka A and Izaki M 2008 *Chem. Mater.* **20** 4 1254-6
[5] Hu X, Hu C and Wang R 2015 *Appl. Cata. B: Environ.* **176-177** 637-45
[6] Liu S, Wang N, Zhang Y, Li Y, Han Z and Na P 2015 *J. Hazard Mater.* **284** 171-81
[7] Liu C, Cao C, Luo X and Luo S 2015 *J. Hazard Mater.* **285** 319-24
[8] Chen J, Che H, Huang K, Liu C and Shi W 2016 *Appl. Catal. B: Environ.* **192** 134-44
[9] Umukoro E H, Peleyeju M G, Ngila J C and Arotiba O A 2016 *Solid State Sci.* **51** 66-73
[10] Jia K, Khaywah M Y, Li Y, Bijeon J L, Adam P M, Déturche R, Guelorget B, François M, Louarn G and Ionescu R E 2014 *ACS Appl. Mater. Interfaces* **6** 219-27
[11] Bhui D K, Bar H, Sarkar P, Sahoo G P, De S P and Misra A 2009 *J. Mol. Liq.* **145** 1 33-7
[12] Zhao T, Sun R, Yu S, Zhang Z, Zhou L, Huang H and Du R 2010 *Colloids Surfaces A Physicochem. Eng. Asp.* **366** 197-202
[13] Ramimoghadam D, Hussein Z M and Taufiq-Yap H Y 2012 *Int. J. Mol. Sci.* **13** 13275-93
[14] Slistan-Grijalva A, Herrera-Urbina R, Rivas-Silva J F, Ávalos-Borja M, Castillón-Barraza F F and Posada-Amarillas A 2008 *Mater. Res. Bull.* **43** 90-6
[15] Chen L J, Wan C C and Wang Y Y 2006 *J. Colloid Interface Sci.* **297** 143-50

[16] Prakoso S P and Saleh R 2012 *Mater. Sci. Appl.* **3** 530-7
[17] Bi H, Cai W, Kan C, Zhang L, Martin D and Träger F 2002 *J. Appl. Phys.* **92** 7491-7
[18] Sze S M and Ng K K 2006 *Physics of Semiconductor Devices* (New Jersey: Wiley-Interscience)
[19] Grovenor C R M 1985 *J. Phys. C Solid State Phys.* **18(21)** 4079-120
[20] Ye J D *et al.* 2005 *Appl. Phys.* A **81** 759-62

The synthesis of 2-(5-(3-methoxyphenyl)-4,5-dihydro-1*H*-pyrazol-3-yl)phenol using sodium impregnated on activated chicken eggshells catalyst

L Mardiana, R Bakri, A Septiarti and B Ardiansah

Department of Chemistry, Faculty of Mathematics and Natural Sciences Universitas Indonesia, Kampus UI Depok, Depok 16424, Indonesia

Corresponding author's e-mail: bakri@ui.ac.id

Abstract. The novel compound of 2-(5-(3-methoxyphenyl)-4,5-dihydro-1*H*-pyrazol-3-yl)phenol as a pyrazoline derivative has been synthesized by two-steps reaction using sodium impregnated on activated chicken eggshells (Na-ACE) catalyst. Na-ACE was primarily prepared by a simple wet impregnation of NaOH solution on activated chicken eggshells solid support. The Na-ACE catalyst produced was characterized using FTIR spectrophotometer, XRD and SEM then applied in pyrazoline synthesis. First, chalcone was prepared from the reaction of 2-hydroxyacetophenone and 3-methoxybenzaldehyde by base-catalyzed aldol condensation. This product was subsequently reacted with hydrazine hydrate to give corresponding pyrazoline. The structure elucidation of the compound using FTIR, UV-Vis, LC-ESI-MS and ^{1}H-NMR indicated the desired product has been successfully synthesized. Furthermore, the potential antioxidant activities of chalcone and pyrazoline have also been studied *in-vitro* using DPPH radical scavenging method. The results revealed that pyrazoline has a greater antioxidant activity than chalcone.

1. Introduction

Chalcone and its derivatives are known for a wide spectrum of applications in the field of pharmaceuticals and other industries [1, 2]. For example, they have been found to exhibit various biological functions as antioxidant, anti-inflammatory, analgesic, anticancer, antitubercular, antitumor, antimicrobial, and antihyperglycemic agents [3-6]. Chemically, chalcone consists of two aromatic rings separated by keto-vinyl chain, and it is a versatile intermediate for synthesizing heterocyclic molecules [7, 8]. Meanwhile, pyrazoline, a type of hydrogenated pyrazole, is a five-membered nitrogen-containing heterocyclic molecule which has broad applications, especially in medicinal chemistry. Nitrogen heterocyclic is of special interest because it constitutes an important group of natural and synthetic products, many of them show valuable biological and pharmacological performances such as antiviral, anti-inflammatory, antifungal, antibacterial, and also demonstrates unique optical properties [9, 10]. Owing to the attractive properties of both chalcone and pyrazole, novel protocols to design different molecular structures have attracted attention of many researchers [11-13].

Several synthetic protocols are available for synthesis of chalcone and its pyrazole derivatives in the absence or presence of catalysts. The catalyst of KOH/EtOH [14] and NaOH/MeOH [15] have been used in chalcones synthesis, whereas some catalysts used in pyrazole synthesis such as triethylamine [16], hydrotalcite [17], *L*-proline in ionic liquid [bmim]BF_4 [18] and in catalyst-free condition using water as green solvent [19]. Many of the previously reported protocols suffer some limitations, such as high reaction temperatures, time consuming, poisonous solvents, and/or high cost reactants [20]. Thus, the demand for the efficient methods using single catalyst that can accommodate two-steps pyrazoline synthesis from aromatic aldehydes/ketones via chalcone has accentuated the demand to develop a novel, green routes and eco-compatible motivated the present investigation. The activated chicken eggshell (ACE) is a simple, low cost and multipurpose material which can be used as catalyst for several types of organic reactions, including transesterification [21, 22], aldol condensation [6] and Schiff base formation [23]. Because of its potential to be used in organic synthetic as heterogeneous catalyst, in this work, we will demonstrate and compare the catalytic ability of Na-modified activated chicken eggshells (Na-ACE) in two-steps reaction during pyrazoline synthesis.

2. Materials and methods

2.1. Materials

Chemicals were from analytical grade (from Sigma-Aldrich, Merck and other commercial suppliers) and used without any further purification. TLC analyses were conducted on Merck silica gel 60 F_{254} aluminium plates. Melting point measurements were performed in an Electrothermal-9100 melting point apparatus and were uncorrected. IR spectra were conducted on a Shimadzu Prestige-21 spectrophotometer in KBr pellets. XRD pattern was recorded on a D8 Bruker diffractometer (40 kV and 30 mA) with Cu-Kα radiation (0.154 nm). Catalyst morphology and its elemental content were measured on SEM-EDAX instrument Hitachi S4160 at accelerating voltage of 20 kV. UV-Vis spectra were recorded in Shimadzu 2450 spectrophotometer. LC-ESI-MS Mariner biospectrometer was used to determine the molecular weight of the products, and the exact organic structure was confirmed using ^{1}H-NMR on a 500 MHz Agilent Technology instrument.

2.2. Catalyst preparation

The Na-ACE catalyst was prepared based on our previous work and re-characterized using IR, XRD, and SEM-EDAX [6]. Briefly, the chicken eggshells were boiled in distilled water for 1 h then dried in an oven at 110ºC for 3 h. Subsequently, the chicken eggshells were activated at 900ºC for 4 h under atmospheric air in a furnace to give activated chicken eggshells (ACE). The Na-ACE catalyst was obtained by impregnating 3% NaOH solution on ACE, and followed by calcination at 700ºC for 2 h. The total basicity of the catalyst was measured by titration method [6,24].

2.3. Chalcone and pyrazoline synthesis

Chalcone synthesis: A mixture of 2-hydroxyacetophenone (1 mmol), 3-methoxybenzaldehyde (1 mmol), solvent (5 mL), and proper amount of Na-ACE were magnetically stirred. The amount of catalyst added, reaction time, temperature and organic solvent were varied to get optimum condition of chalcone synthesis. Purity of the synthesized chalcone was tested by using TLC-scanner. (*E*)-1-(2-hydroxyphenyl)-3-(3-methoxyphenyl)prop-2-en-1-one, a chalcone, was obtained. Pyrazoline synthesis: chalcone (1 mmol), hydrazine hydrate (2 mmol), ethanol (5 mmol) and 30% wt. of Na-ACE were stirred for 24 h at 60ºC. The product was purified with column chromatography with gradient elution to get pure pyrazoline, 2-(5-(3-methoxyphenyl)-4,5-dihydro-1*H*-pyrazol-3-yl)phenol for further analyses.

2.4. Antioxidant activity

Chalcone and pyrazoline were evaluated for their *in vitro* free radical scavenging activity by the 2,2'-diphenyl-1-picrylhydrazyl (DPPH) radical scavenging method [25, 26]. Stock solutions of chalcone and pyrazoline with various concentrations were mixed with DPPH ethanolic solution (0.5 mL, 0.01 mM) in 3 mL of total reaction mixture and allowed to react at room temperature. Absorbance values were measured at 517 nm every 10 min for 40 min and converted to percentage of inhibition. The percentage of inhibition activity was then calculated. The data are summarized in table 1. Percentage of inhibition = [(absorbance of control – absorbance of sample) / absorbance of control] ×100%.

3. Results and discussion

3.1 Catalyst preparation

Chicken eggshells (CE) waste can be considered as material which has more than 98% wt. calcium carbonate by mass [27]. By activating it through calcination, the activated chicken eggshells (ACE) is produced with CaO as major constituent. As the basic properties of the alkaline earth metal oxides [28, 29], ACE is a promising material used for solid support in base-catalysed organic reactions. In addition, the use of ACE has advantages such as less expensive and less toxic compared to other alkaline earth metal oxides [28]. Catalytic performance of metal oxides based heterogeneous catalysts was found to be a function of their basicity. The basic strength of ACE used as solid support in this work was found to be 33.31 mmol/g, and can be enhanced up to 35.96 mmol/g by impregnating 3% Na^+ in ACE. Impregnation of alkali metal ions could increase the basic strength of metal oxides based catalysts, and it is confirmed with earlier published articles [6, 28].

IR spectra of CE in figure 1 shows sharp peak near 1700 cm^{-1} due to the carbonyl stretching of carbonate ions in CE. The disappearance of C=O band in ACE and Na-ACE indicated that calcium carbonate in CE is fully converted to calcium oxide. A broad peak near 500 cm^{-1} is attributed to the Ca-O vibration. Meanwhile, a sharp and strong peak at 3600 cm^{-1} is attributed to the non-hydrogen bonded –OH stretching in $Ca(OH)_2$ formed during adsorpsion of water over Na-ACE. The ACE and Na-ACE are chemically similar, and our data regarding the unactivated and activated chicken eggshells agree with those previously reported work [27, 30].

X-ray diffraction analysis results for CE, ACE and Na-ACE are shown in figure 2. The typical X-ray diffraction patterns of CE are observed at the Bragg angles, $2\theta = 22.93°$, 29.21°, 35.68°, 39.27°, 43.04°, 47.53° and 48.43°, implying the presence of *calcite* structure in chicken eggshells waste. The XRD pattern of ACE (activated at 900°C for 4 h) has peak at $2\theta = 32.09°$, 37.12°, 53.64°, 64.05° and 67.11°, indicating that *lime* (CaO) is the only phase presents in the activated chicken eggshells. The major peak of the *calcite* phase ($2\theta = 29.21°$) is not observed in the XRD pattern of ACE, signifying that the $CaCO_3$ phase was completely changed to the CaO phase. Upon Na^+ impregnation the formation of calcium hydroxide in hexagonal form was depicted due to the presence of peaks at $2\theta = 17.90°$, 28.67°, 34.06°, 47.00°, 50.77° and 62.44°. Impregnation of Na^+ was not found to affect the structure of *lime* phase, so the two major phases in Na-ACE are CaO and $Ca(OH)_2$. XRD resulted in this work was relevant with previous investigations [27, 28, 31].

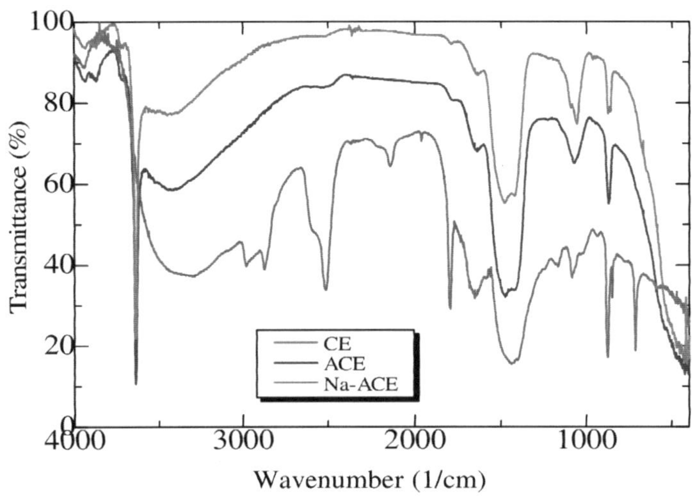

Figure 1. IR spectra of CE, ACE and Na-ACE.

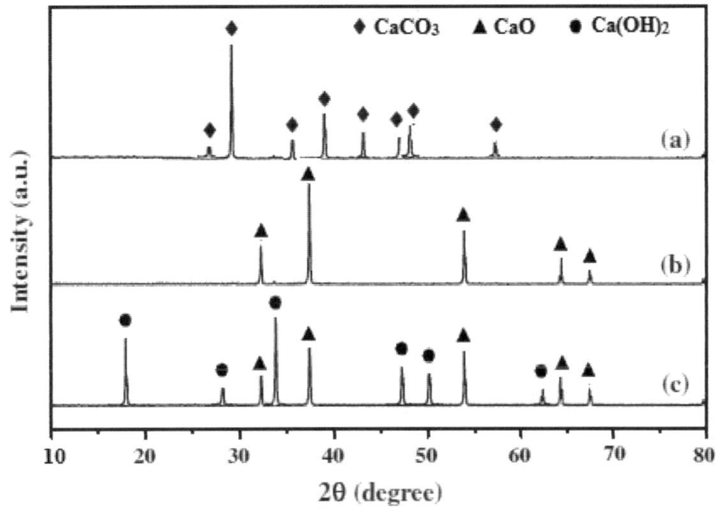

Figure 2. XRD patterns of (a) CE, (b) ACE and (c) Na-ACE.

(a)

(b)

Figure 3. FESEM micrographs of (a) ACE and (b) Na-ACE.

FESEM micrograph of ACE shows that it has a porous structure and irregular shaped cluster of 1-2.5 µm sized particles as shown in figure 3(a). In Na-ACE catalyst, particles size decreased to 40-500 nm. This result is supported by calculation of the crystallite size using Debye-Scherrer equation, and the size of ACE and Na-ACE were found to be 108 and 50 nm, respectively. The data revealed that the 3% Na^+ impregnation can decrease the particles size of the activated chicken eggshells.

3.2. Chalcone and pyrazoline synthesis
Substrate 2-hydroxyacetophenone was reacted with 3-methoxybenzaldehyde using Na-ACE catalyst via base-catalysed aldol condensation, and it produced yellowish crystal of chalcone with melting point of 83-84°C. Structure elucidation of chalcone was performed from its spectroscopic data (IR, LC-ESI-MS and UV). The IR spectrum of chalcone showed an absorption peak at 1642 cm^{-1} associated to the α,β-unsaturated carbonyl group and absorption peaks at 3200 and 2960 cm^{-1} associated to the hydroxyl and C-H sp^3 group, respectively. Meanwhile, absorption at 3016 and 1491 cm^{-1} associated to the C-H sp^2 and C-C aromatic ring. In LC-ESI-MS analysis, there was only one single peak at retention time 7.57 min, and it has m/z values of 255.23 as $[M+H]^+$ and 277.24 as $[M+Na]^+$. The absorption profile of chalcone was screened from 200 to 500 nm and gave maximum wavelength at 310 and 255 nm. Analysis

using TLC-scanner indicated that the chalcone obtained was relatively pure with area chromatogram more than 99%. Chalcone obtained from step 1 was subsequently reacted with hydrazine hydrate via base-catalysed nucleophilic addition at step 2 to give corresponding pyrazoline (yellowish crystal, mp. 90-92°C). Structure elucidation of pyrazoline was performed from its spectroscopic data (IR, UV, LC-ESI-MS and ^{1}H-NMR). The IR spectrum of pyrazoline showed peaks at 3200-3380 cm^{-1} for overlap –OH and secondary –NH functional groups, at 1566 cm^{-1} for imine (C=N) group and at 3020-3050 cm^{-1} for C-H aromatic stretching. The disappearance of carbonyl peak near 1700 cm^{-1} indicated the successive transformation of chalcone to pyrazoline. Regarding the LC-ESI-MS data, there was a single peak at retention time 1.66 min with $[M]^+$ of 268.31 for $C_{16}H_{16}N_2O_2$. Analysis of pyrazoline using ^{1}H-NMR showed a singlet with integration of 3H at 2.89 ppm, assigned to the methyl protons of the methoxy group. Chemical shift between 6.90-7.78 ppm indicated there were 9 protons, including 8 proton of aromatic ring and 1 proton of N-H in pyrazole ring. Hydroxyl proton appeared in the downfield region at 10 ppm due to the hydrogen bond interaction by N atom from pyrazole ring. The UV absorption maximum of pyrazoline was at 300 and 291 nm, slightly lower than chalcone due to hypsochromic effect regarding to the number of its conjugated double bond.

To investigate the catalytic ability of Na-ACE in chalcone synthesis, several optimizations have been carried out. Figure 5a shows the chalcone yield obtained with Na-ACE catalyst, as a function of the reaction time. The reaction was carried out using 30% wt. of catalyst at 60°C in ethanol as solvent. The reaction produced chalcone with low percentage of yield (20%) at 1h, and achieved the optimum at 3h. In figure 5b, the use of 10% wt. of catalyst at 60°C for 3h contact time in ethanol, produced very low percentage of yield (~1%). The increase in the amount of catalyst used, increases the chalcone obtained, with the optimum catalyst of 30% wt. Surprisingly, increasing the catalyst amount to 35% wt. will not affect to the increase of chalcone yield. On the contrary, the chalcone yield decreased from 57% to 33% in this condition due to very strong adsorption of starting materials over Na-ACE catalyst, and will not produce the desirable product. Another reason to this phenomenon is the chalcone produced from the reaction was strongly adsorbed over Na-ACE catalyst and it was hard to handle, so the isolated yield was lower. Furthermore, when large amount of catalyst is added, the sintering and agglomeration of catalyst particles happens, and could decrease the catalytic ability of Na-ACE. Figure 5c shows that 60°C is the most suitable temperature for chalcone synthesis using Na-ACE catalyst. As depicted in figure 5d, we have performed chalcone synthesis in four different solvents. Ethyl acetate, as semi-polar solvent, gave the lowest percentage yield of chalcone by 29%. By changing the medium with more polar solvent, using t-butanol, for example the yield of chalcone increased and reached the optimum when ethanol was used in the reaction. Ethanol can dissolve all starting materials and disperse the catalyst particles well. When methanol was used, a more polar solvent than ethanol, the percentage yield was observed to be lower. The use of very polar solvents can be considered to be unsuitable condition for chalcone synthesis. The reaction scheme of chalcone and pyrazoline synthesis can be seen in figure 4.

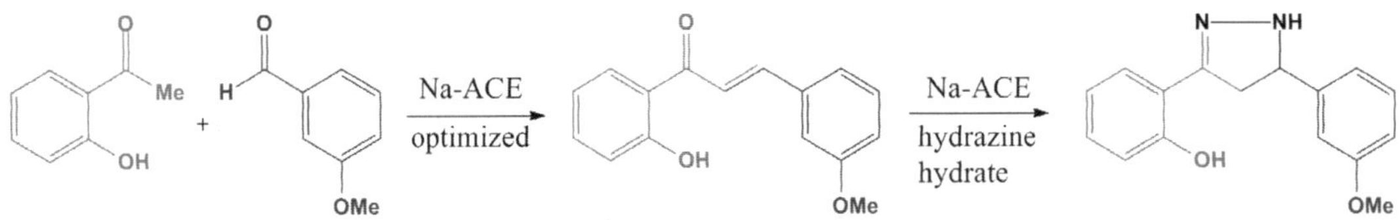

Figure 4. Reaction scheme in pyrazoline synthesis.

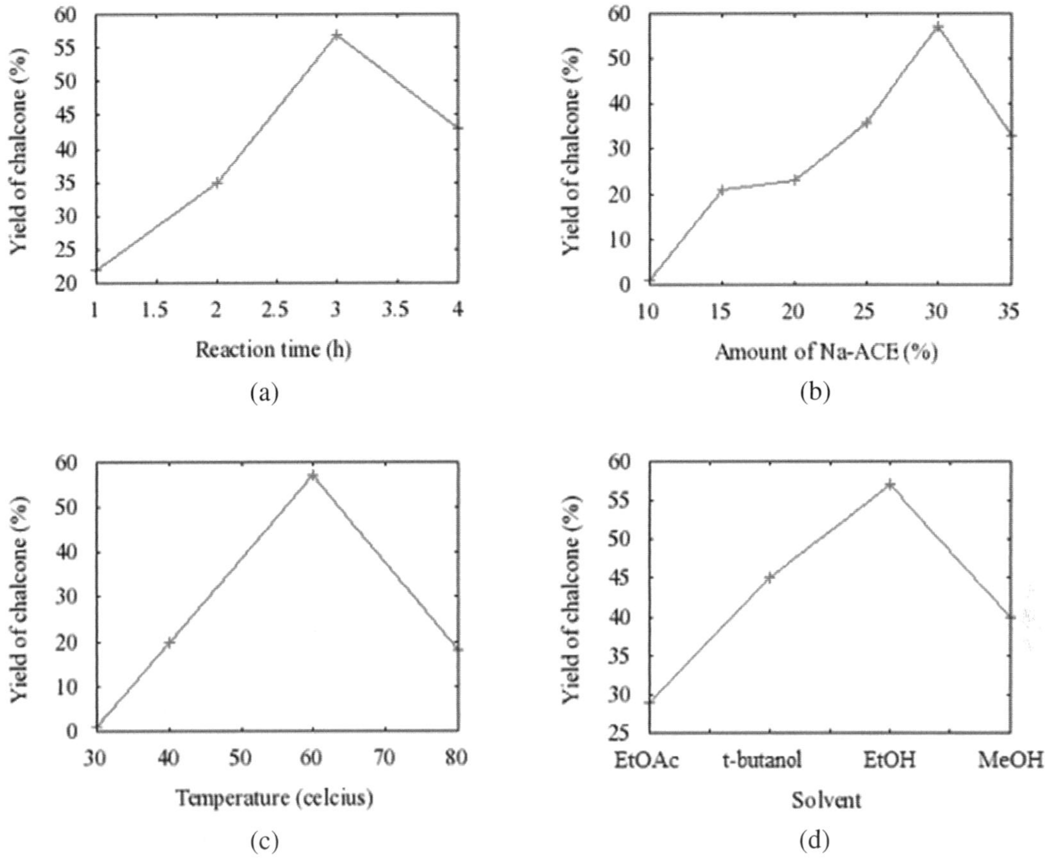

Figure 5. Optimization in chalcone synthesis (a) reaction time (b) amount of Na-ACE catalyst (c) temperature and (d) solvent.

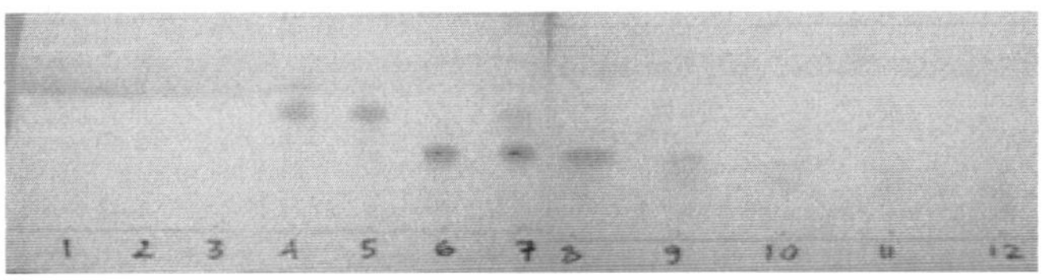

Figure 6. Undesirable by-product in pyrazoline synthesis was observed using TLC analysis. Fraction No. 6-8 is considered contains pyrazoline.

After chalcone was obtained (> 99% purity), we conducted the second step in pyrazoline synthesis. Molar ratio 1:2 of chalcone and hydrazine was used, and the reaction was performed at 60°C for 24 h by using 30% wt. catalyst amount produced very low desirable product, approximately 12% yield. Shortening the reaction time and decreasing both of the catalyst amount and temperature would produce pyrazoline lower than 10% yield. For example, when we used 30% wt. catalyst at 60°C for 10 h, 8% yield pyrazoline was isolated. Pyrazoline was not formed when the reaction was carried out at room temperature using Na-ACE catalyst. From the data, we concluded that the Na-ACE catalyst is less suitable for pyrazoline synthesis via reaction of chalcone and hydrazine hydrate than for chalcone

Table 1. Percentage inhibition by chalcone and pyrazoline.

Chalcone	DPPH inhibition (%)				Pyrazoline	DPPH Inhibition (%)			
	10 min	20 min	30 min	40 min		10 min	20 min	30 min	40 min
150 ppm	18.7	18.8	19.1	19.4	10 ppm	31.4	33.2	33.6	35.0
200 ppm	19.1	19.2	19.3	20.0	20 ppm	33.6	37.3	40.0	43.2
250 ppm	20.0	20.4	20.8	20.9	30 ppm	39.6	45.5	49.1	53.2
300 ppm	21.7	23.5	24.3	25.1	-	-	-	-	-

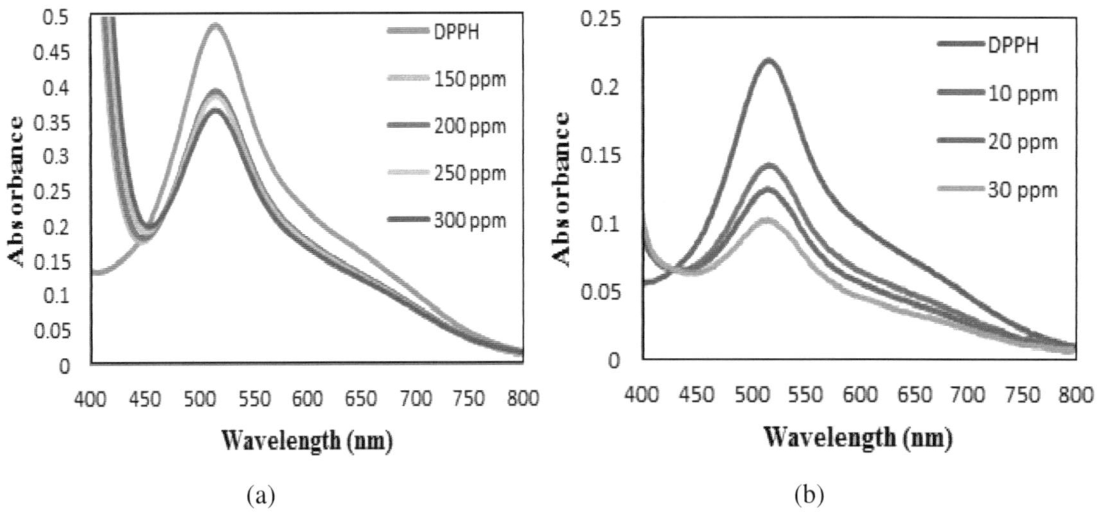

Figure 7. Visible absorption profile of (a) chalcone and (b) pyrazoline in antioxidant test.

synthesis. Another limitation in the pyrazoline synthesis using Na-ACE was separation method of the product mixture. Undesirable product was observed in TLC analyses and need to be purified using gradient concentration column chromatography to get pyrazoline (figure 6).

3.3. Antioxidant activity

The *in vitro* antioxidant and scavenging effects of chalcone and pyrazole were evaluated by using DPPH radical scavenging method, and the results are shown in table 1. The interaction of the chalcone and pyrazoline with the radical DPPH denotes their radical scavenging performance. Both of chalcone and pyrazoline exhibit higher DPPH inhibition ability at higher concentration and longer time exposure. The full visible absorption profiles of DPPH control and sample-containing DPPH solution at different sample concentrations are depicted in figure 7. By using percentage inhibition data, the IC_{50} value for pyrazoline was calculated and found to be 26.8 ppm. On the other hand, the DPPH inhibition value for chalcone at 300 ppm and 40 min was 25.1% only. Hence, the IC_{50} value for chalcone was considered more than 300 ppm and classified as inactive antioxidative agent.

4. Conclusions

A new green approach to the synthesis of chalcone and pyrazoline using the same catalyst, sodium impregnated on activated chicken eggshells (Na-ACE), has been developed. The Na-ACE was found to be more active in catalysing aldol condensation to produce chalcone than in catalysing ring closure pyrazoline formation, with percentage yield of chalcone and pyrazoline of 57% and 12%, respectively. The future research should be focused on the catalytic modification of eggshells-based materials to improve their ability. From the bioassay, it was proven that pyrazoline, 2-(5-(3-methoxyphenyl)-4,5-

dihydro-1H-pyrazol-3-yl)phenol are potent and possessing high antioxidant and DPPH scavenging activities.

References

[1] Pasquale G, Romanelli G P, Autino J C, Garcia J, Ortiz E V and Duchowicz P R 2012 *J. Agric. Food Chem.* **60** 692-7
[2] Prasath R, Bhavana P, Sarveswari S, Seik Weng Ng and Tiekink E R T 2015 *J. Mol. Structure* **1081** 201-10
[3] Dimmock J R, Elias D W, Beazely M A and Kandepu N M 1999 *Curr. Med. Chem.* **6** 1125-49
[4] Nowakowska Z 2007 *Eur. J. Med. Chem.* **42** 125-37
[5] Hussain S and Kaushik D 2015 *J. Saudi Chem. Soc.* **19** 274-81
[6] Mardiana L, Ardiansah B, Bakri R and Cahyana H 2016 *AIP Conference Proceedings* **1729** ID 020051
[7] Li J, Li X and Wang S 2012 *Spectrochim. Acta Part A: Molecular and biomolecular spectroscopy* **88** 31-6
[8] Patil P S, Dharmaprakash S M, Ramakrishna K, Fun H K, Kumar R S S and Rao D N 2007 *J. Cryst. Growth* **303** 520-24
[9] Baraldi P G, Bovero A, Fruttarolo F, Romagnoli R, Tabrizi M A, Preti D, Varani K, Borea P A and Moorman A R 2003 *Bioorg. Med. Chem.* **11** 4161-69
[10] Solomon V R and Lee H 2011 *Curr. Med. Chem.* **18** 1488-508
[11] Shi D, Mou J, Zhuang Q, Niu L, Wu N and Wang X 2004 *Synth. Commun.* **34** 4557-63
[12] Nikpassand M, Mamaghani M, Tabatabaeian K and Abiazi M K 2009 *Mol. Divers.* **13** 389-93
[13] Hossein Nia R, Mamaghani M, Shirini F and Tabatabaeian K 2014 *J. Het. Chem.* **51** 363-7
[14] Bhat B A, Dhar K L, Puri S C, Saxena A K, Shanmugavel M and Qazi G N 2005 *Bioorg. Med. Chem. Lett.* **15** 3177-80
[15] Hans R H, Guantai E M, Lategan C, Smith P J, Wan B, Franzblau S G, Gut J, Rosenthal P J and Chibale K 2010 *Bioorg. Med. Chem. Lett.* **20** 942-4
[16] Litvinov Y M, Shestopalov A A, Rodinovskaya L A and Shestopalov A M 2009 *J. Comb. Chem.* **11** 914-9
[17] Kshirsagar S W, Patil N R and Samant S D 2011 *Synth. Commun.* **41** 1320-5
[18] Khurana J M, Nand B and Kumar S 2011 *Synth. Commun.* **41** 405-11
[19] Vasuki G and Kumaravel K 2008 *Tetrahedron Lett.* **49** 5636-8
[20] Shabalala N G, Pagadala R and Jonnalagadda B 2015 *Ultrasonics Sonochem.* **27** 423-9
[21] Khemthong P, Luadthong C, Nualpaeng W, Changsuwan P, Tongprem P, Viriya-empikul N and Faungnawakij K 2012 *Catal. Today* **190** 112-6
[22] Buasri A, Chaiyut N, Loryuenyong V, Wongweang C and Khamsrisuk S 2013 *Sustain. Energy* **1** 7-13
[23] Patil S, Jadhav S D and Shinde S K 2012 *Organic Chemistry International* article **2012** 153159
[24] Hu K, Wang H, Liu Y and Yang C 2014 *J. Ind. Eng. Chem.* **28** 334-43
[25] Bandgar B P, Gawande S S, Bodade R G, Gawande N M and Khobragade C N 2009 *Bioorg. Med. Chem.* **17** 8168-73
[26] George S, Prameswaran M K, Chakraborty A R and Ravi T K 2008 *Acta Pharm.* **58** 119-29
[27] Mosaddegh E and Hassankhani A 2014 *Chinese J. Catal.* **35** 351-6
[28] Kumar D and Ali A 2012 *Biomass and Bioenergy* **46** 459-68
[29] Watkins R S, Lee A F and Wilson K 2004 *Green Chem.* **6** 335-40
[30] Liao H, Mutvei H, Sjöström M, Hammarström L and Li J 2000 *Biomaterials* **21** 457-68
[31] Degirmenbasi N, Coskun S, Boz N and Kalyon D M 2015 *Fuels* **153** 620-7

IOP Conf. Series: Materials Science and Engineering **188** (2017) 012023 doi:10.1088/1757-899X/188/1/012023

Zinc-porphyrin-imide type sensitized TiO_2 nanotube arrays as photoactive electrode in a dyes sensitized solar cell

F Syafaat and J Gunlazuardi

Department of Chemistry, Faculty of Mathematics and Natural Sciences Universitas Indonesia, Kampus UI Depok, Depok 16424, Indonesia

Corresponding author's e-mail: jarnuzi@ui.ac.id

Abstract. In this work, TiO_2 nanotube was prepared by rapid breakdown electrooxidation of Ti foil in electrolyte containing 0.15 M $HClO_4$. The obtained TiO_2 nanotube bundling powder was calcinated at 450°C for 3 hours, then characterized by SEM, XRD, UV-Vis DRS, and BET. Zinc-porphyrin-imide dyes was deposited into TiO_2 nanotube by immersion of TiO_2 nanotube in zinc-porphyrin-imide solution for 24 hours. Free zinc-porphyrin-imide dyes shows characteristics of absorption spectra in visible region at 439 nm and 620 nm. Meanwhile, when it was immobilized to TiO_2-nanotube, the absorption peak shifted to 421 nm and 640 nm. The zinc-porphyrin-imide/TiO_2 electrode showed an excellent response toward the visible light with the typical photocurrent density of 1.1 mA/cm^2. When the fabricated photoelectrode was assembled in a typical Dyes Sensitize Solar Cell (DSSC), the I-V curve showed photoconversion efficiency of the assembled zinc-porphyrin-imide/TiO_2 DSSC was 1.914% (front side illumination) and 1.147% (backside illumination), respectively.

Keywords: Zinc-porphyrin-imide, TiO_2 nanotube, TiO_2/Zinc-porphiyin-imide

1. Introduction

Titanium dioxide (TiO_2) is great interest for a number of technological areas, including photocatalysis, photovoltaic solar cells, and waste remediation, because of its electronic, optoelectronic, catalytic, and biocompatibility properties. TiO_2 is a wide band gap semiconductor, which has great ability to produce oxidants (holes in the valence band) and reductants (electrons in the conduction band) upon photons absorption [1-3]. Because of this feature and its mechanical strength, TiO_2 has been widely studied for applications in photoelectrochemical systems, such as TiO_2 photoanodes for dye-sensitized photovoltaic solar cells.

Dyes sensitized Solar Cells (DSSC) is one of the photoelectrochemical systems that can convert light energy into electrical energy. Typically, DSSC system so far has achieved energy conversion efficiency of 13% under AM 1.5 lights 1000 Watt [4]. The sensitizer is one important factor in DSSC. Gratzel *et al.* [5] has developed a DSSC with promising energy conversion, using ruthenium complexes compounds. Typical sensitizer that can be used generally is a complex compound that has a cluster of anchoring as a bridge binder and an electron donor of the sensitizer to the semiconductor. The anchoring group used is hydroxyl and carboxyl, which attached to a chromophore groups that can absorb visible light efficiently. The Chromophore group must have a LUMO orbital located above the conduction band of TiO_2, in order to thermodynamically facilitate the injection electrons from the LUMO level of sensitizer into the conduction band of TiO_2. Cluster-based porphyrin complex compounds can be used as an alternative to the ruthenium-polypyridine, wherein porphyrin acts as ligands in conjunction with

IOP Conf. Series: Materials Science and Engineering **188** (2017) 012023 doi:10.1088/1757-899X/188/1/012023

cobalt-based redox mediators [6-9]. Porphyrin group which has π donor-acceptor with a large molar absorptivity can be applied as a sensitizer in a thin layer of TiO_2 [2]. It also provides a highly flexible platform for the development of panchromatic sensitizers [8, 10]. The porphyrin chromophore has intrinsically strong light absorption in the visible region (400-600 nm) [6-9]. Recently, Zn^{II} (5-(2-(4'-dimethylaminophenyl)-ethynyl)-10,20diphenylporphyrinate has been synthesized by Handayani, et al as a two-terminal wired single molecular diodes, not typical sensitizer in DSSC and it has porphyrin as a chromophore and imide as a linker. When the complexes are exposed to light, electron will be transferred from HOMO porphyrin to LUMO imide with the strong character charge transfer [11]. Zinc-porphyrin-imide has characteristicwhen in a polar solvent low-energy absorption bands of the Zn-complex will shift to longer wavelengths [11], therefore it is likely suitable as sensitizer in a DSSC system.

In this study, we constructed a DSSC system which was assembled using TiO_2 nanotubes sensitized by typical zinc-porphyrin-imide as the working photoelectrode. The TiO_2 nanotube bundle was prepared by a Rapid Breakdown Anodization (RBA) method. The developed visible responsive photoanode was characterized and assembled in a typical DSSC system, and the results will be discussed hereafter.

2. Materials and methods

2.1. Preparation of TiO_2 nanotubes powder
Titanium sheet (5 cm × 5 cm; Baoji Jinsheng Metal Material Co., 99.6% purity, 0.3 mm thickness China) was cleaned properly before used. Typically, the Ti sheet was abrased by sand paper, wiped by a clean soft paper, then rinsed subsequently by organic solvent, water, and organic solvent. Then, the clean Ti sheet was clamped onto a copper clip, positioned as anode and immersed directly in electrolytes containing 0.15 M. A titanium foil as a working electrode and a platinum foil as a counter electrode were used for anodization process under a constant potential of 15 V at a temperature of approximately 25°C.

2.2. Coating of TiO_2 to FTO
0.1 gram TiO_2 NT was dispersed in 10 mL of acetic acid 10%, then sonicated for 1 h to form a colloid with the aid of Triton X surfactant. The freshly prepared TiO_2 colloid was then used to coat the fluoride doped tin oxide (FTO) by the doctor blade method, performed on a hot plate set at 60°C. The TiO_2 coated with FTO was then annealed at 350°C for 1 h. The FTO was self- fabricated as reported by Supriyono *et al.* [12].

2.3. Immobilization of Zinc-Phorphyrin-Imide on FTO/TiO_2
Approximately 11 mg of zinc porphyrinwas dissolved in 25 mL mixture of acetonitrile and tert-butanol 1:1. The solution is considered as the stock solution (440 ppm) and then diluted to a working solution at concentration of 110 ppm. The clean FTO/TiO_2 was immersed in the sensitizer solution for 24 h.

2.4. Electrochemical characterization of photoanode working electrode
The FTO/TiO_2 or FTO/TiO_2/Dyes was placed as working electrode in three electrochemical cells, with Ag/AgCl as reference electrode, and platinum foil as counter electrode. The working electrode was illuminated by proper light source. The amperometric measurement was performed at a potential of 500 mV for 100 seconds, with a scan rate of 25 mV/s. The typical photocurrent was measured and evaluated.

2.5. Assembly of DSSC
The FTO/TiO_2/Dyes was arranged in a DSSC system, which the electrolyte (I_3^-/I_2 in a proper solvent) was sandwiched between working electrode (FTO/TiO_2/Dyes) and counter electrode (FTO/Pt). The parafilm was used as spacer to avoid short contact of the electrodes.

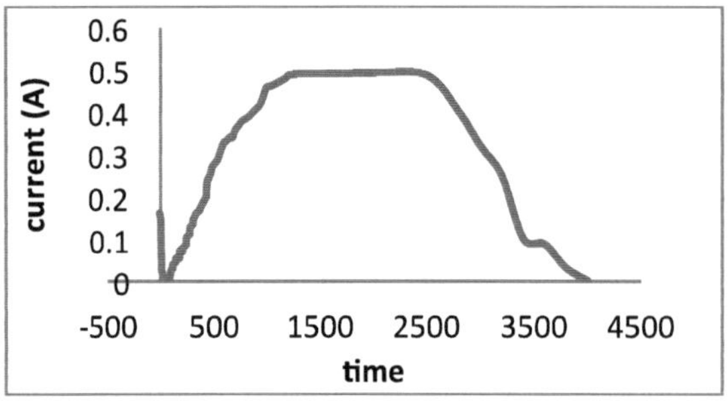

Figure 1. Profile of rapid breakdown anodization.

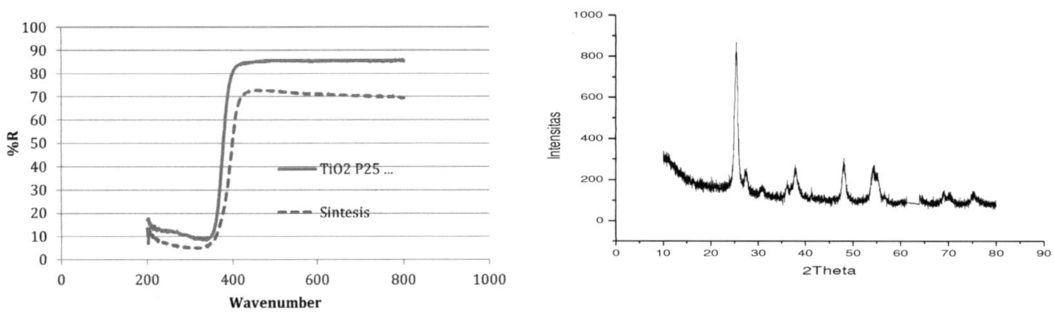

(a) (b)

Figure 2. Spectra UV-Vis DRS and diffractogram XRD TiO_2 nanotube and P25 Degussa.

3. Results and discussion

3.1. Synthesis of TiO_2 nanotube powder

In figure 1, it shows in section I the significant current decrease of 0.158 Å to 0.0019 Å for 80 seconds. It is because on the Ti plate surface oxide layer was formed, when the oxide layer formed it become less conductive than the Ti plate, causing delays of the electron flow which characterized by the decrease in flow during the synthesis process. In section II the flow increased significantly from 0.0019 A to 0.4 A. This is because the oxide layer formed on the Ti plate to form small holes in the oxide layer of TiO_2, and a process of etching by chloride ions, so that the electrolyte can access Ti plate surface to form a new oxide layer. Ti transformation into TiO_2 and TiO_2 etching will continue until the flow is marked by the currents to reach the steady state. The section III shows the currents tend to stabilize at 0.4 A for 27 min. This is because TiO_2 that has been formed, dissolved to form a stable complex compounds with Cl^- ions of ClO_4^- to be $[TiCl_6]^{2-}$.

3.2. Characterization of TiO_2 nanotube powder

Matrix of TiO_2 absorbs light having an equal or greater energy than its energy gap. Light that is not absorbed will be reflected and captured by the detector in the UV-Vis DRS instrument. UV-Vis spectra profile of prepared TiO_2 compared with Degussa P25 TiO_2 can be seen in figure 2. Energy band gap values is determined by changing the value of % R into the value of F using Kubelka-Munk equation, then the value of F is plotted against photon energy in a particular range. The F value of the photon energy in the equivalent area to wavelengths between 250-450 nm will produce linear region in a narrow energy range, and by using the Tauc equation the energy gap value can be determined.

Figure 3 shows the measurement results of prepared TiO_2 nanotubes. The results of XRD measurements indicate the presence of 6 diffractogram peaks corresponding to anatase TiO_2 diffractogram patterns. Measurement was done by comparing the data with the synthesis data

Table 1. Comparison of BET and band gap value of TiO_2 nanotube and P25 Degussa.

Material	Surface area (m^2/g)	Band gap (eV)
TiO_2 *nanotube*	120,971	3,21
TiO_2 P25 Degussa [13]	59	3,31

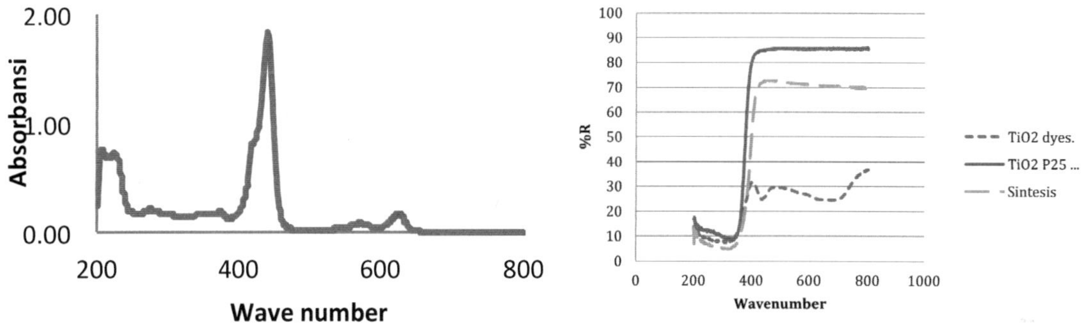

(a) (b)

Figure 3. Spectra UV-Vis of zinc-porphyrin-imide.

of TiO_2 nanotubes interpretation JCPDS 1272. From the results of XRD spectra, it can be seen that the peaks appear at 2θ around 25, 37, 38, 48, 54, and 55. From table 1, it shows that the TiO_2 nanotube synthesis product has a larger surface area compared to Degussa P25 TiO_2. TiO_2 nanotubes synthesis product has a surface area of 120.971 m^2/g, whereas Degussa P25 TiO_2 according to Slamet *et al.* [14] has a surface area of 59 m^2/g. The surface area difference is big enough due to prepared TiO_2 has a nanotube morphology.

3.3. Characteristic of dyes zinc-porphyrin-imide

Figure 4 is a UV-Vis spectrum of Zn-porphyrin dyes. There is a specific absorption in the visible light region (visible) at the maximum wavelength of 439 nm, and 627 nm, as seen in the picture. Absorption occurs at a wavelength of 439 nm, which is characteristic for porphyrin components. Excitation occurs $\pi \rightarrow \pi^*$ intramoleculer charge transfer character, which is the transition of an electron from porphyrin group to imide unit. Excitation occurs in the Zn-porphyrin is involving one electron at HOMO to LUMO character charge transfer [1]. A wavelength of 627 nm and 547 nm are the characteristic of low energy absorption of Zn^{2+}. It can be concluded that the Zn-porphyrin dyes can absorb photons in the visible light region. Free zinc-porphyrin-imide dyes shows characteristic of absorption spectra in the visible region, namely 439 nm and 620 nm. Meanwhile, when it was immobilized to TiO_2-nanotube, the absorption peak shifted to 421 nm and 640 nm.

3.4. Photoelectrochemical of working electrode (FTO-TiO_2- Dyes)

From the graph in figure 4(a), it can be seen that an increase in the current flow was observed when TiO_2-Zn porphyrin irradiated by visible light then reached the steady state at a current of about 1.1 mA/cm^2. While when the TiO_2-Zn porphyrin was illuminated by UV lamp the generated currents reach the steady state at 0.6 mA/cm^2. This phenomenon may indicate the contribution of Zn-porphyrin-imide, to give the absorption region of TiO_2-Zn shifted toward visible light.

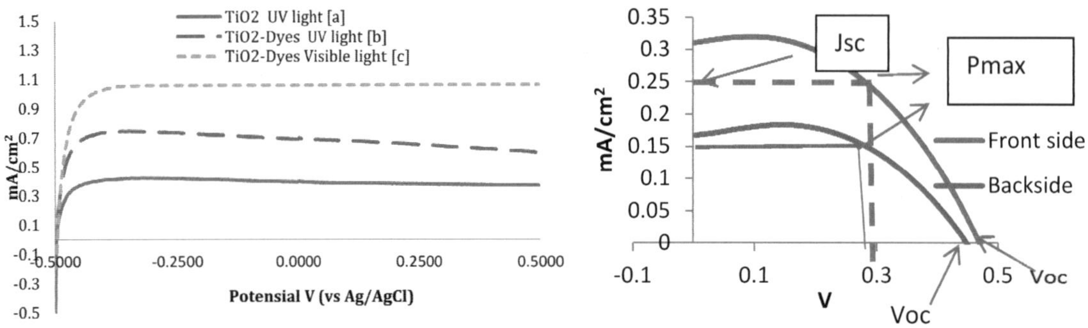

(a) (b)

Figure 4. Voltammogram working electrode and I-V curve DSSC.

The shifting uptake occurs because the TiO_2-Zn porphyrin has a cluster of chromophore as an absorber of visible light, in this case porphyrin, confirmed by the spectrum peak of UV-Vis Zn-porphyrin at 439nm region of visible light. The TiO_2-Zn porphyrin irradiated visible light electrons at HOMO porphyrin moved to the LUMO imide, and going to hole (h +) in the HOMO porphyrin. Hole in the HOMO porphyrin will be filled by an electron from the Zn metal as a donor (metal-ligand charge transfer) to prevent recombination of electrons and holes, due to orbital LUMO on imide is higher compared with ribbon conduction in TiO_2. Then the electronsin the LUMO imide will move into the conduction band of TiO_2 through anchoring group, namely hydroxyl groups and will create a chelating effect on TiO_2-Zn porphyrin. The produced Chelate can be monodentate or bidentate. This chelate will become a bridge that connected the dye with TiO_2.

3.4. DSSC performance
The present study measures the difference of current flow when exposed by backside and front side light illumination. The backside illumination condition mean that the rays do not directly exposed to the TiO_2 layer section but have to passes the Pt film first. At the time of irradiation at the front side position, the generated current is higher than the backside position, the maximum current that occurs at the front side was 1,434 mA, while at the backside flow only reached 0.824 mA. This is caused by the indirect radiation of dyes, since it was absorbed first on CE (Counter Electrode), so the energy of photons reached the dyes must be less. From the measurement results using a potentiostat, DSSC efficiency obtained different results between the front side and the backside illumination. In this study, the efficiency (η) DSSC by irradiating backside illumination was obtained at η = 1.147%, while the value of efficiency (η) DSSC at front side illumination was η = 1.914%.

4. Conclusions
Photoanode of FTO/TiO_2 zinc-porphyrin-imide dyes has been succesfully prepared and characterized, which showed an excellent response toward the visible light. The FTO/TiO_2 zinc-porphyrin-imide dyes photoanode was successfully assembled to a DSSC device, which gave a % light to current conversion of 1.914% and 1.147% upon front side and back side illumination, respectively.

Acknowledgements
Directorate General of Higher Education, Ministry of Research, Technology, and Higher Education, for providing the research funding (PUPT Scheme to JG). Authors convey many thanks to Professor Takuji Ogawa and Murni Handayani, Osaka University, who generously gave their precious Zn-imide-porphyrine dyes type.

References
[1] Stone V F Jr and Davis R J 1998 *Chem. Mater.* **10** 1468-74
[2] Adachi M, Murata Y, Harada M and Yoshikawa S 2000 *Chem. Lett.* **29** 942-3

[3] Fahim N F and Sekino T 2009 *Chem. Mater.* **21** 1967-79
[4] Mathew S, Yella A, Gao P, Humphry-Baker R, Curchod B F E, Ashari-Astani N, Tavernelli I, Rothlisberger U Nazeeruddin M K and Grätzel M 2014 *Nature Chem.* **6** 242-7
[5] Grätzel M 2005 *Inorg. Chem.* **44** 6841-51
[6] Yella A, Lee H W, Tsao H N, Yi C, Chandiran A K, Nazeeruddin M K, Diau E W, Yeh C Y, Zakeeruddin S M and Grätzel M 2011 *Science* **334** 629-34
[7] Feldt S M, Gibson E A, Gabrielsson E, Sun L, Boschloo G and Hagfeldt A 2010 *J. Am. Chem. Soc.* **132** 16714-24
[8] Yum J H *et al.* 2012 *Nature Commun.* **3** 631
[9] Tsao H N, Yi C, Moehl T, Yum, J H, Zakeeruddin S M, Nazeeruddin M K and Grätzel M 2011 *Chem. Sus. Chem.* **4** 591-4
[10] Imahori H, Umeyama T and Ito S 2009 *Acc. Chem. Res.* **42** 1809-18
[11] Handayani M, Gohda S, Tanaka D, Ogawa T 2014 *Chem. Eur. J.* **20** 7655-64
[12] Supriyono, Surahman H, Krisnandi Y K and Gunlazuardi J 2015 *Procedia Environmental Sciences* **28** 242-51
[13] Slamet, Syakur R and Danumulyo W 2003 *Makara Teknologi* **7** 27-32
[14] Slamet, Nasution H W, Purnama E, Kosela S and Gunlazuardi J 2005 *Catalys Communications* **6** 313-9

Superabsorbent nanocomposite synthesis of cellulose from rice husk grafted poly(acrylate acid-co-acrylamide)/bentonite

Helmiyati, G H Abbas and S Kurniawan

Department of Chemistry, Faculty of Mathematics and Natural Sciences Universitas Indonesia, Kampus UI Depok, Depok 16424, Indonesia

Corresponding author's e-mail: helmi-yt@sci.ui.ac.id

Abstract. Superabsorbent nanocomposite synthesis of cellulose rice husk as the backbone with free radical polymerization method in copolymerization grafted with acrylic acid and acrylamide monomer. The cellulose was isolated from rice husk with mixture of toluene and ethanol and then hemicellulose and lignin were removed by using potassium hydroxide 4% and hydrogen peroxide 2%. The obtained cellulose rendement was 37.85%. The functional group of lignin analyzed by FTIR spectra was disappeared at wavenumber 1724 cm^{-1}. Crystal size of the obtained isolated cellulose analyzed by XRD diffraction pattern was 34.6 nm, indicated the nanocrystal structure. Copolymerization was performed at temperature of 70°C with flow nitrogen gas. Initiator and crosslinking agent used were potassium persulfate and N'N-methylene-bis-acrylamide. The swelling capacity of water and urea showed the results was quite satisfactory, the maximum swelling capacity in urea and water were 611.700 g/g and 451.303 g/g, respectively, and can be applied in agriculture to absorb water and urea fertilizer.

Keywords: rice husk, nanocomposite, cellulose, swelling capacity

1. Introduction

Rice husk is an agricultural by products that is biodegradable with three main components, namely cellulose, hemicellulose and lignin [1, 2]. Cellulose is natural polymer that has biodegradable, non-toxic properties, and abundant amounts in nature [3]. Cellulose can be converted to nanocellulose that would be outstanding if combined with inorganic nanoparticles, in order to obtain nanocomposites that have properties of both nanocellulose and nanoparticles [4]. Nanocomposite can be considered as solid structures in nanometer scale dimensions, repeated on the distance between the different forms of constituent structure [5]. Compared with the ordinary cellulose, nanocellulose has a smaller diameter <100 nm which makes nanocellulose has many unique properties, such as high mechanical strength and large surface area [6].

In this study, nanocellulose was used as the backbone to form a superabsorbent nanocomposite which will be used to absorb the urea fertilizer carried by rainwater and irrigation in agriculture. Synthesis of superabsorbent nanocomposite with free radical polymerization method used acrylic acid and acrylamide as monomer, N'N-methylene-bis-acrylamide as crosslinker, and bentonite as inorganic compound.

2. Materials and methods

2.1. Material
Rice husk used was obtained from Bogor, Indonesia. The source of rice husk was similar with the previous research [7]. Toluene-ethanol (Merck) was used to extract the cellulose from rice husk; hydrogen peroxide (Merck), and potassium hydroxide (Merck) were used as impurity removal from rice husk cellulose; acrylic acid, and acrylamide (Nippon Shokubai) were used as monomer; potassium persulphate (KPS; Merck) as initiator; *N,N'*-methylene*bis*-acrylamide (MBA, Sigma) as crosslinking agent, urea (Merck) was used as the absorbate and bentonite as inorganic compound.

2.2. Isolation of cellulose from rice husk
The isolation of rice husk cellulose was performed by the method proposed by Fan *et al.* [7] and Helmiyati and Kurniawan [8]. Rice husk was washed with warm water to remove impurities and water soluble substances.The rice husk was extracted with mixture of toluene: ethanol at ratio of 2:1 for 6 h, then dried at 50°C for 24 h. Hemicellulose and lignin was removed by H_2O_2 4% solution with higher concentration than previous work [9], pH of solution was set at 12 with 4% KOH. The precipitate was washed with aquadest to neutral and dried at 50°C for 24 h.

2.3. Superabsorbent synthesis
Synthesis of polymer was based on experiment performed by Spagnol *et al.* [10]. First, the dissolution process of rice husk cellulose was performed referred to Wu *et al.* [11], in which the rice husk cellulose was dissolved in 7% sodium hydroxide solution and 12% urea. The solution of cellulose was fed into a three neck flask reactor , equipped with reflux condenser and nitrogen gas. The reactor was placed in water bath at 70°C. Then, the potassium persulfate, acrylic acid, acrylamide, MBA, and bentonite were fed into the reactor. The polymerization process was conducted for 2 h at 70°C. The obtained superabsorbent was washed in distilled water, ethanol and acetone for 24 h at room temperature.

3. Results and discussion

3.1. Isolation of cellulose from rice husk
Rice husk consists of cellulose, hemicellulose, lignin, and other extractive substances [8]. After the pretreatment theobtained rendement average of cellulose can be seen in table 1. From table 1, it can be seen the obtained cellulose rendement after isolation as much as 37.85%. The rendement was smaller than the previous rendement [8], due to the H_2O_2 concentration rising during the lignin removing process so that the cellulose will be whiter and bright.

3.2. Characterization of cellulose by FTIR
Figure 1 shows the FTIR spectra of rice husks, isolated cellulose, and pure cellulose. Figure 1a shows infrared absorptions for rice husk, the main absorption band shows the functional groups of rice husk at 1724 cm^{-1} (C=O stretching vibration). This is specific band for lignin [8]. The spectrum 1b does not show an absorption band at wavenumber 1724 cm^{-1}. It indicates that the lignin in the isolated cellulose has been partially removed from rice husk. It can be also observed that spectrum of pure cellulose 1c is similar to the spectrum of rice husk cellulose in spectrum 1b.

Table 1. Average rendement of cellulose isolation.

Rice husk (g)	Dewaxing (g)	Delignification (g)	Rendement(g)	Rendement (%)
10.00	9.50	5.30	3.75	37.5
10.00	9.60	5.20	3.82	38.2
Average	9.55	5.25	3.78	37.85

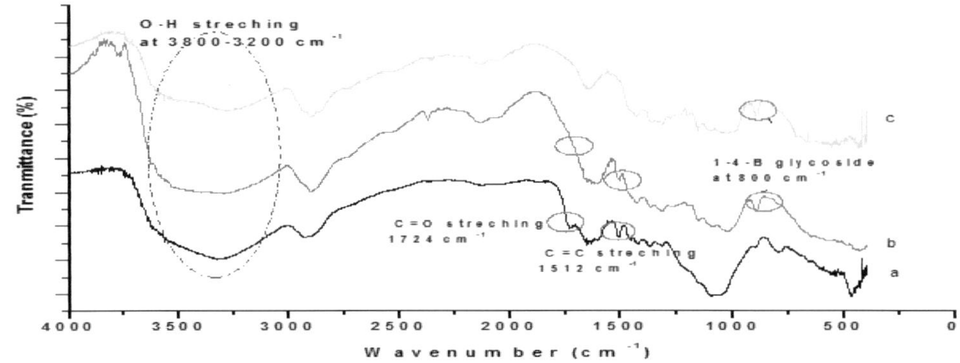

Figure 1. Spectra FTIR of (a) rice husk, (b), isolated cellulose and (c) pure cellulose.

(a) (b)

Figure 2. Micrograph of (a) rice husk and (b) rice husk cellulose.

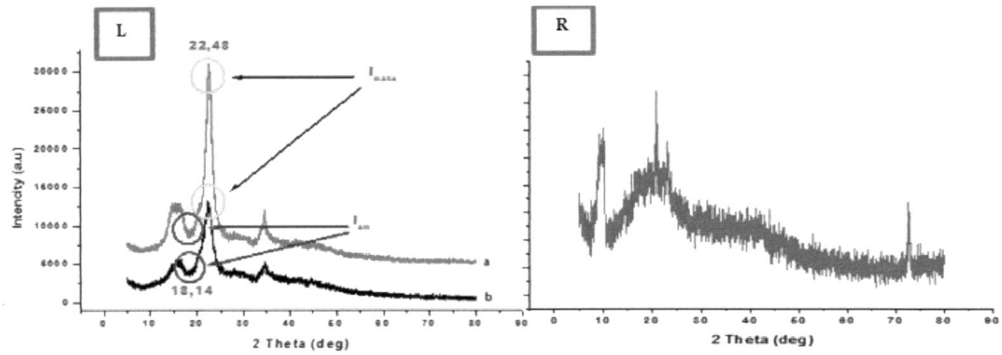

(a) (b)

Figure 3. Diffraction patterns of pure cellulose (a) rice husk cellulose and (b) superabsorbent nanocomposite.

3.3. Characterization of cellulose by SEM

Figure 2 shows the surface morphology. In figure 2a, it can be seen that the surface morphology of rice husk surface is rough and uneven, showing that lignin and other compounds were covered the surface of cellulose. While on the isolated cellulose of rice husk in figure 2b, the fibrils of cellulose can be clearly seen. This is because after the isolation process, lignin, hemicelluloses, and other substances were removed from the isolated cellulose.

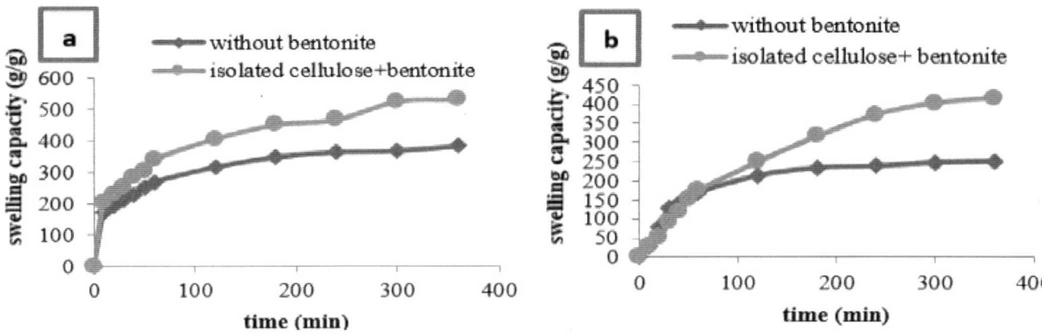

(a) (b)

Figure 4. Swelling capacity of superabsorben in (a) urea solution and (b) water.

3.4. Characterization by XRD

Figure 3 shows diffraction pattern. In figure 3a, it can be seen the diffraction pattern of pure cellulose (a) and rice husk cellulose (b) have similarity with sharp peak 22.48° as intensity of crystalline and the peak at 18.25° as intensity of amorphous. The size of cellulose crystal that calculated by Scherrer's law was 34.6 nm, which indicated nanocrystal size. Figure 3b shows that the diffraction pattern of superabsorbent indicated the amorphous structure. The amorphous structure of the superabsorbent will increase the swelling capacity, due to superabsorbent with the amorphous structure has an irregular composition, so the solution would be easy to get in and bind with the superabsorbent.

3.5. Swelling capacity

The swelling capacity of superabsorbent in urea and water are shown in figure 4. Figure 4 shows the swelling capacity of superabsorbent cellulose-poly (AA-co-AM) with bentonite and without bentonite in urea solution (a) and water (b). The swelling capacity was directly proportional towards the time, where the superabsorbent capacity will increase exponentially with time. The result of swelling capacity shows that superabsorbent of isolated cellulose with bentonite have the highest swelling capacity in urea and water. The maximum swelling capacity of isolated cellulose superabsorbent in urea and water were 611.700 g/g and 451.303 g/g, respectively.

4. Conclusions

Cellulose was successfully isolated from rice husk at 37.85%and has been characterized by FTIR, SEM, and XRD. Superabsorbent of cellulose rice husk graft poly(acrylic acid-coacrylamide)/bentonite has been synthesized successfully and characterized by XRD. The diffraction pattern of XRD indicated that cellulose was nanocrystal with the size of 34.6 nm and the diffraction pattern of superabsorbent indicated the amorphous structure. The result of swelling capacity shows that superabsorbent of isolated cellulose with bentonite have the highest swelling capacity in urea and water. The swelling capacity of superabsorbent with the highest value was the superabsorbent isolated cellulose with bentonite in urea and water with value of 611.700 g/g and 451.303 g/g, respectively.

Acknowledgements

The authors would like to thank Universitas Indonesia for funding this research through PITTA Grant Universitas Indonesia with contract No: 2045/UN2.R12/HKP.05.05/2016.

References

[1] Rosa S M L, Rehman N, de Miranda M I G, Nachtigall S M B and Bica C I D 2012 *Carbohydrate Polymer* **87** 1131-8
[2] Shukla S K *et al.* 2013 *Adv. Mat. Lett.* **4** 714-9
[3] Ma J, Li X and Bao Y 2015 *RSC Adv.* **5** 59745-57
[4] Jabasingh S A, Lalith D, Prabhu M A, Yimam A and Zewdu T 2016 *Carbohydrate Polymers* **136** 700-9
[5] Wang M, Anoshkin I V, Nasibulin A G, Korhonen J T, Seitsonen J, Pere J, Kauppinen E I, Ras

R H A and Ikkala O 2013 *Advanced Materials* **25** 2428-32
[6] Wei H, Rodriguez K, Renneckar S and Vikesland P J 2014 *Environmental Science: Nano* **1** 302-16
[7] Fan G, Wang M, Liao C, Fang T, Li J and Zhou R 2013 *Carbohydrate Polymers* **94** 71-6
[8] Helmiyati and Kurniawan S 2015 *Journal of Chemical and Pharmaceutical Research* **7** 878-85
[9] Helmiyati, Saefumillah A and Yulianti W 2014 *Asian Journal of Chemistry* **26** 7337-42
[10] Spagnol C, Rodrigues F H A, Neto A G V C, Pereira A G B, Fajardo A R, Radovanovic E, Rubira A F and Muniz E C 2012 *European Polymer Journal* **48** 454-63
[11] Wu F, Zhang Y, Liu L and Yao J 2012 *Carbohydrate Polymers* **87** 2519-25

Application of SnO_2 nanoparticle as sulfide gas sensor using UV/VIS/NIR spectrophotometer

N Juliasih, Buchari and I Noviandri

Department of Chemistry, Faculty of Mathematics and Natural Sciences Bandung Institute of Technology, Jl. Ganesha No 10, Bandung, Indonesia

Corresponding author's e-mail: asihjuli@gmail.com

Abstract. Sulfide gas monitoring is required to protect organisms from its toxicity. Nanoparticles of metal oxides have characteristics that applicable as sensors for controlling environmental pollution like sulfide gas. Thin film of SnO_2 as one part of the sulfide gas sensor was synthesized with the chemical liquid deposition method, and characterized by UV/VIS/NIR-Spectrophotometer before and after gas application, also using FTIR, SEM and XRD. Characterization studies showed nanoparticle sizes from the diameters range of 38-71 nm. Application of SnO_2 thin film to sulfide gas detected by UV/VIS/NIR Spectrophotometer with diffuse reflectance showed chemical reaction by the shifting of maximum % R peak at wavelength of 1428 cm. The benefit of measurement of sulfide gas using this SnO_2 nano thin film is that it could be done at the room temperature.

Keywords: metal oxide, SnO_2, sulfide gas sensor, UV/VIS/NIR-Spectrophotometer

1. Introduction

Carbon disulfide, CS_2 is one of gas that produced by many activities in viscose rayon production process, petroleum or natural gas drilling and refining, also pulp production, which is using Kraft method [1]. The toxicity level of this gas can harm organisms. At low concentration it can affect the nervous system, while at high level can cause death. The monitoring of CS_2 concentration becomes very important nowadays, and nanoparticle of metal oxide as one type of material applied as sensors can control this environmental pollution. The research of nanoparticle synthesis for improving the sensor system of CS_2 become increases [2]. The previous research results showed that SnO_2 nanoparticle can be applied as sensor of gas used the electrically method by conductance to change the sample upon exposure of the gas. In this research, it also showed that SnO_2 gave chemical reaction when it applied to adsorb CS_2 detected by UV/VIS/NIR Spectrophotometer. The benefit of measurement of sulfide gas using SnO_2 nano thin film is that it could be done at the room temperature and low cost production.

2. Materials and methods

Thin film of SnO_2 can be produced by deposition of this metal oxide on the glass substrate. Soda lime float glass with 1-1.2 mm of thickness (Yancheng Huida Medical Instruments Co.) was used in this experiment. The substrate was cleaned by rinsing with acetone at 40°C for 15 minutes, then rinsed with ethanol at room temperature, and finally washed with distilled water and dried. $SnCl_2.2H_2O$ and NaOH were used as the precursors, and the deposition was done by spraying. The annealing temperatures were varied from 200ºC - 500ºC. Thin film of SnO_2 as the sulfide gas sensor was synthesized by a simple method of chemical liquid deposition, and characterized using UV/VIS/NIR-Spectrophotometer

(Shimadzu UV3101PC) at λ = 200-2400 nm, FTIR Spectrophotometer, Optic and Scanning Electron Microscope, and also XRD with Cu[(Kα)] with $\lambda\alpha_1$ = 1.54060 nm and $\lambda\alpha_2$ = 1.54439 nm. The application of thin film to sulfide gas was tested using 20 ppm of CS_2 cylinder gas standard with flow of 0.2 L/min.

3. Results and discussion

Infrared spectrum of SnO_2 nano thin film showed specific peaks for annealing temperature variation from 200°C to 500°C (figure 1). It was analyzed by FTIR in the range from 400 to 4500 cm^{-1}. The bands between 400-800 cm^{-1} were correlated to metal oxide bond (SnO_2). Bands around 800-1700 are due to the oxygen stretching and bending frequency. Infrared spectrum of SnO_2 nano thin film showed specific peaks at 3425 cm^{-1}, 1625 cm^{-1}, and 615 cm^{-1} for annealing temperature 500°C. It can be seen from the spectrum that the highest annealing temperature showed results as most clear spectrum. Functional group –OH seems to be wider at the lower annealing temperature, showed at band around 3400 cm^{-1}.

Optical measurement of SnO_2 nanoparticle was carried out with UV/VIS-spectrophotometer, and analysis result showed that the increasing of annealing temperature can decrease the transmittance or increase the absorbance (figure 2). The results were the average spectrum from ten times measurements, and done at the room temperature with the glass substrate slide as the standard.

XRD pattern of SnO_2 variance 500°C were calculated based on Bragg law,

$$n\lambda = 2d \sin\theta \quad (1)$$

where λ is X-ray wavelength, d is the distance between lattice planes, θ is the angle of incidence with lattice plane, and n is integer [3, 4]. The observed peaks of XRD pattern in the figure 3 are indexed assuming SnO_2 phase, giving peaks position, with Miller indices (110), (101), (200), (211), (220), (310) and (301) planes. Characterization studied by SEM showed nanoparticle sizes from the range 38-71 nm of diameters (figure 4).

The application of SnO_2 thin film for CS_2 gas sensor tested using UV/VIS/NIR-spectrophotometer showed a shifting of %Reflectance peak from wavelength 1274 cm^{-1} with 45.21 % of reflectance to wavelength 1428 cm^{-1} with 79.21%. This response was used as a starting point for detecting gas quantitatively at wavelength 1428 cm^{-1} (figure 5). The research will be continued with statistical approach of quantitative analysis by making standard curve and determination of detection limit.

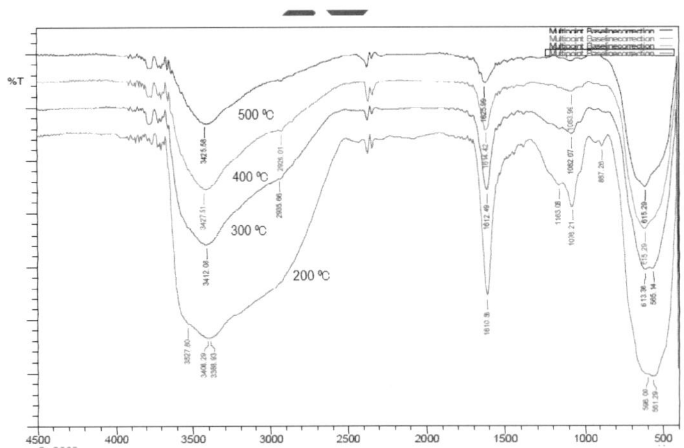

Figure 1. Infrared spectrum of SnO_2 nano thin film at various annealing temperatures of 200°C, 300°C, 400°C, and 500°C.

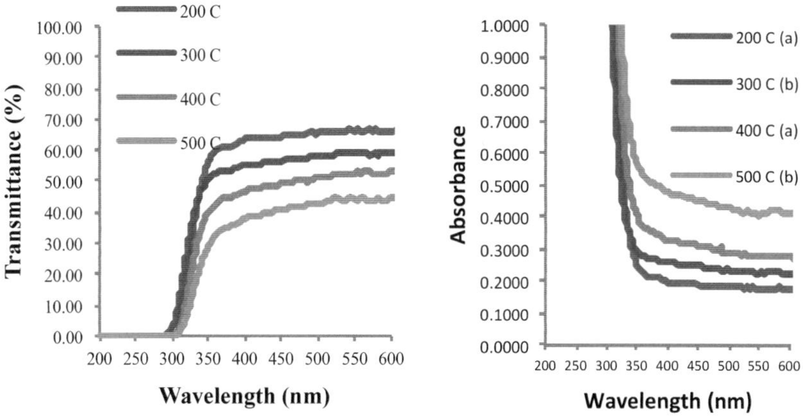

(a) (b)

Figure 2. UV/VIS (a) transmittance and (b) absorbance spectrum of SnO_2 nano thin film.

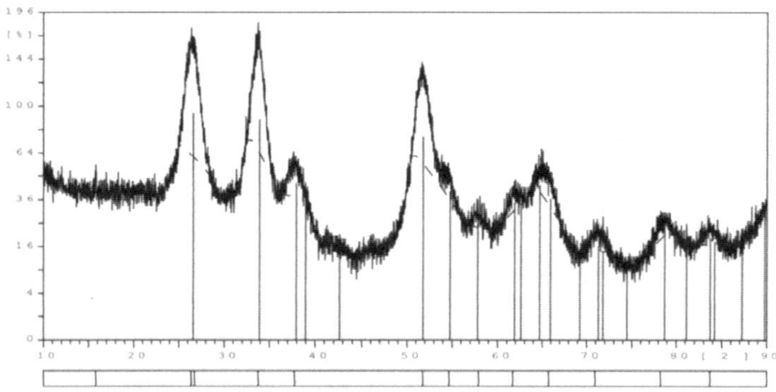

Figure 3. XRD spectrum of SnO_2 nano thin film.

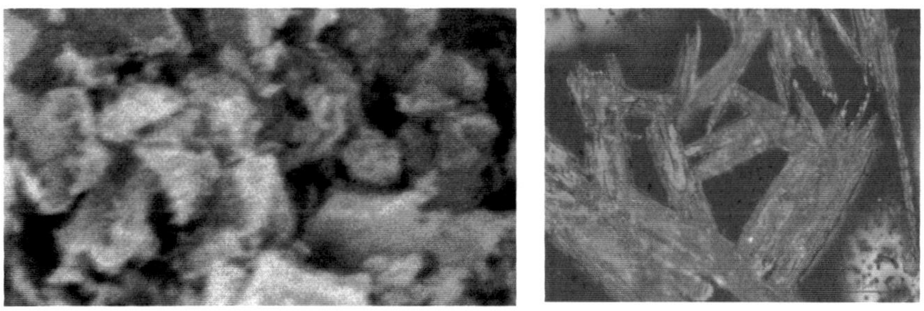

(a) (b)

Figure 4. (a) SEM and (b) optical microscope morphology of SnO_2 nano thin film.

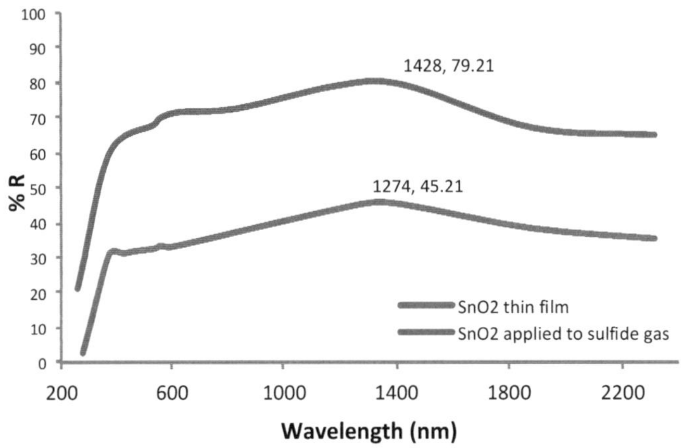

Figure 5. Application of SnO_2 to sense 20 ppm of CS_2 gas standard.

4. Conclusions

SnO_2 nanoparticle can be produced by a simple chemical liquid deposition, and at annealing temperature of 500 °C it can be achieved particle sizes diameter in the range of 38-71 nm. Application of SnO_2 thin film to sulfide gas detected by UV/VIS/NIR Spectrophotometer with diffuse reflectance showed a specific chemical reaction by the shifting of maximum % R peak from wavelength of 1274 to 1428 nm.

Acknowledgements

This work was supported by Department of Chemistry, Bandung Institute of Technology, Indonesia.

References

[1] Pandey S K, Kim K H and Tang K T 2012 TrAC *Trends in Analytical Chem.* **32** 87-99
[2] Zheng Z 2009 *Synthesis and Modification of Metal Oxide Nanostructure and Their Applications* PhD thesis (Queensland: Queensland University of Technology, Australia)
[3] Lee J H, Jang G E, Yoon D H and Son S H 2007 *Journal of Ceramic Processing Research* **8** 59-63
[4] Maleki M and Rozati S M 2013 *Bulletin of Materials Science* **36** 217-21

Preparation of Fe_3O_4/SiO_2-guanidine organobase catalyst for 1,5-diphenylpenta-2,4-dien-1-one synthesis

A H Cahyana, D Fitria, B Ardiansah and D U C Rahayu

Department of Chemistry, Faculty of Mathematics and Natural Sciences Universitas Indonesia, Kampus UI Depok, Depok 16424, Indonesia

Corresponding author's e-mail: herrykim@ui.ac.id

Abstract. A novel heterogeneous organobase catalyst of Fe_3O_4/SiO_2-guanidine was prepared in three stages. First, Fe_3O_4 nanoparticle was obtained by co-precipitation method using seaweed *Sargassum* Sp. as natural reductant. Fe_3O_4 was then coated by SiO_2 using TEOS as silica source, resulting Fe_3O_4/SiO_2. Finally, Fe_3O_4/SiO_2-Guanidine was obtained by modifying Fe_3O_4/SiO_2 with guanidine in the suitable reaction condition. This organobase catalyst was characterized by Scanning Electron Microscope (SEM), Energy Dispersive X-ray Spectroscopy (EDS), and Particle Size Analyzer (PSA). The material was then used as a highly active catalyst in aldol condensation reaction between acetophenone and cinnamaldehyde to produce 1,5-diphenylpenta-2,4-dien-1-one. The structure elucidation of the organic product was confirmed by UV-Vis, FTIR, and LC-MS.

1. Introduction

Recently, the development of heterogeneous catalysts for various organic transformations became a trending topic. Fe_3O_4 based heterogeneous catalysts were widely applied in the synthesis of organic compounds with specific functional groups, such as Fe_3O_4/SiO_2-metformin [1], Fe_3O_4/SiO_2-urea [2], and Fe_3O_4/SiO_2-NH_2 [3] magnetic catalyst. The wide range applications of mostly heterogeneous catalysts are interesting because they can be easily separated from the reaction mixture by an external permanent magnet due to the presence of catalyst's magnetic properties [1].

In this study, Fe_3O_4 nanoparticle was obtained by reducing the ferric chloride solution with seaweed *Sargassum* Sp. as a reducing agent and efficient stabilizer (capping agent) [4] and by adding ferrous chloride which also acts as a reducing agent [1-3]. The reaction was then followed by silica coating and guanidine functionalization of Fe_3O_4 to form Fe_3O_4/SiO_2-guanidine which will be used in the synthesis of 1,5-diphenylpenta-2,4-dien-1-one compound by aldol condensation reaction between acetophenone and cinnamaldehyde.The catalyst has amino group that can be associated with its basic properties. The basic catalyst can extract the H-α atom of acetophenone in Aldol condensation mechanism [5]. From the previous research regarding the basic properties of the catalyst, we hope that Fe_3O_4/SiO_2-guanidine catalysts can be used as an alternative heterogenous catalyst for aldol condensation reaction to replace NaOH as the conventional catalyst for this reaction [5, 6].

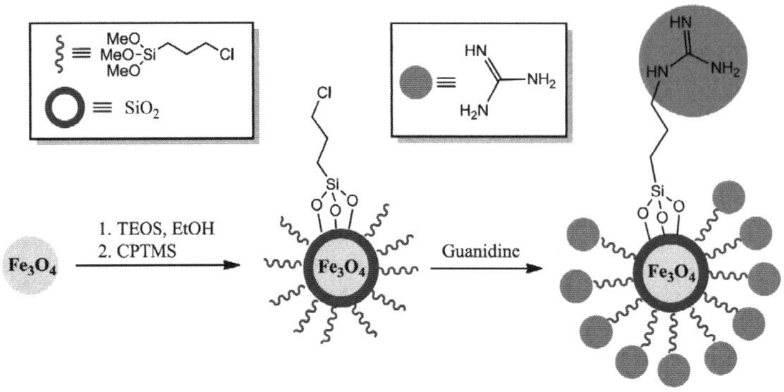

Figure 1. Illustration of the prepared catalyst

2. Materials and methods

2.1. General

Chemicals used were pro-analytical grade. The *Sargassum* Sp. was collected from Binuangeun Sea in Banten, Jawa Barat, Indonesia.

2.2. Preparation of magnetic nanoparticles (Fe_3O_4)

In a 1 L beaker glass, 1 mmol of $FeCl_3.6H_2O$, 0.25 mmol of $FeCl_2.4H_2O$, 5 mL of *Sargassum* extract, and 50 mL of deionized water were stirred at room temperature, while NaOH was added instantaneously to adjust to pH 10. After 2 hours, the product was filtered and washed with deionized water and alcohol, then dried at 60°C [1, 2].

2.3. Preparation of silica-coated magnetic nanoparticles (Fe_3O_4/SiO_2)

In a 1 L beaker glass, 1 g of Fe_3O_4 was dissolved by adding water (20 mL), ethanol (60 mL), and ammonia (2 mL, 25 wt.%). The resulting dispersion was then homogenized by ultrasonic bath. Then, the solution of TEOS (0.5 mL)in ethanol (10 mL) was added to the dispersion. Followed by stirring for 16 h, the obtained solid product was colleted by an external magnetic and washed with ethanol. Finally, the product was dried at 60°C [7].

2.4. Preparation of Fe_3O_4/SiO_2-Guanidine

In a 250 mL of round bottom flask, 1 mL (5 mmol) of 3-chloropropyl-trimethoxysilane (CPTMS) was dissolved in 100 mL of dried toluene. This mixture was added to 1 g of Fe_3O_4/SiO_2 and the solution was stirred for 18 h at 60°C. The obtained solid product (Fe_3O_4/SiO_2-Cl) was washed with toluene, separated by a magnet, and dried at 60°C. Then, it was used in the following step to synthesize Fe_3O_4/SiO_2-guanidine. 1 g of Fe_3O_4/SiO_2-Cl and KI (1.66 g, 10 mmol) were added to a solution of guanidine hydrochloride (0.59 g, 10 mmol) and K_2CO_3 (1.38 g, 10 mmol) in acetonitril (50 mL) in a round-bottom flask and the mixture was stirred under reflux condition for 8 h. The obtained solid product was then separated by a magnet and washed with water followed by drying at 80°C [1,2]. Illustration of the prepared catalyst was depicted in figure 1.

2.5. Synthesis of 1,5-diphenylpenta-2,4-dien-1-one

A mixture of cinnamaldehyde (5 mmol), acetophenone (5 mmol), and Fe_3O_4/SiO_2-guanidine (20% weight) in ethanol was stirred for 10 h at 75°C. After completion of the reaction, the catalyst was separated by an external magnet. The obtained solid product was filtered and washed with water and hexene, then the solid product was purified by recrystallization using hot methanol.

3. Results and discussion

The morphology of Fe_3O_4/SiO_2 was also characterized by using SEM as shown in figure 2(b). The surface differences between these two compounds indicated that the Fe_3O_4 was successfully coated with silica.

The presence of Fe and O elements as the fundamental components of Fe_3O_4 was investigated using EDS spectra as shown in Table 1. EDS analysis for Fe_3O_4/SiO_2 showed the element contained in this compound were Fe, O and Si, indicated that the silica was successfully coated on Fe_3O_4.

Figure 3 shows the particle size of Fe_3O_4, Fe_3O_4/SiO_2 and Fe_3O_4/SiO_2-guanidine characterized by PSA are 35, 159, and 257 nm, respectively. The increasing of particle size was affected by the silica coating and guanidine functionalization of the Fe_3O_4.

The Fe_3O_4/SiO_2-guanidine was then applied in synthesis of 1,5-diphenylpenta-2,4-dien-1-one compound by aldol condensation reaction between acetophenone and cinnamaldehyde. Figure 4 is the reaction scheme for the synthesis. The reaction was carried out with the suitable condition using 20% weight of catalyst at 75°C for 10 hours. The purity of the product was analyzed by using the thin layer chromatography with ethyl acetate:n-hexane(1:4) as the solvent. The yield product obtained was 42 %.

The active site of the Fe_3O_4/SiO_2-guanidine catalyst was $-NH_2$ which acted as a Bronsted-Lowry base that can take the acidic hydrogen from substrates. This site was very suitable for the nucleophilic attack. The other active site of the catalyst was Fe^{3+} which showed a Lewis acid behavior and it can react with the carbonyl groups of cinnamaldehyde to accelerate the conjugation (process) [3].

The structure elucidation of the 1,5-diphenylpenta-2,4-dien-1-one product was confirmed by UV-Vis, FTIR and LC-MS. The UV-Vis spectrum shows the maximum wavelength is 336 nm that confirmed with the appearance of the yellowish crystal product. The FTIR spectrum shows the presence of C-H

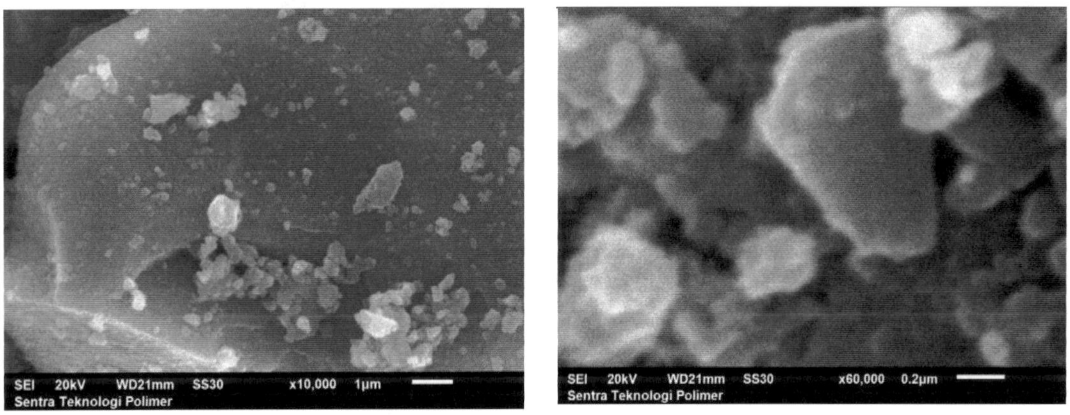

(a) (b)

Figure 2. SEM images of Fe_3O_4/SiO_2 with (a) 10,000 × and (b) 60,000 × magnification.

Table 1. EDS analysis of the catalyst.

Element	Composition (% wt)	
	Fe_3O_4	$Fe_3O_4 - SiO_2$
C	10.29 ± 3.11	9.44 ± 1.91
O	33.24 ± 11.58	28.66 ± 6.71
Fe	54.65 ± 15.60	54.92 ± 8.17
Na	1.32 ± 1.07	0.00 ± 00
Si	0.00 ± 0.00	6.98 ± 0.46

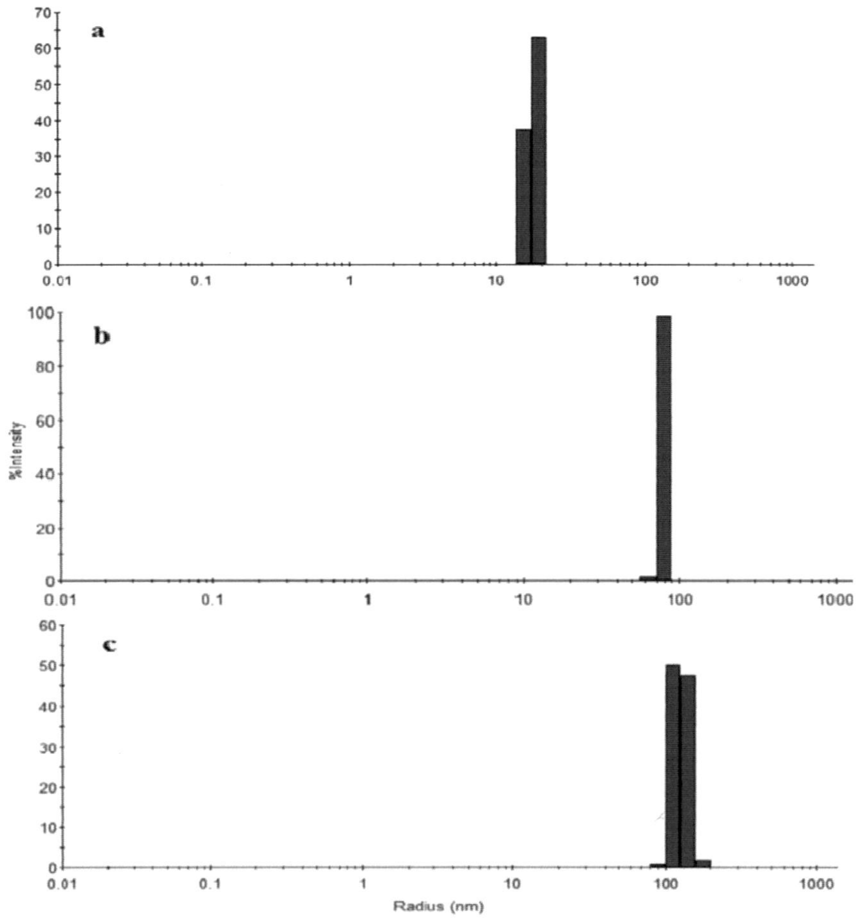

Figure 3. Particle size of (a) Fe_3O_4, (b) Fe_3O_4/SiO_2 and (c) Fe_3O_4/SiO_2-guanidine.

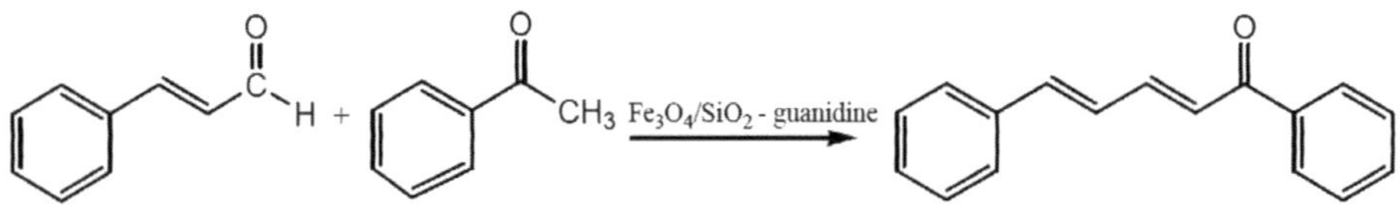

Figure 4. The reaction scheme for the synthesis of 1,5-diphenylpenta-2,4-dien-1-one.

aromatic and olefinic at 3023 and 3065 cm^{-1}, respectively. Furthemore, the peak at 1653 cm^{-1} and 1590 cm^{-1} showed the presence of C=O vibration and C=C groups, respectively. LC-MS spectrum showed the fragmentation at m/z 235 ([M+H]$^+$) and 257 ([M+Na]$^+$), confirming the formation of 1,5-diphenylpenta-2,4-dien-1-one (figure 5).

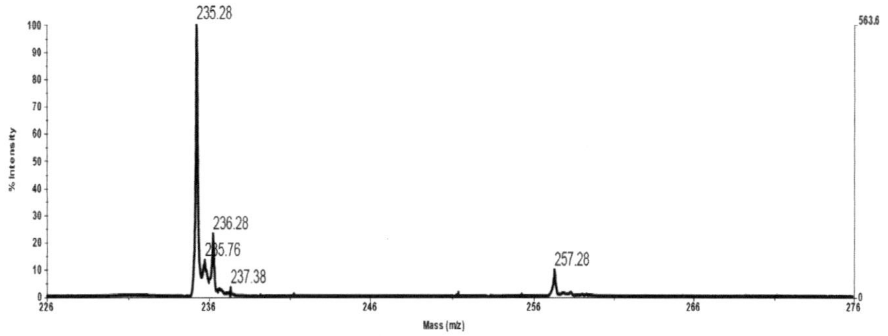

Figure 5. Mass spectrum of the 1,5-diphenylpenta-2,4-dien-1-one.

4. Conclusions

Fe_3O_4 nanoparticle was obtained by co-precipitation method using seaweed *Sargassum* Sp. as reducing agent and stabilizer. The characterization of Fe_3O_4 nanoparticle showed that silica coating and guanidine functionalization of the Fe_3O_4 was performed succesfully. The Fe_3O_4/SiO_2-guanidine organobase catalyst can be used in the synthesis of 1,5-diphenylpenta-2,4-dien-1-one compound by aldol condensation reaction between acetophenone and cinnamaldehyde. The UV-Vis, FTIR, and LC-MS spectrum confirmed that the 1,5-diphenylpenta-2,4-dien-1-one was succesfully synthesized.

Acknowledgements

This work was supported by PITTA 2016 research Grant No: 2035/UN2.R12/HKP.05.05/2016 from Universitas Indonesia.

References

[1] Alizadeh A, Khodaei M M, Beygzadeh M, Kordestani D and Feyzi M 2012 *Bull. Korean Chem. Soc.* **33** 2546-52
[2] Maleki A, Alrezvani Z and Maleki S 2015 *Catal. Commun.* **69** 29-33
[3] Ghasemzadeh M A, Abdollahi-basir M H and Babaei M 2015 *Green Chem. Lett. Rev.* **8** 40-9
[4] Mahdavi M, Namvar F, Ahmad M B and Mohamad R 2013 *Molecules* **18** 5954-64
[5] Mardiana L, Ardiansah B, Bakri R and Cahyana H 2016 *AIP Conf. Proc.* **1729** 020051
[6] Weldon D J *et al.* 2014 *Bioorg. Med. Chem. Lett.* **24** 3381-4
[7] Naeimi H and Nazifi Z S 2013 *J. Nanopart. Res.* **15** 2026-37

Temperature-dependent impedance spectroscopy of $La_{0.8}Sr_{0.2}FeO_3$ nano-crystalline material

C A Kafa, D Triyono and H Laysandra

Department of Physics, Faculty of Mathematics and Natural Sciences Universitas Indonesia, Kampus UI Depok, Depok 16424, Indonesia

Corresponding author's e-mail: djoko.triyono@sci.ui.ac.id

Abstract. $LaFeO_3$ is a material with perovskite structure which electrical properties frequently investigated. Research are done due to the exhibition of excellent gas sensing behavior through resistivity comparison from the p-type semiconductor. Sr doping on $LaFeO_3$ or $La_{1-x}Sr_xFeO_3$ are able to improve the electrical conductivity through structural modification. Using Sr dopant concentration (x) of 0.2, $La_{0.8}Sr_{0.2}FeO_3$ nano-crystal pellet was synthesized. The synthesis used sol-gel method, followed by gradual heat treatment and uniaxial compaction. XRD characterization shows that the structure of the sample is Orthorhombic Perovskite. Topography of the sample by SEM reveals grain and grain boundary existence with emerging agglomeration. The electrical properties of the material, as functions of temperature and frequency, were measured by Impedance Spectroscopy method using RLC meter, for temperatures of 303-373K. Through the Nyquist plot and Bode plot, the electrical conductivity of $La_{0.8}Sr_{0.2}FeO_3$ is contributed by the grain and grain boundary. Finally, the electrical permittivities of $La_{0.8}Sr_{0.2}FeO_3$ are increasing with temperature increase, with the highest achieved when measured at 1 kHz frequency.

1. Introduction

The synthesized perovskite $LaFeO_3$ is often applied as an electrode material for Solid Oxide Fuel Cells (SOFC), and electrocatalytic material as in gas sensors [1, 2]. However, there are possibilities that the compound is able to increase its mediocre electrical conductivity by the giving of doping [3]. Sr^{2+} cation is one well-known doping for increasing $LaFeO_3$ electrical conductivity, by generating oxygen vacancies δ that increasing O^{2-} ion mobility, and/ or causing the Fe cation oxidized from Fe^{3+} to Fe^{4+} [3, 4]. The size and oxidation numbers of Sr^{2+} cation doping towards $LaFeO_3$ perovskite would affect its electrocatalytic behavior, however these values are adjustable by doping composition calculation to achieve the desirable certain degree.

The composition of $La_{0.8}Sr_{0.2}FeO_3$ from the $La_{1-x}Sr_xFeO_3$ group is usually used as the electrode for fuel cells. This is due to its superior electrochemical behavior compared with the other members of the set, even though it did not have the most oxygen vacancies δ [5, 6]. The very small deviation from this composition, $La_{0.75}Sr_{0.25}FeO_3$ is reported to have the excellence on "syngas" (synthesis gas, mainly consist of H_2, CO, and CO_2) production (hence energy production) through catalytic partial oxidation from methane (CH_4), due to its relatively wide BET (Brunauer – Emmett – Teller) surface area and its relatively plentiful H_2 reduction agent consumption [7].

From previous backgrounds, this work focuses on the study of electrical behavior of nano crystalline $La_{0.8}Sr_{0.2}FeO_3$ by using Impedance Spectroscopy (IS) method. The data acquisition through different system temperatures, enables the possibility to calculate the activation energy for relaxation and conduction processes.

2. Materials and methods

The material is synthesized using sol-gel method, with citric acid and aquadest as chelating agents. For synthesizing $La_{0.8}Sr_{0.2}FeO_3$, each 30 mmol of (0.4 ×) La_2O_3, (0.2 ×) $Sr(NO_3)_2$, and $Fe(NO_3)_3.9H_2O$ were mixed with the chelating agents in a beaker glass. Then, the colloid phase sample was heated and stirred on a Magnetic stirrer – Hot plate (WiseStir MSH-20D), with temperature of 120°C and rotational speed of 700 rpm. After the sample transform into the gel phase, it experienced gradual heat treatments (200°C for 5 h, 900°C for 6 h). Finally, the resulting powder was pressed into pellet with pressure of 5 ton/ square inch for 120 s and then sintered at 1300°C for 1 h.

The XRD characterizations of the calcined powder and the sintered pellet were done on room temperature, by Rigaku MiniFlex 600 X-Ray Diffractometer with radiation source of Cu Kα (λ = 1.5418 Å), with the 2θ range from 5° to 90°. The topography of the sample were investigated by SEM, using JEOL JSM-6510 LA electron microscope with acceleration voltage of 20 kV and magnification of 50000 ×. For the Impedance Spectroscopy (IS) characterization, the pellet was modelled as parallel plate capacitor, connected with conductor probe, and inserted on a thermocouple furnace that was connected to a Fluke PM6306 RLC meter. The measurements were done between temperature 303-373K and frequencies 1 MHz, 100 kHz, 10 kHz, and 1 kHz.

3. Results and discussion

Figure 1 shows the XRD pattern of the $La_{0.8}Sr_{0.2}FeO_3$ sample, both on the calcined powder phase and sintered pellet phase. The XRD pattern reveals that $La_{0.8}Sr_{0.2}FeO_3$ have Orthorhombic perovskite structure and space group Pnma. There are no other phases observed in the samples, indicating that the samples are single phase.

Table 1 lists the measured structure identities of the $La_{0.8}Sr_{0.2}FeO_3$ sintered pellet. Its grain size was calculated using Scherrer equation (equation 1) [8], and its tolerance factor follows Goldschmidt equation (equation 2) [9].

$$\tau = \frac{K\lambda}{B\cos\theta} \tag{1}$$

On equation (1), K is the shape factor usually values 0.9, λ is the used X-ray wavelength, B is the Full Width at Half Maximum (FWHM) of the selected peak on radian, and θ is the Bragg angle. As an approach, Scherrer equation may gives different crystallite size value, compared with the electron microscope characterization, if the tested crystal experienced inhomogeneous strain and crystal defect.

$$t = \frac{r_A + r_O}{2^{1/2}(r_B + r_O)} \tag{2}$$

On equation (2), r_A, r_B, r_O are the ionic radius of A-site cation, B-site cation, and oxygen anion, respectively. The less the tolerance factor value from unity, the lower the crystal structure regularity level of the perovskite.

Figure 2 shows the atomic position manifestation on the $La_{0.8}Sr_{0.2}FeO_3$ lattice, and table 2 lists the unit cell position of the related elements. Both information reveal that the Sr^{2+} doping composition still not showing a significant Fe oxidation from Fe^{3+} to Fe^{4+}, and the variation occurs only on the comparison of the number of A-sites atoms (Sr and La).

Figure 3 shows the SEM image of $La_{0.8}Sr_{0.2}FeO_3$. The SEM image shows that the grain distribution of the sample is uniform and quite good, with some grain agglomeration observed. There are no obvious signs of secondary phases observed on the image. Using the measuring bar included in the image, $La_{0.8}Sr_{0.2}FeO_3$ had the grain size of about 0.167 μm, with agglomeration size of about 0.333 μm. This agglomeration condition makes it rather hard to distinguish between its grain and its grain boundaries, indicating that the contribution of the grain boundary in measured impedance would show inhomogeneous values, compared with the grain contribution.

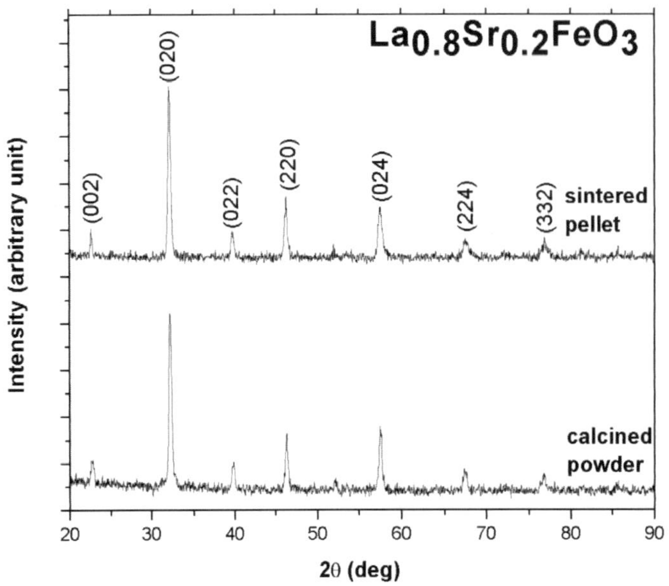

Figure 1. XRD pattern of $La_{0.8}Sr_{0.2}FeO_3$ showing single phase orthorhombic perovskite crystal structure with Pnma space group.

Table 1. Structure identities of the $La_{0.8}Sr_{0.2}FeO_3$ sintered pellet.

Identity	a (Å)	b (Å)	c (Å)	Crystallite size (nm)	Density (gr/cm^3)	Tolerance factor
Value	5.555	7.855	5.514	23	6.42	0.893

Table 2. Unit cell position of the related elements in $La_{0.8}Sr_{0.2}FeO_3$.

Elements	Sr	La	Fe	O1	O2
x	0.4724	0.4724	0.0000	0.2290	0.5060
y	0.2500	0.2500	0.0000	0.0350	0.2500
z	0.0010	0.0010	0.0000	0.2660	0.5700
Oxidation numbers	+2	+3	+3	-2	-2

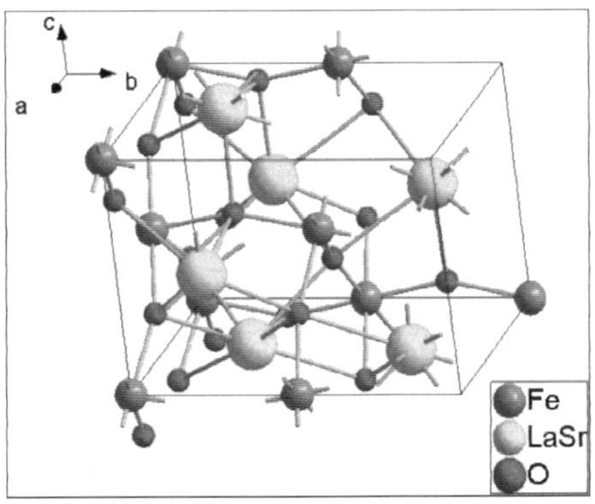

Figure 2. Atomic position on $La_{0.8}Sr_{0.2}FeO_3$ lattice.

Figure 3. SEM images of $La_{0.8}Sr_{0.2}FeO_3$ with the magnification of 50000 ×.

Figure 4 shows the Nyquist plots of the $La_{0.8}Sr_{0.2}FeO_3$ sample on varying temperatures, from RT to 373 K. The Nyquist plot curves have the form of almost a semicircle, however the high-temperature curves have slightly additional shape on small Z' values (high frequencies), which resemble another semicircle that showing the influence of grain and grain boundaries on the electrical properties of the material. The lowering of temperatures would enlarge the radius of the Nyquist plot semicircles, showing that the impedance values are increasing. This temperature-dependence impedance phenomenon indicates that $La_{0.8}Sr_{0.2}FeO_3$, as a member of $La_{1-x}Sr_xFeO_3$, is a semiconductor [10]. The double semicircle Nyquist plots are fitted using ZSimpWin software, that aims to acquire the appropriate equivalent electronic circuits and the component values. The proposed equivalent electronic circuit as the electrical properties model is the (RC)(RC) circuit, where one (RC) part represents grain contribution and the another (RC) part represents grain boundary contribution. Using ZSimpWin software and the Nyquist plots, the component values of this equivalent electronic circuit has been obtained for each temperatures, written in table 3. It can be concluded that the multiplication of R and C (also known as relaxation time τ) from grain contribution is having tendency to be smaller than the ones from grain boundary contributions. That means for the process of conduction in $La_{0.8}Sr_{0.2}FeO_3$, grain conduction is more reliable than the grain boundary conduction, probably due to agglomeration observed that blur the grain boundaries [11].

Figure 5 shows imaginary impedance Z'' ordinate Bode plots of the $La_{0.8}Sr_{0.2}FeO_3$ sample on varying temperatures, from RT to 373 K. The Z'' ordinate Bode plots show that along with frequency increase, the curve increasing until a certain maximum point, and then goes down. Those peaks reveal a distribution of relaxation time of conduction process, which is also temperature-dependence. The peak shift on the Bode plots shows that the relaxation time becoming faster with temperature increase [11]. This phenomenon happens because the peak frequencies on the plots are inversely proportional to the dielectric relaxation time, and increasing temperature caused the peak frequencies to increase.

After dealing with semiconducting behavior of $La_{0.8}Sr_{0.2}FeO_3$, next is the discussion about its dielectric behavior through the complex permittivity quantity ε. Real relative permittivity (dielectric constant ε') represents its ability to store electrical energy, while imaginary relative permittivity (dielectric loss ε'') represents its ability to dissipate electrical energy.

Figure 6 shows $La_{0.8}Sr_{0.2}FeO_3$ real permittivity curves toward the temperatures. The plot of real permittivity versus temperature, shows that the less the frequencies of data acquisition, the more the real permittivity values. Moreover, the real permittivity increase with temperature increase. Because both real and imaginary permittivities are inversely proportional to frequency, there is an assumption that the imaginary permittivity also follow the same trend [11].

Table 3. Electrical properties modeling component values of $La_{0.8}Sr_{0.2}FeO_3$ at various temperature.

T (K)	R1 (Ω)	C1 (F)	R2 (Ω)	C2 (F)
300	148.30	4.424×10^{-06}	1.97	4.322×10^{-20}
323	16.91	9.624×10^{-17}	105.90	4.680×10^{-07}
333	17.01	9.961×10^{-19}	87.24	4.977×10^{-07}
343	18.39	2.179×10^{-16}	68.95	5.508×10^{-05}
353	20.64	1.000×10^{-20}	51.48	6.485×10^{-06}
363	23.66	1.015×10^{-23}	33.14	9.210×10^{-06}
373	1.77	1.795×10^{-05}	26.30	1.328×10^{-20}

IOP Conf. Series: Materials Science and Engineering **188** (2017) 012027 doi:10.1088/1757-899X/188/1/012027

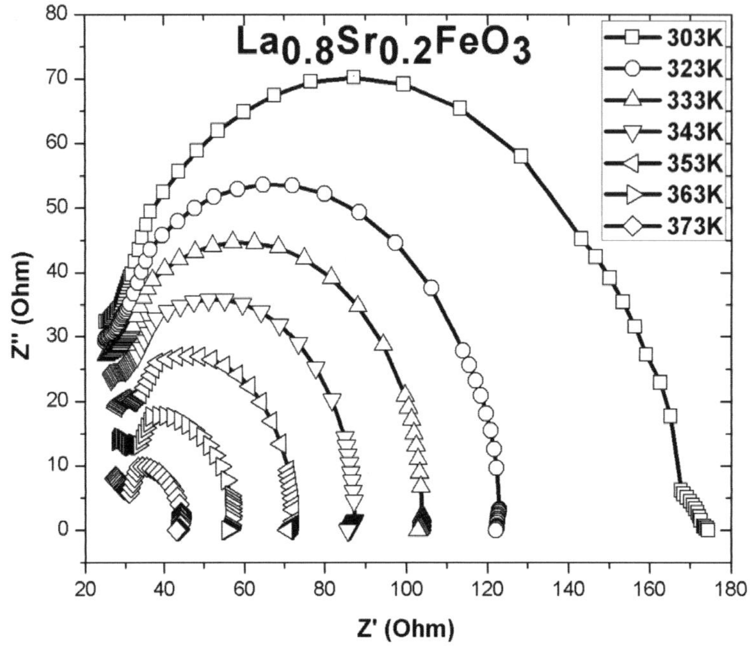

Figure 4. Nyquist plots of the $La_{0.8}Sr_{0.2}FeO_3$ sample on varying temperatures.

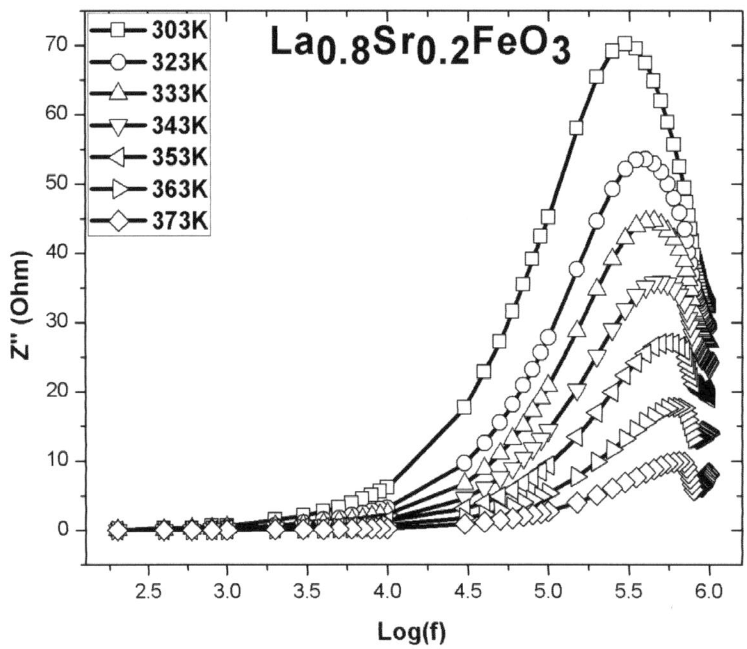

Figure 5. Z'' Ordinate bode plots of the $La_{0.8}Sr_{0.2}FeO_3$ sample on varying temperatures.

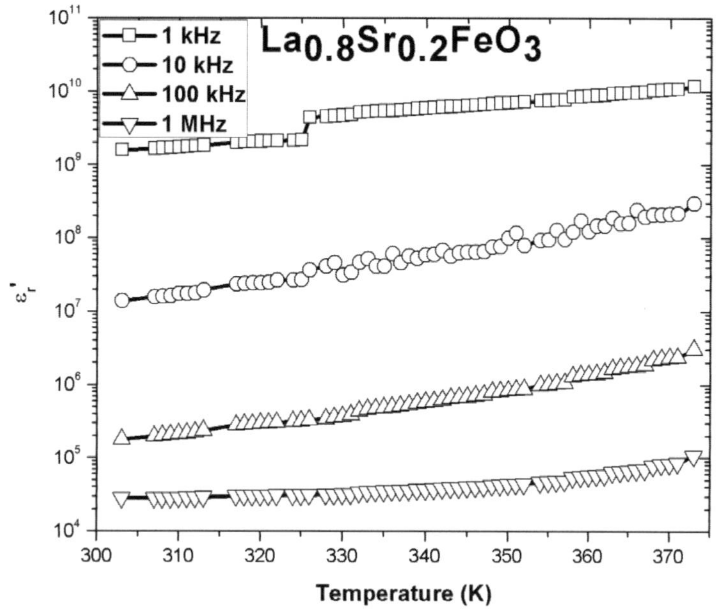

Figure 6. Real permittivity logarithmic plot versus temperatures for $La_{0.8}Sr_{0.2}FeO_3$.

4. Conclusions

$La_{0.8}Sr_{0.2}FeO_3$ was successfully synthesized by sol-gel method. XRD results show single phase Orthorhombic Perovskite crystal structure with Pnma space group. SEM image shows that the synthesized product is uniform and quite good in grain distribution, with a few grain agglomeration observed. Impedance Spectroscopy characterization shows that in general, the sample had semiconducting behavior or impedance decrease with the increase of temperature. The electrical conductivity of $La_{0.8}Sr_{0.2}FeO_3$ is contributed by both grain and grain boundary, and the dielectric relaxation time of the sample is faster with increasing temperature.

Acknowledgements

The authors would like to thank Universitas Indonesia for funding this research through PITTA Grant Universitas Indonesia with contract No: 2007/UN2.R12/HKP.05.00/2016.

References

[1] Qin J, Cui Z, Yang X, Zhu S, Li Z and Liang Y 2015 *Sens. Actuators* B **209** 706-13
[2] Khetre S M, Jadhav H V, Jagadale P N, Kulal S R and Bamane S R 2011 *Adv. Appl. Sci. Res.* **2(4)** 503-11
[3] He F, Li X, Zhao K, Huang Z, Wei G and Li H 2013 *Fuel* **108** 465-73
[4] Evdou A, Zaspalis V and Nalbandian L 2010 *Fuel* **89** 1265-73
[5] Cui X, Li S and Zhu X 2014 *Mater. Lett.* **130** 267-70
[6] Wang J, Zhang Y, Guo W, Hou X, Liang T, Deng C and Xu J 2013 *Ceram. Int.* **39** 5263-70
[7] Khine M S S, Chen L, Zhang S, Lin J and Jiang S P 2013 *Int. J. Hydrogen Energy* **38** 13300-8
[8] Cullity B D 1978 *Elements of X-ray Diffraction 2nd ed* (Boston: Addison-Wesley Publishing Company Inc)
[9] Clemens O, Kuhn M and Haberkorn R 2011 *J. Solid State Chem.* **184** 2870-6
[10] Wang H C, Wang C L, Zhang J L, Su W B, Liu J, Zhao M L, Yin N, Lv Y G and Mei L M 2010 *Curr. Appl. Phys.* **10** 866-70
[11] Macdonald J R 2005 *Impedance Spectroscopy: Theory, Experiment, and Applications* ed E Barsoukov (Hoboken, New Jersey: John Wiley & Sons Inc)

Study of phosphate release from Bogor botanical gardens' sediment into pore water using diffusive gradient in thin film (DGT)

A P Tirta[1], A Saefumillah[1] and Foliatini[2]

[1]Department of Chemistry, Faculty of Mathematics and Natural Sciences Universitas Indonesia, Kampus UI Depok, Depok 16424, Indonesia
[2]Politeknik AKA Bogor, Bogor, 16154, Indonesia

Corresponding author's e-mail: asep.saefumillah@gmail.com

Abstract. Eutrophication is one of the environmental problems caused by the excessive nutrients in aquatic ecosystems. In most lakes, phosphate is a limiting nutrient for algae photosynthesis. Even though the concentration of phosphate from external loading into the water body has been reduced, eutrophication could still be occured due to internal mobilization of phosphate from the sediment pore water into the overlying water. Therefore, the released phosphate from sediments and their interaction in the pore water must be included in the monitoring of phosphate concentration in aquatic system. The released phosphate from sediment into pore water has been studied by DGT device with ferrihydrite as binding gel and N-N'-methylenebisacrylamide as crosslinker. The results showed that DGT with 15% acrylamide; 0.1 % N-N'-methylenebisacrylamide and ferrihydrite as binding gel was suitable for the measurement of the released phosphate from sediment into pore water. The result of the deployed DGT in oxic and anoxic conditions in seven days incubation showed the released phosphate process from the sediment into pore water was affected by incubation time and the existence of oxygen in the environment. The released phosphate from the sediment into pore water in anoxic condition has a higher value than oxic condition. The experimental results of the deployed DGT in natural sediment core at a depth of 1 to 15 cm from the surface of the water for 7 days showed that the sediment has a different phosphate mass profile based on depth. The concentration of phosphate tends to be increased with depth. The maximum C_{DGT} of phosphate released in oxic and anoxic conditions at 7th day period of incubation are 29.23 μg/L at 14 cm depth and 30.19 μg/L at 8 cm depth, respectively.

Keywords: sediments, phosphates release, DGT, pore water

1. Introduction

Eutrophication is one of the environmental problems that causing damage to aquatic ecosystems, especially to freshwater. Eutrophication occurs when the aquatic environment is enriched with nutrients. Nitrates and phosphates are nutrients that needed to grow the plants. In sufficient amounts they are beneficial to many ecosystems, but in excess amounts can cause eutrophication [1]. Agricultural residual waste such as fertilizers is one of the cause of eutrophication, mostly they are dissolved into a nearby water system causing the increase of nutrient content [2]. Eutrophication stimulates an explosive growth of algae (algal blooms), may cause the oxygen content on the water to be consumed by algae, also block the sunlight from photosynthesizing the aquatic plant that lead to the depletion of the dissolved oxygen in water. This bloom disrupts normal ecosystem function and causes many problems. Some algae even produce toxins that are harmful to the higher forms of life. This can cause problems along the food chain and affect any animal which uses them as the food source [3].

In most lakes, phosphate becomes the limiting factor in the algae photosynthesis process because its presence is relatively small compared with the number of aquatic organisms that need them. Phosphate controls the growth of biomass in the overlying water [4], so it necessary to monitor the presence of phosphate in the water. Even though the presence of phosphate has been reduced, eutrophication can still continue to occur due to the mobilization of phosphate from the sediment into pore water through the physics, chemistry and biochemistry processes [5]. Sediments play an important role in the presence of phosphate in the water because it acts as a source and reservoir of phosphate. Therefore, the release of phosphate from the sediment into the pore water must be included in the process of monitoring phosphate pollution. Pore water defines as water that presents between sediment particles, the phosphate is released into pore water and then diffused into overlying water, in which algae could consume orthophosphate as the available phosphate that required by plants [6].

Orthophosphate in the water occurs from the dissolution of apatite mineral and mineralization of organic phosphate [7]. Rumhayati [8], Edlund and Carman [9] and Suzumura and Kamatani [10] found that organic phosphate mineralization in sediments occurs in anoxic condition. It is well known that phosphorous cycle in water is affected by Fe (III) reduction in the sediment and Fe (II) oxidation in the water [11]. The accurate determination of phosphate is difficult to obtain because the dynamic interaction during storage could change the concentration of phosphate. Zhang *et al.* [12] has developed the Diffusive Gradient in Thin Films (DGT) device for in situ measurement of reactive phosphorus species in natural waters, sediments and soils. The principle of phosphate measurement using DGT technique is that phosphate ion diffuses through the membrane filter and diffusive layer to reach the ferrihydrite gel. Mass of phosphate absorbed by binding gel was analyzed by colorimetry method after being eluted by acid. Concentration of phosphate obtained is considered to be equal to the phosphate which may diffuse into algae, therefore DGT can be used for measuring the bioavailability.

DGT is a device that consists of a membrane filter, diffusive gel and adsorbent, packed in a probe. Diffusive gel used was an acrylamide polymer in the form of a hydrogel. Generally, DGT Crosslinker from DGT research used as a crosslinker in the synthesis of acrylamide polymer. According to Ratshania, N-N'-methylenebisacrylamide can be used as a crosslinker in the synthesis of acrylamide polymer for DGT [13]. In this study, the authors used DGT probe technique using ferrihidryte as binding gel and N-N'-methylenebisacrylamide as crosslinker for the measurement of the phosphate release from sediment into pore water sediment in oxic and anoxic conditions. Measurement of the phosphate release was carried out at different depths of 1-15 cm.

2. Materials and methods

2.1. Chemicals

All reagents were of analytical grade. The reagents and materials employed for the preparation of in-house manufactured DGT gels were acrylamide (Sigma–Aldrich,Germany), ammonium persulfate (Sigma–Aldrich,Germany), tetramethylethylenediamine (Sigma–Aldrich,Germany), N,N'-methylenebisacrylamide (Sigma–Aldrich, Germany), celullose nitrate membrane filter (Whatman), Sodium Nitrate from Merck, and $Fe(NO_3)_3.9H_2O$ (Merck, Germany). The reagents for PO_4 determination were potassium hydrogen phosphate (Merck, Germany), sodium hydroxide (Merck, Germany), sulphuric acid (Merck, Germany), ammonium molybdate (Merck, Germany), potassium antimonyl tartrate (Merck, Germany), and ascorbic acid (Merck, Germany).

2.2. Experimental

The preparation of diffusive gel and ferryhidryte was conducted using the procedure by Zhang *et al.* [12]. The available phosphate species was measured using a DGT technique with ferryhydryte as a diffusive gel. The gel contains 15% acrylamide and 1.0% crosslinker. For binding gel, the gel solution was added by ferrihydryte. 70 μL of ammonium persulfate (10% in water) and 20 μL of tetramethylethylenediamine were added to initiate the polymerization reaction. The prepared gel

IOP Conf. Series: Materials Science and Engineering **188** (2017) 012028 doi:10.1088/1757-899X/188/1/012028

solution was then maintained at 45°C for 1 hour. Resin gels were hydrated in demineralized water for 24 h, the diffusive gels disks were stored in the demineralized water and the binding gel was stored in 0.1 M $NaNO_2$ solution. The thickness of the prepared gels was 0.23 cm.

Sediment sample from Bogor Botanical Gardens' lake was sampled as core sediment to attain the undisturbed sample. DGT probe was deployed for 7 days in oxic and anoxic conditions with bubbling of oxygen and nitrogen. For initial data, DGT probe was deployed for 2 hr without bubbling of oxygen and nitrogen. At each retrieval time, the binding gel was removed from the probe and then cut into 1 cm size. The binding gel was eluted by 0.25 M H_2SO_4 solution and the concentration of PO_4 was measured by spectrophotometer UV-Visible with blue molybdenum method.

3. Results and discussion

3.1. Assembly of DGT probe for sediment

DGT is a device that consists of a membrane filter, diffusive gel and adsorbent, packed in a plastic probe. Diffusive gel is the acrylamide polymer and the binding gel is diffusive gel which added with adsorbents. In this research, ferrihydrite was used as adsorbent for phosphate binding. DGT probe sediment used in this research was plastic assembled with an open window of 2.6 cm x 16 cm. The schematic representation of asection of the DGT assembly can be seen in figure 1.

DGT probe sediments were assembled by the following stages, the first stage is to put ferrihydrite gel on the probe, continued with the placement of diffusive gels, cellulose nitrate membrane and finally it was closed using window probe. The DGT probe sediments assembly process can be seen in figure 2.

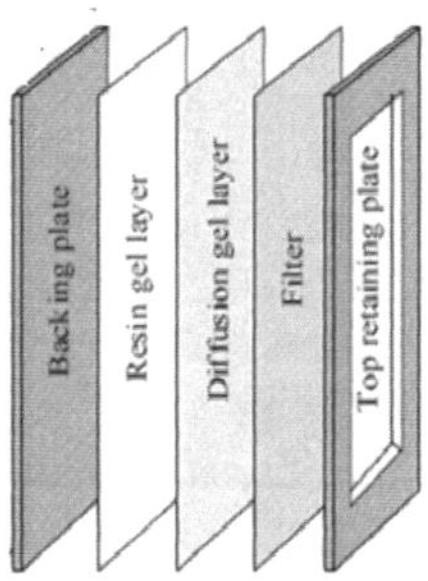

Figure 1. Schematic Representation of asection of the DGT Assembly [14].

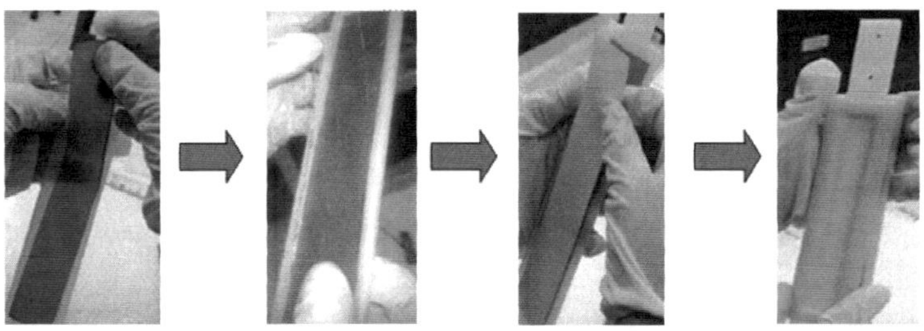

Figure 2. Schematic representation of asection through the DGT assembly.

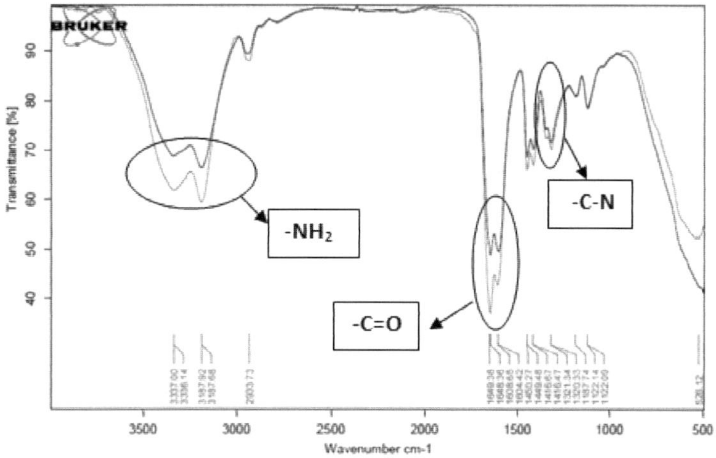

Figure 3. FTIR Spectra diffusive gel (grey line) and ferrihydrite gel (black line).

Table 1. Data of the dissolved Oxygen, temperature and pH.

No.	Parameter	Result		
		Initial (2 hour)	Oxic (7 day)	Anoxic (7 day)
1	Temperatur (°C)	25.0 – 25.5	25.0 – 25.5	25.0 – 25.5
2	Dissolved Oxygen (mg/L)	7.38 – 7.56	6.73 – 7.82	0.12 – 0.90
4	pH	7.05 – 7.10	6.93 – 7.20	7.60 – 7.92

3.2. FTIR characterization of diffusive gel and ferrihydrite gel

FTIR characterization of diffusive gel and ferrihydrite gel can be seen in figure 3. Based on the results, there are two peaks at the wavenumber of 3337 cm^{-1} and 3188 cm^{-1} for sample diffusive gels. For gel sample of ferrihydrite gel, two peaks are also appear at wavenumber 3338 cm^{-1} and 3181 cm^{-1} which is a typical absorption of the primary amine (NH_2) at wavenumber 3300-3600 cm^{-1} with a medium intensity. Peak at wavenumber 1648 cm^{-1} for sample of diffusive gel and 1649 cm^{-1} for sample of ferrihydrite gel indicated an absorption of carbonyl amide (-CONH-) at the wavenumber 1690-1650 cm^{-1} with a strong intensity. Peak at wavenumber 1186 cm^{-1} for sample diffusive gel and 1187 cm^{-1} for sample ferrihydrite gel showed an absorption of aliphatic amine (C-N) at the wavenumber 1250-1020 cm^{-1} with medium intensity. Functional groups that appear in the spectra are the main functional groups of the acrylamide polymer. Functional groups that appear on diffusive gels and ferrihydrite gel showed no significant difference, so it can be concluded that no changes took place in the structure of the polymer in the presence of ferrihydrite. The data also showed that interaction between ferrihydrite and acrylamide polymer is a physical interaction.

3.3. DGT deployment on nature sediment sample

During the deployment of DGT probe, the dissolved oxygen, temperature and pH of the water surface were monitored. The result can be seen in table 1. After seven day period of deployment of DGT probes in sediment, the probe was removed from the DGT system then rinsed using aquabidest. The binding gel was removed from the devices and then sliced for each 1 cm layer of depth to be analyzed for concentration of phosphate using spectrophotometer UV-Visible after eluted with 0.25 M sulphuric acid. Data of phosphate concentrations absorbed by the ferrihydrite binding gel (C_{DGT}) during deployment in oxic condition are showed in table 2 and figure 3.

Table 2. Mass of PO_4 during DGT deployment.

Depth (cm)	Mass of PO_4 (μg) Initial (2 jam)	*Oxic* (7 hari)	*Anoxic* (7 hari)
1	0.00 ± 0.00	0.33 ± 0.09	1.74 ± 2.34
2	0.00 ± 0.00	0.31 ± 0.07	4.43 ± 4.98
3	0.08 ± 0.07	0.32 ± 0.13	9.17 ± 1.45
4	0.22 ± 0.08	6.44 ± 5.05	10.44 ± 2.59
5	0.34 ± 0.58	5.10 ± 4.49	9.79 ± 1.95
6	0.08 ± 0.14	6.93 ± 3.14	8.05 ± 4.92
7	0.00 ± 0.00	6.93 ± 2.35	10.22 ± 3.38
8	0.00 ± 0.00	7.38 ± 1.41	10.84 ± 1.78
9	0.40 ± 0.47	9.59 ± 1.98	9.05 ± 1.95
10	0.29 ± 0.06	5.90 ± 2.27	8.44 ± 5.45
11	0.09 ± 0.15	7.46 ± 1.16	7.18 ± 1.31
12	0.11 ± 0.18	9.46 ± 3.45	7.19 ± 1.80
13	0.00 ± 0.00	8.62 ± 2.41	10.72 ± 3.69
14	0.00 ± 0.00	10.50 ± 1.89	9.25 ± 2.83
15	0.00 ± 0.00	9.14 ± 4.08	8.90 ± 1.98

Note : *0.00 ± 0.00 means phosphate was below detection limit

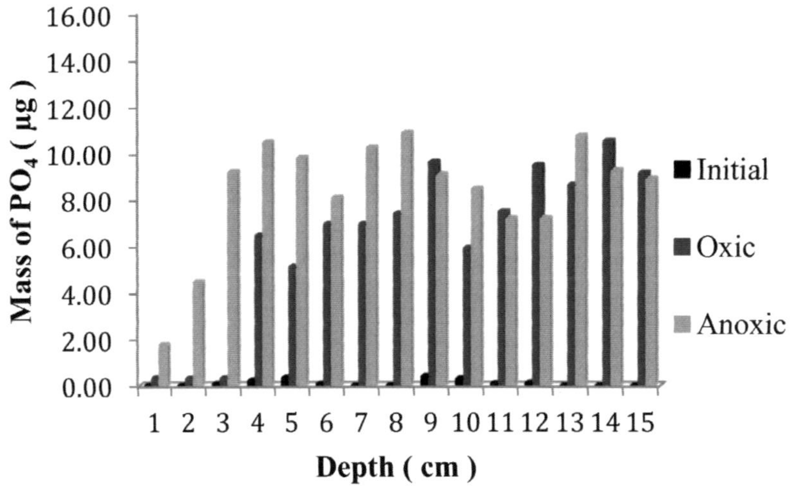

Figure 4. The comparison mass of phosphate absorbed on initial, oxic and anoxic conditions.

Based on the data in table 2, the maximum mass of phosphate after 7 days on oxic and anoxic conditions are 10.50 μg at a depth of 14 cm and 10.84 μg at a depth 8 cm, respectively. Figure 4 shows that mass of phosphate tend to be increased by the depth in both oxic and anoxic treatments. In oxic condition, mass of phosphate increased sharply compared to the increase in anoxic condition. Observation at a depth of 1-8 cm showed that difference mass phosphate absorbed in anoxic condition was much greater than oxic condition.

Phosphate mobilization from sediment into pore water is strongly influenced by environmental redox condition [15]. According to Baken *et al.* [11] the dynamics of phosphate and iron in the aquatic environment is affected by the Fe (III) reduction in the sediment and Fe (II) oxidation in the water. In anoxic condition, the sediment will release phosphate which bind with iron hydroxides [10] and increase the concentration of phosphate in the water. While in oxic condition, Fe (II) will be oxidized to Fe (III) to form iron oxyhydroxide (FeOOH) which adsorbs phosphate from the water. Therefore, in oxic condition the phosphate concentration became lower than phosphate in anoxic condition. Moreover, the process of release of phosphate from the sediment into the water was not only caused by dissolution and precipitation of apatite mineral, but also occurred by mineralization of organic phosphate compounds. Phosphorus bound by the organic phase is released from microbial activity during the degradation of organic matter [16]. Suzumura and Kamatani [10] found that at 60 days of incubation, myo-inositol hexaxis phosphate (IP_6) are mineralized in oxic and anoxic conditions. In oxic condition, the IP_6 mineralization was at 48% and in anoxic condition at 95%.

The concentration of phosphate absorbed by binding gel (C_{DGT}) is calculated using the following equation (1) bellow:

$$C_{DGT} = \frac{M \times \Delta g}{D \times t \times A} \quad (1)$$

C_{DGT}	=	concentration of phosphate absorbed by binding gel
M	=	mass of phosphate absorbed by binding gel
Δg	=	the thickness of the diffusive gel plus the thickness of the filter membrane
D	=	diffusion coefficient of phosphate in the gel
t	=	deployment time
A	=	exposure area

Data of the concentration of phosphate absorbed by binding gel (C_{DGT}) during seven days of deployment can be seen in table 3. The concentration profile of C_{DGT} of phosphate at various depths after seven days period of deployment can be seen in figure 5.

Based on the data in table 3, the maximum of C_{DGT} of phosphate after 7 days period of oxic and anoxic conditions are 29.23 g/L at a depth of 14 and 30.19 g/L at a depth of 8 cm, respectively. Based on profile of C_{DGT} Phosphate at various depths after seven days of deployment in oxic and anoxic conditions, it showed that CDGT phosphate tends to increase by the depth. Observation at depth of 1-8 cm showed

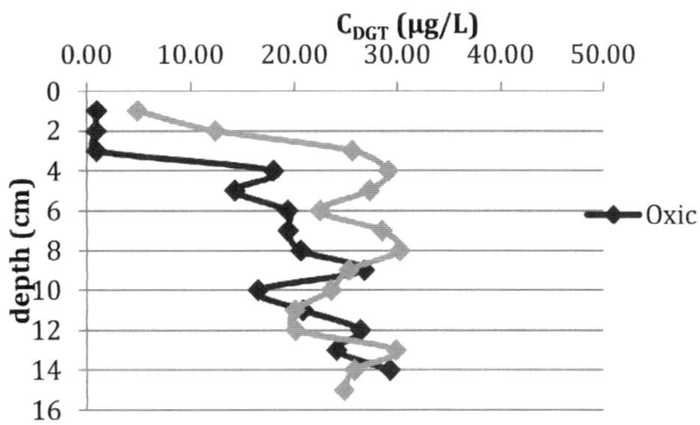

Figure 5. Profile C_{DGT} phosphate at various depths after seven days period of deployment.

Table 3. C_{DGT} during seven days period of deployment.

Depth (cm)	C_{DGT} PO_4 (µg/L)	
	Oxic	Anoxic
1	0.92 ± 0.25	4.85 ± 6.52
2	0.86 ± 0.21	12.35 ±13.86
3	0.90 ± 0.37	25.53 ± 4.05
4	17.94 ± 14.07	29.08 ± 7.22
5	14.20 ± 12.51	27.25 ± 5.43
6	19.29 ± 8.74	22.42 ± 13.71
7	19.29 ± 6.53	28.45 ± 9.41
8	20.55 ± 3.92	30.19 ± 4.96
9	26.71 ± 5.50	25.21 ± 5.43
10	16.43 ± 6.32	23.49 ± 15.16
11	20.78 ± 3.22	20.00 ± 3.64
12	26.35 ± 9.60	20.02 ± 5.01
13	24.00 ± 6.71	29.84 ± 10.27
14	29.23 ± 5.26	25.76 ± 7.89
15	25.44 ± 11.36	24.77 ± 5.51

that CDGT phosphate in anoxic condition was greater than oxic condition. This is due to at the increasing depth the oxygen concentration decreases, so both two core sediment samples are in anoxic condition. According to Effendi, at the water surface, oxygen levels will be higher due to the diffusion process of water with free air. Furthermore, the surface layer exposed to sunlight allows the process of photosynthesis to produce oxygen. At the increasing depth, the level of dissolved oxygen is decreased, caused by the decreasing sunlight that would inhibit the process of photosynthesis. The dissolved oxygen is used by the aquatic biota for breathing process and the oxidation of organic and inorganic materials [17]. At a depth of 9-15 cm the absorbed mass phosphate was relatively the same, it could be due to the influence of oxygen and nitrogen bubbling was reduced at the higher depths, therefore the oxygen cannot diffuse to the higher depths. This causes the environmental conditions on both treatments to be relatively equal.

Redox potentials in natural waters range from about -400 mV to +800 mV at pH 7 to 8. They are bounded in the negative range by the reduction of H_2O to H_2 (g) and in the positive range by the oxidation of H_2O to O_2 (g). According to Wetzel, Stumm and Morgan have classified the representative ranges of redox potentials in water and sediment at pH 7 to 8. The first range redox potentials are from about 710 to 800 mV the second range redox potentials are from about –100 to 710 mV, and the third range redox potentials are from about-240 to –100 mV. In the first range, where saturated with oxygen, the soluble iron II (Fe^{2+}) was oxidized to solid iron III (Fe^{3+}) which adsorb phosphate. In the second range, where some oxygen has been consumed, the solid iron III (Fe^{3+}), manganese III and IV (Mn^{3+}, Mn^{4+}) were reduced to soluble iron II (Fe^{2+}) and manganese II (Mn^{2+}) in which organic matter was mineralized. Phosphorus, which initially bound to Fe^{3+} hydroxide, was released to the pore water and diffused to overlying water as a result of the reduction of Fe^{3+} solid phases. The third range, under even more reduce conditions, the concentration of Fe^{2+} and Mn^{2+} were further increased; SO_4^{2-} was started to be reduced, accompanied by the precipitation of Manganous Sulfide (MnS) and Ferrous Sulfide (FeS) and the formation of pyrite [18].

4. Conclusions

DGT with 15% acrylamide; 0.1 % N-N'-methylenebisacrylamide and ferrihydrite as binding gel were suitable for the measurement of the phosphate release from sediment into pore water. The result of the deployed DGT in oxic and anoxic conditions in seven days period of incubation showed the released phosphate process from the sediment into pore water was affected by incubation time and the existence of oxygen in the environment. The released phosphate from the sediment to the water in anoxic condition has a higher value than oxic condition. The experimental results of the deployed DGT in natural sediment core at a depth of 1 to 15 cm from the surface of the water in 7 days period of incubation showed that the sediment has a phosphate mass profile difference based on depth. The concentration of phosphate tends to be increased with depth. The maximum C_{DGT} of phosphate release in oxic and anoxic conditions at 7th days are 29.23 μg/L at 14 cm depth and 30.19 μg/L at 8 cm depth, respectively.

Acknowledgements
The authors would like to thank Universitas Indonesia for funding this research through PITTA Grant Universitas Indonesia with contract No: 1992/UN2.R12/HKP.05.00/2016.

References
[1] Boulton A J and Brock M A 1999 Australian Freshwater Ecology: Processes and Management (Adelaide: Gleneagles Publishing, Australia)
[2] Boesch D F, Brinsfield R B and Magnien R E 2002 *J. Environ. Qual.* **30** 303-20
[3] Smith V H, Tilman G D and Nekola J C 1999 *Environmental Pollution* **100** 179-96
[4] Thompson P A, Waite A M and McMahon K 2003 *Marine and Freshwater Research* **54** 27-37
[5] Bostrom B, Andersen J M, Fleischer S and Jansson M 1988 *Hydrobiologia* **170** 229-44
[6] Hesslein R H 1976 *Limnology and Oceanography* **21** 912-4
[7] Golterman Han L 2004 The Chemistry of Phosphate and Nitrogen Compounds in Sediments (Dordrecht: Kluwer Academic Publishers, The Netherlands)
[8] Rumhayati B 2010 *Jurnal ILMU DASAR* **11** 160-6
[9] Edlund G and Carman R 2001 *Chemosphere* **45** 1053-61
[10] Suzumura M and Kamatani A 1995 *Geochim. Cosmochim. Ac.* **59** 1021-6
[11] Stijn B, Verbeeck M, Verheyen D, Diels J and Smolders E 2015 *Water Research* **71** 160-70
[12] Zhang H, Davison W, Gadi R and Kobayashi 1998 *Anal. Chim. Acta* **370** 29-38
[13] Ratshania 2014 *Pengembangan Metode DGT (Diffusive Gradients in Thin Film) dengan Binding Gel Fe-Al-Oksida dan Pengikat Silang N,N'-methylenebisacrylamide untuk Penyerapan Fosfat dalam Air* BSc projects (Depok: Departemen Kimia, FMIPA Universitas Indonesia)
[14] DGT Research 1997. Technical Info-DGT (Lancaster: DGT Research)
[15] Froelich P N, Klinkhammer G P, Bender M L, Luedtke N A, Heath G R, Cullen D, Dauphin P, Hammond D, Hartman B and Maynard V 1979 *Geochim. Cosmochim. Ac.* **43** 1075-90
[16] Hupfer M, Gtichter R and Ruegger R R 1995 *Limnol. Ocenogr.* **40** 610-7
[17] Effendi H 2003 *Telaah Kualitas Air, Bagi Pengelolaan Sumber Daya dan Lingkungan Perairan* (Jakarta: Kanisius, Indonesia)
[18] Wetzel R G 2001 *Limnology: Lake and River Ecosystems 3rd ed* (San Diego, California: Academic Press, United States)

IOP Conf. Series: Materials Science and Engineering **188** (2017) 012029 doi:10.1088/1757-899X/188/1/012029

Ag/Ag_2O/TiO_2 nanocomposites: microwave-assisted synthesis, characterization, and photosonocatalytic activities

S P Prakoso[1], A Taufik[1,2] and Rosari Saleh[1,2]

[1]Integrated Laboratory of Energy and Environment, Faculty of Mathematics and Natural Sciences Universitas Indonesia, Kampus UI Depok, Depok 16424, Indonesia
[2]Department of Physics, Faculty of Mathematics and Natural Sciences Universitas Indonesia, Kampus UI Depok, Depok 16424, Indonesia

Corresponding author's e-mail: rosari.saleh@gmail.com

Abstract. Nanocomposites of Ag/Ag_2O/TiO_2 were synthesized using a microwave-assisted method. As-prepared samples were systematically characterized by X-ray diffraction, electron spin resonance, and diffuse reflectance spectroscopy. The photosonocatalytic performance was investigated under both UV and visible light and ultrasonic irradiation in terms of degradation of methylene blue (MB) dye in aqueous solution. Compared with the pure TiO_2 nanoparticles and Ag/Ag_2O nanocomposites, the Ag/Ag_2O/TiO_2 nanocomposites presented more responsive photosonocatalytic activity in the wide range of light wavelengths. It is believed that defect states in TiO_2 nanoparticles, which were observed by ESR measurements, played an important role in trapping the charge carrier in the photosonocatalytic process. In addition, the degradation process was studied in terms of the influence of several experimental parameters, and a reasonable mechanism of photosonocatalysis with Ag/Ag_2O/TiO_2 nanocomposites was proposed by correlating the active radical species involved with the physical properties of the as-synthesized samples.

1. Introduction

In recent years, photocatalytic technology has greatly aided the environment by its use in wastewater treatment [1, 2]. The efficiency of solar conversion in the photocatalytic process depends on several factors, including light absorption, charge separation and migration, and surface reactions [3, 4]. Along with low cost and high catalytic efficiency, physical and chemical stability are the primary reasons that titanium dioxide is chosen. However, it also has limitations, including that it works only with ultraviolet light [5] and, therefore, cannot be used with visible light. These limitations, along with its high rate of charge recombination and separation difficulties [6], make large-scale application impossible.

The presence of a metal element can provide efficient charge separation and migration, thus helping to facilitate photocatalytic activity and stability. Several studies have exhibited metal-semiconductor properties for capturing, storing, and discharging photo-generated electrons [7, 8]. One widely used material, Ag_2O, exhibits active photocatalysts and good stability while forming Ag-Ag_2O during photocatalytic reaction [9]. Deposition of silver nanoparticles to TiO_2 has attracted growing interest because this process can extend titanium dioxide excitation under visible irradiation [10-12]. This effect is due to suppressing recombination by trapping electrons using a Schottky barrier at the interface.

However, photocatalytic activity has several limitations due to its low ability to penetrate a nontransparent medium, including highly concentrated dye in wastewater. Therefore, it is important to find a new method that effectively treat highly concentrated dye. Sonocatalytic process is an alternative method for degrading highly concentrated organic pollutant in wastewater because its ultrasonic wave has the ability to penetrate any water medium [13]. Therefore, the present study used a microwave-assisted method to synthesize a photosonocatalytic process that synthesized $Ag/Ag_2O/TiO_2$ as a catalyst for removing organic dye from an aqueous solution.

2. Materials and methods

2.1. Materials

Both $AgNO_3$ and TiO_2 were obtained from Merck Millipore. $AgNO_3$ was used as a precursor to synthesize Ag/Ag_2O. TiO_2 was used to obtain the composite material. All chemicals were used without further purification.

2.2. Ag/Ag_2O synthesis

The synthesis step included producing the Ag/Ag_2O nanoparticles and adding theTiO_2 nanoparticles. The Ag/Ag_2O nanoparticles were synthesized using the developed microwave-assisted method.

First, the silver nitrate precursor was mixed with sodium hydroxide by constant stirring at room temperature for 30 min. This mixture was heated in a microwave oven (Electrolux). Because beaker glass was used, the microwave heating was accomplished in power steps to avoid drying out the inside of the mixture. Total time inside the microwave oven was about 5 min.

The mixture was removed from the microwave oven and allowed to cool slowly in the open air for 30 min. Then, it was stirred for 1 h to ensure better particle growth. The resulting precipitate was washed with deionized (DI) water and alcohol to ensure there were no impurities in the sample and then dried in a 120°C furnace to yield Ag/Ag_2O nanoparticles.

2.3. $Ag/Ag_2O/TiO_2$ formation

Commercially available TiO_2 (Merck) was added to Ag/Ag_2O at ratios of 0.25:1 (25%), 0.5:1 (50%), and 0.75:1 (75%). Silver nanoparticles were immersed in DI water, and titanium dioxide was dissolved in sodium hydroxide. The mixing process was conducted under constant stirring and temperatures (80°C) for 1 h. Precipitate washing and drying were applied, resulting in the $Ag/Ag_2O/TiO_2$ composite. The $Ag/Ag_2O/TiO_2$ with 25%, 50%, and 75% Ag/Ag_2O were denoted as A 25-TiO_2, A 50-TiO_2 and A 75-TiO_2, respectively.

2.4. Characterization

X-ray diffraction (XRD) analysis was performed using a Rigaku Miniflex 600. The patterns were recorded from 10°-90° at 30 kV with an intensity of 15 mA. The samples' UV-Vis diffuse reflectance spectra (DRS) were obtained using a Shimadzu Spectrophotometer. Electron spin resonance (ESR) was performed using an X-band spectrometer equipped with a 9.1-GHz field modulation unit and an X-band JEOL JES-RE1X at room temperature.

2.5. Photosonocatalytic experiment

Phtosonocatalytic activity was tested for its ability to remove methylene blue as a model organic pollutant. Using the typical process, 0.03 g of catalyst was immersed in 100 mL of methylene blue solution with a concentration of 40 mg/L and placed in a dark place for 30 min to reach adsorption-desorption equilibrium. Then, the solution was placed in an ultrasonic bath operated at 150 W and 40 kHz frequency and irradiated with a UV-C lamp and a Xenon (visible-light source) lamp separately. Every 15 min, the samples were investigated using UV-Vis spectroscopy to analyze the decolorization of the methylene blue.

3. Results and discussion

Figure 1 shows the XRD spectra of the $Ag/Ag_2O/TiO_2$ with various Ag/Ag_2O wt.%. For comparison, the figure also shows the XRD spectra of TiO_2 and Ag/Ag_2O. The peak at $2\theta = 25.24°$ was attributed to the anatase structure of the TiO_2, and the peak at $2\theta = 44.12°$ was attributed the presence of the silver nanoparticles. In addition, cubic structures from the Ag_2O were detected at $2\theta = 25.24°$, 32.57°, 37.88°, and 54.91°. No undesired peaks were detected, which would indicate the existence of impurity structures. The lattice parameter and grain size of the samples were calculated using the Rietveld refinement method and Debye Scherrer equations, and table 1 shows the results. As can be seen, the value of the sample's lattice parameters differed slightly from those of its single materials. This was attributed to the intra-granular coupling of the composites, which provided intimate contact between the two oxide phases and improved the efficiency of the semiconductor-coupling effects of the crystal structures [14]. The TiO_2 grain size did not differ significantly from that of its single materials.

Figure 2 shows the UV-Vis DRS of the $Ag/Ag_2O/TiO_2$ composite for each TiO_2 concentration. All samples had broad absorption in the 400-700 nm range. A comparison was made with pure TiO_2, which had similar absorption characteristics. This broad absorption was attributed to the SPR effect of the sample's Ag nanoparticles [15]. The absorption edge of the $Ag/Ag_2O/TiO_2$ was slightly red-shifted and had enhanced photoactivity under visible radiation. Less absorption was seen at higher concentrations of TiO_2, and absorption dropped even lower with pure TiO_2. The samples' band-gap energies were calculated using the Kubelka-Munk equation, and the results are shown in table 1. The band-gap energy shifted slightly lower with increased Ag/Ag_2O.

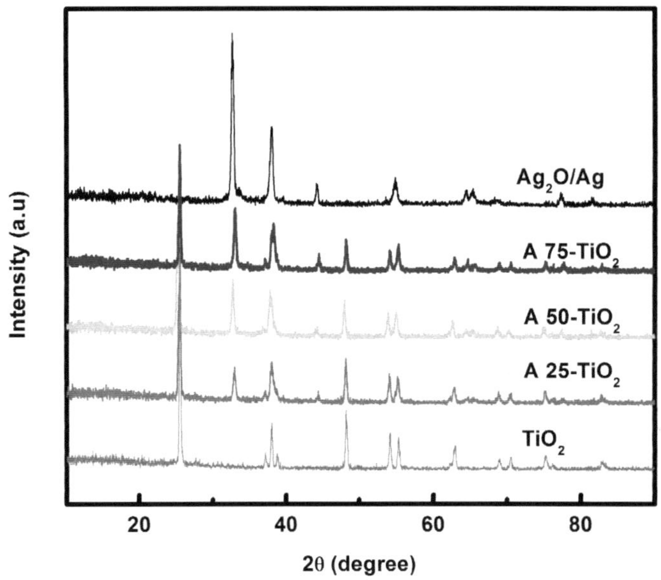

Figure. 1 XRD spectra of TiO_2 nanoparticles, Ag/Ag_2O, and $Ag/Ag_2O/TiO_2$ with different Ag/Ag_2O loading.

Table 1. Lattice parameter, grain size and band gap energy of TiO_2 nanoparticles, Ag/Ag_2O, and $Ag/Ag_2O/TiO_2$ with different Ag/Ag_2O loading.

Sample	Lattice Parameter (Å)				Grain Size (nm)			Energy Gap (eV)
	Ag	Ag_2O	TiO_2		Ag	Ag_2O	TiO_2	
	a=b=c	a=b=c	a=b	c				
Ag/Ag_2O	4.088	4.733	-	-	22	21	-	-
A 25 - TiO_2	4.094	4.736	3.791	9.529	20	19	31	2.87
A 50 - TiO_2	4.090	4.734	3.788	9.524	21	19	29	2.71
A 75 - TiO_2	4.071	4.734	3.762	9.526	21	20	27	2.51
TiO_2	-	-	3.791	9.529	-	-	45	3.32

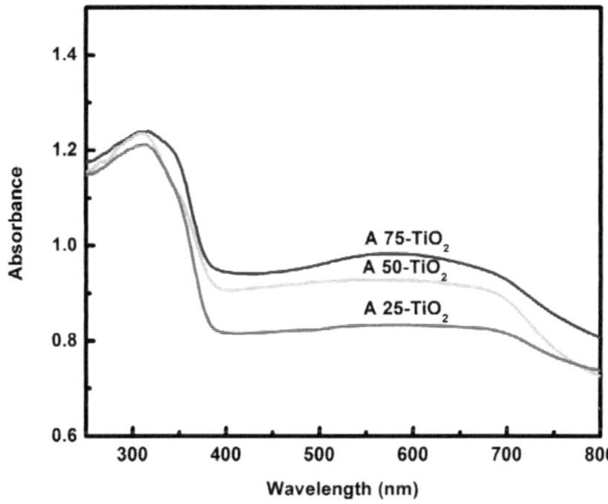

Figure 2. UV Vis Absorbance spectra of $Ag/Ag_2O/TiO_2$ with different Ag/Ag_2O loading.

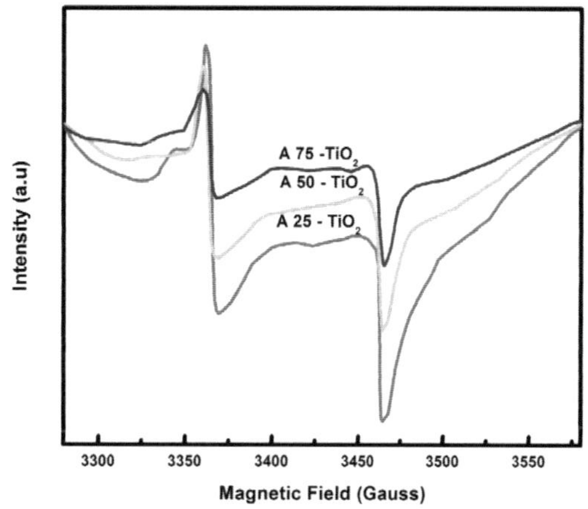

Figure 3. ESR spectra of $Ag/Ag_2O/TiO_2$ with different Ag/Ag_2O loading.

Figure 3 shows the ESR spectra of $Ag/Ag_2O/TiO_2$ with various Ag/Ag_2O loadings. As can be seen, there were two main signals, in the ranges of 3361 and 3465 gauss, which were attributed to the Ti^{3+} and oxygen-vacancy signals and which had g values of 1.95 and 2.04, respectively [16]. The intensity of both signals increased gradually as Ag/Ag_2O decreased. Both the Ti^{3+} and the oxygen vacancy acted as charge carrier traps that enhanced the samples' catalytic performance.

Figure 4 shows the photosonocatalytic activity of $Ag/Ag_2O/TiO_2$ with various Ag/Ag_2O loadings under UV and visible-light irradiation. As can be seen, under both UV and visible light, the samples' photosonocatalytic activity gradually increased with decreased Ag/Ag_2O. In addition, to understand the role of the photosonocatalytic process, the samples' photocatalytic and sonocatalytic activities were investigated. Figure 5 shows the comparison of photocatalytic, sonocatalytic, and photosonocatalytic activities under both UV and visible-light irradiation. As can be seen, the samples' degradation ability was enhanced by using simultaneous photocatalytic and sonocatalytic activity. This might be due to the existence of ultrasound irradiation that enhanced the production of OH radicals used to decompose the methylene blue in the aqueous solution. Ultrasound irradiation also has the ability to create spontaneous light, called sonoluminescence, which has a wide range of wavelengths. Therefore, formation of electron and hole pairs may increase with the addition of ultrasound irradiation [13]. Both photocatalytic and photosonocatalytic activities were higher in the $Ag/Ag_2O/TiO_2$ nanocomposites than in the TiO_2 nanoparticles. This may have been due to a retardation of electron-hole recombination caused by formation of heterojunctions among Ag, Ag_2O, and TiO_2.

Figure 6 shows the effects of scavenger addition on the photosonocatalytic activity of $Ag/Ag_2O/TiO_2$ nanocomposites. The addition of scavengers was tested to identify the main active species involved in the photocatalytic process. Three type of scavengers were used: ammonium oxalate, sodium sulfate, and tert-butyl ethanol, which detected hole, electron, and OH radical main active species. The addition of scavengers reduced the photocatalytic activity because these main active species in the photosonocatalytic process were captured by the scavengers. As can be seen, under both UV and visible-light irradiation, the addition of hole scavengers caused the lowest photosonocatalytic activity, indicating that holes are the main active species in the degradation process.

The possible mechanisms of both photocatalytic and photosonocatalytic activity can be described as follows: Under visible-light irradiation, TiO_2 cannot create electron and hole pairs due to insufficient visible-light energy. However, Ag_2O with a narrow band gap can create electron and hole pairs due to its suitable energy [17]. Electrons from the Ag_2O valence band were excited into the conduction band and left holes in the valence band. These electrons did not readily recombine with holes in the valence band but were captured by Ag and trapping state from TiO_2, including Ti^{3+} and oxygen vacancy.

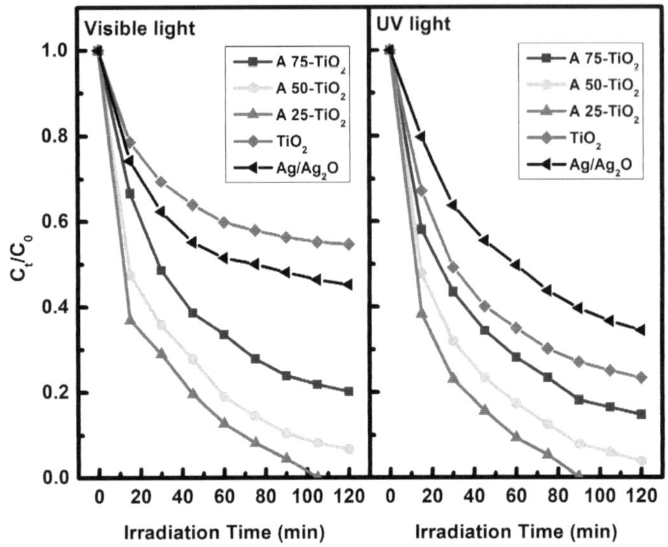

Figure 4. Photosonocatalytic activity of $Ag/Ag_2O/TiO_2$ under UV and visible light irradiation.

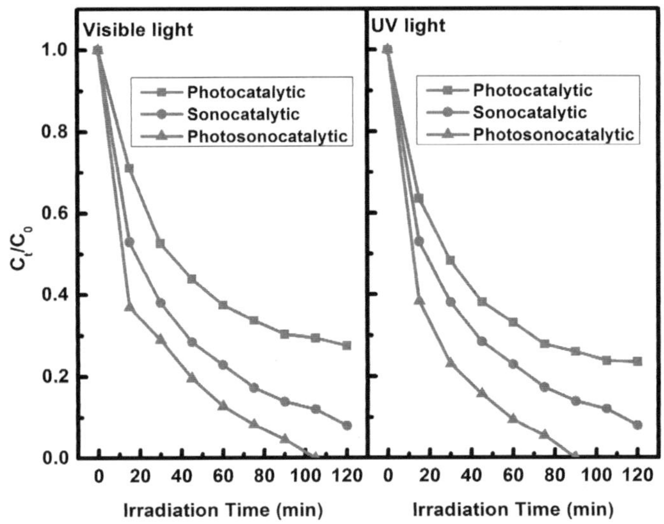

Figure 5. Photocatalytic, sonocatalytic, and photo-sonocatalytic activity of $Ag/Ag_2O/TiO_2$ nano-composites under visible and UV light irradiation

Therefore, the recombination of electrons and holes was prevented, and photocatalytic activity under visible-light irradiation increased. Under UV light irradiation, only TiO_2 electrons and holes. The addition of ultrasound irradiation to the photocatalytic process (photosonocatalytic) created acoustic cavitation, which emitted light (i.e., sonoluminescences), which has a wide wavelength range (from UV to visible-light irradiation). Under these conditions of local high temperature and pressure, water molecules can be thermally dissociated into hydroxyl radicals (OH) to some extent [13, 18]. The sonoluminescence from ultrasound irradiation could create electron and hole pairs neither for TiO_2 nor for Ag_2O at the same time. Electrons from the conduction band of TiO_2 migrated into the conduction bands of Ag_2O and metallic Ag, while electrons from the conduction band of Ag_2O were captured by native defects, including Ti^{3+} and oxygen vacancy. Based on the ESR results, the Ti^{3+} and oxygen-vacancy signals increased with decreased Ag_2O/Ag content, indicating that decreased Ag_2O/Ag content resulted in higher oxygen vacancy and Ti^{3+} in the samples. Therefore, the highest photosonocatalytic activity was achieved by 25% Ag_2O/Ag. When electrons migrated from one state to another, holes

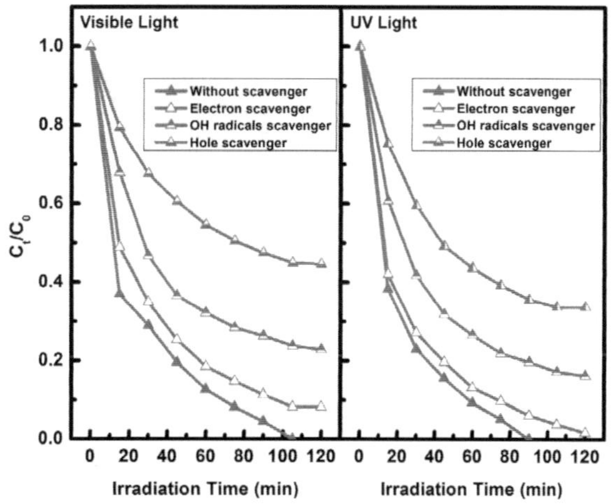

Figure 6. The effet of scavenger addition on photosonocatalytic activity of $Ag/Ag_2O/TiO_2$ nanocomposites under visible and UV light irradiation

reacted with water molecules to form OH radicals and decompose the methylene blue in the aqueous solution. This result was confirmed by a scavenger experiment showing that holes and OH radicals were the main active species in the degradation process.

4. Conclusions

$Ag/Ag_2O/TiO_2$ composite was successfully synthesized using a microwave-assisted method. Composition of the composite was confirmed by XRD measurement. Adding silver to the TiO_2 enhanced absorption into the visible region and extended usability for photosonocatalytic applications. The existence of native defects, including Ti^{3+} and oxygen vacancy, improved the photosonocatalytic experiment. Photosonocatalytic activity exhibits faster degrading activity than the photocatalytic or sonocatalytic processes.

References

[1] Chen C, MaW and Zhao J 2010 *Chem. Soc. Rev.* **39** 4206-19
[2] Li Y, Liu J R, Jia S Y, Guo J W, Zhuo J and Na P 2012 *Chem. Eng. J.* **191** 66-74
[3] Wu N, Wang J, Tafen D N, Wang H, Zheng J-G, Lewis J P, Liu X, Leonard S S and Manivannan A 2010 *J. Am. Chem. Soc.* **132** 6679-85
[4] Cushing S K, Li J, Meng F, Senty T R, Suri S, Zhi M, Li M, Bristow A D and Wu N 2012 *J. Am. Chem. Soc.* **134** 15033-41
[5] Asahi R, Morikawa T, Ohwaki T, Aoki K and Taga Y 2001 *Science* **293** 269-71
[6] Fujishima A, Rao T N and Tryk D A 2000 *J. Photochem. Photobiol. C Photochem. Rev.* **1** 1-21
[7] Choi H, Chen W T and Kamat P V 2012 *ACS Nano* **6** 4418-27
[8] Takai A and Kamat P V 2011 *ACS Nano* **5** 7369-76
[9] Wang X, Li S, Yu H, Yu J and Liu S 2011 *Chem. A Eur. J.* **17** 7777-80
[10] Awazu K, Fujimaki M, Rockstuhl C, Tominaga J, Murakami H, Ohki Y, Yoshida N and Watanabe T 2008 *J. Am. Chem. Soc.* **130** 1676-80
[11] Chuang H Y and Chen D H 2011 *Int. J. Hydrogen Energy* **36** 9487-95
[12] De Souza M L and Corio P 2013 *Appl. Catal. B Environ.* **136** 325-33
[13] Zhou M, Yang H, Xian T, Li R S, Zhang H M and Wang X X 2015 *J. Hazard. Mater.* **289** 149-57
[14] Mageshwari K, Nataraj D, Pal T, Sathyamoorthy R and Park J 2015 *J. Alloys Compd.* **625** 362-70
[15] Hu X, Hu C and Wang R 2015 *Appl. Catal. B Environ.* **176** 637-45

[16] Coronado J M and Soria J 2007 *Catal. Today* **123** 37-41
[17] Wei N, Cui H, Song Q, Zhang L, Song X, Wang K, Zhang Y, Li J, Wen J and Tian J 2016 *Appl. Catal. B Environ.* **198** 83-90
[18] Zhou M, Yang H, Xian T, Yang Y and Zhang Y 2015 *Chinese Journal of Catalysis* **36** 1987-94

Preparation of boron doped diamond modified by iridium for electroreduction of carbon dioxide (CO_2)

A M Ichzan, J Gunlazuardi and T A Ivandini

Department of Chemistry, Faculty of Mathematics and Natural Sciences Universitas Indonesia, Kampus UI Depok, Depok 16424, Indonesia

Corresponding author's e-mail: ivandini.tri@sci.ui.ac.id

Abstract. Electroreduction of carbon dioxide (CO_2) at iridium oxide-modified boron-doped diamond (IrOx-BDD) electrodes in aqueous electrolytes was studied by voltammetric method. The aim of this study was to find out the catalytic effect of IrOx to produce fine chemicals contained of two or more carbon atoms (for example acetic acid) in high percentage. Characterization using FE-SEM and XPS indicated that IrO_2 can be deposited at BDD electrode, whereas characterization using cyclic voltammetry indicated that the electrode was applicable to be used as working electrode for CO_2electroreduction.

Keywords: Electroreduction, carbon dioxide, iridium oxide, boron-doped diamond

1. Introduction

CO_2 produced by human's activity offers a great value to overcome the global warming issues. CO_2 can be recycled to O_2 naturally, during the photosynthesis process, however this process is limited due to the decrease of plant population.The remaining CO_2 then continously increases at the surface of the earth. This problem prompted some researchers to investigate the alternative process to decrease CO_2 content in the atmosphere. One of the alternative method to reduce CO_2 is to convert the gas electrochemically to industrial chemicals such as formic acid [1], methanol [2], and acetic acid [3, 4]. Le *et al.* [2] reported that direct reduction of CO_2 to form methanol can be performed by using copper as the working electrode, while several other studies reported the reduction of CO_2 to formic acid with cadmium, indium, tin and lead as the working electrodes [5]. In addition, theuse of gold or silver as the working electrode with KOH-methanol as the electrolyte was also reported [6] whereas methane can be produced electrochemically through CO_2 reduction using copper electrode with lithium as a supporting electrolyte [7].

Furthermore, Nakata *et al.* [8] reported that the use of boron-doped diamond (BDD) electrodes in the reduction of CO_2 can produce formaldehyde. BDD is reported to be abetter electrode compared to the previous electrode used in this aplication, because it generates higher products percentage. BDD is also known to have a wide potential window as well as a high electrochemical stability [9, 10]. The use of a single element for the working electrode such as copper, BDD and other working electrode also showed the tendency to form organic compounds consists only one carbon atom, such as methane, methanol, formic acid and formaldehyde. Meanwhile, Panglipur *et al.* [4] reported an indication of the formation of acetic acid from CO_2 reduction process using BDD modified by copper at its surface [3, 4]. It was proposed that the use of composite consisted of BDD and copper correlates to the formation of organic compounds with two carbon atoms.

IOP Conf. Series: Materials Science and Engineering **188** (2017) 012030 doi:10.1088/1757-899X/188/1/012030

This research studies the preparation of iridium-modified BDD electrodes for an application in electrochemical reduction of CO_2. Iridium was used as a composite metal at the BDD surface due to its electrocatalytic effect. The use of iridium oxide nanoparticles composite with cobalt oxide electrode on water oxidation and mercury detection have been reported [11, 12]. Characterization with SEM, XPS, EDX and cyclic voltammetry was performed to confirm that the iridium-modified electrode is ready to be employed as the working electrode in CO_2 electroreduction.

2. Materials and methods

2.1. Chemicals

H_2IrCl_6. xH2O, KNO_3, NaCl, and 2-propanol were supplied from Sigma and used in this study without purification.

2.2. Experimental

Boron doped diamond (BDD) was characterized by cyclic voltametry in 0.1 M KNO_3 prior to be modified by Iridium. As the modified solution, a solution consisted of 0.5 M H_2IrCl_6 and 0.1 M KNO_3 (1:1) wasaged by heating for a specified time at 80°C in closed vials, and cooled to room temperature [9]. Then, the solution was used as a modified solution to BDD by performing cyclic voltammetry for several times from a potential of +0.2 to 1.20 V (vs. Ag/AgCl). The iridium-modified BDD was characterized using SEM, EDX, XPS and cyclic voltammetry.

3. Results and discussion

3.1. Electrochemical behavior of iridium at Boron Doped Diamond (BDD)

Before the BDD electrode was modified by iridium, cyclic voltammetryusing KNO_3 as the electrolyte was performed in the potential range of -1.0 to +1.5 V with a scan rate of 100 mV/s (figure 1). No oxidation and reduction peak were observed in the applied potential indicated that the BDD was electrochemically stable and no contamination of other analytes that can go through redox reactions in the applied potential range on the cyclic voltammetry system was observed.

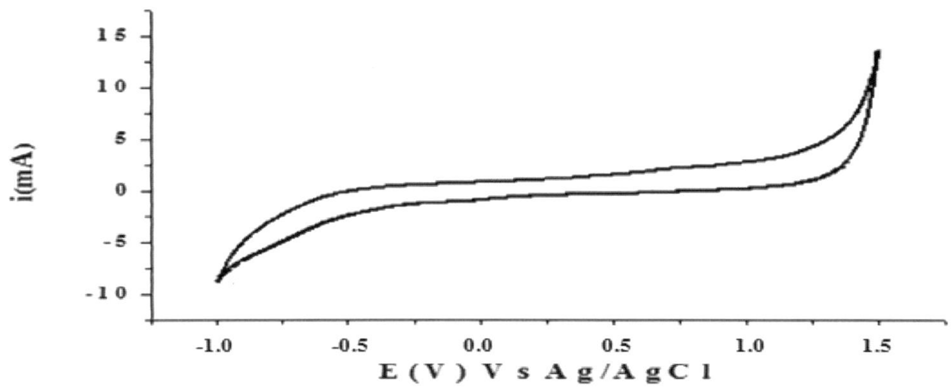

Figure 1. Voltammogram of BDD in 0.1 $MKNO_3$

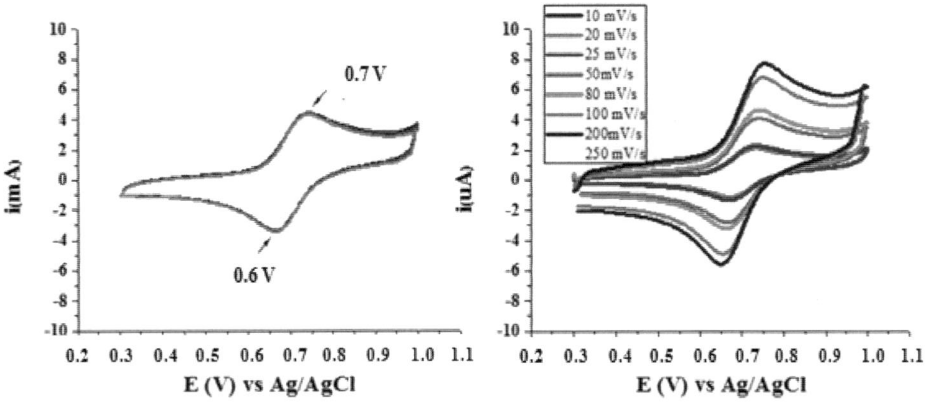

(a) (b)

Figure 2. (a) Repeating cyclic voltammetry process (10 Cycles) of 0.3 mM Ir in 0.1 M KNO_3 (1:1) and (b) variation of scan rate on Ir 0.3 mM in 0.1 M KNO_3 (1:1)

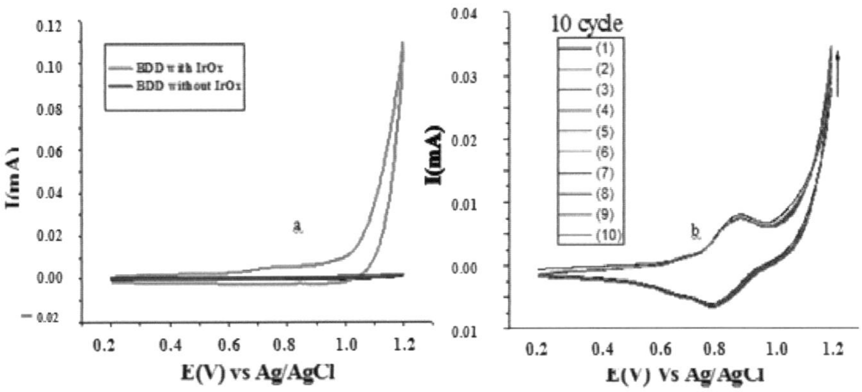

(a) (b)

Figure 3. The voltammograms of (a) the unmodified and the iridium-modified BDD and (b) the iridium-modified BDD after 10 cycles ofvoltammetryin the solution consisted of 0.5 mM Ir and 0.1 M KNO_3 (1:1)

In the presence of 0.3 mM Ir in 0.1 M KNO_3 (1: 1), the cyclic voltammograms (figure 2a) shows an oxidation-reduction couple peaks in the potential range of +0.7 V and +0.6 V. The peaks shows a good reproducibility, which indicates the stability of the working electrode as well as the electrolyte solution and analytes during the measurements. Meanwhile, characterization using the same solution by varying the scan rate (figure 2b) also shows that the current increases proportionally, scan rate suggested a good performance of the working electrode to generate the current [13].

3.2. Deposition IrO_X-BDD

Deposition of iridium oxide on the surface of BDD was performed using chronoamperometry with applied potential of +1.2 V for 10 minutes. As the depositing solution, a solution consisted of 0.5 mM Ir and 0.1 M KNO_3 (1: 1) was used after aging by heating for 1 hour at 80°C.

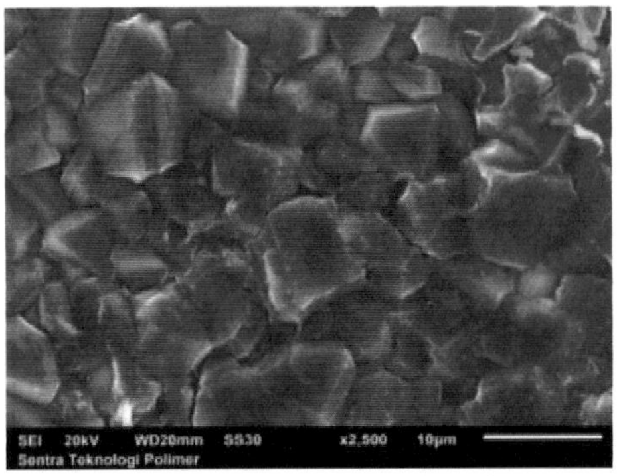

Figure 4. Characterization of BDD by FE-SEM

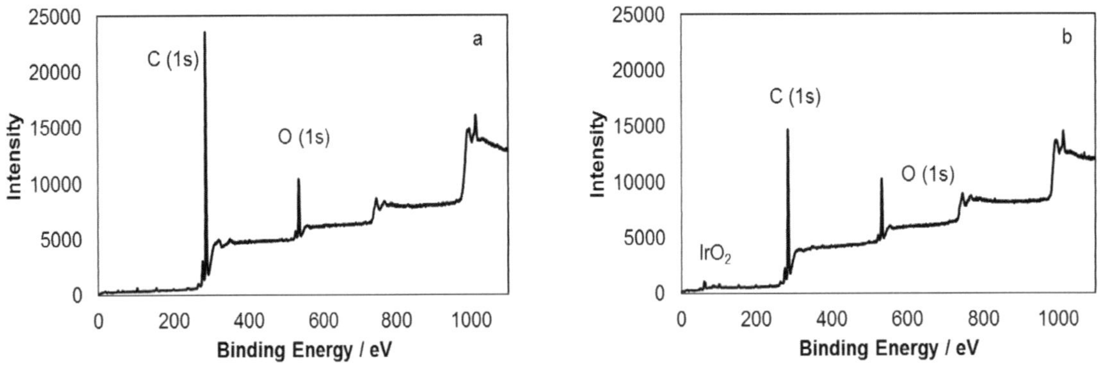

Figure 5. XPS characterization of BDD (a) before and (b) after deposition with iridium

The characterization using cyclic voltammetry was performed in 0.1 M KNO_3 (figure 3a). Whereas the unmodified BDD generates very low background current, a higher current at a potential 1.2 V was observed for the modified BDD, suggested the formation of an electroactive compound at the surface of BDD. The current becomes higher when more of cyclic voltammetry process was performed and then saturated after 10 cycles, indicated that the deposition was saturated at the surface of BDD (figure 3b).

3.3. Characterization of IrOx-BDD by using FE-SEM
The electrode was also characterized by using FE-SEM with a magnification of 10000 × (figure 4). In BDD which deposited by Ir, Ir particles on the surface of BDD marked with the brighter dotted area than the BDD.

3.4. XPS characterization
Elements composition at the BDD surface was characterized with XPS. The characterization before and after modification (figure 5) shows a peak at 284 eV attributtable to C 1s and another peak at 532 eV attributed to O 1s. On the other hand, at the Ir-modified BDD the peak at 86 eV was observed, while it was not appeared at unmodified BDD. This peak was identified as IrO_2 which was formed on the surface of BDD after the deposition.

4. Conclusions
Modification of boron-doped diamond by iridium was successfully performed by chronoamperometry technique. Characterization of the electrode by FE-SEM and XPS indicated the formation of IrO_2 on

the surface of BDD, whereas cyclic voltammetry suggested that the electrode is promising to be applied for electrochemical reduction of CO_2.

References

[1] Alvarez-Guerra M, Quintanilla S and Irabien A 2012 *Chem. Eng. J.* **207-208** 278-84
[2] Le M, Ren M, Zhang Z, Sprunger P T, Kurtz R L, and Flake J C 2011 *J. Electrochem. Soc.* **158** E45-9
[3] Panglipur H S 2016 *Elektroreduksi CO_2 menjadi Bulk Chemichal Menggunakan Boron- Doped Diamond (BDD) Termodifikasi Ir* BSc projects (Depok: Departemen Kimia, FMIPA Universitas Indonesia)
[4] Panglipur H S, Ivandini T A, Wibowo R and Einaga Y 2016 *AIP Conference* Proceedings **1729** 020047
[5] Hori Y, Kikuchi K and Suzuki S 1985 *Chem. Lett.* **14** 1695-8
[6] Kaneco S, Iiba K, Ohta K, Mizuno T and Saji A 1998 *J. Electroanal. Chem.* **441** 215-20
[7] Kaneco S, Iiba K, Yabuuchi M, Nishio N, Ohnishi H, Katsumara H, Suzuki T and Ohta K 2002. *Ind. Eng. Chem. Res.* **41** 5165-70
[8] Nakata K, Ozaki T, Terashima C, Fujishima A and Einaga Y 2013 *Angew. Chem. Int. ed.* **53** 871-4
[9] Shibano S, Ivandini T A, Terashima C, Nakata K and Einaga Y 2014 *Chem. Let.* **43** 1292-3
[10] Ivandini T A and Einaga Y 2013 *Electrocatal.* **4** 367-74
[11] Tae E L, Song J, Lee A R, Kim C H, Yoon S, Hwang I C, Kim M G and Yoon K B 2015 *ACS Catal.* **5** 5525-9
[12] Pikulski M and Gorski W 2000 *Anal. Chem.* **72** 2696-702
[13] Jitaru M 2007 *Journal of University of Technology and Metallurgy* **42** 333-44

Synthesis of ZSM-5 zeolite from coal fly ash and rice husk: characterization and application for partial oxidation of methane to methanol

Y K Krisnandi[1], F M Yanti[1,2] and S D S Murti[2]

[1]Department of Chemistry, Faculty of Mathematics and Natural Sciences Universitas Indonesia, Kampus UI Depok, Depok 16424, Indonesia
[2]Agency for Assessment and Application of Technology, Jl. M.H Thamrin 8, Jakarta, Indonesia

Corresponding author's e-mail: yuni.krisnandi@sci.ui.ac.id

Abstract. Indonesian fly ash (SiO_2/Al_2O_3 mole ratio = 3.59) was used together with rice husk (SiO_2 92%) as raw material for mesoporous ZSM-5 zeolite synthesis. Prior being used, coal fly ash and rice husk were subjected to pre-treatment in order to extract silicate (SiO_4^{4-}) and aluminate (AlO_4^{5-}) and to remove the impurities. Then the ZSM-5 zeolite were synthesized through hydrothermal treatment using two types of templates (TPAOH and PDDA). The as-synthesized ZSM-5 was characterized using FTIR, XRD, SEM-EDX, and BET. The result of FTIR showed peaks at 1250-950 cm^{-1} ($\nu_{asymetric}$ T-O), 820-650 cm^{-1} ($\nu_{symetric}$ T-O), and at 650-500 cm^{-1} confirming the presence of the five number ring of the pentasil structure. The result of XRD showed the appearance of certain peaks in the position 2 theta between 7-9° and 22-25° indicative of ZSM-5 structure, but also showed the pattern of low intensity magnetite and hematite. The SEM image showed the rough surface of hexagonal crystals from ZSM-5 structure, indicative of mesoporosity in the structure. EDX result showed Si/Al ratio of 20, while surface area analysis gave SA of 43.16.The ZSM-5 zeolites then was modified with cobalt oxide through impregnation method. The catalytic activity as heterogeneous catalysts in partial oxidation of methane was tested. The result showed that hence the catalytic activity of ZSM-5 and Co/ZSM-5 from fly ash and rice husk were still inferior compared to the pro-analysis sourced-counterpart, they were potential to be used as catalyst in the partial oxidation of methane to methanol.

Keyword: coal fly ash, rice husk ash, synthesis of ZSM-5, partial oxidation, methane to methanol

1. Introduction

Coal is the world's most abundant and widely distributed fossil fuel, with global proven reserves totaling nearly 1000 billion [1]. The major problem in coal-fired power generation is that the plenty of solid waste so-called fly ash (of about 30% of raw coal) is produced [2]. Previously, almost all the fly ash was disposed by landfill, which became increasingly expensive and caused an environmental problem. Therefore, research on utilization of the by-product coal fly ash requires an urgent attention. This fly ash contains at least 50% of alumina and silicate which is composed of amorphous solid and hollow sphere [3]. Due to these similarities in the composition between coal fly ash and zeolite, coal fly ash has been viewed as a potential candidate for conversion to zeolite [4].

IOP Conf. Series: Materials Science and Engineering **188** (2017) 012031 doi:10.1088/1757-899X/188/1/012031

Chareopanich *et al.* [5] has been success to synthesis of ZSM-5 zeolite from lignite coal fly ash and rice husk ash which is produced microporous ZSM-5 zeolite. Microporous zeolite makes large molecules cannot react effectively over these microporous materials because of the limitation of their small pore sized [6]. Panpa *et al.* [7] has been success to synthesis of ZSM-5 from rice husk. Mohammed *et al.* [8] also has been success to synthesis and characterization of ZSM-5 zeolite from rice husk ash and their implication for photocatalytic degradation catalyst.

The focus of this work is the synthesis of ZSM-5 zeolite using fly ash the main raw material. However, the SiO_2/Al_2O_3 mole ratio in the raw fly ash is too low for ZSM-5 synthesis (SiO_2/Al_2O_3 mole ratio = 3.59). In order to obtain the appropriate SiO_2/Al_2O_3 mole ratio, the silicate was added to adjust the mole ratio of fly ash. From economic point of view, the silicate prepared from rice husk ash was used instead of commercial ones because there is abundant supply of rice husk in Indonesia.

2. Materials and methods

2.1. Materials

The materials used were: Coal Fly Ash from Paiton, Indonesia, rice husk from Bogor, Indonesia, sodium hydroxide (NaOH) pellet, Chloride acid (HCl) 1M, Tetrapropylammonium hydroxide, TPAOH 1M (Sigma Aldrich), Poly Diallyl Dimethyl Ammonium Chloride (PDDA 35%) (Aldrich), Cobalt (II) Nitrate ($Co(NO_3)_2.6H_2O$) (Sigma Aldrich), methane gas (99.99% BOC), N_2 gas (99.99% UHP) and deionized water.

2.2. Pretreatment of coal fly ash and rice husk ash

Coal fly ash was dried and sieved using a 200 mesh sieve, and then activated in a furnace at a temperature of 800°C for 2 hours, then refluxed with 1M HCl for 3 hours at 90°C. After this, the mixture was washed with distilled water to pH neutral. Then the product was filtered with filter paper, and then sample was dried in an oven with a temperature of 105°C for 6 hours.

Rice husk sample was washed, then dried in an oven at 105°C for 1 hour, then the sample was crushed and sieved 60 mesh, then burned with a furnace at 700°C for 4 hours, then characterized by AAS, FTIR and XRD.

2.3. Synthesis of ZSM-5

Synthesis of ZSM-5 was prepared with a molar ratio: 12NaOH: $30SiO_2$: $0.75Al_2O_3$: 6TPAOH: $1800H_20$. In this experiments, 2.0286 g of coal fly ash was mixed with 30 mL of 0.4 M NaOH solution (mixture A) at 100°C. Then 5.7187 g rice husk ash was added with 70 mL of 0.4 M NaOH and then stirred at 100°C (mixture B). Furthermore, mixture's B was slowly added to the mixture A, then 4.0552 g TPAOH was added to the mixture, and glacial acetic acid was added to adjust pH up to11. Stirring was continued for 3 hours at a temperature of 100°C. After that, added 1 g of PDDA (10 wt.%) to the mixture and stirred again for 15 hours at room temperature. Then the reaction mixture was transferred into the autoclave and heated for 144 hours at a temperature of 150°C in the oven. The reaction product was be cooled, washed with aquademin, filtered, and dried at room temperature for one night. After that, the product was calcined for 8 hours at a temperature of 500°C. Synthesis of ZSM-5 from pro-analysis materials as references also was prepared with raw material from $NaAlO_2$ and TEOS. The as-synthesized crystals of mesoporous ZSM-5 were characterized by XRD instruments, FTIR, SEM, and pore size analyzer (BET).

Post-synthesis modification was carried out to obtain Co/ZSM5, test catalytic partial oxidation of methane to methanol in a vessel of stainless steel as a batch reactor, following the procedure have been reported previously [9] by mixing 0.2495 M cobalt solution (from $Co(NO_3)_2$) with the zeolite. After the impregnation process, the mixture was calcined at 500°C to obtain Co/ZSM-5.

2.4. Application for partial oxidation of methane to methanol

Test catalytic partial oxidation of methane to methanol in a vessel of stainless steel as a batch reactor, following the procedure have been reported previously [9]. The catalyst used for each reaction is 0.5 g, which has been activated at 500°C. The reaction was performed at 150°C with methane pressure of 0.75

bar and a nitrogen pressure of 2 bar with a variety of reaction time of 30 minutes and 60 minutes then extracted with ethanol, and characterized using Gas Chromatography-FID.

3. Results and discussion

3.1. X-Ray Diffaction (XRD)

Diffraction patterns of ZM-5 from fly ash dan rice husk ash (nZSM-5) and ZSM-5 from pro-analysis materials (pZSM-5) showed the appearance of certain peaks at 7° to 9° and 22° to 25° 2 theta (figure 1), indicative of successfully synthesized ZSM-5 structure. Figure 1, also shows the peak intensity of nZSM-5 is lower than pZSM-5. It means that the crystallinity of pZSM -5 is higher than nZSM-5. Besides that, figure 1 also shows other pattern in position 2 theta 33°; 36°; 49°; 57°, indicative of hematite and magnetite.

3.2. Fourier transform infa red (FTIR)

Figure 2 shows the FTIR spectra of nZSM-5 and pZSM-5. The characteristic absorption bands of ZSM-5 were found at around 1250-950 cm^{-1} ($\nu_{asimetri}$ T-O), 1100-700 cm^{-1} ($\nu_{simetri}$ T-O), and also were found at round 650-500 cm^{-1} indicated the presence of double 5 ring in ZSM-5 [10].

3.3. Thermal gravimetry analysis (TGA)

TG analyses of ZSM-5 were perfomed on nZSM-5 and pZSM-5. The TG curve (figure 3) of ZSM-5 particles shows weight loss up to three steps which is weight loss in the region <150°C showed a loss of water mass in the ZSM-5, while the area was around the 200-440°C shows the decomposition and weight loss of PDDA. Then in the region 440-510°C showed decomposition of TPAOH template.

3.4. Chemical content

Figure 4 shows the chemical content of nZSM-5 and pZSM-5 and the Si/Al ratio of pZSM-5 and nZSM-5 from are 16.73 and 20.10 respectfully.

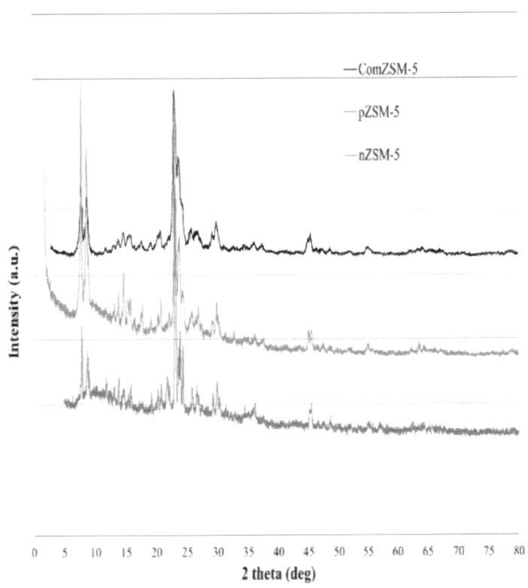

Figure 1. XRD Pattern of ZSM-5 synthesized (pZSM-5 and nZSM-5) and Com-ZSM-5.

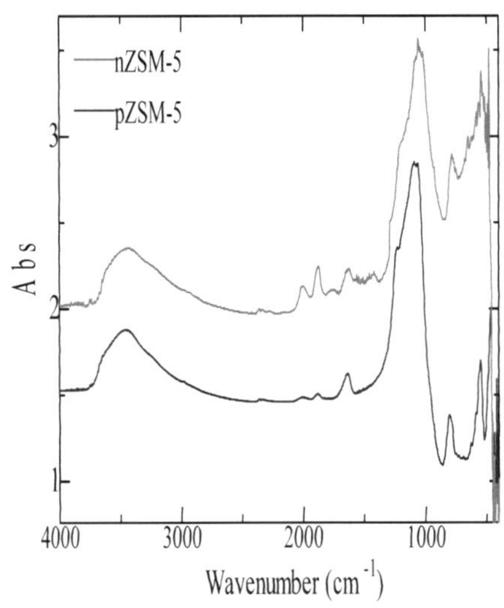

Figure 2. FTIR Spectra of ZSM-5 synthesized (pZSM-5 and nZSM-5).

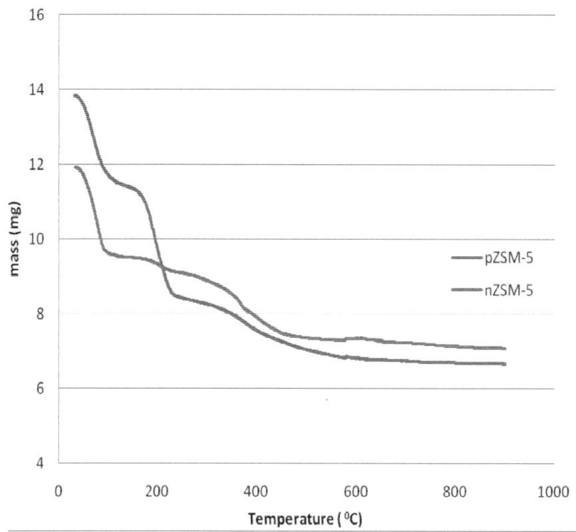

Figure 3. Thermagravimetry of as-synthesized ZSM-5.

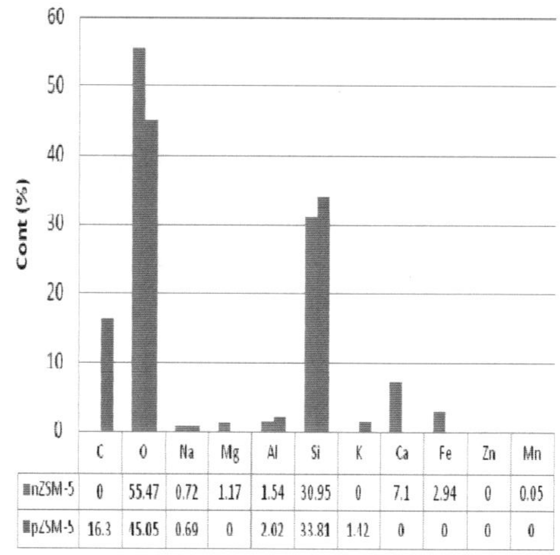

Figure 4. Analysis of chemical content of as-synthesized ZSM-5.

Figure 5a. The morphology of zeolite pZSM5 with magnification sequentially (a) 3.000 ×, (b) 4.000 ×, and (c) 7.000 ×.

Figure 5b. The morphology of zeolite nZSM5 with magnification sequentially (a) 3.000 ×, (b) 4.000 ×, and (c) 7.000 ×.

3.5. Scanning electron microscopy (SEM)

The hexagonal morphology of nZSM5 and pZSM5 crystal were formed with unsmooth or rough surface, indicating the formation of mesoporous feature on the surface of the crystal (figure 5). SEM image also showed there were still many impurities in as-synthesized nZSM-5. Futhermore, the nZSM-5 has relatively shorter length compared to the pZSM-5.

3.6. Surface area analysis

The results of surface area analyses of nZSM-5 and pZSM-5 are summarized in table 1. It can be seen that the surface area of nZSM-5 from fly ash and rice husk ash is lower than the pZSM-5 from pro-analysis materials, due to ZSM-5 from fly ash and rice husk ash still has many impurities especially Fe which could act as inhibitor in zeolititation process. The pore volume of nZSM-5 is also lower than that of pZSM-5.

3.7. Catalytic test: partial oxidation of methane to methanol

Catalytic test on partial oxidation of methane to methanol in a vessel of stainless steel as a batch reactor, following the procedure have been reported previously [9]. The catalytic activity of Co/ZSM-5 also was tested. Prior to the catalytic reaction, the ZSM-5 catalyst must be first activated at a temperature of 500 °C with the aim to form cobalt oxide species which constitute the active site of the catalyst Co/ZSM-5. Furthermore, flushing was carried out using nitrogen gas to remove organic impurities, water vapour and excess oxygen in the reactor. The condition of the reaction was adapted from the work reported by Krisnandi *et al.* [9] and Bezniz *et al.* [11]. The first set of experiments was carried out at 150 °C with reaction time of 30 min and 60 min, and CH_4 pressure (Pmethane) of 0.75 Bar in 200 cm^3 stainless steel vessel. The results of this experiment are summarized in figure 6.

Figure 6 showed that hence the catalytic activity of ZSM-5 and Co/ZSM-5 from fly ash and rice husk was still inferior (7.08% and 12.43% conversion of methane) compared to the pro-analysis materials (28.33% and 37.65%) and commercial one (9.92% and 21.36%), they were potential to be used as catalysts in partial oxidation of methane to methanol. The activity of ZSM-5 and Co/ZSM-5from fly ash and rice husk ash are inferior compared to the pro-analysis materials because ZSM-5 from fly

Table 1. Surface area analyses of ZSM-5 catalyst calculated from BET measurement.

Sample	S BET (m^2/g)	S ext (m^2/g)	V tot (cm^3/g)	V meso (cm^3/g)	V micro (cm^3/g)	Average diameter pore (nm)
pZSM-5	294.75	104.44	0.594	0.495	0.099	8.11683
nZSM-5	43.759	43.759	0.287	0.287	-	26.2618

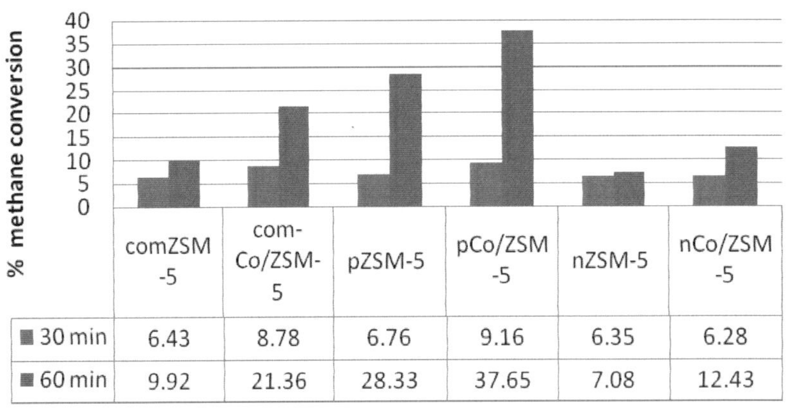

Figure 6. % methane conversion to methanol from reaction carried out at 150°C, t = 30 and 60 min and methane partial pressure of 0.75 bar.

ash and rice husk has lower surface area than ZSM-5 pro-analysis materials. The cobalt-loaded ZSM-5 catalysts gave higher conversion than the ZSM-5 without cobalt-loaded impregnated Co/ZSM-5, because it contains cobalt oxidic spesies at the zeolite outer surface, which becomes active sites that play important role in methane partial oxidation [11].

4. Conclusions

ZSM-5 zeolite catalysts from fly ash and rice husk have been prepared and their catalytic activities were tested in the partial oxidation of methane to methanol. The result showed that hence the catalytic activity of ZSM-5 and Co/ZSM-5 from fly ash and rice husk was still inferior (7.08% and 12.43% conversion of methane) compared to the pro-analysis sourced-counterpart (28.33% and 37.65%) and commercial one (9.92% and 21.36%), they were potential to be used as catalyst in the partial oxidation of methane to methanol.

Acknowledgments

This research was funded by Hibah PUPT BOPTN No. 0545/UN2.R12/HKP/05.00/2015 and BPPT SK No. No.211a/2014.

References

[1] Bukhari S S, Behin J, Kazemian H and Rohani S 2015 *Fuel* **140** 250-266
[2] Kouprianov V I 2002 *Fuel Process. Technol.* **76** 187-9
[3] Zacco A, Borgese L, Gianoncelli A, Struis R P W J, Depero L E and Bontempi E 2014 *Environmental Chem Lett.* **12** 153-75
[4] Zhao X S, Lu G Q and Zhu H Y 1997 *J. Porous Mater.* **4** 245-51
[5] Chareonpanich M, Namto T, Kongkachuichay P and Limtrakul J 2004 *Fuel Processing* **85** 1623-34
[6] Groen J C, Moulijin J A and Perz-Ramirez J 2005 *Microporous and Mesoporous Materials* **87** 153-61
[7] Panpa W and Jinawath S 2009 *Applied Catalyst B: Enviromnental* **90** 389-94
[8] Mohamed M M, Zidan F I and Thabet M 2008 *Microporous and Mesoporous Materials* **108** 193-203
[9] Krisnandi Y, Putra Bimo A P, Bahtiar M, Zahara, Abdullah M and Howe R F 2015 *Procedia Chemistry* **14** 508-15
[10] Naskar M K, Kundu D and Chatterjee M 2012 *J. American Ceramic Society* **95** 925-30
[11] Beznis N V, Weckhuysen B M and Bitter J H 2010 *Catalysis Lett.* **136** 52-6

Synthesis and characterization of noble metal nanocomposites: $Ag/Fe_3O_4/ZnO$ and $Ag/Fe_3O_4/CuO/ZnO$ for better photocatalytic activity under visible light irradiation

H Tju[1,2], S P Prakoso[2], A Taufik[1,2] and R Saleh[1,2]

[1]Department of Physics, Faculty of Mathematics and Natural Sciences Universitas Indonesia, Kampus UI Depok, Depok 16424, Indonesia
[2]Integrated Laboratory of Energy and Environment, Faculty of Mathematics and Natural Sciences Universitas Indonesia, Kampus UI Depok, Depok 16424, Indonesia

Corresponding author's e-mail: rosari.saleh@gmail.com

Abstract. Nobel metal such as silver (Ag) nanoparticles have been proven could enhance photocatalytic activity under visible light irradiation and prevent recombination of electron and hole. Fe_3O_4/ZnO and $Fe_3O_4/CuO/ZnO$ has been investigated in our previous study. So in this work, a magnetic $Ag/Fe_3O_4/ZnO$ and $Ag/Fe_3O_4/CuO/ZnO$ nanocomposites were successfully synthesized using sol-gel method. The as-synthesized products were characterized by X-ray diffraction and ultraviolet–visible (UV-Vis) spectroscopy. The results showed that the nanocomposites were the combination of the desired nanoparticles. From the UV-Vis absorption spectra, we found the surface plasmon resonance (SPR) to be around ~440 nm. Under visible light irradiation, the $Ag/Fe_3O_4/ZnO$ and $Ag/Fe_3O_4/CuO/ZnO$ nanocomposites exhibited much higher photocatalytic activity than the Fe_3O_4/ZnO and $Fe_3O_4/CuO/ZnO$. The effect of catalyst dosage and initial concentration of methylene blue (MB) were also tested. To understand the mechanism in photocatalytic activity, several scavengers were tested. The reusability study suggested that the prepared nanocomposites can still maintain the degradation efficiency after four cycles, showing great potential for water purification.

Keywords: $Ag/Fe_3O_4/CuO/ZnO$, Ag, photocatalytic, methylene blue.

1. Introduction

In recent years, the use of semiconductor photocatalysts in the field of environmental purification have been intensively studied. Many researchers have found that by using semiconductors, especially metal-oxide semiconductors materials, in photocatalytic experiments, could degrade the organic pollutants efficiently without causing secondary pollutants [1, 2]. Zinc oxide, a II-IV compound semiconductor with a wide band gap, has been widely used in photocatalytic experiments due to its high photocatalytic efficiency, photosensitivity, low cost, high stability, and environmental friendly [3, 4].

However, ZnO as a direct wide band-gap (~3.24 eV [5]) semiconductor is only responsive to ultraviolet (UV) light, which reduces its efficiency in visible light, and ZnO has a high recombination rate of electron–hole [3–5]. Those limitations hinder the practical photocatalytic applications of ZnO. Therefore to overcome these limitations, we could modified ZnO by using several methods such as doping [6], coupled with another semiconductor [7], and deposition of noble metal [5]. Our research group has reported the synthesis and characterization of coupled ZnO with another semiconductor, CuO and Fe_3O_4, which created a $Fe_3O_4/CuO/ZnO$ and also Fe_3O_4/ZnO nanocomposites [8, 9]. We found that coupling ZnO with another metal oxide could enhance the degradation efficiency of methylene blue (MB) under UV light irradiation [10]. However, the degradation efficiency in the visible light region is still low.

Coupling semiconductors with noble metals such as silver (Ag) has also attracted a lot attention. Silver (Ag) has good physical chemical properties and high electrical conductivity [11]. Furthermore, the presence of Ag nanoparticles in semiconductors could create Schottky junctions, which could hinder the recombination of electrons and holes [12]. Additionally, the surface plasma resonance (SPR) effect of Ag nanoparticles could further increase the degradation efficiency in the visible light region [13].

Herein, we investigate the coupling of Fe_3O_4/ZnO and Fe_3O_4/CuO/ZnO nanocomposites with silver (Ag) nanoparticles to create Ag/Fe_3O_4/ZnO and Ag/Fe_3O_4/CuO/ZnO nanocomposites. The as-synthesized nanocomposites are used as photocatalysts for the degradation of MB under visible light irradiation. The results indicate the increase of degradation efficiency, which may be due to the incorporation of Ag in the nanocomposites.

2. Experimental

2.1. Materials

Iron (II) sulfate heptahydrate ($FeSO_4{\cdot}7H_2O$, 99%), copper sulfate pentahydrate ($CuSO_4{\cdot}5H_2O$, 99%), zinc sulfate heptahydrate ($ZnSO_4{\cdot}7H_2O$, 99%), silver nitrate ($AgNO_3$), sodium hydroxide (NaOH), ethanol, ethylene glycol (EG), and methylene blue, were purchased from Merck. All chemical reagents (Merck) were of analytical grade and used without further purification.

2.2. Synthesis of nanomaterials

The preparation of Fe_3O_4/ZnO and Fe_3O_4/CuO/ZnO nanocomposites were discussed in the previous study [8–11]. With slight modification, the preparation of Ag/Fe_3O_4/ZnO nanocomposites are given as follows: $ZnSO_4{\cdot}7H_2O$ and NaOH were mixed in distilled water. This suspension was further stirred well under 80°C using a magnetic stirrer. Then an amount of Fe_3O_4 and Ag nanoparticles were dispersed into a solution of ethanol and distilled water with particular ratio were added into the suspension. After 2h stirring and heating, the mixed solution was collected, washed, aging, and then dried at 125°C for 1h under vacuum condition. The synthesis method for Ag/Fe_3O_4/CuO/ZnO is similar to that for Ag/Fe_3O_4/ZnO but an amount of CuO nanoparticles were added. We vary the amount of Ag in Fe_3O_4/ZnO and Fe_3O_4/CuO/ZnO by 5 wt.% and 25 wt.%, and the products were denoted as FZ–05 wt.% Ag, FZ–25 wt.% Ag, FCZ–05 wt.% Ag, and FCZ–25 wt.% Ag, where FZ and FCZ represent Fe_3O_4/ZnO and Fe_3O_4/CuO/ZnO nanocomposites. The materials were structurally characterized using X-ray diffractometer (Rigaku Miniflex 600), additionally the optical property was studied using UV-Vis spectroscopy.

2.3. Photocatalytic experiments

Methylene blue (MB) solution (20 mg/L) was used as a model of organic pollutants. In a typical photocatalytic experiment, 0.03 g of the as-synthesized sample was dispersed with vigorous stirring in 100 mL MB solution for 30 min under dark conditions to ensure adsorption–desorption equilibrium. Then the as-prepared solution was irradiated with visible light for 2 h. In every certain times interval (15 min), the absorbance of the MB solutions were measured using UV-Vis spectroscopy. The degradation efficiency was presented as C_t/C_0 versus irradiation time, where C_0 = initial concentration of MB before irradiation and C_t = concentration of MB after a certain irradiation time.

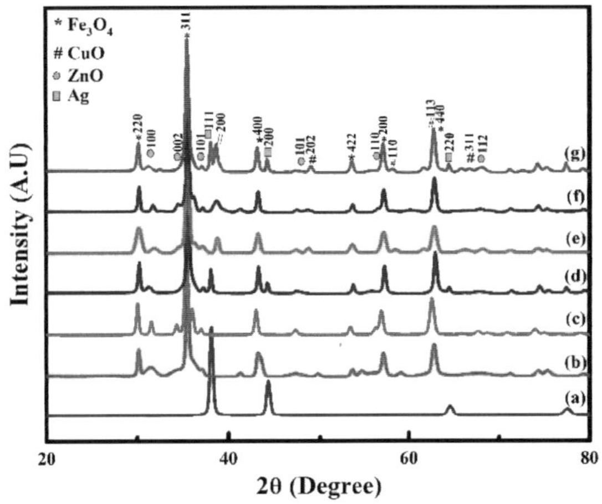

Figure 1. XRD patterns of (a) pure Ag, (b) FZ, (c) FZ-05 wt.% Ag, (d) FZ-25 wt.% Ag, (e) FCZ, (f) FCZ-05 wt.% Ag, and (g) FCZ-25 wt.% Ag.

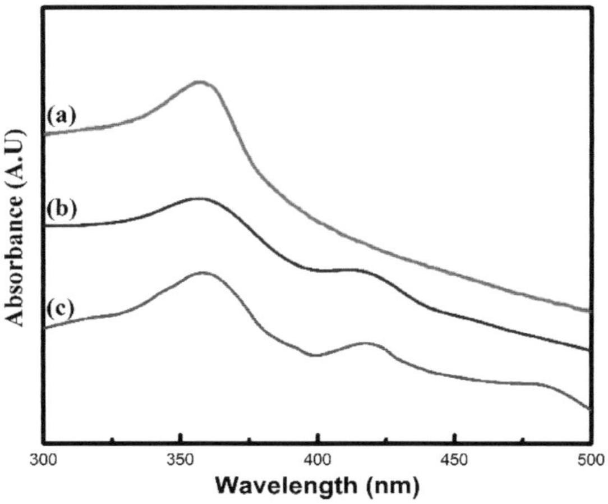

Figure 2. UV-Vis absorbance spectra of (a) FCZ, (b) FCZ-05 wt.% Ag, and (c) FCZ-25 wt.% Ag.

Table 1. Lattice Constant and grain size of FZ, FZ-05 wt.% Ag, FZ-25 wt.% Ag, FCZ, FCZ-05 wt.% Ag, and FCZ-25 wt.% Ag.

Samples	Lattice Constant (Å)							<D> (nm)			
	Ag	ZnO		CuO			Fe_3O_4				
	a=b=c	a=b	C	a	b	c	a=b=c	Ag	ZnO	CuO	Fe_3O_4
Fe_3O_4/ ZnO	–	–	–	–	–	–	8.374	–	25	–	42
Ag/F_3O_4/ZnO	4.085	3.242	5.214	–	–	–	8.349	13	20	–	39
Fe_3O_4/CuO/ZnO	–	3.242	5.214	4.692	3.425	5.136	8.374	–	26	15	42
Ag/Fe_3O_4/CuO/ZnO	4.085	3.242	5.214	4.692	3.425	5.136	8.374	13	18	15	40

3. Results and discussion

3.1. Structural and optical properties

As we can see in figure 1, the XRD patterns of all prepared nanocomposite samples exhibit a series peaks that can be characterized for cubic spinel Fe_3O_4 (2θ = 30.14°, 35.49°, 43.28°, 53.76°, 57.20°, and 62.83°), hexagonal wurtzite ZnO (2θ = 34.47°, 36.26°, 47.50°, 56.70°, 68.06°, and 69.21°), and monoclinic CuO (2θ = 38.81° and 48.7°) in the samples coupled by CuO. With the addition of 25 wt.% Ag in Fe_3O_4/ZnO and Fe_3O_4/CuO/ZnO, new diffraction peaks were found at 2θ = 38.14°, 44.26°, 64.42°, and 77.44°, which can be assigned to the (111), (200), (220), and (311) planes of face-center-cubic (fcc) Ag. However, no new diffraction peaks could be found on addition of 5 wt.% Ag, which may be due to the low Ag content [14]. Absence of another peaks in the patterns indicated no impurities in prepared samples. The XRD results confirm the presence of Fe_3O_4, ZnO, CuO, and Ag nanoparticles in the FZ-Ag and FCZ-Ag nanocomposites. The grain size and lattice parameters of all prepared samples were analyzed using the Debye–Scherrer equation and Rietveld refinement model; the results are shown in table 1.

Figure 2 shows the UV-Vis absorbance spectra of FCZ, FCZ–05 wt.% Ag, and FCZ–25 wt.% Ag. As can be seen, the absorbance spectrum of the sample with the incorporation of Ag nanoparticles displays typical SPR around 400-450 nm [15], indicating the presence of Ag in the prepared samples.

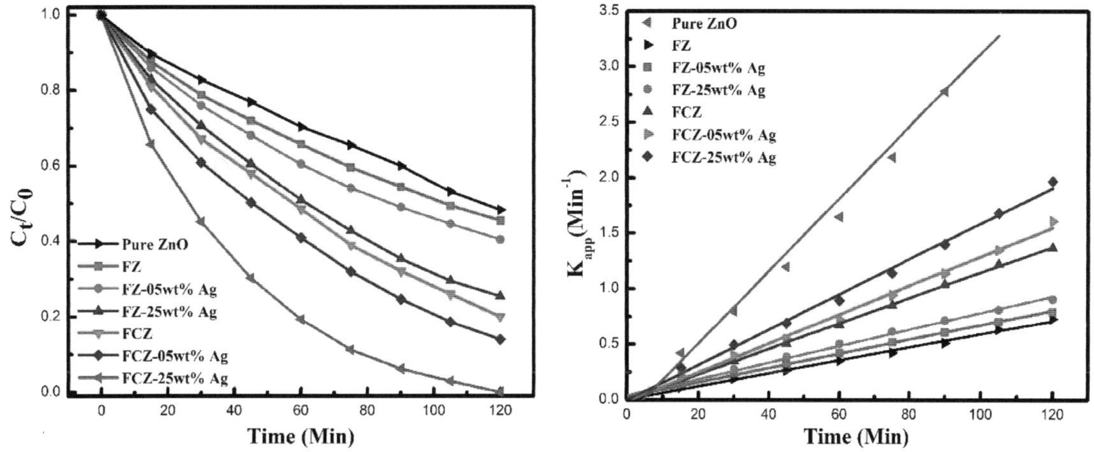

(a) (b)

Figure 3. The degradation of MB in the presence of different photocatalyst and pseudo first order linear plots ln C_0/C_t vs irradiation time.

3.2. Photocatalytic performance

To study the photocatalytic performance of all prepared samples, MB was used as an organic pollutant. The photocatalytic experiment was tested under visible light irradiation. The MB solution containing the catalyst was kept in the dark for 30 min to reach adsorption–desorption equilibrium. As shown in figure 3a, after 120 mins the degradation efficiency of MB reached about 51%, 55%, 60%, 75%, 80%, 86%, and 100%, respectively, for pure ZnO, FZ, FZ–05 wt.% Ag, FZ–25 wt.% Ag, FCZ, FCZ–5 wt.% Ag, and FCZ–25 wt.% Ag. The incorporation of Ag nanoparticles into Fe_3O_4/ZnO and $Fe_3O_4/CuO/ZnO$ nanocomposites was found to enhance the degradation efficiency under visible light irradiation. And with the increased amount of Ag, the degradation was further increased.

The pseudo first-order kinetic rate reaction is expressed by equation (1) bellow:

$$\ln\left(\frac{C_0}{C_t}\right) = kt + A \tag{1}$$

where C_0 is the initial concentration at $t = 0$, C_t is the concentration at t, t is irradiation time, and k is the pseudo first-order kinetic rate constant. The kinetics of all prepared samples are presented in figure 3b. The k values of FZ and FCZ increased with the addition of Ag nanoparticles. The rate order was determined as follows : FCZ–25 wt.% Ag → FCZ–05 wt.% Ag → FCZ → FZ–25 wt.% Ag → FZ–05 wt.% Ag → FZ → pure ZnO. The FCZ–25 wt.% Ag displays the best photocatalytic performance, with a k value of 0.03261. These results further proof the increased in catalytic performance with the introduction of Ag in the nanocomposites.

In order to obtain the optimum catalyst dosage (FZ–25 wt.% Ag and FCZ–25 wt.% Ag), photocatalytic experiments were carried out by varying the dosage of catalyst from 0.01g to 0.04g with the concentration of MB = 20 mg/L. The results are shown in figure 4a; with the increased amount of catalyst from 0.01 g to 0.03 g, the degradation efficiency was also increased, but further increment in catalyst dosage lead to a decrease in degradation efficiency. The increase in degradation efficiency may be due to the increased number of active sites, which increases the formation of oxygen species [13]. However, further increase of dosage could result in severe catalyst aggregation, which leads to light scattering and reduced light penetration through the MB solution [16].

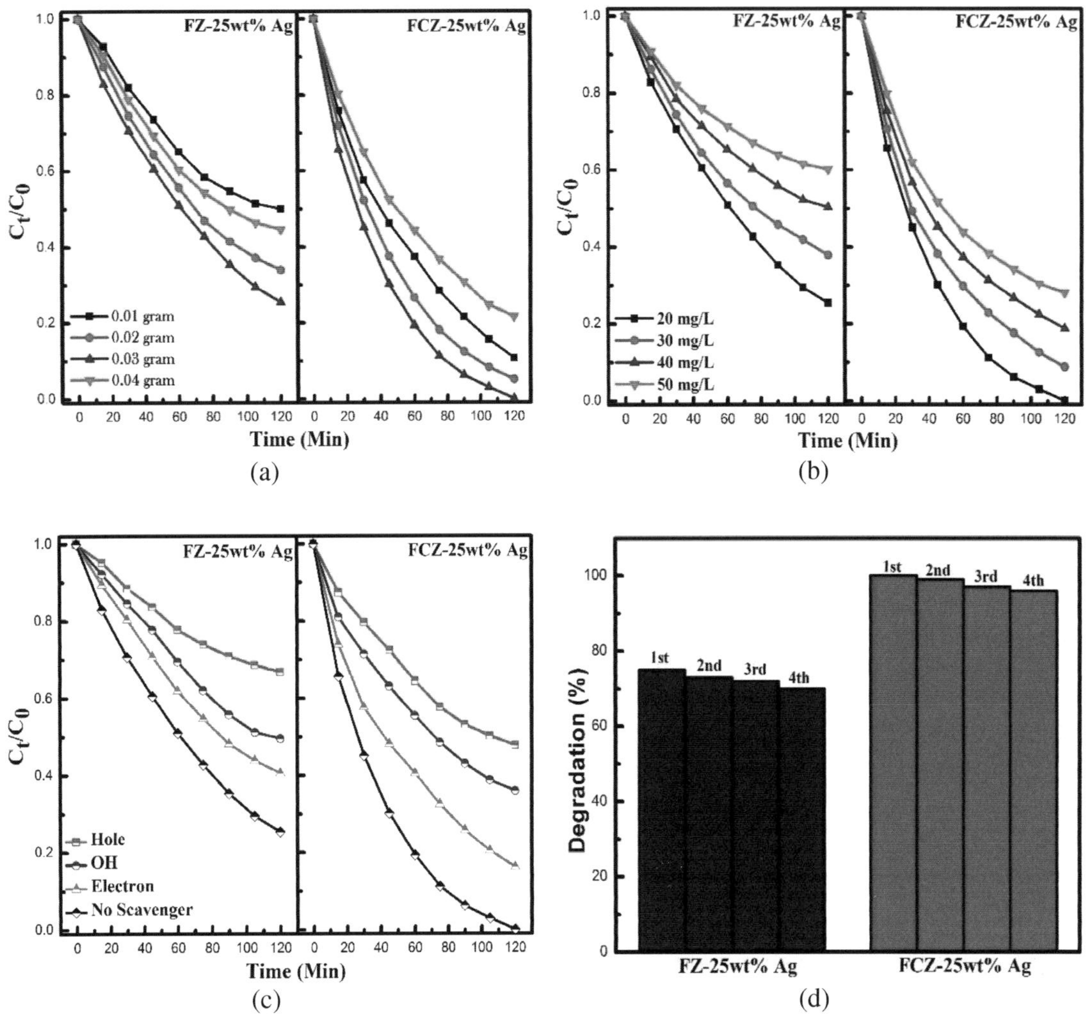

Figure 4. (a) The effect of dosage on the degradation efficiency of MB, (b) effect of initial concentration of MB on the degradation efficiency ([Catalsyt] = 0.03 g), (c) effect of various scavenger and (d) the recyling experiment.

The effect of initial MB concentration on photocatalytic degradation was also examined in the presence of 0.03 g FZ–25 wt.% Ag and FCZ–25 wt.% Ag, varying the concentration of MB from 20 mg/L to 50 mg/L. The results (figure 4b) indicate that with the increase of the MB initial concentration decreased the degradation efficiency. The decreased in degradation efficiency may due to the shielding of light by dye molecules [17].

To further understand the mechanism in the photocatalytic activity process, we controlled the experiment by adding several trapping agents. In this experiment, sodium sulfate, di-ammonium oxalate, and tert-butyl alcohol were used to trap electrons, holes, and OH, respectively. Figure 4c shows that the addition of any trapping agent caused the degradation efficiency of MB to decrease by several percent. With the addition of di-ammonium oxalate (hole scavenger) the degradation efficiency was greatly inhibited; therefore, it can be inferred that the main active species in the photocatalytic experiment with FZ–25 wt.% Ag and FCZ–25 wt.% Ag nanocomposites are holes followed by OH and electrons.

Separation and reusability of the nanocomposite materials are also important issues. To examine reusability, the prepared nanocomposites were separated from the MB solution using an external magnet and reused for a second, third, and fourth photocatalytic cycle. In figure 4d we can see that the degradation efficiency is not significantly decreased after four cycles. This result shows that FZ–25 wt.% Ag and FCZ–25 wt.% Ag show great promise for use in industrial processes.

The enhanced photocatalytic activity of Ag/Fe_3O_4/ZnO and Ag/Fe_3O_4/CuO/ZnO nanocomposites under visible light irradiation is may caused by SPR of the Ag nanoparticles [12, 18], which was confirmed using UV-Vis absorbance spectroscopy (figure 2). The obtain main active species, holes, may due to the fact that hole could interact directly to the dyes pollutants whereas the electrons are trapped in metal region (Ag), in which SPR effect taken places [12, 15, 17].

4. Conclusions

Ag/Fe_3O_4/ZnO and Ag/Fe_3O_4/CuO/ZnO nanocomposites with different added amounts of Ag (5 and 25 wt.%) were successfully synthesized by using sol-gel method. The XRD pattern exhibited all desired peaks, confirm the existence of Ag, Fe_3O_4, CuO, and ZnO. UV-Vis spectroscopy could confirm the existence of SPR effect. These prepared samples exhibit highly efficient photocatalytic activity for the degradation of MB in aqueous solution under visible light irradiation. The maximum degradation efficiency and k values follow the order FCZ–25 wt.% Ag > FCZ–05 wt.% Ag > FCZ > FZ–25 wt.% Ag > FZ–05 wt.% Ag > FZ > pure ZnO. The maximum photocatalytic degradation is achieved with a reaction time of 120 min, catalyst dosage of 0.03 g, and initial concentration of 20 mg/L. The main active species follow the order: hole > OH > electron. In addition, the photocatalysts are found to be stable without loss of activity after four cyclic experiments.

References

[1] Wang Y, Shi R, Lin J and Zhu Y 2010 *Appl. Catal.* B **100** 179-83
[2] Khataee A, Sheydaei M, Hassani A, Taseidifar M and Karaca S 2015 *Ultrason. Sonochem.* **22** 404-11
[3] Cozzoli P D, Fanizza E, Comparelli R, Curri M L, Agostiano A and Laub D 2004 *J. Phys. Chem.* B **108** 9623-30
[4] Xu L P, Hu Y L, Pelligra C, Chen C H, Jin L, Huang H, Sithambaram S, Aindow M, Joesten R and Suib S L 2009 *Chem. Mater.* **21** 2875-85
[5] Mun S Lam, Chung J Sin, Zuhairi A Abudllah and Rahman A Mohamed 2014 *Sep. Purif. Technol.* **132** 378-87
[6] Saidani T, Zaabat M, Aida M S and Boudine B 2015 *Superlattices Microstruct.* **88** 315-22
[7] Xu T, Zhang L, Cheng H and Zhu Y 2011 *Appl. Catal.* B **101** 382-7
[8] Taufik A, Susanto I K and Saleh R 2015 *Mater. Sci. Forum* **827** 37-42
[9] Susanto I K, Taufik A and Saleh R 2015 *Adv. Mater. Res.* **1123** 227-32
[10] Taufik A, Tju H and Saleh R 2016 *J. Phys. Conf. Ser.* **710** 012004
[11] Yang Y, Li H, Hou F, Hu J, Zhang X and Wang Y 2016 *Mater. Lett.* **180** 97-100
[12] Zhang X, Wang L, Liu C, Ding Y, Zhang S, Zeng Y, Liu Y and Luo S 2016 *J. Hazard. Mater.* **313** 244-52
[13] Xia D, Hu L, Tan X, He C, Pan W, Yang T, Huang Y and Shu D 2016 *Appl. Catal.* B **185** 295-306
[14] Zhang X, Wang Y, Hou F, Li H, Yang Y, Hang X, Yang Y and Wang Y 2016 *Appl. Surf. Sci.* **391** 476-83
[15] Patil S S *et al.* 2016 *Catal. Today* **260** 126-34
[16] Tao H, Liang X, Zhang Q and Chang C 2015 *Appl. Surf. Sci.* **324** 258-64
[17] Mohamed R M and Aazam E S 2014 *J. Ind. Eng. Chem.* **20** 4377-81
[18] Chen J, Che H, Huang K, Liu C and Shi W 2016 *Appl. Catal.* B **192** 134-44

Theoretical investigation on the magnetization enhancement of Fe_3O_4-reduced graphene oxide nanoparticle system

M A Majidi[1], Y Wicaksono[1], A D Fauzi[1], A Taufik[1], R Saleh[1] and A Rusydi[2,3]
[1]Department of Physics, Faculty of Mathematics and Natural Sciences Universitas Indonesia, Kampus UI Depok, Depok 16424, Indonesia
[2]NUSNNI-NanoCore, Department of Physics, Faculty of Science, National University of Singapore, Singapore 117576, Singapore
[3]Singapore Synchrotron Light Source, National University of Singapore, Singapore 117603, Singapore

Corresponding author's email: aziz.majidi@sci.ui.ac.id

Abstract. We present a theoretical study on the enhancement of magnetization of Fe_3O_4 nanoparticle system upon addition of reduced graphene oxide (rGO). Experimental data have shown that the magnetization of Fe_3O_4-rGO nanoparticle system increases with increasing rGO content up to about 5 wt%, but decreases back as the rGO content increases further. We propose that the enhancement is due to spin-flipping of Fe ions at the tetrahedral sites assisted by oxygen vacancies at the Fe_3O_4 particle boundaries. These oxygen vacancies are induced by the presence of rGO flakes that adsorb oxygen atoms from Fe_3O_4 particles around them. To understand the enhancement of the magnetization, we construct a tight-binding based model Hamiltonian for the Fe_3O_4 nanoparticle system with the concentration of oxygen vacancies being controlled by the rGO content. We calculate the magnetization as a function of the applied magnetic field for various values of rGO wt%. We use the method of dynamical mean-field theory and perform the calculations for a room temperature. Our result for rGO wt% dependence of the saturated magnetization shows a very good agreement with the existing experimental data of the Fe_3O_4-rGO nanoparticle system. This result may confirm that our model already carries the most essential idea needed to explain the above phenomenon of magnetization enhancement.

1. Introduction

Fe_3O_4, informally known as magnetite, is a magnetic material with ferrimagnetic ordering. Due to its half-metallic characteristic with 100% spin polarization at the Fermi level [1, 2] and high Curie temperature ($\sim$ 851 K), magnetite becomes a prospective material for spintronic devices [3–5]. To some extent, Fe_3O_4 nanoparticles still preserve the magnetic properties as those of bulk Fe_3O_4. Due to high Curie temperature and biocompatibility [6], Fe_3O_4 nanoparticle system also becomes promising for use in photo-catalysis and biomedical applications [7, 8].

A recent study by Taufik *et al.* suggests that the magnetization of Fe_3O_4 nanoparticle system enhances when mixed with reduced graphene oxide (rGO) [9]. The mechanism behind this magnetization enhancement has still not been fully understood. However, a hint to explanation of this phenomenon may be given by another recent experimental study by Herng *et al.*, suggesting that the magnetization enhancement occurs due to spin flipping of Fe ions at tetrahedrally coordinated sites, which is suspected

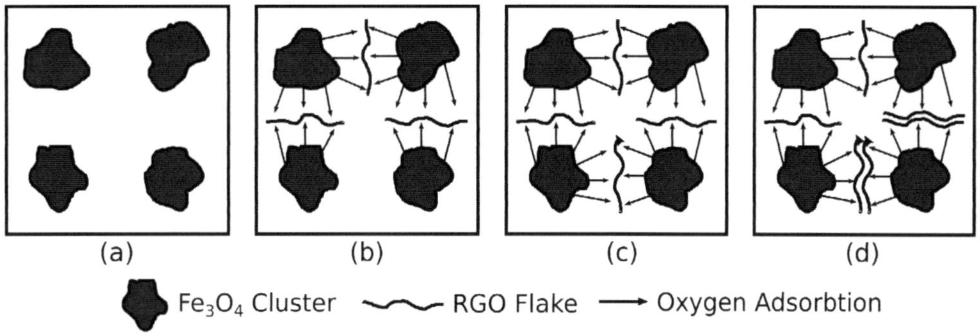

Figure 1. The system of Fe_3O_4 nanoparticle clusters (a) without additional rGO, (b) with additional rGO less than 5 wt%, (c) with additional rGO of $\sim$ 5 wt% and (d) with additional rGO more than 5 wt%

to be induced by oxygen vacancies. These spin flippings lead the system to transform from ferrimagnetic to ferromagnetic ordering [10]. Motivated by that experimental hint, we develop the idea and propose a theoretical study to explain the magnetization enhancement phenomenon of the Fe_3O_4-rGO nanoparticle system. We hypothesize that the magnetization enhancement of Fe_3O_4-rGO nanoparticle system is induced by oxygen vacancies formed on the surface of Fe_3O_4 nanoparticle clusters due to the presence of rGO flakes nearby the Fe_3O_4 clusters. This argument is based on a previous experimental study by Santoso *et al.* [11], supported by a theoretical study by Majidi *et al.* [12], indicating that graphene can easily adsorb oxygens. Our main goal is to calculate magnetization of Fe_3O_4 nanoparticle system with and without oxygen vacancies, and use this to mimic the experimental data of magnetization of the combined Fe_3O_4-rGO nanoparticle system by connecting the concentration of oxygen vacancies with the rGO content.

2. Model of Fe_3O_4-rGO nanoparticle system

The experimental data of reference [9] shows that the magnetization of Fe_3O_4-rGO nanoparticle system increases with increasing rGO content up to about 5 wt%, but decreases back as the rGO content increases further. We propose a model to explain this situation as illustrated in figure 1. When there is no rGO content (figure 1(a)), all the Fe_3O_4 clusters are assumed to have no oxygen vacancies. As the system is added with rGO content less than 5 wt% (figure 1(b)), rGO flakes start to fill spaces in between neighboring Fe_3O_4 clusters. At this low concentration of rGO, we suppose that each space between two neighboring Fe_3O_4 clusters can only be filled by at maximum of one layer of rGO flake. The insertion of an rGO flake induces the extraction of oxygen atoms from the nearby Fe_3O_4 clusters, which then get adsorbed onto the rGO surface. As a result, oxygen vacancies are formed on the surfaces of Fe_3O_4 clusters, inducing spin flippings of the Fe ions at tetrahedral sites in those clusters, thus increasing the total magnetization of the whole system by some amount. We suppose that when the rGO content is $\sim$ 5 wt% (figure 1(c)), each space between neighboring Fe_3O_4 clusters has been filled with one rGO layer. At this rGO concentration, the adsorption of oxygen atoms by the rGO flakes is optimum, hence the enhancement of magnetization of the system becomes maximum. When the rGO content is exceeding 5 wt% (figure 1(d)), there are some spaces between Fe_3O_4 clusters that are filled with rGO flakes of more than one layer. The increase of number of layers of an rGO flake does not increase its capability of adsorbing oxygen atoms since the area of its open surfaces remains the same. However, the addition of more rGO content without more formation of oxygen vacancies in Fe_3O_4 clusters causes the total magnetization per unit mass of the system to decrease. Therefore, adding rGO flakes of more than 5 wt% decreases the total magnetization per unit mass of the system monotonically.

The rGO content added to Fe_3O_4 nanoparticles is experimentally measured using weight percent unit (wt%). This weight percentage can be understood as ratio of the total mass of rGO flakes to the total mass of Fe_3O_4 clusters. Mathematically, we can formulate it as follows

$$x = \frac{N_G M_G}{N_F M_F}, \tag{1}$$

where N_F and N_G, respectively, are the number of Fe_3O_4 clusters and rGO layers in the system, M_F and M_G are the averaged mass of Fe_3O_4 clusters and the averaged mass of rGO layers, respectively, while x is wt% of rGO layers added to the Fe_3O_4 nanoparticles.

Now, let the mass ratio of Fe_3O_4 clusters to rGO flakes be α. As an Fe_3O_4 cluster is likely heavier than an rGO layer, α should be larger than unity. In our model, a space between neighboring Fe_3O_4 clusters can be filled by an rGO flake or one or more layers. How rGO layers fill such spaces affects the magnetization enhancement of the system. Hence, we include this into our model. Let n be the maximum number of rGO layers filling a space between two Fe_3O_4 clusters while the system still alows a vacant space. This means, no space will get filled with an rGO flake of more than n layers as long as there is still a vacant space. Hence, ratio of the total number of rGO layers in the system to the maximum number of rGO layers that can be accomodated by the system with no vacant spaces can be formulated as

$$\beta \equiv \frac{N_G}{N_{G_{max}}} = \frac{N_G}{nN_F} = \frac{x\alpha}{n}. \tag{2}$$

Based on the definition of n stated above, the probability of finding a space between two Fe_3O_4 clusters with no rGO flake is

$$P_0 = \begin{cases} (1-\beta)^n, & \text{if } \beta < 1 \\ 0, & \text{if } \beta \geq 1, \end{cases} \tag{3}$$

while the probability of finding an rGO flake containing one or more layers between two Fe_3O_4 clusters is $1 - P_0$. When rGO flakes are added to the Fe_3O_4 nanoparticles with random distribution, we expect that the resulting net magnetization of the system is contributed by the magnetic moments of the unit cells inside of Fe_3O_4 clusters that have no oxygen vacancies and those on the surfaces of the Fe_3O_4 clusters some of which have oxygen vacancies. For simplicity, let us assume that each Fe_3O_4 cluster has the same number of unit cells making up the total mass of M_F for each cluster. Likewise, let us assume each rGO layer contains the same number of graphene unit cells making up the total mass of M_G for each layer. Further, let μ_0 be the magnetic moment of an Fe_3O_4 cluster with no oxygen vacancies, and μ_v be the magnetic moment of an Fe_3O_4 cluster assuming all the Fe_3O_4 unit cells in it has an oxygen vacancy. Lastly, let γ be the portion of unit cells in an Fe_3O_4 cluster that have oxygen vacancies. Thus, we may construct the relation between the total magnetization $\mathscr{M}$, that depends on the rGO content x, with the quantities just described as

$$\mathscr{M}(x) = \frac{P_0 N_F \mu_0 + (1-P_0) N_F \gamma \mu_v}{N_F M_F + N_G M_G}, \tag{4}$$

with $\mathscr{M}_0 = \frac{\mu_0}{M_F}$. By dividing both the numerator and the denominator with $N_F M_F$, the equation can be rearranged to be

$$\mathscr{M}(x) = \frac{P_0 \mathscr{M}_0 + (1-P_0)\gamma \mathscr{M}_v}{1 + n\frac{\beta}{\alpha}}. \tag{5}$$

To use equation (5), we first specify the n value corresponding to the model we choose. Next, for every x value we take β according to equation (2), and then P_0 according to equation (3). Along with n, γ and α can be adjusted to obtain the best fit to the experimental data. As for $\mathscr{M}_0$, we take the empirical value of the saturation magnetization from the experimental data, that is ~ 80 emu/g. Whereas, $\mathscr{M}_v$ is the theoretical value of the saturation magnetization if spin flipping occurs in all unit cells in the

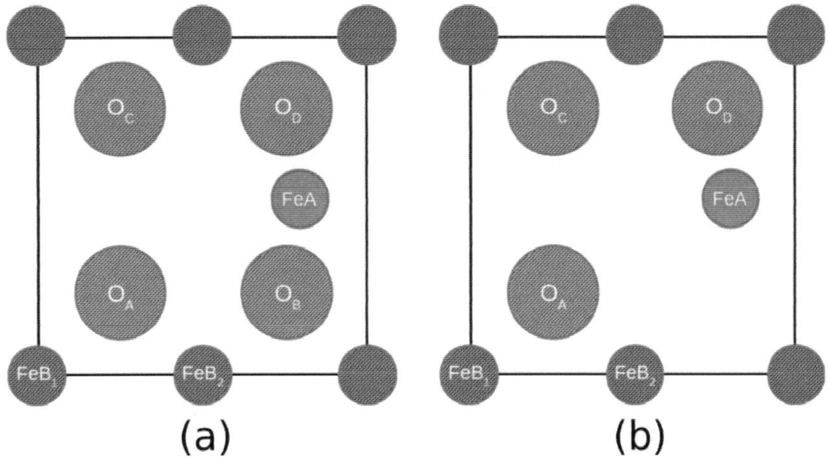

Figure 2. Model unit cell of Fe_3O_4 (a) without oxygen vacancy and (b) with oxygen vacancy.

cluster induced by oxygen vacancies, that is $\mathscr{M}_V = 3\mathscr{M}_0$. The expected realistic value of the saturation magnetization in an Fe_3O_4 cluster if spin flipping occurs in some portion, γ, of all unit cells on the surface of the cluster, would be $\gamma\mathscr{M}_V$.

3. Model to calculate the magnetization as a function of temperature and external magnetic field

To calculate the magnetization of the system as a function of temperature we model the Fe_3O_4 nanoparticle cluster as if it is a bulk system. The model has been proposed in our recent study [13]. Following reference [13], we use (001) Fe_3O_4 surface and simplify the crystal structure by taking only the bottom-most two-layers of the actual unit cell. The half metallic property along the (001) Fe_3O_4 surface has been confirmed by first-principle calculation [14]. We further simplify the unit cell by restricting only the square block containing one unit of Fe_3O_4 as shown in figure 2. With this model unit cell, for the system without oxygen vacancies (referring to figure 2(a)), we choose 11 basis atomic orbitals to construct our Hilbert space, that we order as follows: $|O_A - p_x\rangle$, $|O_A - p_y\rangle$, $|O_B - p_x\rangle$, $|O_B - p_y\rangle$, $|O_C - p_x\rangle$, $|O_C - p_y\rangle$, $|O_D - p_x\rangle$, $|O_D - p_y\rangle$, $|FeA - e_g\rangle$, $|FeB_1 - e_g\rangle$ and $|FeB_2 - e_g\rangle$. The 11 basis orbitals along with the two spin orientations make up a set of 22 basis states. Whereas, for the system with oxygen vacancies, we remove one oxygen atom, i.e. O_B, hence removing also $|O_B - p_x\rangle$ and $|O_B - p_y\rangle$ orbitals, leaving only 9 basis orbitals or 18 basis states remaining. Using the above sets of basis states, we propose our model Hamiltonian as

$$H = \sum_{\mathbf{k}} [a_{\mathbf{k}}]^{\dagger} [H_0(\mathbf{k})] [a_{\mathbf{k}}] - J_H \sum_i \mathbf{S}_i . \mathbf{s}_i + U \sum_i n_{\uparrow,i} n_{\downarrow,i}. \quad (6)$$

The first term in the Hamiltonian is the kinetic term which is constructed using tight-binding approximation. Here, $[a_{\mathbf{k}}]$ is a column vector containing the corresponding fermion annihilation operators, while $[a_{\mathbf{k}}]^{\dagger}$ is its hermitian conjugate, with $\mathbf{k}$ index denotes the $\mathbf{k}$-points in the Brillouin zone. $[H_0(\mathbf{k})]$ is a 22×22 (18×18) matrix in momentum space for the system without (with) oxygen vacancies, constructed using tight-binding approximation, which consists of four 11×11 (9×9) blocks corresponding to their spin directions as

$$[H_0(\mathbf{k})] = \begin{bmatrix} H_0(\mathbf{k})_{\uparrow} & 0 \\ 0 & H_0(\mathbf{k})_{\downarrow} \end{bmatrix}. \quad (7)$$

See reference [13] for the detailed content of $[H_0(\mathbf{k})]$ for both systems without and with oxygen vacancies. The second term represents the magnetic exchange interaction between the collective spin

S of the t_{2g} electrons and the spin **s** of an e_g electron in each Fe ion, with i denoting the site of the Fe ion, and J_H being Hund's coupling constant. The last term represents the Coulomb repulsive interaction between two electrons occupying the same Fe-e_g orbital, with $n_{i\sigma}$ being the electron occupation number operator at site i with spin component σ,and U the Hubbard parameter. To simplify our calculations, we replaced this term with the mean-field approximated version, that is

$$U\sum_i n_{\uparrow,i}n_{\downarrow,i} = U\sum_i \langle n_{i\uparrow}\rangle n_{i\downarrow} + \langle n_{i\downarrow}\rangle n_{i\uparrow} - \langle n_{i\uparrow}\rangle\langle n_{i\downarrow}\rangle, \tag{8}$$

with $\langle n_{i\sigma}\rangle$ to be calculated self-consistently.

4. Calculation Methods

We use dynamical mean-field theory (DMFT) algorithm to calculate magnetization of the system. We begin the algorithm by defining the Green function matrix for our system as

$$[G(\mathbf{k},z)] = [z[I] - [H_0(\mathbf{k})] - [\Sigma(z)]]^{-1}, \tag{9}$$

where z is the complex frequency variable defined as $z = \omega + i0^+$ for the real-frequency domain and $z = i\omega_n + \mu$ for the Matsubara-frequency domain, with $\omega_n = (2n+1)\pi T$ being fermion Matsubara frequency, T temperature and μ chemical potential. Note that throughout this paper, we set $\hbar = k_B = 1$. Here, the self-energy matrix $[\Sigma(z)]$, along with $\langle n_{i\sigma}\rangle$, are taken from initial guess. We then average the Green function matrix over all **k**-points in the Brillouin zone as

$$\overline{[G(z)]} = \frac{1}{N}\sum_{\mathbf{k}}[G(\mathbf{k},z)], \tag{10}$$

where N is the number of **k**-points in the Brillouin zone. Next, we extract the mean-field Green function matrix through

$$[G(z)]_{\mathrm{MF}} = \left[\overline{[G(z)]}^{-1} + [\Sigma(z)]\right]. \tag{11}$$

We use this mean-field Green function matrix to calculate the local interacting Green function matrix that depends on the orientation of the three Fe spins

$$[G(z,\theta_1,\theta_2,\theta_3)]_{\mathrm{loc}} = \left[[G(z)_{\mathrm{MF}}]^{-1} - [\Sigma(\theta_1,\theta_2,\theta_3)]_{\mathrm{loc}}\right]^{-1}, \tag{12}$$

where $[\Sigma(\theta_1,\theta_2,\theta_3)]_{\mathrm{loc}}$ is the local self-energy matrix manifesting the dynamics of the Coulomb repulsive interactions and magnetic exchange interactions at a local unit cell, containing the dependence on the angles θ_1,θ_2 and θ_3, of the three Fe spins with respect to the direction of net magnetization (taken as the direction of the positive z axis), and the averaged occupancies of electrons at Fe-e_g orbitals, $\langle n_{i\sigma}\rangle$. The datailed form of $[\Sigma(\theta_1,\theta_2,\theta_3)]_{\mathrm{loc}}$ can be seen in reference [13].

Now $[G(z,\theta_1,\theta_2,\theta_3)]_{\mathrm{loc}}$ must be averaged over all possible orientations of the three Fe spins in a unit cell. The averaging process is implemented using the Boltzmann weight defined as

$$P(\theta_1,\theta_2,\theta_3) = \frac{1}{Z}e^{-S_{\mathrm{eff}}+hS(\cos\theta_1+\cos\theta_2+\cos\theta_3)/T}, \tag{13}$$

where

$$S_{\mathrm{eff}} = -\sum_n \ln\det\left[[G_{\mathrm{MF}}(i\omega_n)][G^{-1}_{\mathrm{loc}}(i\omega_n,\theta_1,\theta_2,\theta_3)]\right]e^{i\omega_n 0^+}, \tag{14}$$

is the effective action, which is calculated only in Matsubara-frequency domain and

$$Z = \int\int\int d\theta_1 d\theta_2 d\theta_3 e^{-S_{\text{eff}}+hS(\cos\theta_1+\cos\theta_2+\cos\theta_3)/T}, \tag{15}$$

serves as the partition function. The term $hS(\cos\theta_1 + \cos\theta_2 + \cos\theta_3)/T$ is added to accomodate the effect of external magnetic field h (defined in unit of energy) at a temperature T. Thus, the averaged interacting Green function matrix is obtained through

$$G_{\text{ave}}(z) = \int\int\int d\theta_1 d\theta_2 d\theta_3 \left[G_{\text{loc}}(z,\theta_1,\theta_2,\theta_3)\right] P(\theta_1,\theta_2,\theta_3). \tag{16}$$

This averaging process is also used to calculate the magnetization through

$$M = \int\int\int P(\theta_1,\theta_2,\theta_3) S(\cos\theta_1 + \cos\theta_2 + \cos\theta_3)\, d\theta_1 d\theta_2 d\theta_3. \tag{17}$$

The new self-energy matrix can then be extracted as

$$[\Sigma(z)] = [G_{\text{MF}}(z)]^{-1} - [G_{\text{ave}}(z)]^{-1}. \tag{18}$$

Along with the self-energy matrix, we also need to update the chemical potential, μ, by imposing a fixed electron filling per unit cell, that is $n_{\text{filling}} = 20$ for the system without oxygen vanancies, such that

$$n_{\text{filling}} = \int d\omega\, \text{DOS}(\omega) f(\omega,\mu,T), \tag{19}$$

where

$$\text{DOS}(\omega) = -\frac{1}{\pi}\text{ImTr}[G_{\text{ave}}(\omega + i0^+)], \tag{20}$$

is the electron density of states, and $f(\mu,\omega,T) = (e^{(\omega-\mu)/T} + 1)^{-1}$ is the Fermi-Dirac distribution function. The calculated chemical potential is then used to update the averaged electron occupancy of each e_g orbital of Fe ion at site i in a unit cell with spin component σ through

$$\langle n_{i\sigma}\rangle = \int d\omega\, \text{PDOS}_{i\sigma}(\omega) f(\omega,\mu,T), \tag{21}$$

with $\text{PDOS}_{i\sigma}(\omega) = -(1/\pi)\ \text{Im}\ G_{\text{ave}\ i\sigma}(\omega + i0^+)$ being the corresponding projected density of states. Finally, we iterate the algorithm using the updated self energy matrix, chemical potential and averaged electron occupancy, and the self-consistency process is repeated until convergence is achieved.

5. Results and discussion

Since in this paper we use DMFT to obtain mainly the magnetization as a function of temperature and magnetic field, we restrict our calculations only for the system with oxygen vacancies. While, for the system with oxygen vacancies, we take for granted the result of reference [13] from which the system with oxygen vacancies has a ferromagnetic ordering with magnetization being three times greater than that of the system without oxygen vacancies that has a ferrimagnetic ordering. Further, we normalize the magnetization such that the saturation magnetization of the pure Fe_3O_4 nanoparticle system is ~ 80 emu/g, as obtained from the experiment. For our calculations we use the following parameters: $\varepsilon_{O_A} = \varepsilon_{O_C} = -1.5$ eV, $\varepsilon_{O_B} = \varepsilon_{O_D} = 1.8$ eV, $\varepsilon_{\text{FeA}} = 0.5$ eV, $\varepsilon_{\text{FeB}_1} = \varepsilon_{\text{FeB}_2} = 0.4$ eV, $t_{\text{O-O}_{(\text{near})}} = 0.4$ eV, $t_{\text{O-O}_{(\text{far})}} = 0.3$ eV, $t_{\text{FeA-O}} = 1.8$ eV, $t_{\text{FeB}_1\text{-O}} = 0.9$ eV, $t_{\text{FeB}_2\text{-O}} = 0.9$ eV, $t_{\text{FeB}_1\text{-FeB}_2} = 0.825$ eV, $t_{\text{FeA-FeB}_1} = t_{\text{FeA-FeB}_2} = 1.8$ eV, $J_H = 1$ eV, $U = 4$ eV and $n_{\text{filling}} = 20$ electrons per unit cell.

Figure 3 shows our result of spin-resolved density of states (DOS) for a room temperature (~ 300 K). The density of states shows that our model reveals its half-metallic characteristic, where in the majority spin channel, the chemical potential (or the Fermi level) lies inside of the band, while in the minority spin channel, the chemical potential lies outside the band. This result demonstrates that our model captures

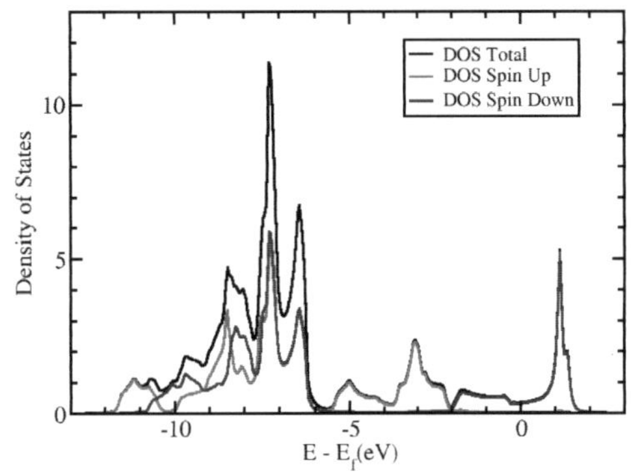

Figure 3. Density of states (DOS) of our Fe_3O_4 model at a room temperature showing the half metallic characteristic.

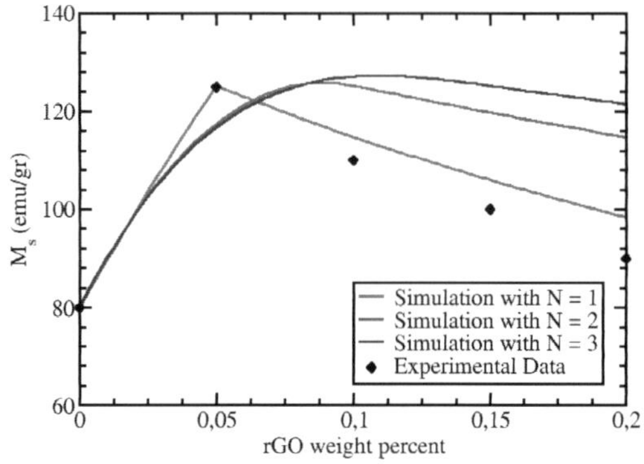

Figure 4. Saturation magnetization of Fe_3O_4-rGO nanoparticle system with rGO wt% variation, for $\alpha = 20$ and $\gamma = 0.574$.

one of the most relevant characteristics of Fe_3O_4 as observed in experiment [2]. Having this result, we argue that our model is adequate to be used for further calculations.

In calculating the saturation magnetization as a function of rGO wt% (x), we tune the parameters α, γ and n, and find that the best fit to experimental data is obtained using $\alpha = 20$, $\gamma = 0.574$ and $n = 1$. Figure 4 shows our results of the saturation magnetization for $n = 1, 2$, and 3, as compared to the experimental data. By varying n, we can track how the distribution of number of layers of the rGO flakes influences the result. For $n = 1$, the result shows the linear increase of the saturation magnetization with increasing x up to 5 wt%, after which the additional rGO flakes make the saturation magnetization decreases also almost linearly. A little discrepancy is shown by our $n = 1$ result, that is, the saturation magnetization values above $x = 5\%$ are obtained to be slightly higher than the experimental data. For higher values of n, the maximum saturation magnetization shifts to the right. That is because there is still a probability of finding spaces unoccupied by rGO flakes when some other spaces are already filled by rGO flakes with two (for $n = 2$) or three (for $n = 3$) layers. This mechanism also implies that the peak of saturation magnetization for $n = 2$ and $n = 3$ become broader. Meanwhile, the fact that our saturation magnetization calculation result is in agreement with experimental data for $\gamma = 0.574$ indicates that not all Fe_3O_4 unit cells have oxygen vacancies. This is because the capability of the rGO flakes to adsorb more oxygens from the Fe_3O_4 clusters are limited by the oxygen adatoms that are already precedingly attached on the graphene surface. As for the value of α which is about 20, it shows us that the Fe_3O_4 nanoparticle cluster is much heavier than the rGO layer which is consistent with our presumption.

The results of applied magnetic field dependent magnetization for all rGO wt% variation are shown in figure 5. The magnetization of the model for Fe_3O_4 nanoparticle system with no additional rGO flakes is represented by the black curve. While, the red curve represents the magnetization of the system with the optimum rGO content ($x = 5\%$), leading to highest saturation magnetization. For the other rGO wt% values ($x = 0.10, 0.15$ and 0.20) the calculations are performed using equation 5. The overall trend shows by our calculation results is in agreement with the experimental data, in which the highest magnetization occurs at 5 wt% of rGO, and the magnetization subsequently decreases until the content of rGO flakes achieving 20 wt%, which is still higher than that of the pure Fe_3O_4 nanoparticle system.

In contrast to the saturation magnetization results, our calculated remanent magnetization values do not really fit with the experimental data, instead they are significantly higher than experimental data, and are rather closer to the saturation magnetization values. This is related to the fact that the Curie temperature (T_C) for our model of Fe_3O_4 without oxygen vacancies shown in figure 6 is much higher than

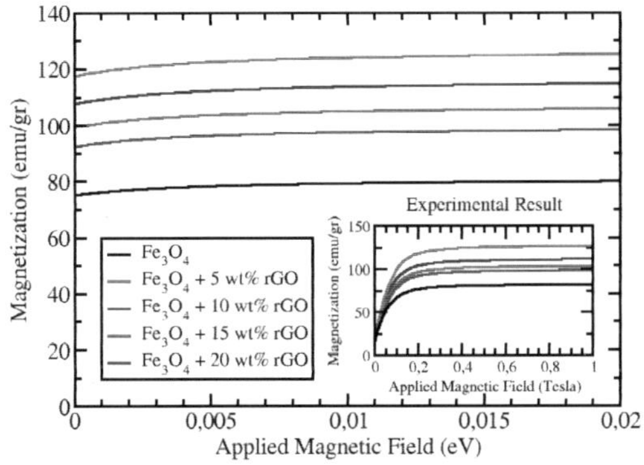

Figure 5. Magnetization as a function of applied magnetic field of our model for Fe_3O_4-rGO nanoparticle system with rGO wt% variation. Inset shows the experimental data for comparison.

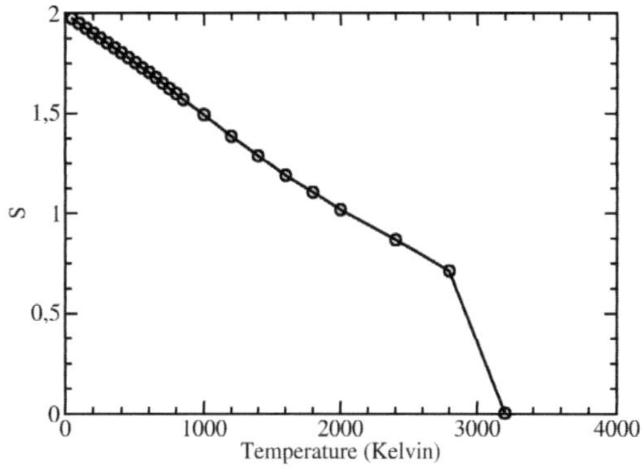

Figure 6. Averaged total spin per unit cell as a function of temperature of our model for pure Fe_3O_4 without oxygen vacancies. The over-estimated T_C may be attributed to the over-simplification of the model unit cell.

the experimental value of 858 Kelvin. This may partly arise as a consequent of the over-simplification of our model unit cell. Another possible source of discrepancy may come from the mean-field treatment of the Hubbard term in our model Hamiltonian, which commonly overestimates physical quantities such as T_C. Nevertheless, overall, our model is able to qualitatively confirm the magnetization enhancement of Fe_3O_4 nanoparticle system with addition of rGO flakes.

6. Conclusions

In conclusion, our present study supports our previous hypothesis that the enhancement of magnetization of Fe_3O_4 nanoparticle system upon addition of reduced graphene oxide (rGO) is due to spin-flipping of the Fe ions in the tetrahedral sites and that oxygen vacancies at the Fe_3O_4 nanoparticle clusters assist the occurrence of spin-flipping. The presence of rGO flakes nearby Fe_3O_4 clusters induces the formation of oxygen vacancies in the Fe_3O_4 clusters, leading to the occurrence of spin-flipping and thus enhancing the magnetization of the system. Our model for Fe_3O_4 nanoparticle cluster can describe the saturation magnetization enhancement of Fe_3O_4 nanoparticle system upon addition of reduced graphene oxide (rGO). However, our calculation method overestimates the Curie temperature and the remanent magnetization of the Fe_3O_4 nanoparticle system. This may be caused by the over-simplification of our model unit cell and the mean-field approximation used in our method.

Acknowledgements

We acknowledge and thank Universitas Indonesia, Indonesia, for the full funding provided for the completion of this project through PITTA Research Grant No. 2022/UN2.R12/HKP.05.00/2016.

References

[1] Yanase A and Siratori K 1984 *J. Phys. Soc. Jpn.* **53** 312-7
[2] Zhang Z and Satpathy S 1991 *Phys. Rev. B* **44** 13319-31
[3] Matsuda H, Takeuchi M, Adachi H, Hiramoto M, Matsukawa N, Odagawa A, Setsune K and Sakakima H 2002 *Jpn. J. Appl. Phys.* **41** L387-90
[4] Hu G and Suzuki Y 2002 *Phys. Rev. Lett.* **89** 276601

[5] Seneor P, Fert A, Maurice J-L, Montaigne F, Petroff F and Vaurs A 1999 *Appl. Phys. Lett.* **74** 4017
[6] Wang J, Wu W, Zhao F and Zhao G 2011 *Appl. Phys. Lett.* **98** 083107
[7] Shariati S, Faraji M, Yamini Y and Rajabi A A 2011 *Desalination* **270** 1605
[8] Laurent S, Forge D, Port M, Roch A, Robic C, Vander Elst L and Muller R N 2008 *Chem. Rev.* **108** 2064-110
[9] Taufik A and Saleh R 2016 *AIP Conf. Proc.* **1725** 020089
[10] Herng T S *et al.* 2015 *Nano Res.* **8** 2935-45
[11] Santoso I, Singh R S, Gogoi P K, Asmara T C, Wei D, Chen W, Wee A T S, Pereira V M and Rusydi A 2014 *Phys. Rev. B.* **89** 075134
[12] Majidi M A, Siregar S and Rusydi A 2014 *Phys. Rev. B.* **90** 195442
[13] Fauzi A D, Majidi M A and Rusydi A 2017 *J. Phys.: Condens. Matter* **29** 135802
[14] Cheng C 2005 *Phys. Rev. B* **71** 052401

IOP Conf. Series: Materials Science and Engineering **188** (2017) 012034 doi:10.1088/1757-899X/188/1/012034

The effect of twisting motion dependent diagonal hopping on the I-V characteristics of poly(dG)-poly(dC) DNA molecule

Efta Yudiarsah

Department of Physics, Faculty of Mathematics and Natural Sciences Universitas Indonesia, Kampus UI Depok, Depok 16424, Indonesia

Corresponding author's e-mail: e.yudiarsah@sci.ui.ac.id

Abstract. The effect of twisting motion dependent diagonal hopping on the I-V characteristic of Poly(dG)-Poly(dC) DNA molecule has been studied. Twisting angle dependent hopping constant is model using semi-empirical Slater-Koster theory. The I-V characteristic of DNA molecule is calculated from transmission probability using Landauer-Buttiker formalism by assuming symmetric voltage drop at the contacts. The transmission probability of charge on the molecule is calculated using transfer and scattering matrix methods, simultaneously. The calculation is carried out on 32 base pairs doubled-stranded DNA molecule model sandwiched in between two metallic electrodes. The results show that at low voltage the magnitude of current is not influenced much by the change of diagonal hopping term caused by base pair twisting motion. Larger influenced is observed at higher voltages, after decreases by increasing the coupling parameter of vibration and diagonal hopping constant up to some value, current increases. This trend is observed in the I-V characteristic of DNA molecule for all frequencies and temperatures used. Lower twisting motion frequency shows larger influences of diagonal hopping term on the current than the one at higher frequencies.

1. Introduction

In 1962, Eley and Spivey [1] showed that DNA can conduct current. The electrical conduction through DNA may involve in several biological processes, such as DNA mutation [2], damage detection and self-repairing [3, 4], as well as sensor and signaling system [5, 6], as shown by several experimental results. These results shed the light on the possibility of using DNA molecules as an active element in electronic devices, such as molecular transistor [7] and biosensor [8]. Moreover, DNA molecule can be used as a template in fabricating nanoscale wire [9] for its long nature.

The possibility of using DNA in molecular scale technology increases the interest of physicists in studying the transport property of DNA molecules. However, experimental results show that DNA can be an insulator up to proximity-induced superconductor [10]. This controversy may be due to the fact that different groups measuring conductivity of DNA molecule using different types of DNA [11, 12]. It has been shown that base pairing mismatch [13] also affect the transport property of DNA. Several other factors may also lead to this controversy, *e.g.* medium or substrate [14, 15] and internal motion of DNA [16, 11].

The effect of internal motion of DNA in the form of base pair twisting motion has been studied by modelling electron hopping term in single channel effective tight binding Hamiltonian model changes as cosine the twist angle [14]. In this paper, two channels model is used, with electron hopping terms following semi-empirical theory of Slater-Koster [10]. This paper will be focused on the effects of twisting motion dependent diagonal hopping on the I-V characteristic of 32 base pair long Poly(dG)-Poly(dC) DNA molecule. Twisting motion dependent diagonal hopping is tuned by varying the linear coupling parameter of vibration and diagonal hopping constant. I-V characteristic is calculated from transmission probability using Landaur-Büttiker formalism [17] by assuming potential drops at contact symmetrically. The transfer matrix technique and scattering matrix method are employed simultaneously in calculating transmission probability [18]. The results show the influence of the linear coupling parameter of vibration and diagonal hopping constant is strong at high voltage. It is observed that by increasing the coupling parameter up to some value, current decreases, then it increases. The molecule model used in the present work will be discussed in section 2 along with the methods of calculation. Then, calculation results will be discussed in section 3. In the last section, the conclusion of discussion will be drawn.

2. Materials and methods

In the model we used, electron can hop from a base to a sugar-phosphate as well as from a sugar-phosphate backbone to a base. Electron can also hop from a base to other base in the same strand (intra-strand hopping), in Watson-Crick base pair (Watson-Crick pair hopping). The model also considers electron can hop diagonally between two bases in different strands (diagonal hopping). Semi-empirical theory of Slater-Koster [10] gives angle dependent intra-strand, diagonal, and Watson-Crick pair hopping constant, consecutively, as shown in equation (1)

$$t_{i,i+1}^{j,j} = t_{i,i+1}^{pa}\left[1-\alpha_{i,i+1}^{pa}\sin^2\left(\Delta_{i,i+1}^{j,j}/2\right)\right]\exp\left[-\beta_{i,i+1}^{pa}\left(\cos\varphi_{eq}-\cos\Delta_{i,i+1}^{j,j}\right)\right]/\left[1-\alpha_{i,i+1}^{pa}\sin^2\left(\varphi_{eq}/2\right)\right],$$

$$t_{i,i+1}^{j,k} = t_{i,i+1}^{sl}\left[1-\alpha_{i,i+1}^{sl}\cos^2\left(\Delta_{i,i+1}^{j,k}/2\right)\right]\exp\left[\beta_{i,i+1}^{sl}\left(\cos\varphi_{eq}-\cos\Delta_{i,i+1}^{j,k}\right)\right]/\left[1-\alpha_{i,i+1}^{pa}\cos^2\left(\varphi_{eq}/2\right)\right], \text{ and}$$

$$t_{i,i}^{j,k} = t_{i,i}^{pp}\left[1+\alpha_{i,i}^{pp}\sin^2\left\{\left(\varphi_i^j-\varphi_i^k\right)/2\right\}\right]\exp\left[\beta_{i,i}^{pp}\sin^2\left\{\left(\varphi_i^j-\varphi_i^k\right)/2\right\}\right], \quad (1)$$

where $\alpha^{pa} = 6.73$, $\alpha^{pp} = 1.00$, $\beta^{pa} = 6.26$, $\beta^{sl} = 2,85$, and $\beta^{pp} = 11.5$. α^{sl} , the linear coupling parameter of vibration and diagonal hopping constant, will be varied.

Twisting angle of base measured relative to its position at equilibrium is φ. Its value is assumed following Gaussian distribution with deviation standard $\sqrt{k_bT/\left(I_n\omega^2\right)}$ and zero average. T is temperature, k_b is Boltzmann constant, I_n is base moment of inertia, which assumed to be in the order of theoretical value, i.e. 1.0×10^{-44} kgm^2 [19], and ω is twisting motion frequency. 36° is taken as φ_{eq}, the angle between two neighbouring base pairs at equilibrium. The twisting motion dependent on-site energy according to [20] can be written as $\varepsilon_i^j = \phi_i^j + b_{i-1,1}^j\left(\cos\Delta_{i-1,i}^{j,j}-\cos\varphi_{eq}\right)+b_{i,1+1}^j\left(\cos\Delta_{i,i+1}^{j,j}-\cos\varphi_{eq}\right)$, where $\Delta_{i,i+1}^{j,k} = \varphi_{eq}-\left(\varphi_i^j-\varphi_{i+1}^k\right)$, ϕ is base on-site energy at equilibrium, 7.84 eV for G and 8.85 eV for C and b is coupling parameter between charge and base pair twisting motion.

In tight binding Hamiltonian form, the DNA molecule model can be written as follow.

$$\begin{aligned} H_{dna} = &\sum_{i=1}^{N}\sum_{j=1}^{2}\left\{\varepsilon_i^j c_{i,j}^+ c_{i,j} + t_{i,i}^{j,j+1}\left(c_{i,j}^+ c_{i,j+1} + c_{i,j+1}^+ c_{i,j}\right)\right\} + \sum_{i=1}^{N}\sum_{j=1}^{2}\sum_{k=u,d}\left\{B_i^k c_{i,k}^+ c_{i,k} + t_{i,i}^{j,k}\left(c_{i,j}^+ c_{i,k} + c_{i,k}^+ c_{i,j}\right)\right\} \\ &+ \sum_{i=1}^{N-1}\sum_{j,k=1}^{2} t_{i,i+1}^{j,k}\left(c_{i,j}^+ c_{i+1,k} + c_{i+1,k}^+ c_{i,j}\right) \end{aligned} \quad (2)$$

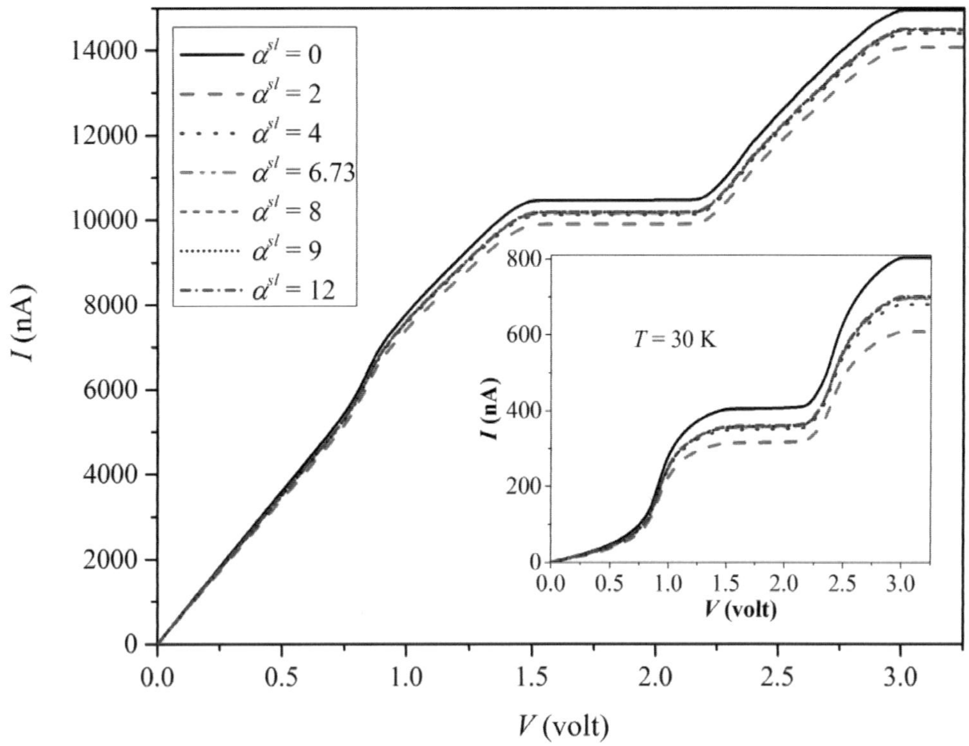

Figure 1. The I-V characteristic of 32 base pairs long poly(dG)-poly(dC) DNA molecule at 4.2 K, $W = 0.1$ eV, and frequency 0.51 meV for various linear coupling parameter of vibration and diagonal hopping constant α^{sl}. In the inset is the I-V characteristic calculated at temperature 30 K.

Table 1. Parameter for electron hopping constants ($t_{x,y}$) and coupling constant between charge and base pair twisting motion ($b_{x,y}$) for tight binding Hamiltonian [20, 21].

Base (X,Y)	$t^{pa}_{x,y}$ **(eV)**	$t^{sl}_{x,y}$ **(eV)**	$t^{pp}_{x,y}$ **(eV)**	$b_{x,y}$ **(eV)**
G,G	0.30	0.14	-	0.0334
C,C	0.38	0.28	-	0.0341
C,G	0.14	0.27	0.57	0.0299

In equation 2, ε, B, and t, consecutively, represent base on-site energy, sugar-phosphate backbone on-site energy, and electron hopping constant from one base to the other or to sugar-phosphate backbone. N is the length of DNA molecule, in this case $N = 32$ base pairs. The values of b and t can be seen in table 1.

In this study, back bone on-site energy B follows uniform distribution with average value of 9.36 eV and deviation W, where W is energy disorder due to medium and temperature. Other parameters used in this study are metal on-site energy 8.45 eV and electron hopping constant in metal 0.75 eV. Moreover, in calculating current, Fermi energy of 8.45 eV has been used.

3. Results and discussion

The I-V characteristic of 32 base pairs poly(dG)-poly(dC) DNA molecule calculated at 4.2 K, $W = 0.1$ eV, and frequency 0.51 meV for linear coupling parameter of vibration and diagonal hopping constant is shown in figure 1. This result and other calculation results is the average of 400 twisting angle configurations. The figure shows that the variation of the linear coupling parameter of vibration and diagonal hopping constant almost does not change the current magnitude at low voltage. At higher voltage, the influence of coupling parameter is getting larger. As the coupling parameter increases up to a certain value, the current magnitude decreases. After that, the current magnitude increases and seems to saturate. The plateau in the range of 1.5 and 2.0 volt emerges due to gap on transmission probability as function of energy. Even though I-V characteristic profile is different, similar trends are observed at higher temperature as shown in the inset.

Calculated I-V characteristic at frequency 1.02 meV shows similar trend as the one calculated at 0.51 eV. However, the influence of linear coupling parameter is lesser at 1.02 meV than 0.51 meV. Calculated I-V characteristic on 32 base pairs poly(dG)-poly(dC) DNA molecule at 4.2 K, $W = 0.1$ eV, and frequency 1.02 meV for various linear coupling parameter of vibration and diagonal hopping constant is shown in figure 2. The influence of linear coupling parameter on I-V characteristic seems stronger at higher temperature as shown in the inset. The trend that the influence of linear coupling parameter on the current at higher voltage is larger is observed at all frequencies and temperature used. However, the change in current lessens with the increment of frequency as well as temperature.

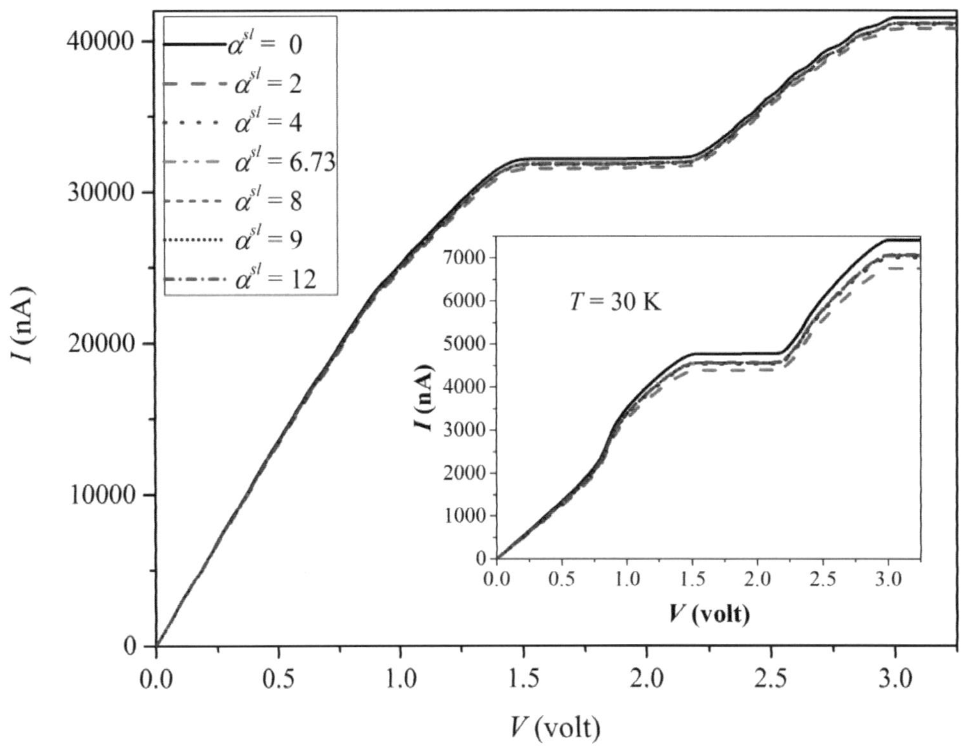

Figure 2. The I-V characteristic of 32 base pairs long poly(dG)-poly(dC) DNA molecule at 4.2 K, $W = 0.1$ eV, and frequency 1.02 meV for various linear coupling parameter of vibration and diagonal hopping constant α^{sl}. In the inset is the I-V characteristic calculated at temperature 30 K.

4. Conclusions

At low voltage, the magnitude of current is not influenced much by the change of the linear coupling parameter of vibration and diagonal hopping constant. At higher voltage, at first current decreases by increasing the linear coupling parameter up to some value, and then it increases. Higher twisting motion frequency shows weaker influences of linear coupling parameter on the current. These trends are observed at all temperature used in the study.

Acknowledgements

The author would like to thank UPP-IPD Universitas Indonesia for laboratory support and Mr. Daniel K. Suhendro for the parameters used in the calculation. This study is funded by Hibah Penelitian Unggulan Perguruan Tinggi (PUPT) fiscal year 2016 No.1122/UN2.R12/HKP.05.00/2016.

References

[1] Eley D D and Spivey D I 1962 *Trans. Far. Soc.* **58** 411-5
[2] Miller J H Jr, Villagrán M Y S, Maric S and Briggs J M 2015 *Phys.* B **460** 119-25
[3] Dandliker P J, Holmlin R E and Barton J K 1997 *Sci.* **275** 1465-8
[4] Boon E M, Livingston A L, Chmiel N H, David S S and Barton J K 2003 *Proc. of the Nat. Ac. of Sci.* **100** 12543-7
[5] Sontz P A, Muren N B and Barton J K 2012 *Acc. of chem. Res.* **45** 1792-800
[6] Genereux J C, Boal A K and Barton J K 2010 *J. the Am. Chem. Soc.* **132** 891-905
[7] Nokhrin S, Baru M and Lee J S 2007 *Nanotech.* **18** 095205-5
[8] Zaffino R L, Mir M and Samitier J 2014 *Nanotech.* **25** 105501-8
[9] Braun E, Eichen Y, Sivan U and Ben-Yoseph G 1998 *Nature* **391** 775-8
[10] Endress R G, Cox D L and Singh R R P 2004 *Rev. of Mod. Phys.* **76** 195-214
[11] Jo Y S, Lee Y and Roh Y 2003 *Mat. Sci. and Eng.* C **23** 841-6
[12] Klotsa D, Römer R A and Turner M S 2005 *Biophys.* J **89** 2187-98
[13] Kratochvílová I *et al.* 2010 *J. Phys. Chem.* B **114** 5196-205
[14] Roche S 2003 *Phys. Rev. Lett.* **91** 108101
[15] Guo A M, Xiong S J, Yang Z and Zhu H J 2008 *Phys. Rev.* E **78** 061922
[16] Muren N B, Olmon E D and Barton J K 2012 *Phys. Chem. Chem. Phys.* **14** 13754-71
[17] Maciá E, Triozon F and Roche S 2005 *Phys. Rev.* B **71** 113106
[18] Taniyama H and Yoshii A 1996 *Phys. Rev.* B **53** 9993
[19] Yakushevich L V 2004 *Nonlinear physics of DNA* 2nd ed (Weinheim: Wiley-VCH)
[20] Suhendro D K, Yudiarsah E and Saleh R 2010 *Physica B: Cond. Mat.* **405** 4806-11
[21] Suhendro D K 2014, private communication

Fabrication and characterization of micro-band boron-doped diamond electrode for an application in adenosine phosphates sensor

P U Prayikaputri, J Gunlazuardi and T A Ivandini

Department of Chemistry, Faculty of Mathematics and Natural Sciences Universitas Indonesia, Kampus UI Depok, Depok 16424, Indonesia

Corresponding author's e-mail: ivandini.tri@sci.ui.ac.id

Abstract. Micro-band electrode was successfully fabricated by lamination method through sealing a piece of boron-doped diamond film inside a sandwich of two insulating plates, namely Teflon and silicon rubber as the gaskets. Characterization was performed using Raman and XPS spectra of the BDD film, while the fabricated micro-band was characterized by analyzing its SEM image. The electrode was examined for cyclic voltammetry of adenosine triphosphate solution, where an oxidation peak at +0.9 V vs. Ag/AgCl can be observed. The influence of scan rate and pH was also studied, in which pH 2 was selected as the optimum pH. The diffusion coefficient of 0.1 mM ATP at micro-band electrode was 3.84 x 10^{-8} m^2/s, while the effective surface of the micro-band BDD electrode was 8.72 x 10^{-14} m^2.

Keywords: micro-band electrode, boron-doped diamond, characterization, ATP, cyclic voltammetry

1. Introduction

Adenosine triphosphate (ATP) is clinically applied for the medication of progress wasting of muscles tissue (amyotrophia), intracerebral hemorrhage, a myocardial infarction and hepatitis. During the process of production and storage, ADP and AMP are often generated as impurities. Therefore, ATP, ADP and AMP determination is substantially significant for the product quality checking, production process monitoring, and clinical effectiveness evaluation of the ATP medicine[1, 2]. In this research, microelectrode was prepared for a detector in capillary electrophoresis in order to achieve an extremely fast scan rates and tiny areas that are necessary for high resistive media without considerable IR drops. The microelectrode is prepared from the cross section of boron-doped diamond (BDD) film, which generally provide around 5 μm thickness. Therefore, such type of electrode can be named as a micro-band electrode since the area of electrode can be considered as the band. Meanwhile, BDD exhibits interesting electrochemical properties, such as low and good stability of background current, a wide potential window in the aqueous media, high endurance towards deactivation, and highly stable for a long period [3-5]. Accordingly, modification of BDD as a micro-band is expected to provide a high performance of BDD microelectrode. Furthermore, electrochemistry of the micro-band electrode was examined for ATP detection.

2. Materials and methods

2.1. Apparatus and chemicals
Apparatus and chemicals used in this research were BDD, Teflon, silicon, platinum spiral, Ag/AgCl (saturated KCl), fused silica capillary, syringe, magnetic stirrer, stir bar, hotplate, ultrasonic bath, screw vial, cable, ATP, K_2HPO_4, KH_2PO_4, H_3PO_4, CH_3COOH, H_3BO_4, aqua bides, and methanol.

2.2. Fabrication of micro-band electrode
Micro-band electrode was prepared with lamination method. First, silicon substrate in BDD was chemically etched by the mixture of HF (48%) : HNO_3(60%) 1:1 for 12 h [6]. Then, BDD was cut to 1 cm x 1 cm size, and insulated by Teflon and 1 cm x 1cm silicon rubber to form layers imitating a sandwich with 4 mm in width (figure 1). Stainless steel was applied to connect the electrode with electric current.

2.3. Procedure
Cyclic voltammetry of ATP in phosphate buffer solution (PBS) pH 6.8 with divers scan rates was performed at potential range of -1600 mV to 2000 mV (vs. Ag/AgCl). Influence of pH was investigated using Britton-Robinson buffer containing 0.04 M phosphoric acid, 0.04 M acetic acid, and 0.0404 M boric acid. NaOH was added to adjust the pH.

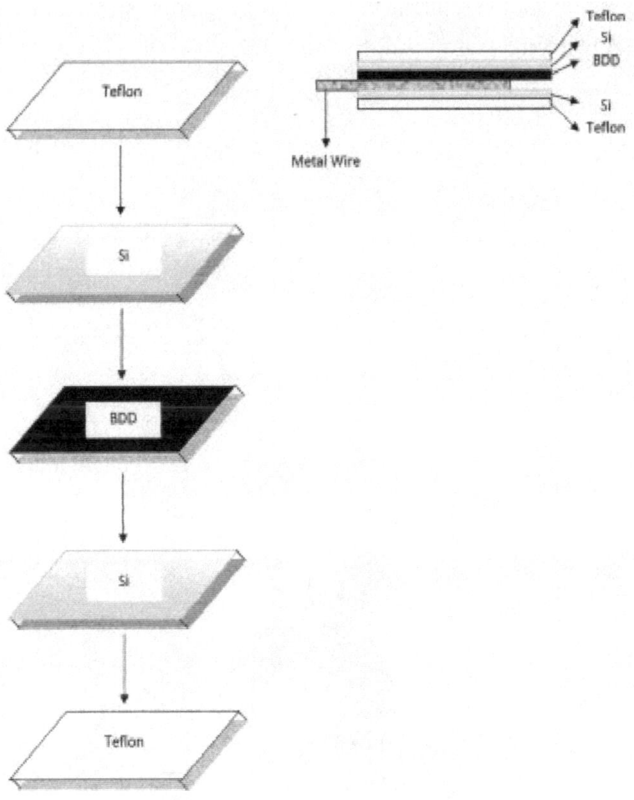

Figure 1. Longitudinal section of micro-band electrode

3. Results and discussion

3.1. Characterization of micro-band boron-doped diamond electrode

Characterization of BDD micro-band was performed by Raman and XPS spectroscopy as well as SEM method. Raman spectroscopy (figure 2a) provided the information concerning carbon phase, such as amorf, graphite, diamond, indicated by a peak at 1332 cm^{-1} attributted to diamond with C-C structure in sp^3 hybridization and a peak at 500 cm^{-1} suggested the presence of the disordered diamond formed due to the influence of the boron doping. Furthermore, XPS spectra (figure 2b) displays a C (1^s) peak in hybridization sp^3 with a binding energy at 284.5 eV and an O(1^s) peak at binding energy of 532.5eV resulted from the bonding between O and H [2]. In addition, SEM image (figure 3c) shows that the BDD film in the micro-band electrode with a 10 μm thickness.

3.2. Electrochemistry of ATP at BDD micro-band electrode

Cyclic voltammetry was performed for 0.1 mM ATP in PBS pH 6.8. A peak at +0.6 V was observed in figure 3a. High resistive backgrounds of cyclic voltammograms were mainly consist of radial diffusion in which the voltammetric response is directly match towards the electrode length and is rather insensitive against width, was identified due to delamination defects. Then, +0.9 V vs Ag/AgCl was selected as the optimum detection potential for adenosine, because when the oxidation process was applied at exactly +0.6 V, the result suggested that the process was just started at that potential and not all species could be oxidated. In figure 3b, cyclic voltammetry was performed with various scan rates from 25 mV/s to 250 mV/s, and the linear correlation of current and square root of scan rate was obtained, therefore the detection of adenosine triphosphates using micro-band electrodes can be performed as per Randles-Sevcik equation and depends on mass transfer only.

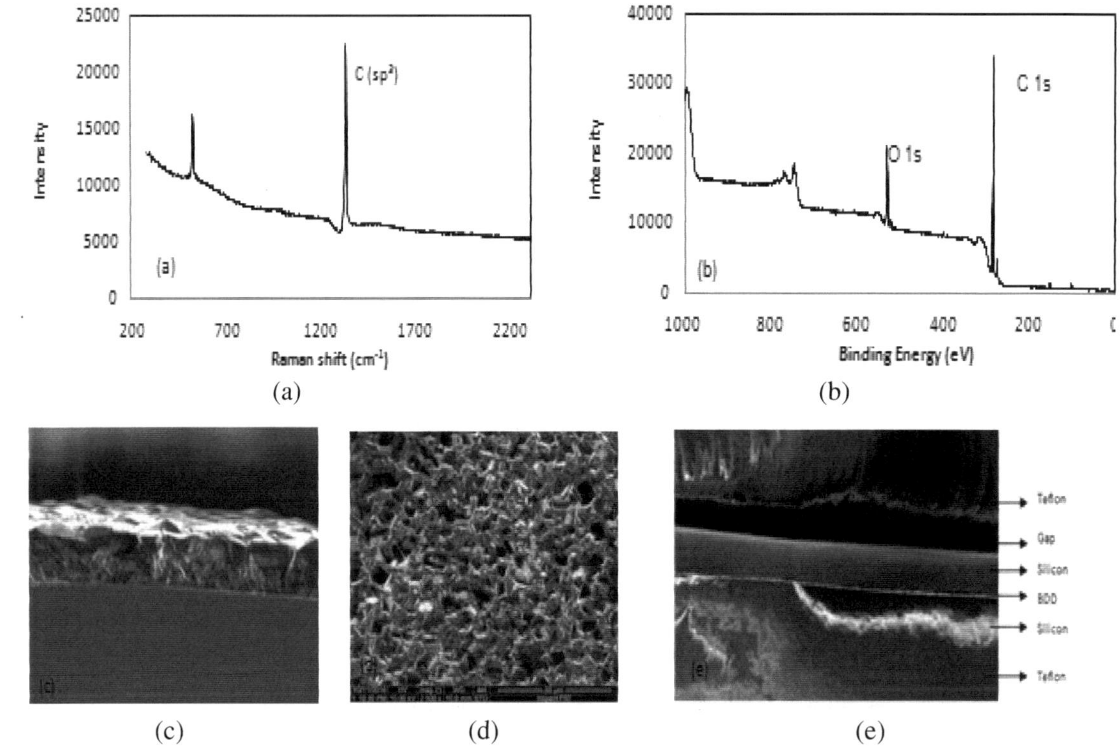

(a) (b)

(c) (d) (e)

Figure 2. (a) Raman and (b) XPS spectra of BDD as well as the SEM image of micro-band electrode in (c) longitudinal section, (d) surface, (e) layer

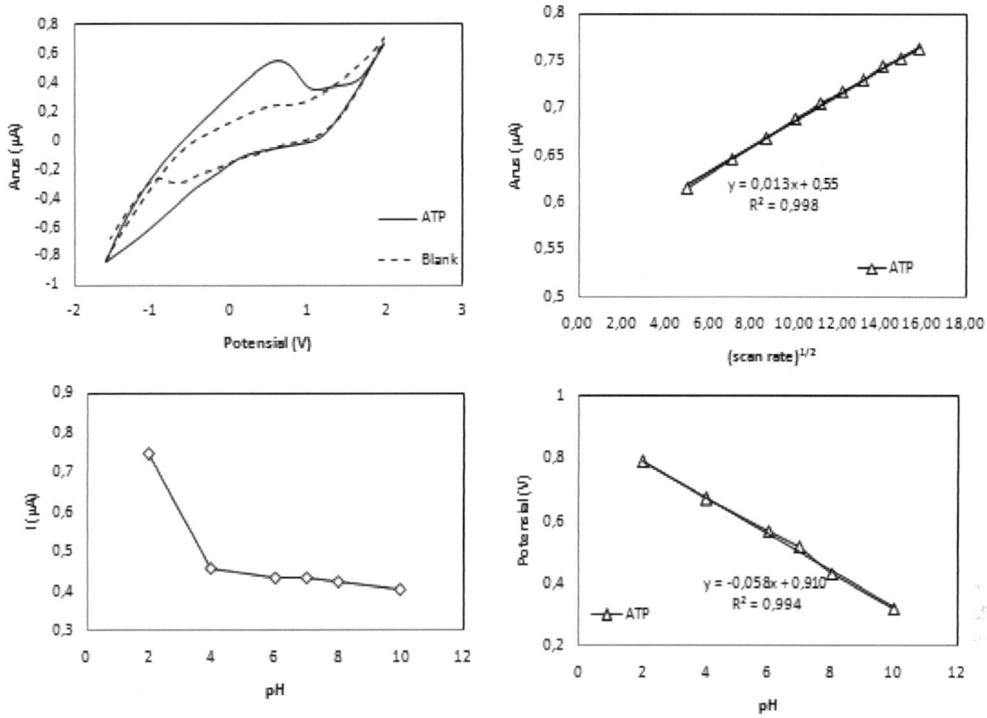

Figure 3. (a) Cyclic voltammograms of blank and ATP (b) plot I vs. (scan rate)^{½}, (c) plot I vs. pH in ATP, (d) plot pH vs. potential in ATP

Influence of pH was investigated using Britton-Robinson buffer as universal buffer from pH 2 to 10. A mixture containing 0.04 M H_3BO_3, 0.04 M H_3PO_4 and 0.04 M CH_3COOH titrated to the desired pH with 0.2 M NaOH. In figure 3c, it was discovered that the highest current response was pH 2 in acid solution due to more phosphate groups with more negative charges are available in ATP. In figure 3d, the determination of adenosine triphosphate was conducted as per Nernst equation, at 25°C; the slope of the pH electrode was close to-59 mV/pH units, indicated that the oxidation and reduction in half reaction occurred at the same amount of electrons. When PBS pH 6.8 and Britton-Robinson buffer pH 2 was applied, the same responses in the reaction was observed, i.e the higher response of current in potential range from +0.6 V to +0.8 V due to the same ionic strength.

Diffusion defined as the net motion of molecules or atoms from a high chemical potential to a low chemical potential one. Diffusion will keep continue until all the species are uniformly diversed and reached the equilibrium, although there was no gradient concentration. Diffusion velocity is influenced by temperature, particle size, and viscosity. Investigation of diffusion coefficient for 0.1 mM ATP in Britton-Robinson buffer pH 2 with micro-band BDD electrode was performed using Cottrell equation, which conducted by amperometry at +0.9 V along 5 minutes. Diffusion coefficient of this process was 3.84 x 10^{-8} m^2/s and the effective surface of the micro-band BDD electrode was 8.72 x 10^{-14} m^2.

4. Conclusions

Micro-band electrode was successfully fabricated by lamination method through sealing boron doped-diamonds inside a sandwich of two insulating plates, namely Teflon and silicon. The thickness was about 10 µm and delivered the ideal condition for the surface of BDD. Under voltammetry experiment, +0.9 V vs. Ag/AgCl was selected as optimum detection potential for adenosine. Additionally, the calculation result demonstrated that this electrode follows the theory of Randles-Sevcik and Nernst equation, with pH 2 as the optimum pH. Furthermore, the diffusion coefficient of 0.1 mM ATP at micro-band electrode was 3.84 x 10^{-8} m^2/s and the effective surface of the micro-band BDD electrode was 8.72 x 10^{-14} m^2.

Acknowledgements
This work was partly funded by Hibah PUPT Kemenristekdikti fiscal year 2016 No. 1127/UN2.R12/HKP.05.00/2016.

References

[1] Chiku M, Watanabe T and Einaga Y 2010 *Diamond Relat. Mater.* **19** 673-9
[2] Fu C G, Song L N and Fang Y Z 1999 *Anal. Chim. Acta* **399** 259-63
[3] Honda Y, Ivandini T A, Watanabe T and Einaga Y 2013 *Diamond Relat. Mater.* **40** 7-11
[4] Asai K, Ivandini T A, Falah M M and Einaga Y 2016 *Electroanal.* **28** 177-82
[5] Shibano S, Ivandini T A, Terashima C, Nakata K and Einaga Y 2014 *Chem. Lett.* **43** 1292-3
[6] Caston S L and McCarley R L 2002 *J. Electroanal. Chem.* **529** 124-34

IOP Conf. Series: Materials Science and Engineering **188** (2017) 012036 doi:10.1088/1757-899X/188/1/012036

Transition metal complexes of 4'-(p-nitrophenyl)-2,2':6',2" -terpyridine as fluorosensor sodium ion

A Zulys and D M A Ovika

Department of Chemistry, Faculty of Mathematics and Natural Sciences Universitas Indonesia, Kampus UI Depok, Depok 16424, Indonesia

Corresponding author's e-mail: zulys@ui.ac.id

Abstract. Ligand 4'-(p-nitrophenyl)-2,2':6',2"-terpyridine has been synthesized using Kröhnke method by aldol condensation reaction. The solid brown precipitate was collected and resulted 40% yield. The ligand was characterized by ^{1}H-NMR spectrometer, IR spectrophotometer, elemental analyzer, UV-Vis spectrometer, and spectrofluorometer. The complexation of ligand was done by chelating method with transition metal (Cu^{2+}, Zn^{2+}). The complex compounds have been confirmed by UV-Vis spectrophotometer from absorbance spectrum data. The data showed a new peak at wavelength 236 nm for Cu^{2+} complex and 233 nm for Zn^{2+} complex. The application of ligand 4'-(p-nitrophenyl)-2,2':6',2"-terpyridine and complex compounds is as fluorosensor of sodium ion. The intensity of fluorescence was determined by spectrofluorometer. The addition of various concentrations of sodium ion into ligand and complex compounds (Cu^{2+}, Zn^{2+}) changed the intensity of fluorescence. The fluorescence data of ligand showed the on-off type while Cu and Zn complex showed the off-on type. The on-off type fluorescence was indicated by decreasing intensity and the off-on by increasing intensity. Both ligand and complex compound showed a wavelength shift when sodium ion was added into solution.

1. Introduction

Recently, the progress of fluorescence based on organic compound has been developed [1-3]. The recent research about fluorescence was used to detect some heavy metals [4]. Fluorescence is one of the types of luminescence which based on excited electromagnetic wavelength. When the substance absorbs some energy at a specific wavelength and releases energy in terms of light in short lifetime, it produces fluorescence that used as sensor.

Most organic compounds which have aromatic or heterocyclic system with π-conjugated system, for example derivatives pyridine, azatriphenylene, and terpyridine, show a higher possibility for a fluorescence characteristic [4]. Fluorescence detection in complex compound is considered to be effective for sensing, owing to the complex compounds' longer lifetime than organic compound [5]. The longer lifetime is able to reduce the interference that comes from another compound, and therefore increases the fluorescence sensitivity [6].

In fluorescence mechanism, the ligand acts as fluorophore, therefore, the type of ligand affects the chemo sensor. The most common ligand used for chemosensor are 2,2':6',2"- terpyridine and its derivatives, because terpyridine have high binding affinity [7, 8]. The photophysical properties of 2,2':6',2"- terpyridine are unique and have strong binding affinity to some metals. In order to achieve these goals, the terpyridine must be substituted to C-4' position [7-9]. This research used ligand

4'-(p-nitrophenyl)-2,2':6',2"-terpyridine that complexed with transition metals Cu^{2+} and Zn^{2+}. The previous research about this ligand had been conducted by Bo *et al.* [10]. This ligand has phi-conjugated system that helps to form a good connection between donor and acceptor electron.

2. Materials and methods

2.1. Materials

The 4'-(p-nitrophenyl)-2,2':6',2"-terpyridine ligand was prepared according to the literature, with some modification, while Cu^{2+} and Zn^{2+} complex was done by chelating process. All chemicals were obtained from commercial sources without further purification. The chemicals used were 2-acetylpyridine (Sigma-Aldrich), 4-nitrobenzaldehyde (Sigma-Aldrich), KOH (Merck), $NH_3.H_2O$ (Merck), Ethanol p.a (Merck), Methanol p.a (Merck), $Cu(NO_3)_2.6H_2O$ (Merck), $Zn(NO_3)_2.4H_2O$ (Merck), and NaCl (Merck). Elemental analysis (C, H, and N) were performed with an Elemental TruSpec LECO. ^1H NMR spectra were recorded at room temperature using JEOL 500 MHz apparatus and tetramethylsilane (TMS) as reference and $CDCl_3$ as solvent. Infrared spectra were obtained from KBr pellets on Shimadzu 8400-S in the range 400-4000 cm^{-1}. Emission spectra were recorded on a Hitachi F-2000 spectrophotometer at room temperature.

2.2. Synthesis of the ligand 4'-(p-nitrophenyl)-2,2':6',2"-terpyridine [10]

Synthesis of ligand 4'-(p-nitrophenyl)-2,2':6',2"-terpyridine: solution of 2-acetylpuridine (5 g, 41.5 mmol) and nitrobenzaldehide (3.17 g, 21 mmol) in methanol (175 mL) was stirred vigorously with 15% KOH aqueous solution (15 mL) and NH_4OH (150 mL) concentrate for 3 days at room temperature. The dark brown precipitate was collected by vacuum filtration, washed with water until pH = 7 and dissolved in CH_2Cl_2. The organic fraction was washed with 1% $NaHCO_3$ aqueous solution, dried with anhydrous $MgSO_4$ and evaporated to give 2.933 g crude ligand 4'-(p-nitrophenyl)-2,2':6',2" -terpyridine (yield 40%). Analytic calculation for $C_{21}H_{14}N_4O_2$ (354.11) are C, 71.11; H, 4.77; N, 15.09. The results are C, 74.37; H, 5.39; N, 11.96. IR (KBr, cm^{-1}): 1315 (s), 1500 (s), 1494 (m), 1401(m), 1388 (w), 1118 (m), 870 (w), 769 (w), 702 (w), 617 (w). ^{1}HNMR ($CDCl_3$): δ (ppm) = 8.66 (d, 2H), 8.62 (s, 2H), 8.59 (d, 2H), 7.80 (d, 2H), 7.71 (td, 2H), 7.27 (dd, 2H), 6.73 (d, 2H). ESI-MS:m/z = 325.2 [MH^+].

3.3. Synthesis of Cu^{2+} complex

$Cu(NO_3)_2.6H_2O$ (0.234 g, 0.64 mmol) in methanol (15 mL) was poured to solution of 4'-(p-nitrophenyl)-2,2':6',2"-terpyridine (0.2 g, 0.64 mmol) in CH_2Cl_2 (10 mL). The color of the reaction mixture turned to dark green immediately. After stirring for 24 h at room temperature, a dark green precipitate was obtained. The solid was then washed with cold methanol (5 mL) and cold diethyl ether (10 mL), and finally dried over silica gel. Recrystallization in warm DMF produced the cuprum complex, [Cu(4'-(p-nitrophenyl)-2,2':6',2"-terpyridine)$(NO_3)_2$], 0.14 g (64%), as pale dark green crystals. IR (KBr, cm^{-1}): 1315 (s), 1500 (s), 1494 (w), 1401(w), 1388 (w), 1118 (m), 870 (w), 769 (w), 702 (w), 617 (w).

3.4. Synthesis of Zn^{3+} complex

$Zn(NO_3)_2.4H_2O$ (0.146 g, 0.64 mmol) in methanol (15 mL) was added to solution of 4'-(p-nitrophenyl)-2,2':6',2"-terpyridine (0.2 g, 0.64 mmol) in CH_2Cl_2 (10 mL). The color of the reaction mixture turned to brown immediately. After stirring for 24 h at room temperature, a brown precipitate was obtained. The solid was then washed with cold methanol (5 mL) and cold diethyl ether (10 mL), and finally dried over silica gel. Recrystallization in warm DMF generated the zinc complex, [Zn(4'-(p-nitrophenyl)-2,2':6',2"-terpyridine)$(NO_3)_2$], 0.14 g (64%), as pale brown crystals. IR (KBr, cm^{-1}): 1315 (s), 1500 (s), 1494 (w), 1401(w), 1388 (w), 1118 (m), 870 (w), 769 (w), 702 (w), 617 (w).

3. Results and discussion

3.1. Synthesis and characterization of ligand 4'-(p-nitrophenyl)-2,2':6',2"-terpyridine and complex Cu^{2+} and Zn^{2+}

Based on the graph above, new peaks at wavelength 239 nm, 236 nm, 233 nm, and 230 nm were

IOP Conf. Series: Materials Science and Engineering **188** (2017) 012036 doi:10.1088/1757-899X/188/1/012036

observed respectively for Fe complex, Zn complex, Cu complex, and Ni complex. It caused by electron transition from its metals that have d-d orbital. The result was a shifting wavelength and decreasing molar absorptivity. The functional group of NO_2 caused a shifting wavelength to the blue shift. It happened because of the functional group does not have an electron lone pair.

3.2. Fluorescence properties

Both ligand and complexes $ML(NO_3)_2$ have concentration of 500 μM and dissolved in methanol-dichloromethane by ratio 3:1. While sodium solution concentration ranged from 1 to100 μM. Before the fluorescence intensity of ligand and complexes were determined, the first step was scanning the wavelength to determine the wavelength number with the highest excitation and emission intensity of ligand and complexes. The scanning measurement was done in the range 220-600 nm. The fluorescence data was plotted between wavelength and emission intensity or excitation intensity (figure 1).

3.3. Mechanism detection of Na analyte was based on PET process

It could be interpreted that NO_2 was as receptor and terpyridine was ligand, while benzene as σ-linker. There were two interaction possibilities when analyte interacts with ligand or complexes compound. The possibilities were turn on or turn off process. Based on the data above, it showed that the addition of sodium ion caused a fluorescence intensity change and wavelength shift. The shift of wavelength was

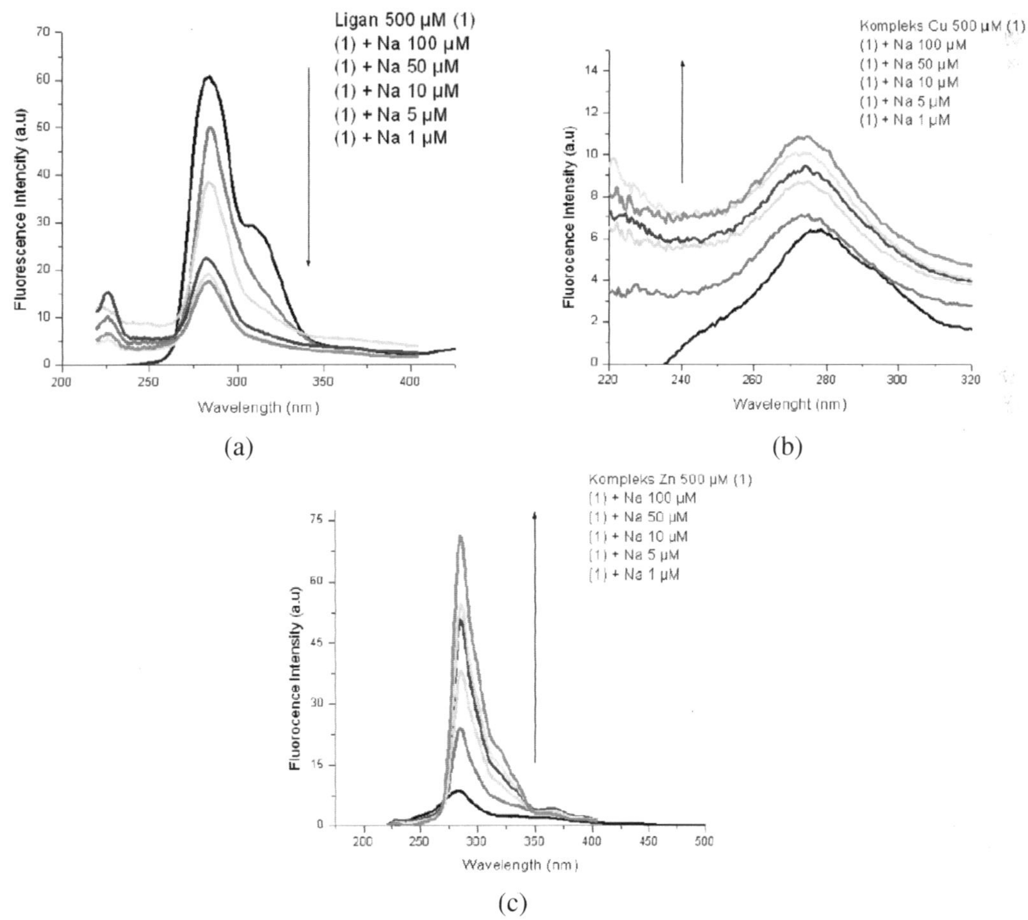

Figure 1. The fluorescence data of (a) ligand, (b) Cu and (c) Zn complexes by adding heavy metals in various concentrations.

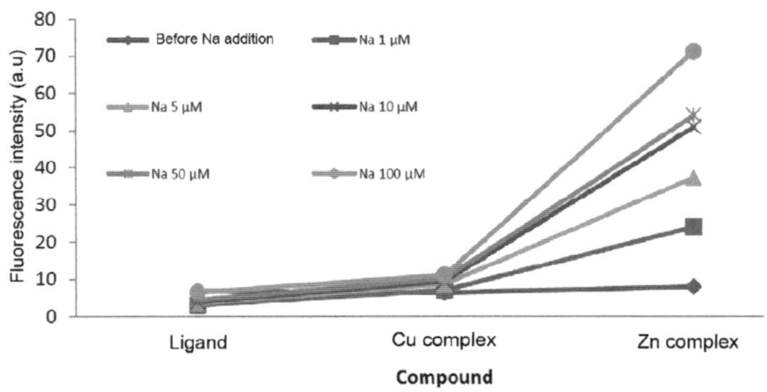

Figure 2. The fluorescence data of ligand and complex by adding heavy metals in various concentrations.

from 283 nm to 285 nm and fluorescence intensity decrease caused a quenching effect or turn off fluorescence.

Cu and Zn complex shifted the wavelength from 278 nm to 273 nm and 282 nm to 287 nm, respectively. Both complexes showed an increase of fluorescence intensity. Based on that data, it showed an enhancing fluorescence or turn on mechanism. By addition of 1 μM sodium ion, the response of fluoroscence intensity was still observed (figure 2). Based on that data of fluorescence, it can be concluded that both ligand and complexes were sensitive for sodium ion detection.

4. Conclusions

Ligand 4'-(p-nitrophenyl)-2,2':6',2"-terpyridinehad has been synthesized using Kröhnke method by aldol condensation reaction. The solid brown precipitate was collected and resulted 40% yield. The fluorescence data of ligand showed the on-off type while Cu and Zn complex showed the off-on type. The on-off type fluorescence was indicated by decreasing intensity and the off-on by increasing intensity. Both ligand and complex compound shifted the wavelength when sodium ion was added into solution. By addition of 1 μM of sodium ion, the response of fluorescence intensity was still observed. Based on the data of fluorescence, it can be concluded that both ligand and complexes were sensitive to detect sodium ion.

Acknowledgements

This research was supported by LPDP and University of Indonesia, for fiscal year 2016.

References

[1] Wolf C, Mei X and Rokadia H K 2004 *Tetrahedron Letters* **45** 7867-71
[2] Staneva D, Grabchev I, Soumillion J P and Bojinov V 2007 *J. Photochemistry and Photobiology A: Chemistry* **189** 192-7
[3] Dalapati S, Paul B K, Jana S and Guchhait N 2011 *Sensor and Actuators* B **157** 615-20
[4] Megan E A F, Sherryllene S P, Catherine N M and Wayne E J J 2012 *Journal of Inorganica Chimica Acta* **381** 78-84
[5] Lakowicz J R 2006 *Principles of Fluorescence Spectroscopy 3rd Ed* (New York: Springer Verlag)
[6] Yongsuk J and Juyoung Y 2012 *Journal of Inorganica Chimica Acta* **381** 2-14
[7] Wang P, Li Z, Lv G C, Zhou H P, Hou C, Sun W Y and Tian Y P 2012 *Inorganic Chemistry Communications* **18** 87-91
[8] Ali N K, Abolghasem B and Sara Z 2011 *Journal of Polyhedron* **30** 1134-42
[9] Naseri Z, Kharat A N, Banavand A, Bakhoda A and Foroutannejad S 2012 *Polyhedron* **33** 396-403
[10] Bo S, Liu X and Zhen Z 2008 *Journal of Luminescence* **128** 1725-30

IOP Conf. Series: Materials Science and Engineering **188** (2017) 012037 doi:10.1088/1757-899X/188/1/012037

Preparation of platinum-modified boron-doped diamond for electroreduction of CO_2

Jasril, J Gunlazuardi and T A Ivandini

Department of Chemistry, Faculty of Mathematics and Natural Sciences Universitas Indonesia, Kampus UI Depok, Depok 16424, Indonesia

Corresponding author's e-mail: ivandini.tri@sci.ui.ac.id

Abstract. Metal-modified boron-doped diamond has been prepared for preliminary study of CO_2 electroreduction. Pt was electrodeposited at boron-doped diamond (BDD) by using chronoamperometry technique. The precursor metal solution concentration of 6 mM was applied with deposition potentials of -0.3 V (vs Ag/AgCl). Characterization by using FESEM and XPS confirmed the presence of Pt on the surface of BDD. Cyclic voltammetry was applied to obtain an optimum condition for electroreduction of CO_2. CO_2 dissolved in 0.1 M NaCl and 0.1 M Na_2SO_4 solutions were applied. A reduction peak, attributable to CO_2, appeared at a potential of -0.7 V (vs Ag/AgCl) in NaCl solution, while no peak was observed in Na_2SO_4 solution. The result indicated that the metal-modified electrodes has successfully prepared as a working electrode for CO_2 electroreduction.

Keywords: platinum, boron-doped diamond, CO_2 electroreduction

1. Introduction

Research on electrochemical reduction of CO_2 has been performed since a long time ago. Various products of the CO_2 reduction into a more useful substances are widely discovered, such as formaldehyde, formic acid, methanol, methane, etc. [1, 2]. Electrochemical reduction of CO_2 employing metal electrodes has been carried out for Pt [2, 3]. Pt belongs to d group metals with an incompletely filled d orbital [2].

On the other hand, boron-doped diamond (BDD) electrodes provide a wide potential window and high electrochemical stability, which establish BDD as an attractive candidate for the electrochemical reduction of CO_2 [1, 4]. Nakata *et al.* [1] reported that electroreduction of CO_2 using BDD generates formaldehyde in a high percentage· Moreover, the electroreduction was affected by the solvent or electrolyte as well as the type of electrodes [1, 5]

In this work, platinum-modified BDD electrodes were prepared in order to improve the catalytic properties of the BDD electrodes for CO_2 electroreduction.

2. Materials and methods

2.1. Materials

H_2PtCl_6 were purchased from WAKO chemicals (Japan), Pt wire was obtained from Nilaco (Japan), and Ag/AgCl (saturated KCl) system was supplied from BAS Inc. (Japan). Double distilled water was obtained from Simply Lab water system (DIRECT-Q 3 UV, Milipore).

2.2. Preparation of platinum-modified BDD (Pt-BDD)

Preparation of Pt-BDD were performed following the previous methods [6]. Briefly, BDD was cleaned in 2-propanol, followed by double distilled water for about 10 min. Then, amperometry technique was applied at BDD by using 6 mM H_2PtCl_6 in 0.1 m H_2SO_4. A deposition potential of -0.3 V (vs Ag/AgCl) was then selected to deposit platinum at BDD electrodes. This treatment was carried out using single cell with a spiral platinum counter electrode and Ag/AgCl as a reference electrode.

2.3. Electrochemical behavior study of CO_2 at Pt-BDD

Cyclic voltammetry of 0.1 M NaCl solution with and without the presence of CO_2 was performed at BDD and Pt-BDD electrodes. Prior to bubbling with CO_2, the solution was bubbled with N_2 gas to remove other gases contained in the solution. After bubbling with CO_2 for 30 minutes, the cyclic voltammetry measurement was performed at a potential range from 0 to -2.0 V (vs. Ag/AgCl) with a scan rate of 100 mV/s.

3. Results and discussion

3.1. Characterization of Pt-BDD electrode

Characterization to confirm the existence of Pt deposited on BDD was performed by FESEM and XPS data of BDD and Pt-BDD. SEM Images of BDD (figure 1a) shows grain crystals of about 5 μm size. In the SEM image of Pt-BDD, the presence of particles on the surface of BDD (figure 1b) indicated that Pt has been deposited on BDD. Since Pt provided a higher conductivity than BDD, the particles of Pt were observed to be lighter than BDD. The shiny color corresponds to the metallic properties of d group metals, which emits light when hit by electrons.

Figure 2 shows the XPS spectra of BDD and Pt-BDD. Without modification, only the peaks of C 1s and O 1s were observed at BDD. On the contrary, in the presence of Pt particles peaks at 71 and 75 eV attributed to Pt were observed. The data confirmed that Pt particles were successfully deposited at BDD [6, 7].

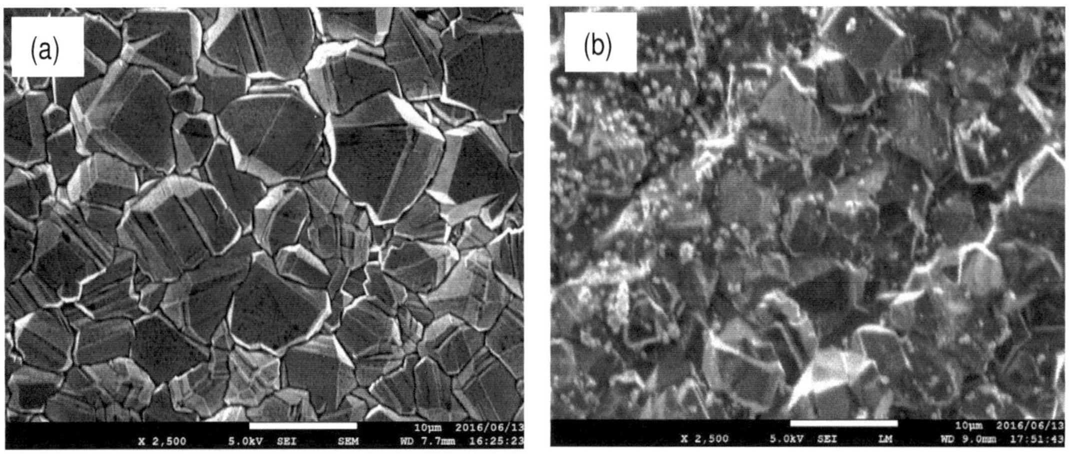

(a) (b)

Figure 1. (a) SEM images of unmodified BDD and (b) Pt-BDD electrode surfaces.

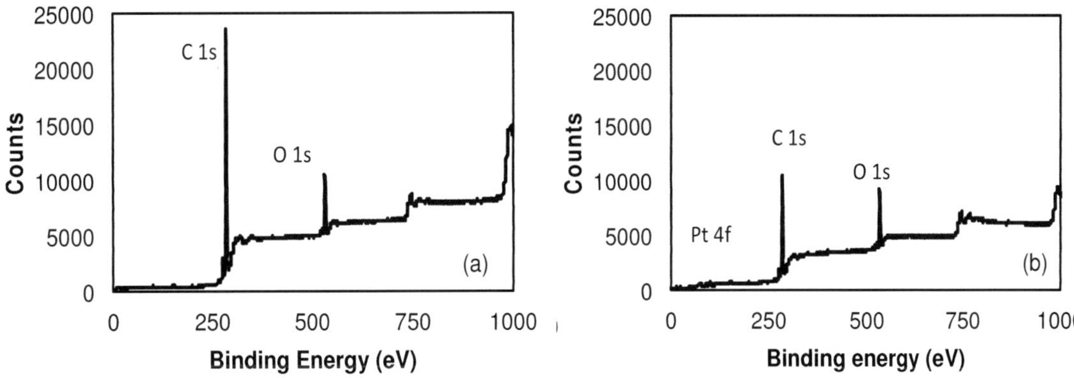

(a) (b)

Figure 2. XPS spectra (a) BDD and (b) Pt-BDD electrode

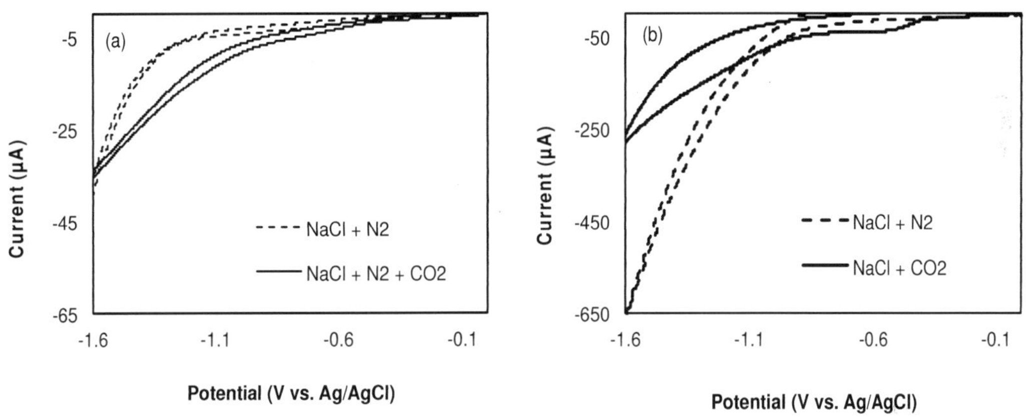

(a) (b)

Figure 3. Cyclic voltammograms of CO_2 in 0.1 M NaCl solution at (a) unmodified BDD and (b) Pt-BDD electrodes

3.2. Cyclic voltammetry of CO_2

Cyclic voltammograms (CVs) of CO_2 in 0.1 M NaCl solutions were performed to obtain the suitable solution to reduce CO_2 using Pt-BDD electrode. Figure 3 shows CVs of CO_2 in 0.1 M NaCl at unmodified BDD (figure 3a) and at Pt-BDD (figure 3b) electrodes. Since the solution was bubbled with N_2 gas to remove other gasses before it was saturated by CO_2 gas, the observed reduction peak can be attributed to the reduction of CO_2. The peak was appeared at a potential of -0.7 V (vs Ag/AgCl) at BDD electrodes, while it was observed at -0.5 V at Pt-BDD, suggested that reduction of CO_2 at Pt-BDD can be performed at lower potential.

4. Conclusions

Preparation of Pt-BDD as a working electrode for CO_2 electrochemical reduction has been successfully performed. It was confirmed by the characterization of electrodes with FESEM and XPS data. Additionally, electrolyte determination suggested that reduction potential of CO_2 at Pt-BDD was occurred at a lower potential than at unmodified BDD.

Acknowledgements
This work was funded by Hibah Desentralisasi PUPT Universitas Indonesia, contract No. 1127/UN2.R12/HKP.05.00/2016.

References

[1] Nakata K, Ozaki T, Terashima C, Fujishima A and Einaga Y 2014 *Angew. Chemie - Int. Ed.* **53** 871-4
[2] Jitaru M, Lowy D A, Toma M, Toma B C and Oniciu L 1997 *J. Appl. Electrochem.* **27** 875-89
[3] Saeki T, Hashimoto K, Kimura N, Omata K and Fujishima A 1996 *J. Electroanal. Chem.* **404** 299-302
[4] Yang N, Gao F and Nebel C E 2013 *Anal. Chem.* **85** 5764-9
[5] Lee J and Tak Y 2001 *Electrochim. Acta* **46** 3015-22
[6] Ivandini T A, Sato R, Makide Y, Fujishima A and Einaga Y 2004 *Chem. Lett.* **33** 1330-1
[7] Ivandini T A, Sato R, Makide Y, Fujishima A and Einaga Y 2005 *Diamond Relat. Mater.* **14** 2133-8

Different heat treatment of CeO_2 nanoparticle composited with ZnO to enhance photocatalytic performance

A Taufik[1,2], H Shabrany[1,2] and R Saleh[1,2]

[1]Department of Physics, Faculty of Mathematics and Natural Sciences Universitas Indonesia, Kampus UI Depok, Depok 16424, Indonesia
[2]Integrated Laboratory of Energy and Environment, Faculty of Mathematics and Natural Sciences Universitas Indonesia, Kampus UI Depok, Depok 16424, Indonesia

Corresponding author's e-mail: rosari.saleh@gmail.com

Abstract. In this study, ZnO/CeO_2 nanocomposites were prepared with four variations of the molar ratio of ZnO to CeO_2 nanoparticles. Both ZnO and CeO_2 nanoparticles were synthesized using the sol-gel method at low temperature, followed by different heat treatments for CeO_2 nanoparticles. Thermal phase transformation studies of the CeO_2 nanoparticles were observed at annealing temperatures of 400-800°C. The complete crystalline structure of CeO_2 nanoparticles was obtained at an annealing temperature of 800°C. The structural and optical properties of all samples were observed using several characterization techniques, such as X-ray diffraction (XRD), ultraviolet-visible diffuse reflectance spectroscopy, and Brunauer, Emmett, and Teller (BET) surface area analysis. The structural characterization results revealed that the prepared CeO_2 nanoparticles were quite crystalline, with a cubic structure. The photocatalytic activities of all samples were tested under visible irradiation. The obtained results showed that ZnO/CeO_2 nanocomposites with a molar ratio 1:0.3 exhibited the highest photocatalytic activity. Further understanding of the role of primary active species underlying the reaction mechanism involved in photocatalytic activity were carried out in controlled experiments by adding several scavengers. The detailed mechanism and its correlation with the properties of ZnO/CeO_2 nanocomposites were discuss.

Keywords: CeO_2, ZnO, photocatalytic, methylene blue

1. Introduction

Photocatalytic semiconductors have been widely used in organic dye removal due to their good ability to degrade organic dye [1]. Among multiple semiconductors, zinc oxide (ZnO) is the favored photocatalyst due to its good photocatalytic activity, eco-friendliness and low cost [2-3]. However, the ZnO photocatalyst still has limitations due to the high rate of electron recombination and holes; moreover, it can only be activated using ultraviolet (UV) light irradiation [4].

To overcome these problems, combining a ZnO semiconductor with another semiconductor has proven to increase the photocatalytic activity [5-7]. Cerium (IV) oxide (CeO_2) is an n-type semiconductor that has been commonly used as a supporting photocatalyst to increase photocatalytic activity [8-9]. The coupled CeO_2 semiconductor with ZnO is able to inhibit the electron recombination and hole rate through the process of electron and hole transfer from one semiconductor to the others [10].

The photocatalytic process is also influenced by the crystal structure of the photocatalytic material. Some researchers have stated that the better the crystal structure the photocatalyst material produces, the better the photocatalytic activity will be [11-12]. This is due to the increasing charge carrier mobility of the photocatalyst [13]. Therefore, in this research, the photocatalytic activity that results from merging a ZnO semiconductor with CeO_2 under irradiating UV light and visible light is tested. In addition, the influence of the crystallized CeO_2 is analyzed through variations in calcination from 400°C to 800°C.

2. Experimental details

Cerium sulfate pentahydrate ($CeSO_4.5H_2O$), zinc sulfate heptahydrate ($ZnSO_4.7H_2O$), sodium hydroxide (NaOH), ethanol, and water were used as the starting material without further purification. CeO_2 nanoparticles were synthesized using the sol-gel method. Typically, $CeSO_4.5H_2O$ were dissolved into the water using a magnetic stirrer. Then, NaOH solution was added to the $CeSO_4.5H_2O$ solution. Next, the solution was heated to 80°C for 4 h. After that, the precipitates were separated from the solution using the centrifugation method. The precipitates were dried under vacuum condition, and finally, the dried sample was calcinated at temperatures of 400°C, 600°C, and 800°C.

ZnO/CeO_2 nanocomposites were synthesized using the sol-gel method. Typically, $ZnSO_4.7H_2O$ was dissolved in water using a magnetic stirrer, then NaOH solution was added dropwise into $ZnSO_4.7H_2O$ solution. Following this, the solution was heated to 80°C. After reaching a temperature of 80°C, the CeO_2 nanoparticles were dispersed in water using an ultrasonic bath for 30 min; dispersed CeO_2 nanoparticles were then added to the solution. The solution was stirred for 2 h. After the mixing process, the precipitates were separated using a centrifuge and washed using ethanol and water several times. The precipitates were aged overnight and then dried at a temperature of 100°C for 1 h under vacuum conditions.All samples were characterized using X-ray diffraction (XRD; Rigaku Miniflex 600), Brunauer, Emmett, and Teller (BET) surface area analysis (Quantachrome 2000), and UV-visible (UV-vis) diffuse reflectance spectroscopy (Shimadzu spectrophotometer).

The photocatalytic activity was tested under visible light irradiation with methylene blue as a model of organic pollutant. Typically, 0.3 g of the photocatalyst was dispersed in a 100 mL glass beaker containing 20 mg/L of methylene blue solution for 30 min under dark conditions to reach adsorption and desorption equilibrium. Following this, the samples were irradiated using visible light for 120 min. At 15-min intervals, the methylene blue solution was taken and analyze using UV-Vis spectroscopy. The methylene blue degradation was calculated based on the following equation (1) bellow:

$$\text{Degradation} = \frac{C_t}{C_0} \tag{1}$$

where C_0 is the methylene blue concentration before the photocatalytic process (mg/L) and C_t is the concentration of methylene blue for every 15-min span of data retrieval (mg/L).

In the photocatalytic activity experiment, the influence of additional scavengers was also analyzed; this analysis functioned to detect the species with the greatest roles in the photocatalytic activity. The scavenger used included sodium sulfate as an electron scavenger, ammonium oxalate as a hole scavenger, and tert-butyl alcohol as a hydroxyl radical scavenger.

3. Results and discussion

Figure 1(a) shows the methylene blue degradation using the CeO_2 photocatalyst with different annealing temperatures. The results show that CeO_2 nanoparticles with an annealing temperature of 800°C exhibited the best photocatalytic performance under visible light irradiation. The methylene blue degradation rate was analyzed using the pseudo–first order equation, as seen in figure 1(b); the results showed that the methylene blue degradation rate using CeO_2 photocatalyst before the annealing process was 0.00636 min^{-1}, and with the increasing of the annealing temperature to 400°C, 600°C, and 800°C, the degradation rate of methylene blue increased to 0.00719 min^{-1}, 0.00855 min^{-1}, and 0.00939 min^{-1},

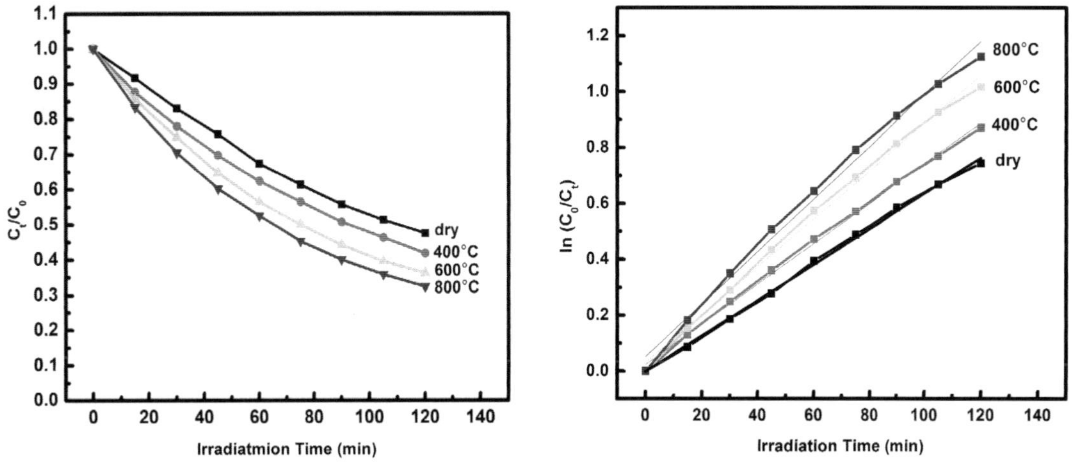

(a) (b)

Figure 1. (a) Photocatalytic and (b) linear plot ln (C_0/C_t) vs time of CeO_2 nanoparticles with different annealing temperatures

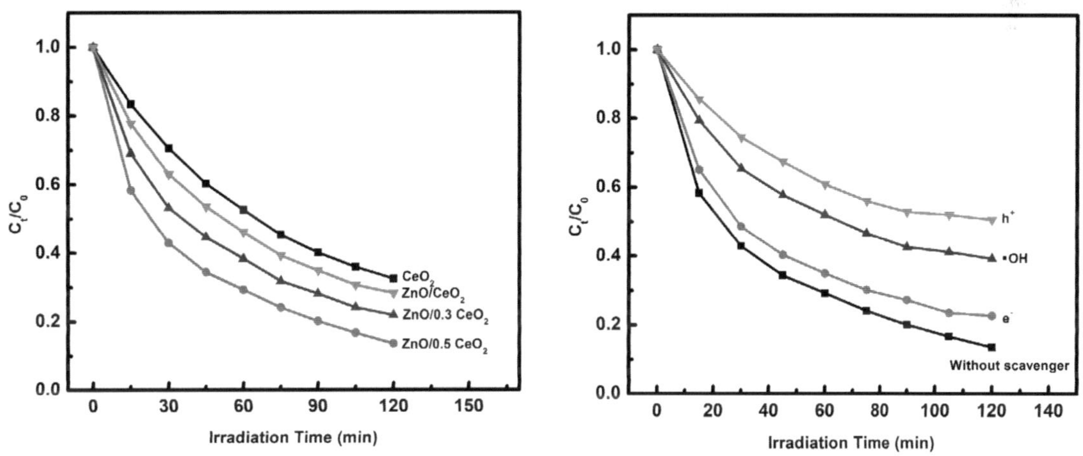

(a) (b)

Figure 2. (a) Photocatalytic activity of ZnO/CeO_2 nanocomposites with different molar ratio and (b) Effect of scavenger addition on photocatalytic performance of ZnO/CeO_2 nanocomposites.

respectively. Therefore, the CeO_2 nanoparticles with an annealing temperature of 800°C were used as composite material with the ZnO semiconductor.

The methylene blue degradation efficiency of the ZnO/CeO_2 (1:0.3, 1:0.5, 1:1) nanocomposite samples under visible light irradiation are shown in figure 2(a). The findings show that the usage of ZnO/CeO_2 nanocomposite with a molar ratio 1:0.5 resulted in the best degradation ability under visible light irradiation.

Figure 2(b) shows the influence of scavenger addition to the photocatalytic activity of the ZnO/CeO_2 sample with molar variation of 1:0.5 under visible light irradiation. The results show that the addition of scavengers decreased the photocatalytic activity of the sample. This occurred due to the scavenger addition captured electron, hole, and OH radicals, which played a role in the photocatalytic process in association with each scavenger used. It can be seen that the addition of ammonium oxalate resulted in the lowest photocatalytic activity under visible light irradiation. This finding indicates that the hole was the most active species in terms of the degradation of methylene blue.

Figure 3(a) shows the XRD measurement of CeO_2 nanoparticles with different annealing temperatures. The XRD result showed that the CeO_2 nanoparticles had a cubic structure with planes of diffraction of (111), (200), (220), (311), (222), (400), (331), (420), and (422). It can be seen that with an increasing annealing temperature, the diffraction pattern CeO_2 became sharper. Thus, improvement was observed in the crystallized sample. The grain size and lattice parameter of the CeO_2 nanoparticles for all annealing temperatures were analyzed using the Debye–Scherrer equation [14] and Rietveld refinement model. The obtained result is shown in table 1. The grain size of the sample increased from 2 nm to 25 nm.

Figure 3(b) shows the XRD curve of the ZnO/CeO_2 nanocomposites with molar ratios of 1:0.3, 1:0.5, and 1:1. For comparison, the XRD curves of ZnO and CeO_2 are also shown in the figure. It can be seen that the existence of the hexagonal wurtzite phase of ZnO and the cubic phase of CeO_2 could be detected from the XRD measurement. The ZnO structures were successfully detected at $2\theta = 31°$, 34°, and 36°, which indicates the existence of planes (100), (002), and (101), while $2\theta = 28°$, 32°, 47°, 56°, and 59° indicated the existence of planes (111), (200), (220), (311), (222), and (400), corresponding to the cubic phase of CeO_2. It can be observed that the diffraction peak of the cubic phase of CeO_2 increased with greater CeO_2 content in the ZnO/CeO_2 nanocomposite. The lattice parameter and grain size are also shown in table 1. The results show that the lattice parameters of ZnO and CeO_2 did not change significantly when they were formed into a nanocomposite. However, the measurement value grain size of ZnO decreased along with the addition of molar CeO_2.

The measurement test of surface area of all samples is shown in table 1. The result obtained shows that along with the increase of annealing temperature, the value of surface area of CeO_2 sample decreased from 76 m^2/g until 13 m^2/g. The decrease of surface area value when annealing temperature was improved caused by the increase of grain size measurement of CeO_2 sample. The value of surface area sample increase from 23 m^2/g until 105 m^2/g when it was formed into ZnO/CeO_2 nanocomposite.

The UV-Vis absorbance of ZnO/CeO_2 with molar ratios of 1:0.3, 1:0.5, and 1:1 are shown in figure 4. The figure demonstrates the light absorbance ability of samples from wavelengths of 200-800 nm.

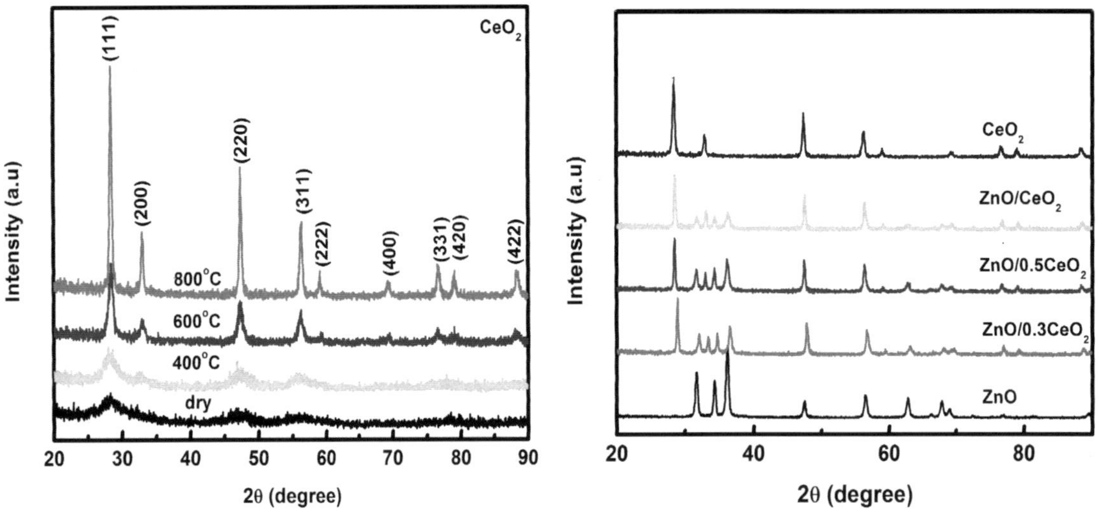

(a) (b)

Figure 3. (a) XRD spectra of CeO_2 nanoparticles and (b) ZnO/CeO_2 nanocomposites with different molar ratio.

Table 1. Lattice parameter, grain size and BET surface area of CeO_2 nanoparticles with different annealing temperatures and ZnO/CeO_2 nanocomposites with different molar ratio.

Sample	Lattice Parameter (Å) CeO_2	Lattice Parameter (Å) ZnO		Grain Size (nm)		Surface Area (m^2/g)
	a	b	a	CeO_2	ZnO	
dry	5.397	-	-	2	-	76
CeO_2 400°C	5.416	-	-	4	-	57
CeO_2 600°C	5.419	-	-	9	-	34
CeO_2 800°C	5.421	-	-	25	-	13
$ZnO/0.3\ CeO_2$	5.422	3.259	5.220	23	25	105
$ZnO/0.5\ CeO_2$	5.414	3.253	5.216	21	20	45
ZnO/CeO_2	5.413	3.253	5.212	21	19	23

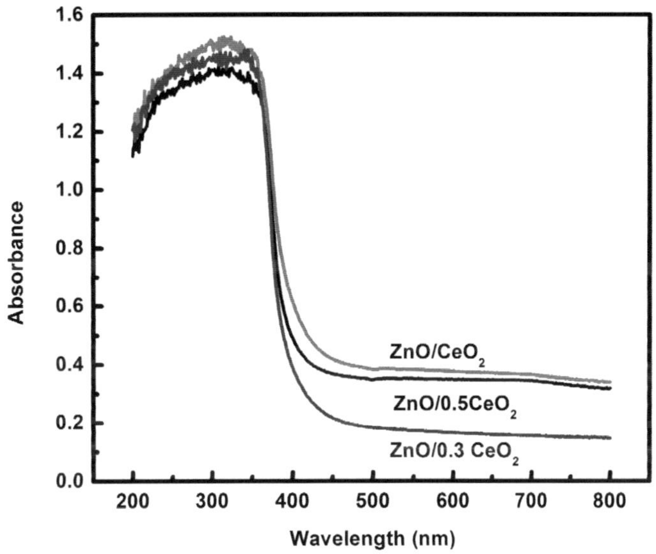

Figure 4. UV Vis absorbance spectra of ZnO/CeO_2 nanocomposites.

The results showed that the light adsorption ability of ZnO/CeO_2 nanocomposite in the visible light range increased with greater CeO_2 loading. The band gap energy of all the nanocomposite samples was calculated with the Kubelka–Munk equation [15]. The results showed that the band gap energy of the sample exhibited a slight decrease with increased CeO_2 loading. This indicates that the ZnO/CeO_2 nanocomposite sample may be activated under visible light irradiation.

Based on the photocatalytic measurement, the photocatalytic activity of the CeO_2 sample increased with the addition of annealing temperature. This was because the higher crystallinity of the sample plays an important role in the photocatalytic process. As could be seen in the XRD measurement, the annealing temperature of 800°C resulted in the highest crystallization. With the higher crystallinity of the sample, the mobility of the charge carrier was also improved, and the photocatalytic activity could be enhanced [13]. In this case, the surface area of the samples did not have a significant influence on the photocatalytic process.

The formation of ZnO/CeO_2 nanocomposites can increase the methylene blue degradation ability due to the charge transfer from each band structure of ZnO and CeO_2. Some studies showed that the conduction and valence bands of the ZnO and CeO_2 positions allow the electron and holes to transfer from one semiconductor to the other. The charge transfer process can inhibit the recombination of electrons and holes; thus, the photocatalytic activity increases [10].

4. Conclusions

CeO_2 nanoparticles with an annealing temperature of 800°C showed the highest crystallinity and photocatalytic performance. The surface area value did not have much influence on the photocatalytic process. The formation of ZnO/CeO_2 nanocomposites is able to increase the photocatalytic activity capacity under the irradiating visible light caused by recombination electron and hole inhibition.

References

[1] Chong R, Cheng X, Wang B, Li D, Chang Z and Zhang L 2016 *Int. J. Hydrogen Energy* **41** 2575-82
[2] Bao Y, Wang C and Ma J Z 2016 *Mater. Des.* **101** 7-15
[3] Zhu C, Lu B, Su Q, Xie E and Lan W 2012 *Nanoscale* **4** 3060-4
[4] Yin H, Yu K, Song C, Huang R and Zhu Z 2014 *ACS Appl. Mater. Interfaces* **6** 14851-60
[5] Shirzadi A and Ejhieh A N 2016 *J. Mol. Catal. A: Chem.* **411** 222-9
[6] Mageshwari K, Nataraj D, Pal T, Sathyamoorthy R and Park J 2015 *J. Alloys Compd.* **625** 362-70
[7] Kalpana K and Selvaraj V 2015 *Ceram. Int.* **41** 9671-9
[8] García M E C *et al.* 2014 *Mater. Sci. Eng.,* B **183** 78-85
[9] Lv Z, Zhong Q and Ou M 2016 *Appl. Surf. Sci.* **376** 91-6
[10] Lamba R, Umar A, Mehta S K and Kansal S K 2015 *J. Alloys Compd.* **620** 67-73
[11] Li D, Cheng X, Yu X and Xing Z 2015 *Chem. Eng. J.* **279** 994-1003
[12] Chong R, Cheng X, Wang B, Li D, Chang Z and Zhang L 2016 *Int. J. Hydrogen Energy* **41** 2575-82
[13] Zhang S, Gu X, Zhao Y and Qiang Y 2015 *Mater. Sci. Eng., B* **201** 57-65
[14] Srivastava S, Sinha R and Roy D 2004 *Aquat. Toxicol.* **66** 319-29
[15] Hapke B 1993 *Theory of Reflectance and Emittance Spectroscopy* (Cambridge: University Press)

IOP Conf. Series: Materials Science and Engineering **188** (2017) 012039 doi:10.1088/1757-899X/188/1/012039

Effect of Fe_3O_4 addition on dielectric properties of $LaFeO_3$ nano-crystalline materials synthesized by sol-gel method

H Laysandra and D Triyono

Department of Physics, Faculty of Mathematics and Natural Sciences Universitas Indonesia, Kampus UI Depok, Depok 16424, Indonesia

Corresponding author's e-mail: djoko.triyono@sci.ui.ac.id

Abstract. Dielectric properties of nano-crystalline material $LaFeO_3.xFe_3O_4$ with x = 0, 0.1, 0.2, 0.3, and 0.4 at.% have been studied by impedance spectroscopy method. $LaFeO_3$ was synthesized by sol-gel method resulting nano-particle. Then, it was mixed with Fe_3O_4 powder. The mixture powder was pressed to form pellet and then sintered at 1300°C for 1 h to form nano-crystalline of $LaFeO_3.xFe_3O_4$. X-ray diffraction characterization at room temperature for all samples show two phases i.e. perovskite $LaFeO_3$ (orthorhombic) as a main phase and Fe_3O_4 (cubic) as second phase. It is found that the crystallite size of main phase increases with addition of Fe_3O_4 until 0.3 at.%. The electrical properties as a function of temperature (300-500 K) and frequency (100 Hz - 1 MHz) are presented in Nyquist and Bode plots. It is observed that from equivalent circuit and their parameters, dielectrical properties are contributed by grain and grain boundary. The dielectric constant, ε' were calculated by parallel plate method and their values reach up to 10^7 exhibiting typical colossal dielectric constant (CDC) material like behavior.

1. Introduction

$LaFeO_3$ is one of the most important multifferoic perovskite materials. It has been wide range in applications, such as electrodes materials for fuel cells and chemical sensors [1]. $LaFeO_3$ shows electrical and magnetic behaviour with complex correlation. $LaFeO_3$ is antiferromagnetic with Neel temperature of 740 K [2]. The electrical properties such as dielectric constant and conductivity can be improved by substitution on A and/or B sites [1]. $LaFeO_3$ ceramics have been reported to exhibit colossal dielectric constant (dielectric constant typically more than 10^4) with weak temperature-dependence [3]. Chen *et al.* [4] reported colossal dielectric behaviour of $LaFeO_3$ in the temperature range 80-315K. Bhargav *et al.* [3] reported colossal dielectric behaviour $La_{1-x}Pb_xFeO_3$ with x = 20 at.% Pb^{2+} with frequency-dependence behavior in the range of 300-500K. Effect perovskite materials composites with Fe_3O_4 have been reported to structural and electrical transport properties. Z. C. Xia *et al.* [5] found that electrical transport properties of $La_{0.67}Ca_{0.33}MnO_3/Fe_3O_4$ are directly related to the Fe_3O_4 content. In this paper, we have synthesized $LaFeO_3$ by sol-gel method and added Fe_3O_4 to form composites of $LaFeO_3.xFe_3O_4$ with x = 0.0, 0.1, 0.2, 0.3, and 0.4 at.%. We focused on the effect of additional Fe_3O_4 on the structural and dielectric properties of composites of $LaFeO_3.xFe_3O_4$ materials.

2. Materials and methods

In this work, we mainly analyze the effects of Fe_3O_4 compound in nano-crystalline $LaFeO_3.xFe_3O_4$ on their electrical properties. The samples were prepared by three steps: First, powder samples of $LaFeO_3$ was synthesized by sol-gel method, and then characterized by XRD to make sure that its phase is formed. The so-obtained $LaFeO_3$ powders were calcined at 900°C for 6 hours. Second, $LaFeO_3$ powder were mixed with Fe_3O_4 powder and grind to form homogeneous powders of $LaFeO_3.xFe_3O_4$ with x = 0, 0.1, 0.2, 0.3, and 0.4 at.%, respectively. Finally, the powders were pressed into pellet with diameter of 10 mm at pressure of 5kN and sintered at 1300°C for 1 h. The phase identification has done by using X-Ray Diffractometer (XRD) with Cu K (λ = 1.5406 Å) radiation source in the 2θ range of 10°- 90° with a step size of 0.02°. Rietveld refinement of the diffraction pattern was carried out using the HighScore Plus software. Electrical properties of $LaFeO_3.xFe_3O_4$ were characterized by *RLC-Meter FLUKE-PM6306* at frequency between 100 Hz until 1 MHz with the temperature range of 300-500 K.

3. Results and discussion

3.1. Structure analysis

Figure 1 shows XRD pattern of nano-crystalline $LaFeO_3.xFe_3O_4$ with x = 0.1, 0.2, 0.3, and 0.4 at.% at room temperature. HighScore Plus software was used to analyze all of XRD pattern. All samples exhibit two phases i.e. $LaFeO_3$ orthorhombic structure as a main phase and Fe_3O_4 cubic structure as a second phase. The Bragg angle, 2θ were observed shifted insignificantly for addition of Fe_3O_4 but the crystal structure of $LaFeO_3$ is still remain. The refined structural parameters are tabulated in table 1.

Table 1. Refined lattice parameters and crystallite size of $LaFeO_3.xFe_3O_4$ bulks sintered at 1300°C for 1 h in air.

Samples	$LaFeO_3.xFe_3O_4$				
	x=0	x=0.1	x=0.2	x=0.3	x=0.4
Lattice parameters of $LaFeO_3$					
a	5.5578	5.5686	5.5621	5.5660	5.5770
b	7.8500	7.8515	7.8649	7.8572	7.8713
c	5.5467	5.5538	5.5469	5.5574	5.5540
Crystallite size of $LaFeO_3$ (nm)	25.0	30.1	33.1	34.3	23.8

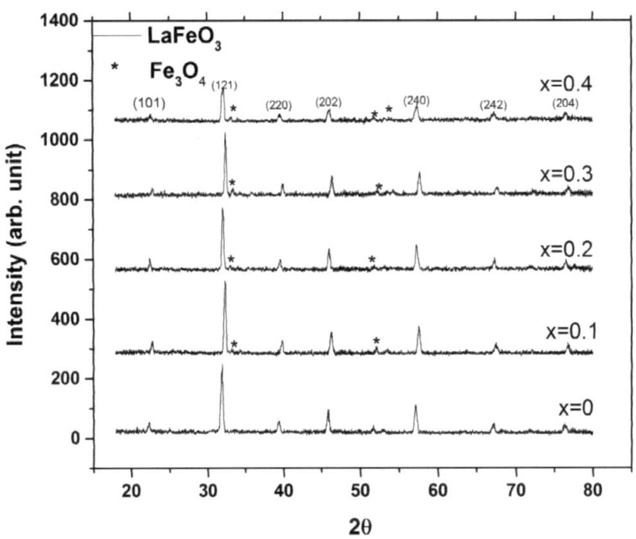

Figure 1. The XRD patterns of $LaFeO_3.xFe_3O_4$ bulks sintered at 1300°C for 1 h in air.

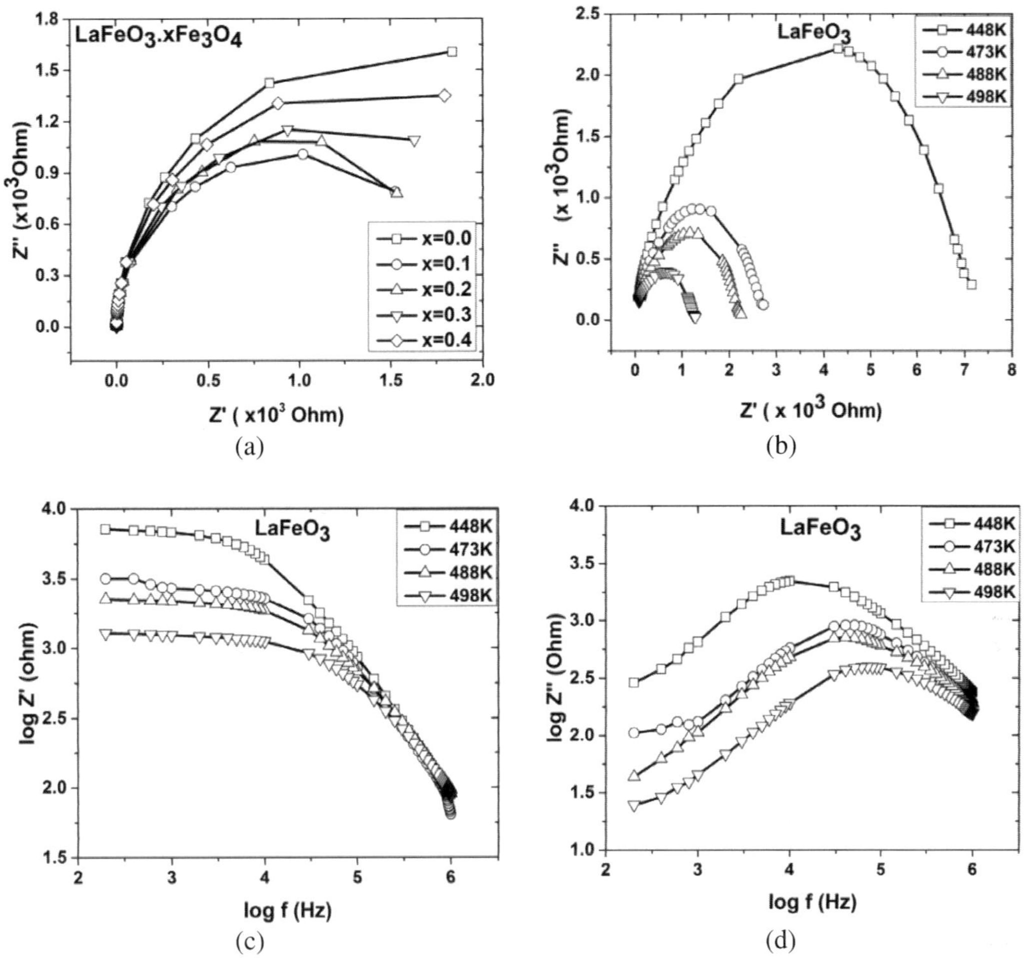

Figure 2. Nyquist plots of (a) $LaFeO_3.xFe_3O_4$ at room temperature and (b) $LaFeO_3$ as a function of temperature; and (c) real and (d) imaginary parts of impedance with frequency at different temperatures for $LaFeO_3$ compounds.

As shown in the table 1, the crystallite size of $LaFeO_3$ is significantly increased with the increasing of Fe_3O_4 content until x = 0.3 (at.%). The similar phenomenon is found by Xia *et al.* [5] for $La_{0.67}Ca_{0.33}MnO_3$ composited with Fe_3O_4. This behavior is due to the Fe_3O_4 composition on composite $LaFeO_3.xFe_3O_4$ (x) that exceeds 0.3, which is able to cause the slowing on the crystal growth process. With the higher composition of Fe_3O_4 to the composite, it would increase the blockade of crystal growth of $LaFeO_3$, lead to decrease of crystallite size [5].

3.2 Electrical properties

The Nyquist plots of $LaFeO_3.xFe_3O_4$ are shown in figure 2. The impedance spectrum is showed by the semicircular arcs, whose changes with increasing of temperature. Nyquist plot in figure 2a and 2b show similar equivalent circuit. The high frequency semicircle can be attributed to the bulk (grain) property, and the low frequency semicircle can be attributed to the grain boundary property of the material that arising due to equivalent circuits of two parallel resistance-capacitance (RC)(RC) elements similar as reported by Triyono *et al.* [6] for $Sr_2(Fe,Ti)O_6$ compounds. Figure 2c and 2d show the bode plots of the $LaFeO_3$ ceramics at different temperatures. Figure 2c shows the *Z'* value of *Z'* (*f*) decrease in increasing

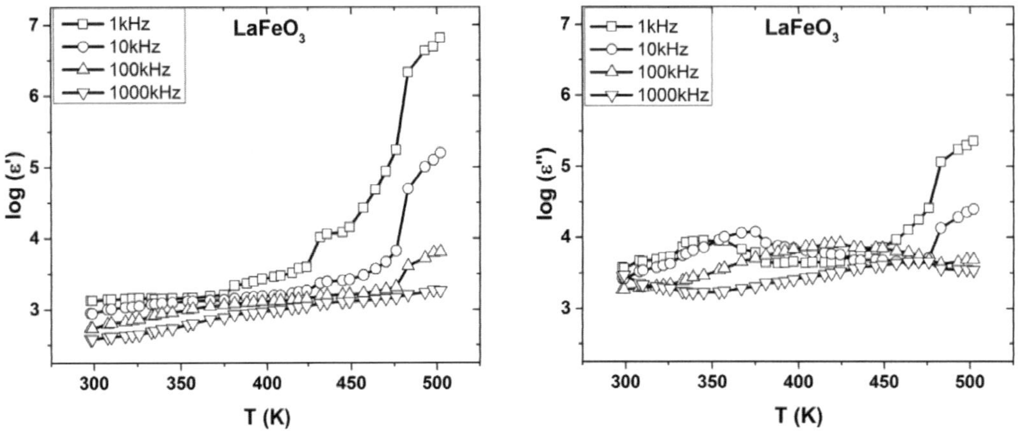

(a)

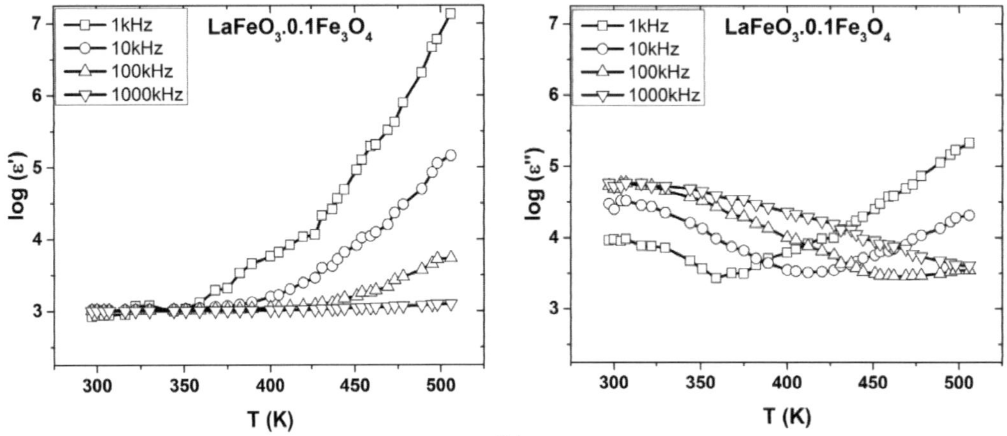

(b)

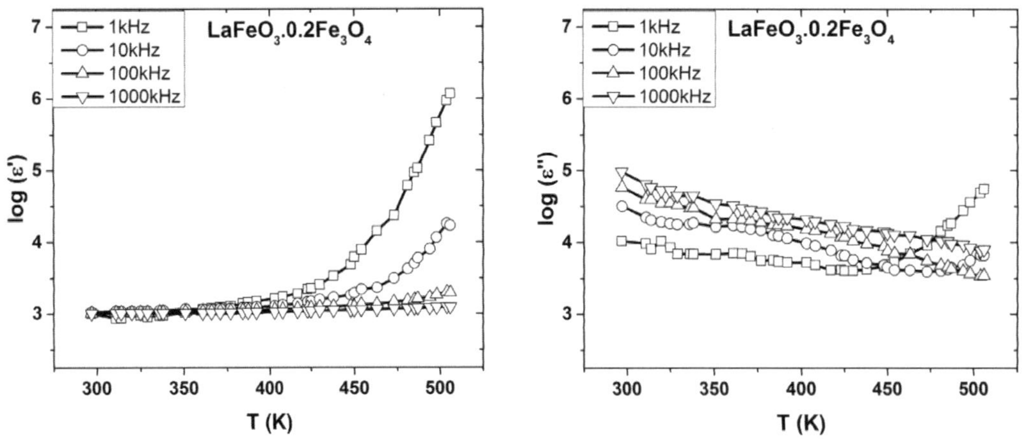

(c)

Figure 3. Variation of real ε' and imaginary ε'' part of dielectric constant with temperature at different frequencies for $LaFeO_3.xFe_3O_4$ composites (a) x = 0, (b) x = 0.1 and (c) x = 0.2.

of temperature. This directly related to the increasing of ac conductivity. Figure 2d shows the value of *Z"* increased with increasing of frequency until exhibited a maximum value before starts to decrease rapidly. The maximum values shift to higher frequency with increase of temperature indicate there is occuring temperature dependent electrical relaxation phenomenon [7].

Figure 3 shows the temperature dependent real (ε') and imaginary (ε'') part of dielectric constant at different frequencies for $LaFeO_3.xFe_3O_4$ from 300 K to 500 K. We have identified that the dielectric constant is decreasing with decreasing of temperature, which is exhibited in all materials. At higher temperatures, the real part of dielectric constant of $LaFeO_3.xFe_3O_4$ shows increasing order of 10^4 and it decreased at low temperature with relatively weak temperature dependence. Interestingly, the temperature-dependent behavior of dielectric observed on all samples is indicated as an observed behavior of typical colossal dielectric constant (CDC) ceramics [3]. The highest dielectric constant at high temperature is achieved for the sample with x = 0.1 i.e. until 10^7. In the high temperature region, higher value of dielectric constant may be related polarization which comes from mobility of ions, Fe^{3+}-Fe^{3+} and imperfections from this material [8]. This combined effects cause a sharp increase in the dielectric constant when temperature increases. We have measured the frequency dependence of dielectric constant especially for x = 0.1 in the temperature range of 300-500 K as presented in figure 4. Figure 4 shows the frequency dependence of dielectric constant of $LaFeO_3.0.1Fe_3O_4$. The dielectric spectra as a function of temperature of all samples are similar. At low frequency, the increasing of dielectric constant at x = 0.1 is observed, where the dielectric constant reaches 10^4. At high temperature, the dielectric constant increases significantly as the frequency decreases. And the dielectric constant changes very little with the frequency at room temperature. The larger dielectric constant at low frequencies is related to stronger polarization at low frequencies of this material [9].

The dielectric constant (real and imaginary parts) decrease when increase in frequency and become independent in higher frequency. This phenomenon can be explained by space charge polarization which is related by presence of fairly conducting grains in poorly conducting grain boundaries [10]. From figure 4, we observe that increase of dielectric constant with temperature may be due to thermal fluctuation [11]. The similar phenomenon is found by Kumari *et al.* [10] for Cr^{3+} substituted Fe_3O_4 and El-Mallah [11] for (Pb,Ca)TiO_3 ceramics. However, in our samples show much higher in the values of dielectric constant.

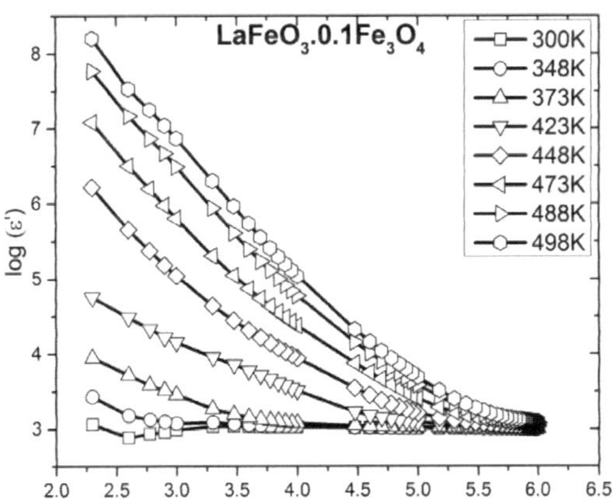

(a)

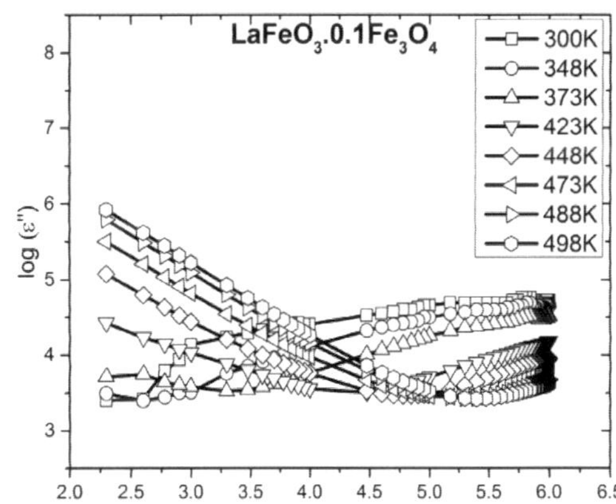

(b)

Figure 4. Real (ε') and imaginary (ε") part of dielectric constant as a function of frequencies at 300-500K for $LaFeO_3.0.1Fe_3O_4$ composites.

4. Conclusions

$LaFeO_3.xFe_3O_4$ with x = 0.0, 0.1, 0.2, 0.3, and 0.4 at.% compounds has been successfully synthesized by sol-gel method. XRD characterization for all samples show two phases i.e. $LaFeO_3$ orthorhombic structure as a main phase and Fe_3O_4 cubic structure as a second phase. The electrical properties are indicated by grain and grain boundary contribution. The dielectric constant, ε' values reach up to 10^7 exhibiting typical colossal dielectric constant (CDC) material like behaviour.

Acknowledgements
The authors would like thanks to HIBAH PITTA (2007/UN2.R12/HKP.05.00/2016) from Universitas Indonesia for financial support.

References

[1] Phokha S, Hunpratup S, Pinitsoontorn S, Putasaeng B, Rujirawat S and Maensiri S 2015 *Mater. Res. Bull.* **67** 118-25
[2] Devi Chandrasekhar K, Mallesh S, Krishna Murthy J, Das A K and Venimadhav A 2014 *Physica* B **448** 304-11
[3] Bhargav K K, Ram S and Majumder S B 2015 *J. Alloys and Compounds* **638** 334-43
[4] Chen C, Xu K B, Cui Y M and Wang C C 2012 *Mater. Lett.* **89** 153-5
[5] Xia Z C *et al.* 2003 *Solid State Comm.* **126** 567-71
[6] Triyono D and Laysandra H 2016 *AIP Conf. Proc.* **1729** 20-2
[7] Dhak P, Dhak D, Das M and Pramanik K 2009 *Mater. Sci. and Eng.* B **164** 165-71
[8] Sutar B C, Choudhary R N P and Das P R 2014 *Ceramics International* **40** 7791-8
[9] Wang J, Wu F Q, Shi K H, Wang X H and Sun P P 2004 *Sensors and Actuators* B **99** 586-5
[10] Kumari N, Kumar V and Singh S K 2014 *Ceramics International* **40** 12199-205
[11] El-Mallah H 2012 *Acta Physica Polonica* A **122** 174-9

The composition of maja leaf extract (*Aegle marmelos*) and silica gel for urea biosensor

Y Triana[1], T P H Hutapea[2] and F Kurniawan[3]

[1]Material and Metallurgical Engineering Department, Kalimantan Institute of Technology, Jl Sukarno Hatta KM. 15, Karang Joang, Balikpapan, Kalimantan Timur 76127, Indonesia
[2]Faculty of Fisheries and Marine, University of Borneo Tarakan, Jl Amal Lama No 1, Tarakan, Kalimantan Utara, Indonesia
[3]Chemistry Department, Institut Teknologi Sepuluh Nopember, Surabaya, Jawa Timur 60111, Indonesia

Corresponding author's e-mail: nita@itk.ac.id

Abstract. In this work, we used Indonesian-naturalized plant, Maja (*Aegle marmelos*) leaf extract instead of urease enzyme for urea sensor. Maja leaf extract, silica gel, and paraffin were mixed until homogenous, and then pasted to the tip of the silver wire, which was tightly covered by glass tube to assemble a silica gel-Maja leaf extract modified silver electrode. Cyclic voltammetry showed that cathodic and anodic current signals were observed at 0.4 V and 0.625 V, respectively. The optimum composition of silica and Maja leaf extract were also investigated. The best response was observed at the paste composition ratio of silica gel and paraffin of 3:2, with 0.05 gr total weight and 0.03 gr Maja leaf extract.

Keywords: silica gel, urea biosensor, Maja leaf extract, urea, Aegle marmelos

1. Introduction

Urea plays an important role in many biological processes. Human body produces 20 to 30 grams of urea each day [1]. The detection of urea is important in clinical diagnosis. Various methods to detect urease enzyme were established, including optical methods [2] and colorimetry methods [3]. However, some of these methods need a complicated instrument, relatively expensive process, less sensitive in detecting urea, and unavailability of the portable equipment. Therefore, in this research, another alternative method is used, that is electrochemical nonenzymatic biosensor using the Maja leaf extract (*Aegle marmelos*).

Coarse extract from Maja leaf (*Aegle marmelos*) have been used since ancient period to cure various illnesses like asthma, inflammation of mucous membrane, acute bronchitis, and as antidiabetic formula [4]. Elements contained in the concentrated extract of Maja leaf and ethanol are zinc, chromium, iron, selenium, magnesium, copper, and phosphor [5]. One of these elements is expected to be able to decompose urea, following the urease enzyme mechanism. Method used in this research is electrochemical technique using silica gel and Maja (*Aegle marmelos*) extract modified silver electrode. Electrochemical performanceof the electrode and optimum condition for glucose biosensor also studied in this research.

2. Materials and methods

2.1 Equipments
The equipments used in this research were measuring pipette, Ohaus® analytical scale, Thermo Scientific® oven, Oakton® pH meter, hot plate, micropipette, spatula, 25 mL measuring flask, watch glass, glass bottle, *ultrasonic bath*, a set of evaporation equipment, a set of distillation equipment, silver wire (1 mm diameter and 50 mm length) and Autolab Metrohm® potentiostat type AUT84948 with three electrodes system. The three electrode cells were silica gel/Maja leaf extract modified electrode for working electrode, Ag|AgCl (KCl 3M) for reference electrode, and platinum wire for auxiliary electrode.

2.2 Materials
Maja leaves were taken from the tree which was cultivated in Institut Teknologi Sepuluh Nopember (ITS) Surabaya, Indonesia. Silica, glucose, urea, and ethanol p.a wet were purchased from sigma aldrich. Paraffin was purchased from Bratachem. Silicon carbide sandpaper grade 1200, N_2 gas, and HVS paper were obtained from local market. Demineralized water was used for dilution in all experiment.

2.3 Preparation of Maja leaf extract
Maja leafs were dried for 5 days at room temperature. The 150 g dried Maja leaf was extracted in 1 L ethanol using maceration technique for 72 hours. The remaining Maja leaf was discarded. The Maja extract solution was evaporated using rotavapor at temperature of 60°C until no more solvent is evaporated (about 3 hours). The residue is Maja leaf extract. This Maja leaf extract was kept in the refrigerator to maintain the activity of the Maja leaf extract. This Maja leaf extract will be used as the electrode active material for further experiments.

2.4 Preparation of silver electrode
Silver wire with purity of 99.97% was used as electrode. The diameter is 1 mm with 5 cm length. The electrode was subsequently washed with 70% ethanol and aqua DM. Then it was dried at room temperature. The electrode was cleaned using ultrasonic apparatus with operational time of 20 minutes, and then dried at room temperature.

2.5 Preparation of silica gel /Maja leaf extract modified silver electrode
Silica gel, solid paraffin, and Maja leaf (*Aegle marmelos*) extract were mixed with certain variation of mass. The mixture was stirred inside a watch glass until a paste-like mixture is formed. Then, the paste was inserted manually into the glass of the electrode using spatula until homogenous. Finally, the surface of the electrode was polished until smooth and flat using a piece of sand paper.

2.6 Characterization of silica gel/Maja leaf modified silver electrode
The silver electrode modified with silica gel/Maja leaf extract paste as in the section 1.5 was tested on glucose and urea solution. The test was performed by comparing the silica gel/Maja leaf modified silver electrode with silica gel modified silver electrode in both solutions. The concentrations of both solutions were 5 μM at potential of -1 V to +1 V with a scan rate of 100 mV/seconds.

3. Results and discussion
Figure 1 and figure 2 show the response of the silica and silica/Maja leaf extract modified electrodes on glucose and urea, respectively. It shows that the silica/Maja leaf extract modified electrodes give a good response for urea but not for glucose. The comparison of both voltammograms of silica and silica/Maja leaf extract modified electrode can be seen in figure 3.

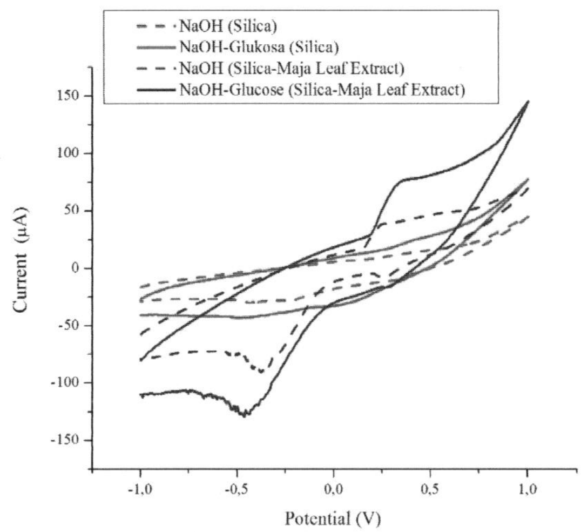

Figure 1. Voltammogram of silica gel and silica gel/Maja leaf extract modified electrode in glucose solution at pH 13

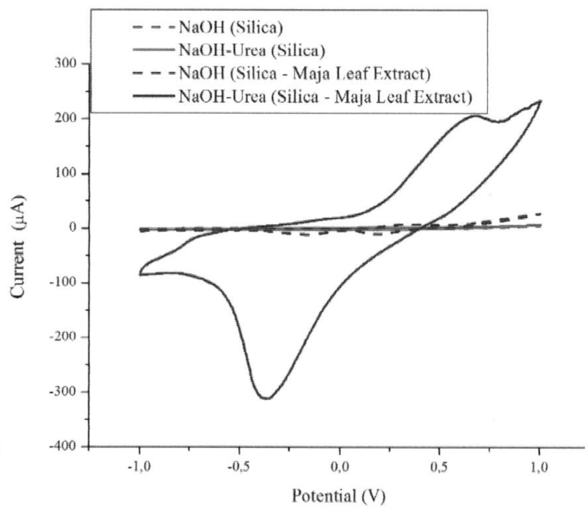

Figure 2. Voltammogram of silica gel and silica gel/Maja leaf extract modified electrode in urea solution at pH 13

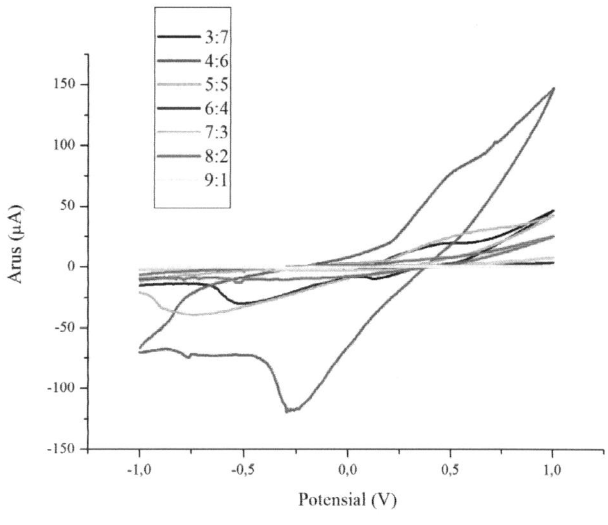

Figure 3. Voltammogram of silica gel and paraffin with variations of pasta composition in urea solution

Based on the voltammogram in figure 3, the best compositions are 0.03 grams silica and 0.02 grams paraffin. From the anodic and cathodic peak current, it can be seen that the peak is higher compared to other compositions. The more silica and the less paraffin composition generated a shifting in potential in the detection of urea. While, the less silica composition and the more paraffin cause the disappearance of anodic and cathodic peak current. This is because the redox process does not take place due to the composition of paraffin is blocking the process.

Based on the voltammogram results in figure 4 the best composition of Maja leaf extract (*Aegle Marmelos*) is 0.03 gram. More than 0.03 grams of the composition causes a potential shifting of the signals. This is because a high content of Maja leaf extract which consists of various compounds will

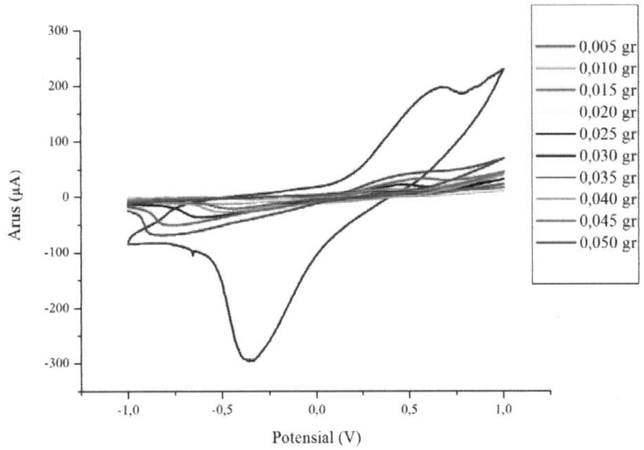

Figure 4. Voltammogram of silica gel and paraffin with variations of leaf extract Maja (*Aegle Marmelos*) composition in urea solution

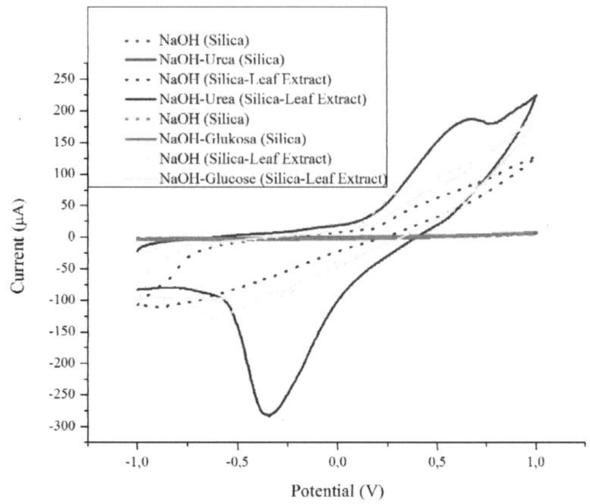

Figure 5. Voltammogram of silica gel and silica gel/Maja leaf extract modified electrode in glucose and urea solution at pH 13

affect the redox process at the electrode surface. In this test, silver electrode was modified with Maja leaf extract and silica gel in paste form. Silica gel has semiconductor property, so it could assist the process of electron transfer, thus the measured electrical current is increased. This matrix change was expected to be able to increase the redox response in electrocatalytic activity of Maja leaf. The test was done by comparing the electrode that only modified with silica gel and the electrode that modified with silica gel and Maja leaf extract. Performance of both electrodes were observed by electrical current responses generated from the blank and tested solution with the same concentration of 5 μM by cyclic voltammetry technique at the range potential of -1 V to +1 V and scan rate of 100 mV/second. This test was performed at pH 13 by using 0.1 M NaOH solution. The obtained voltammogram is shown in figure 5 for all measurement. The detection of glucose using silica gel modified silver electrode can be seen as purple solid line (with the blank in purple dotted line). The generated current is very low in comparison with other electrode. The detection of glucose using silica/Maja leaf extract modified silver electrode can be seen as yellow solid line (the blank is yellow dotted line). The signal current is higher than glucose detection using silica gel modified silver electrode. In the case of urea detection using silica gel modified silver electrode, it can be seen as brown solid line with the blank in brown dotted line. The current response of this electrode on urea is similar to glucose. No significant current is generated. Conversely, the detection of urea using silica/Maja leaf extract modified silver electrode gives a significantly different voltammogram. The voltammogram shows the highest anodic and cathodic peak current response. The voltammogram shows that the cathodic and anodic current signals are 0.4 V and 0.625 V, respectively. The voltammogram also shows that the electrode indicates a good selectivity for urea in the present of glucose. Meanwhile silica modified silver electrode shows no peak at anodic and cathodic current.

4. Conclusions

The good result was obtained on silver electrode modified with silica gel and Maja leaf extract in the detection of urea. The voltammogram showed that this electrode was succeeded to be applied on both anodic and cathodic current, indicated by the generated higher peak compared to the blank solutions. Meanwhile, the silver electrode that modified only with silica gel did not show any higher peak at anodic and cathodic currents than its blank solution. Urea detection response showed a relatively high current without interference current from the glucose. Maja leaf extract which was used as the main active component, acted as catalyst of redox reaction against urea.

References

[1] Meessen J H 2012 *Urea in Ullmann's Encyclopedia of Industrial Chemistry* ed W Gerhartz (Weinheim: Wiley-VCH, Germany)
[2] Krysteva M and Al Hallak M 2003 *Scientific World Journal* **3** 585-92
[3] Reddy K R C, Srivastava P K, Dey P M and Kayastha A M 2004 *Biotechnol. Appl. Biochem.* **39** 323-7
[4] Kurian J C 2001 *Plants That Heal* ed 2 (Pune: Oriental Watchman Publishing House)
[5] Umadevi K, Vanitha V and Vijayalakshmi K 2012 *World Journal Pharmaceutical Research* **1** 813-37

Study of CO_2 adsorption capacity of mesoporous carbon and activated carbon modified by triethylenetetramine (TETA)

I Sulistianti, Y K Krisnandi and I Moenandar

Department of Chemistry, Faculty of Mathematics and Natural Sciences Universitas Indonesia, Kampus UI Depok, Depok 16424, Indonesia

Corresponding author's e-mail: yuni.krisnandi@sci.ui.ac.id

Abstract. Mesoporous carbon was synthesized by soft template method using phloroglucinol and formaldehyde as a carbon source; and Pluronic F-127 as a mesoporous template. The synthesized mesoporous carbon and commercial activated carbon were modified with triethylenetetramine (TETA) to increase CO_2 adsorption capacity. Based on FTIR characterization, the synthesized mesoporous carbon and the activated carbon without modification process has similarity pattern. After the modification, both of them showed absorption peaks in the area around 1580 to 1650 cm^{-1} which is known as N-H bending vibration and absorption peaks in the area around 3150 to 3380 cm^{-1} which is known as N-H stretching vibration. The XRD results showed two peaks at $2\theta = 24.21°$ and $2\theta = 43.85°$, according to JCPDS index No. 75-1621 those peak are the typical peaks for hexagonal graphite carbon. In BET analysis, the synthesized mesoporous carbon and activated carbon modified TETA have surface area, pore volume and pore diameter lower than without modification process. In carbon dioxide adsorption testing, the synthesized mesoporous carbon showed better performance than the commercial activated carbon for CO_2 adsorption both without modification and by modification. The synthesized mesoporous carbon obtained CO_2 adsorption of 9.916 mmol/g and the activated carbon of 3.84 mmol/g for on 3.5 hours of adsorption. It is three times better than activated carbon for adsorption of carbon dioxide. The modified mesoporous carbon has the best performance for adsorption of gas CO_2 if compared by unmodified.

1. Introduction

The industrial revolution was the starting point for environmental issues such as greenhouse effect that leads to global warming. The industrial revolution is an increase in resource utilization will increase industrial activity. Carbon dioxide (CO_2) acts as a blanket to trap infrared radiation coming from the earth's atmosphere, causing warming of the earth and the temperature increase [1].

Carbon Dioxide Capture and Storage (CCS) can stop global warming through the reduction of greenhouse gas emissions, mainly capture process [2]. Absorption amine is advanced commercial technology to remove CO_2 [3]. However, this method is toxic, causing corrosion of equipment, and requires high energy for solvent regeneration. Therefore developed physical adsorption of CO_2 using a solid adsorbent such as zeolite, a metal-organic, porous silica, clay incandescent, and a carbon material such as activated carbon. The adsorption process does not cause corrosion of equipment and environmental pollution, but has the lower capacity of CO_2 adsorption at high temperatures [3].

To search for a suitable adsorbent developed mesoporous materials such as mesoporous carbon material for the adsorption of CO_2 because it has a high porosity, have the large pores, have a large surface area, have a good design for pore structure, and requires low energy. Amine functional groups such as Triethylenetetramine (TETA) is introduced to make it easier to interact with CO_2 [3]. Amine group adsorbs CO_2 and reversibly react with CO_2 to form carbamates and bicarbonate. Adsorbent modified amine is a promising way to capture the CO_2 because it has a high adsorption selectivity and the diffusion rate of CO_2 in high mesoporous.

Pamungkas [4] have successfully demonstrated that mesoporous carbon modified with 50 wt.% EDA has the best CO_2 adsorption capability for more amine groups which can cover the mesoporous carbon pores thus blocking the adsorption of CO_2. Faisal [5] have successfully demonstrated that mesoporous carbon modified with 30 wt.% TETA has the best adsorption capability and comparing the ability of the mesoporous carbon adsorption and activated carbon for the adsorption of CO_2 gas.

This research will be conducted synthesis of mesoporous carbon with soft-template method are then modified amine groups by impregnation method. Then will be compared to commercial activated carbon adsorption ability to CO_2.

2. Materials and methods

2.1. Materials

The materials used were Activated Carbon from Bratachem, *Phloroglucinol*, Pluronic F-127, Formaldehyde 37%, Ethanol, Chloride Acid (HCl) 37%, Triethylenetetramine (TETA), Sodium Hydroxide (NaOH) pellet, Potassium Hydrogen Phthalate (KHP), Borax ($Na_2B_4O_7$. $10H_2O$), Indicator Phenolphthalein, Indicator Methyl Orange, Indicator Methyl Red, CO_2 gas (99.99% UHP), N_2 gas (99.99% UHP) and deionized water.

2.2. Synthesis of mesoporous carbon

Synthesis of mesoporous carbon was prepared by mixing 1.25 g ploroglucinol and 1.25 g Pluronic F-127 into the ethanol-water mixture 10:9 (w/w) of 9.7 g. The solution was added 0.08 mL of HCl 37% (w/w) as a catalyst. After that 1.25 mL of formaldehyde 37% (w/w) solution was added. Then do the hydrothermal process at a temperature of 100°C for 24 hours. The resulting material is carbonized with a tubular furnace under the stream of N_2 gas. Synthesized mesoporous carbon is then characterized using FTIR, XRD, SEM, and BET.

Synthesized mesoporous carbon and activated carbon was modification with Triethylenetetramine (TETA) by impregnation method. Mesoporous carbon and activated carbon modified TETA characterized by a FTIR and BET.

2.3. Application for adsorption carbon dioxide (CO_2)

Test adsorption of carbon dioxide (CO_2) is performed in a reactor by stream of CO_2 and N_2 gas to 0.2 g adsorbent. Then the CO_2 does not adsorbed is collected in 0.1 M NaOH solution and analyzed quantitatively by titrimetric method.

3. Results and discussion

3.1. Synthesis and characterization of mesoporous carbon

The mesoporous carbon were synthesized by soft-template method through hydrothermal treatment using copolymer Pluronic F-127 as a mesoporous template; using phloroglucinol and formaldehyde as a carbon source [6]. The hydrothermal process aims to form a pore structure, which is the decomposition of template. After hydrothermal process, the mesoporous polymer will be formed subsequently carbonized to polymer decomposition and increase the surface area of the synthesized mesoporous carbon [7].

3.2. X-Ray diffraction (XRD)

X-ray diffraction pattern of carbon material according to JCPDS index, No. 75-1621, that there are two distinctive peaks at 2θ area around 22-26° and 42-46° [8]. Two peaks at 2θ = 24.21° and 2θ = 43.85°

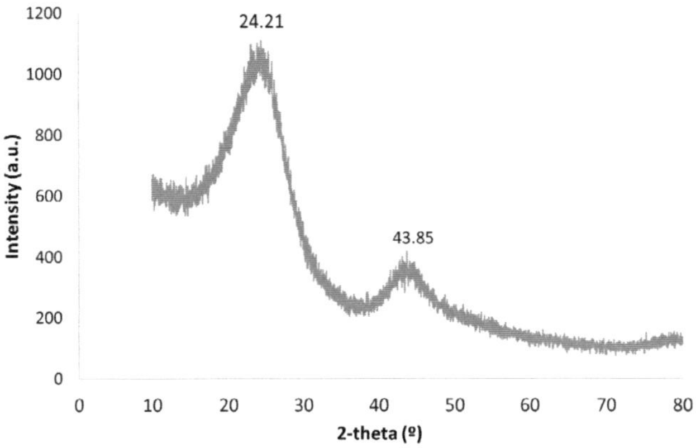

Figure 1. X-ray diffraction pattern of synthesized mesoporous carbon.

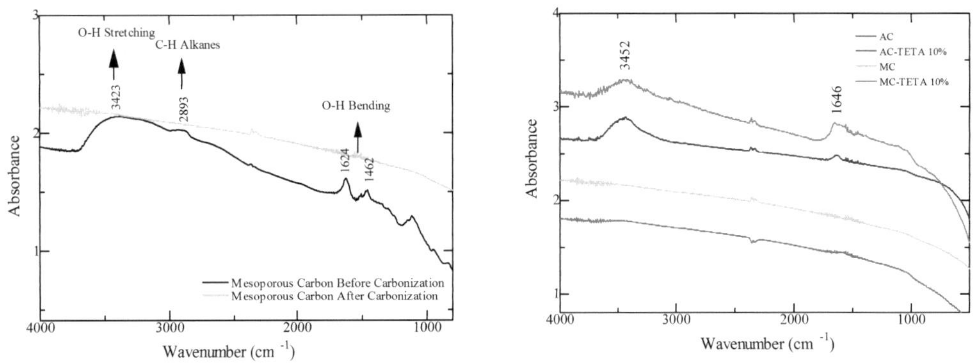

(a) (b)

Figure 2. (a) FTIR spectra of mesoporous carbon synthesized (before and after carbonization process) and (b) FTIR spectra of MC, MC-TETA 10%, AC and AC-TETA 10%.

are diffraction planes of hexagonal graphitic carbon [9]. A broad peak indicates that the mesoporous carbon synthesized contains amorphous carbon (figure 1).

3.3. Fourier transform infrared (FTIR)

The FTIR spectrum carbon mesoporous synthesis product before carbonization (figure 2a) shows the absorption peaks at wave numbers around 3500 cm^{-1} which is the peak of the vibration of OH stretching phloroglucinol and absorption peaks at wave number 2900 cm^{-1} which is the absorption peak of the vibration of symmetric and asymmetric methylene group (CH_2). The FTIR spectrum synthesized mesoporous carbon after carbonization (figure 2a) indicates the absence of absorption peak at wave number around 3500 cm^{-1} and 2900 cm^{-1} which is an absorption peak of O-H stretching vibration and the vibration of symmetric and asymmetric methylene. This indicates that the mesoporous carbon have been synthesized because there are happened a decomposition of Pluronic F-127, phloroglucinol and formaldehyde.

The FTIR spectrum of activated carbon (figure 2b) has similarities with the FTIR spectrum synthesized mesoporous carbon. This indicated that both have the same functional groups. After modified by TETA appears absorption peaks in the area around 1580-1650 cm^{-1} which is N-H bending

IOP Conf. Series: Materials Science and Engineering **188** (2017) 012041 doi:10.1088/1757-899X/188/1/012041

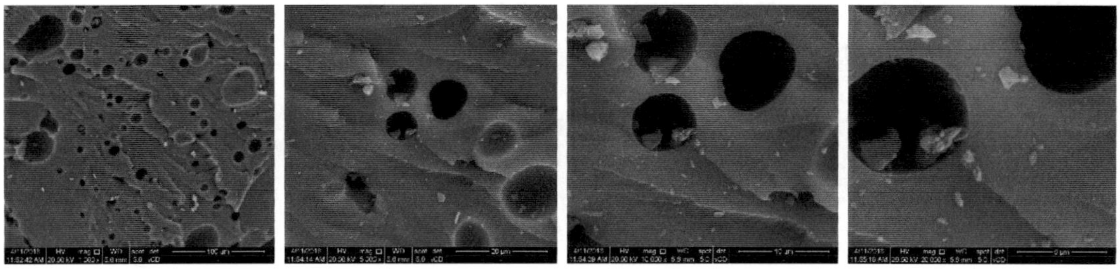

(a) (b) (c) (d)

Figure 3. The surface morphology of synthesized mesoporous carbon with a magnification of (a) 1000 ×; (b) 5000 ×; (c) 10000 ×; and (d) 20000 ×.

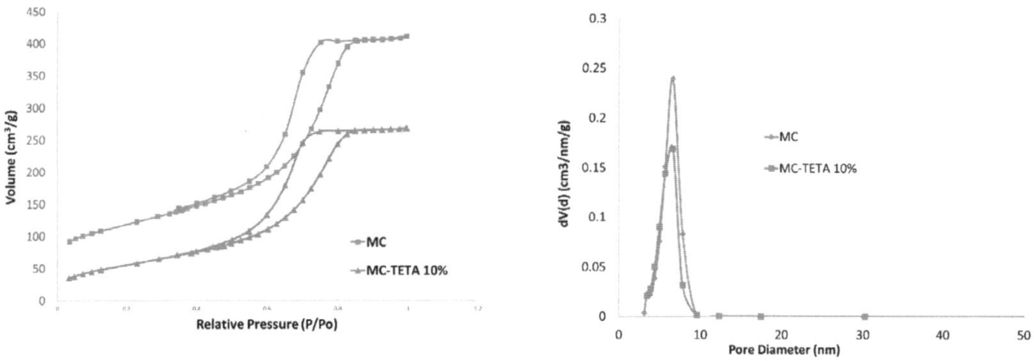

(a) (b)

Figure 4. (a) Adsorption-desorption isotherms curve of MC and MC-TETA 10% and (b) pore size distribution curve of MC and MC-TETA 10%.

vibration of TETA and absorption peaks in the area around 3150-3380 cm^{-1} which is N-H stretching vibration of TETA (figure 2b).

3.4. Scanning electron microscopy (SEM)

Analysis by Scanning Electron Microscopy (SEM) was used to determine the size and morphology of carbon materials synthesis product. Synthesized mesoporous carbon surface morphology is smooth (smooth), by holes in its surface (figure 3). Surface with holes like this are very support to be used as an adsorbent gas carbon dioxide (CO_2).

3.5. Surface area analysis

Modifications by TETA does not change the structure of mesoporous carbon mesoporous (figure 4a). In figure 4a it is shown that the mesoporous carbon without modification and by modification of 10% TETA has adsorption-desorption isotherms curve similar (adsorption-desorption isotherms type IV) [10]. The difference is that the mesoporous carbon has a larger volume than mesoporous carbon modified TETA. This shows that modifications by TETA decrease the pore volume and decrease surface area (table 1). Figure 4b shows that the mesoporous carbon and mesoporous carbon modified TETA had a similar BJH desorption pattern. It is explained that the mesoporous carbon and mesoporous carbon modified TETA has similar a pore size distribution.

Based on the data, mesoporous carbon has the CO_2 adsorption capability better than activated carbon (figure 5a). Mesoporous carbon was able adsorbed CO_2 as much as 9.916 mmol/g adsorbent and activated carbon was able adsorbed CO_2 as much as 3.84 mmol/g adsorbent. It is very possible

Table 1. Surface area analyses of MC and MC-TETA 10% calculated from BET measurement.

Sampel	S_{BET} [a] (m^2/g)	S_{ext} [b] (m^2/g)	V_{total} [c] (cm^3/g)	V_{micro} [b] (cm^3/g)	D_{pore} [d] (nm)
MC	407.278	337.622	0.6355	0.036	6.2409
MC-TETA 10%	205.559	205.559	0.4149	0	6.5691

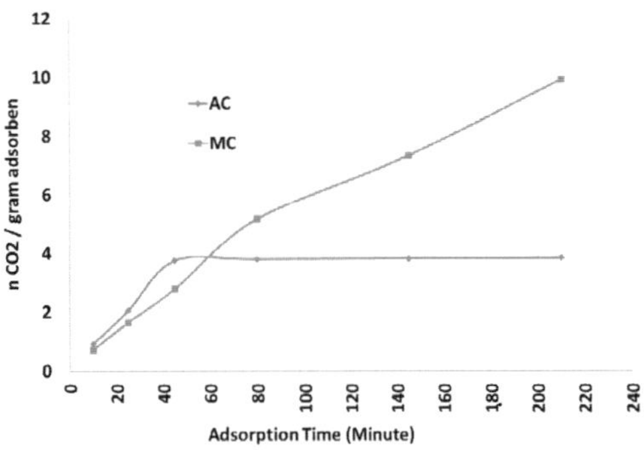

(a)

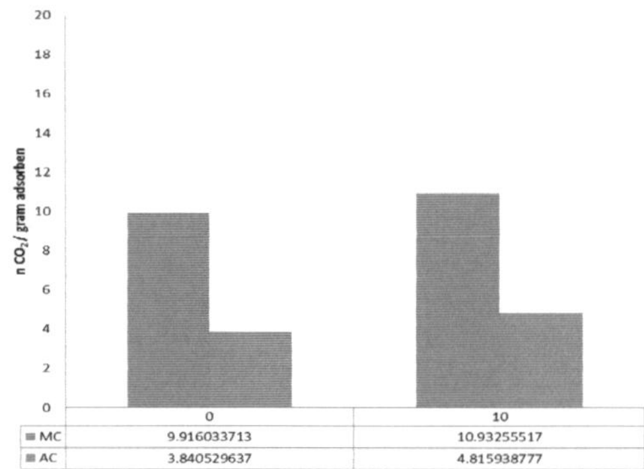

(b)

Figure 5. (a) Curve of adsorption CO_2 by mesoporous carbon and activated carbon and (b) curve of adsorption CO_2 by mesoporous carbon, mesoporous carbon modified TETA, activated carbon and activated carbon modified TETA.

because mesoporous carbon's pore is more uniform than activated carbon. In addition, the activated carbon dominated of micropores, thus activated carbon will be saturated more quickly than the mesoporous carbon. Modifications by TETA able to increase the ability CO_2 adsorption (figure 5b). Mesoporous carbon modified TETA was able adsorbed CO_2 as much as 10.932 mmol/g adsorbent and activated carbon modified TETA was able adsorbed CO_2 as much as 4.816 mmol/g adsorbent.

4. Conclusions

Adsorbent from mesoporous carbon and activated carbon have been prepared and their adsorption capacities were tested. The result showed that adsorption capacity of mesoporous carbon (9.916 mmol/g adsorbent) was better if compared with activated carbon (3.84 mmol/g adsorbent) and modification by 10% TETA could increase CO_2 adsorption capacity.

Acknowledgements

This work was funded by PITTA Grant No. 1753/UN2.R12/PPM.00.00/2016 from Universitas Indonesia.

References

[1] Chatti R, Bansiwal A K, Thote J A, Kumar V, Jadhav P, Lokhande S K, Biniwale R B, Labhsetwar N K and Rayalu S S 2009 *Microporous and Mesoporous Materials* **121** 84-9
[2] Rabensteiner M, Kinger G, Koller M, Gronald G and Hochenauer C 2014 *International Journal of Greenhouse Gas Control* **27** 1-14
[3] Chew T L, Ahmad A L and Bhatia S 1996 *Journal of Chemical Engineering* **153** 43-57
[4] Pamungkas A Z 2014 *Sintesis Karbon Mesopori Termodifikasi Gugus Amina sebagai Adsorben Karbon Dioksida* BSc projects (Depok: Departemen Kimia, FMIPA Universitas Indonesia)
[5] Faisal M 2015 *Studi Pengaruh Modifikasi Trietilentetraamina pada Karbon Mesopori terhadap Kemampuan Adsorpsi Gas CO_2* BSc projects (Depok: Departemen Kimia, FMIPA

Universitas Indonesia)
[6] Górka J , Zawislak A , Choma J, Jaroniec M 2008 *Carbon* **46** 1159-61
[7] Tanaka S, Doi A, Nakatani N, Katayama Y and Miyake Y 2009 *Carbon* **47** 2688-98
[8] Priantini T 2014 *Sintesis Material Karbon Mesopori Melalui Metode Soft Template dan Aplikasinya Sebagai Adsorben Zat Warna Acid Red 119* BSc projects (Depok: Departemen Kimia, FMIPA Universitas Indonesia)
[9] Chen J, Xia N, Zhou T, Tan S, Jiang F and Yuan D 2009 *Electrochem* **4** 1063-73
[10] Landers J, Gor GY and Neimark A V 2013 *Colloids Surf.* A **437** 3-32

Improve sonocatalytic performance using modified semiconductor catalyst SnO_2 and ZrO_2 by magnetite materials

V Paramarta[1,2], **Y Kristianto**[1,2], **A Taufik**[1,2] **and R Saleh**[1,2]

[1]Department of Physics, Faculty of Mathematics and Natural Sciences Universitas Indonesia, Kampus UI Depok, Depok 16424, Indonesia
[2]Integrated Laboratory of Energy and Environment, Faculty of Mathematics and Natural Sciences Universitas Indonesia, Kampus UI Depok, Depok 16424, Indonesia

Corresponding author's e-mail: rosari.saleh@gmail.com

Abstract. Modifying semiconductor catalysts using magnetic materials could enhance the efficiency of wastewater removal and the separation efficiency demonstrated in our previous studies. The catalytic performance of wide band gap ZrO_2 and SnO_2 semiconductors has not been studied with regard to the ultrasonic radiation catalytic process. Therefore, the ultrasound-assisted sonocatalytic performance of ZrO_2 and SnO_2 semiconductors combined with Fe_3O_4 (magnetite) nanoparticles has been studied using methylene blue as a model organic pollutant. The nanocomposites were synthesized using sol-gel methods. The structural properties of the samples were characterized using X-ray diffraction (XRD). Sample morphology was obtained by transmission electron microscopy (TEM). The magnetic properties of the samples were characterized using a vibrating sample magnetometer (VSM). The cubic spinel structure of Fe_3O_4 is successfully identified, as are the tetragonal structures from SnO_2 and ZrO_2. The samples exhibit ferromagnetic behavior at room temperature. Both ZrO_2 and SnO_2 combined with Fe_3O_4 show better sonocatalytic efficiency than without Fe_3O_4 and even better efficiency than the pioneering wide band gap semiconductor TiO_2. The combination of magnetite nanoparticles with SnO_2 exhibit the highest degradation efficiency. The incorporation of magnetic material into catalysts has been proved to enhance the reusability of catalysts with efficient separation process.

Keywords: SnO_2, ZrO_2, Fe_3O_4/SnO_2, Fe_3O_4/ZrO_2, sonocatalytic, methylene blue.

1. Introduction

Environmental pollution is a topic of interest because of its negative impact on society. Dye wastes from industrial processes, including textiles and pharmacies among others, are a particularly large source of environmental pollution. The problem of dye wastes must be overcome by using appropriate methods and technologies [1]. Methods using advanced oxidation processes (AOPs) have been widely used because they have the capacity to degrade organic compounds in solutions [2]. Sonolysis is one of the most effective AOP methods for decomposing pollutants in transparent solutions. Sonolysis uses the formation of hydroxyl radicals (•OH), which result from the splitting of water molecules using an ultrasonic wave. The formation of hydroxyl radicals by acoustic cavitation bubbles consists of three stages, i.e., formation, growth, and implosive collapse [3, 4]. In this process, increased pressure and temperature in the acoustic cavitation bubbles causes thermal dissociation within the solution to form hydroxyl radicals (•OH) and other reactive oxygen species such as $\bullet O_2^-$, $\bullet O$, and $\bullet OOH$, which can oxidize the organic pollutant in the solutions [5, 6]. However, the efficiency of pollutant degradation using the sonolysis process is very low due to the limited amount of •OH available upon ultrasonic radiation. To overcome this issue, catalysts must be added to the sonolysis process, increasing the efficiency of the degradation of dye wastes [7, 8].

Metal-oxide semiconductors have attracted much attention due to their wide application to processes such as water purification, water splitting to produce hydrogen, biological and biomedical sensing, and photovoltaics among others [9]. Wang *et al.* [10] researched the degradation of hydrogen peroxide solutions by sonocatalyst activity of TiO_2, and the results showed that the semiconductor is an efficient catalyst. Zirconium dioxide (ZrO_2) and tin oxide (SnO_2) are also potentially ideal candidates for semiconductor suitable for use as catalysts due to their high photostability, high thermodynamic stability, lack of toxicity, good adsorption capacity, unique physical and chemical properties, and ease of production [11, 12]. However, the high electron–hole recombination rate of those semiconductors could limit the efficiency of degradation. Combining these semiconductors with other materials is an effective way to inhibit the electron–hole recombination rate and thus increase the efficiency of the catalyst to degrade organic pollutants in solutions [13]. Other shortcomings include the process of separating the catalyst material from the dye waste, which is still difficult, limiting the reusability of catalysts for the sonocatalytic process (recycling). Other shortcomings of ZrO_2 and SnO_2 semiconductors could be overcome by adding magnetic material such as Fe_3O_4 (magnetite). Fe_3O_4 has several advantages such as low cost, good adsorption capacity, and chemical stability in solutions [14, 15]; it also makes it possible to magnetically separate the catalyst from the solutions so it can be reused [16]. Therefore, the combination of Fe_3O_4 with SnO_2 and ZrO_2 is expected to increase the efficiency of these catalysts to degrade pollutants.

In this study, Fe_3O_4/ZrO_2 and Fe_3O_4/SnO_2 nanocomposites were prepared using the sol-gel method. The sample was characterized using X-ray diffraction (XRD) and vibrating sample magnetometer (VSM) techniques. Sonocatalytic activity of the two nanocomposites was analyzed using methylene blue (MB) as a model organic pollutant. The sonocatalytic activity of the two nanocomposites is studied by comparing the results to the catalytic activity of other semiconductor catalysts, applied in advance.

2. Experimental

2.1. Materials

All reagents used were analytical grade and were used without further purification. Zirconium (IV) chloride ($ZrCl_4$), anhydrous tin chloride ($SnCl_2$), iron (II) sulfate heptahydrate ($FeSO_4 \cdot 7H_2O$), titanium dioxide (TiO_2), sodium hydroxide (NaOH), methylene blue (MB), ethanol, ethylene glycol (EG), ammonium oxalate, sodium sulfate, and tert-butyl alcohol (TBA) were purchased from Merck (Kenilworth, NJ, USA).

2.2. Synthesis of nanomaterials

The Fe_3O_4 nanoparticles were synthesized using the same method used in our previous study [17]. The ZrO_2 nanoparticles was synthesized using the sol-gel method. First, $ZrCl_4$ and NaOH were each dissolved in aqueous solutions using magnetic stirring. Then, the $ZrCl_4$ solution was mixed slowly into the NaOH solution, and the mixture was stirred using magnetic stirring at 80°C for 3 h until a homogenous solution formed. The mixture solutions were centrifuged to separate out the precipitate, which was then washed using aqueous solution and ethanol several times to remove unexpected compounds. The resulting product was heated at 120°C for 12 h. ZrO_2 nanoparticles were obtained by calcination for 5 h at 550°C.

The SnO_2 nanoparticles were synthesized using a modification of the method reported by Li *et al.* [18]. First, $SnCl_2$ was dissolved in a mixture of ethanol and aqueous solution, which was then added into the NaOH solutions using magnetic stirring. Then, the mixed solutions were heated at 180°C for 3 h, then chilled to room temperature. The precipitate was obtained by centrifugation and washed using aqueous solutions and ethanol several times. The precipitated particles were dried under vacuum at 80°C, and SnO_2 particles were obtained by calcination for 3 h at 700°C.

The Fe_3O_4/ZrO_2 and Fe_3O_4/SnO_2 nanocomposites were synthesized using the sol-gel method. First, ZrO_2 and SnO_2 nanoparticles were mixed with Fe_3O_4 in a mixture of ethanol and aqueous solution. Each mixture was ultrasonicated for 2 h and then centrifuged to obtain the precipitate. The resulting product was then dried under vacuum at 80°C for 12 h to obtain Fe_3O_4/ZrO_2 and Fe_3O_4/SnO_2 nanoparticles. The Fe_3O_4/TiO_2 nanocomposites were synthesized using our previous method [19].

2.3. Characterization
The sample was characterized by XRD using a Rigaku Miniflex 600 (Rigaku, Tokyo, Japan) with a Cu K-α radiation source (λ =1.5406 Å), and the magnetic character of the sample was measured with an Oxford Type 1.2T VSM.

2.4. Sonocatalytic experiments
The sonocatalytic experiment was done by mixing each sample of SnO_2, ZrO_2, Fe_3O_4/SnO_2, and Fe_3O_4/ZrO_2 into 100 mL solutions of MB with concentrations of 20 mg/L; pH was adjusted using NaOH. The solutions were allowed to stand until reaching adsorption–desorption equilibrium. The solutions were put in an ultrasonic bath with frequency and tension of 40 kHz and 150 W, respectively, and ultrasonicated for 2 h. After every 15 min, the concentration of the MB solution was analyzed using a UV-Vis spectrometer. Sonocatalytic activity of TiO_2 nanoparticles and Fe_3O_4/TiO_2 nanocomposites were tested in the same way.

2.5. Scavanger experiments
Different radical-scavengers were used in MB solutions to determine the most influential species in sonocatalytic activity. Ammonium oxalate, sodium sulfate, and TBA were each used for scavenging holes, electrons, and hydroxyl radicals. The measurements were done as described above.

3. Results and discussion

3.1. Structural and optical properties
Figure 1 shows the XRD spectrum of Fe_3O_4, ZrO_2, and SnO_2 nanoparticles and Fe_3O_4/ZrO_2 and Fe_3O_4/SnO_2 nanocomposites. The diffraction pattern of the Fe_3O_4 sample observed at the values 2θ = 30.14°, 35.49°, 43.28°, 57.20°, 62.83°, and 74.8° shows each area (220), (311), (400), (511), (440), and

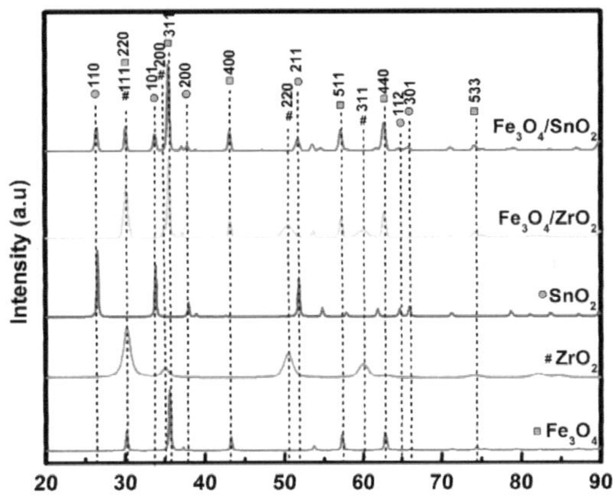

Figure 1. XRD patterns of Fe_3O_4, ZrO_2, and SnO_2 nanoparticles and Fe_3O_4/ZrO_2, Fe_3O_4/SnO_2 nanocomposites.

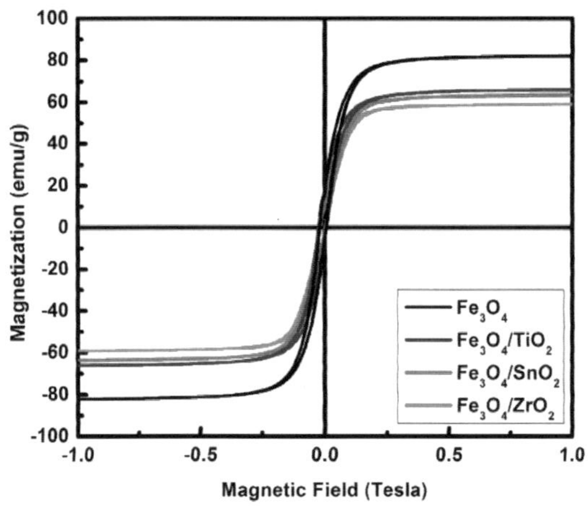

Figure 2. VSM spectra of Fe_3O_4, Fe_3O_4/TiO_2, Fe_3O_4/ZrO_2 and Fe_3O_4/SnO_2 nanocomposites

Table 1. Lattice parameters and grain size of Fe_3O_4, ZrO_2, SnO_2 nanoparticles and Fe_3O_4/ZrO_2, Fe_3O_4/SnO_2 nanoparticles.

Sample	Lattice Parameter					<D> (nm)			M-S
	Fe_3O_4	ZrO_2		SnO_2					
	a=b=c	a=b	c	a=b	c	Fe_3O_4	ZrO_2	SnO_2	(emu/g)
Fe_3O_4	8.36	-	-	-	-	45	-	-	82
ZrO_2	-	3.60	5.14	-	-	-	13	-	-
SnO_2	-	-	-	4.73	3.18	-	-	38	-
Fe_3O_4/ZrO_2	8.37	3.6	5.14	-	-	25	13	-	59
Fe_3O_4/SnO_2	8.38	-	-	4.76	3.19	28	-	21	62

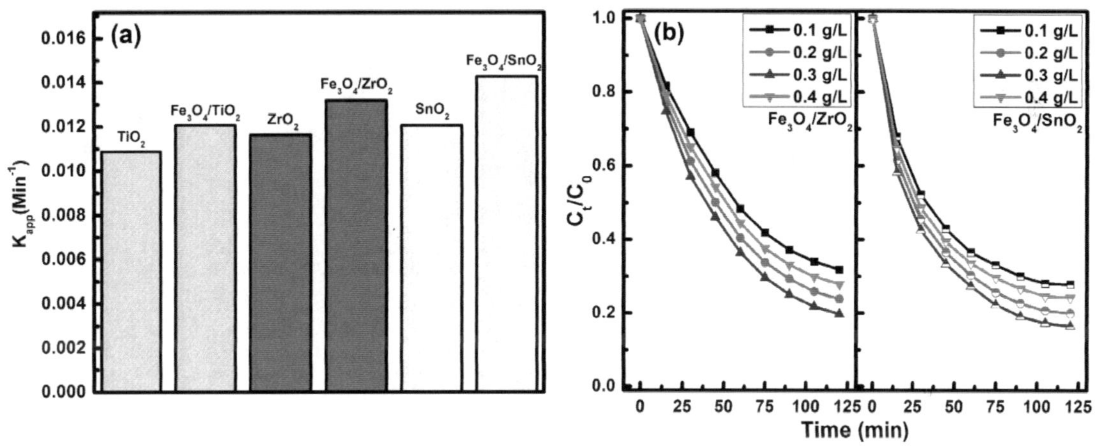

(a) (b)

Figure 3. Sonocatalytic degradation (a) rate constant of all samples (b) effect of dosage nanocomposites Fe_3O_4/TiO_2, Fe_3O_4/ZrO_2 and Fe_3O_4/SnO_2.

(533) of the cubic spinel structure. The diffraction pattern of the ZrO_2 sample observed at the values $2\theta = 30.18°$, $35.03°$, $50.36°$, and $59.94°$ shows the reflection index of each area (111), (200), (220), and (311) of the tetragonal structure of ZrO_2, while the diffraction pattern of the SnO_2 nanoparticle sample observed at the values $2\theta = 26.5°$, $33.8°$, $38°$, $39°$, $51.8°$, $54.8°$, $58°$, $62°$, $64.7°$, $65.8°$, $71.2°$, $78.2°$, $81.2°$, and $83.7°$ shows the existence of areas (110), (101), (200), (111), (211), (220), (002), (310), (112), (301), (202), (321), (400), and (222) of the tetragonal structure of SnO_2 nanoparticles. The XRD patterns of Fe_3O_4/ZrO_2 and Fe_3O_4/SnO_2 nanocomposites show the existence of the cubic spinel phase of Fe_3O_4, followed by the addition of tetragonal SnO_2 and ZrO_2 phases for Fe_3O_4/SnO_2 and Fe_3O_4/ZrO_2 nanocomposites. The grain size was calculated using Scherrer's equation and lattice parameters were calculated using the Rietveld refinement method. The results are shown in table 1.

3.2. Magnetic properties

Figure 2 shows the magnetic hysteresis curve of the Fe_3O_4/ZrO_2 and Fe_3O_4/SnO_2 nanocomposites. As a comparison, the magnetic hysteresis curve of Fe_3O_4 nanoparticles is also shown in the picture. Loop

Table 2. Rate constants and maximum sonocatalytic degradation of MB in presence of prepared samples.

Samples	Rate constant (min^{-1})	Maximum Degradation (%)
TiO_2	0.01088	72.01
Fe_3O_4/TiO_2	0.01206	76.75
ZrO_2	0.01163	74.75
Fe_3O_4/ZrO_2	0.01318	80.29
SnO_2	0.01205	77.19
Fe_3O_4/SnO_2	0.01427	83.50

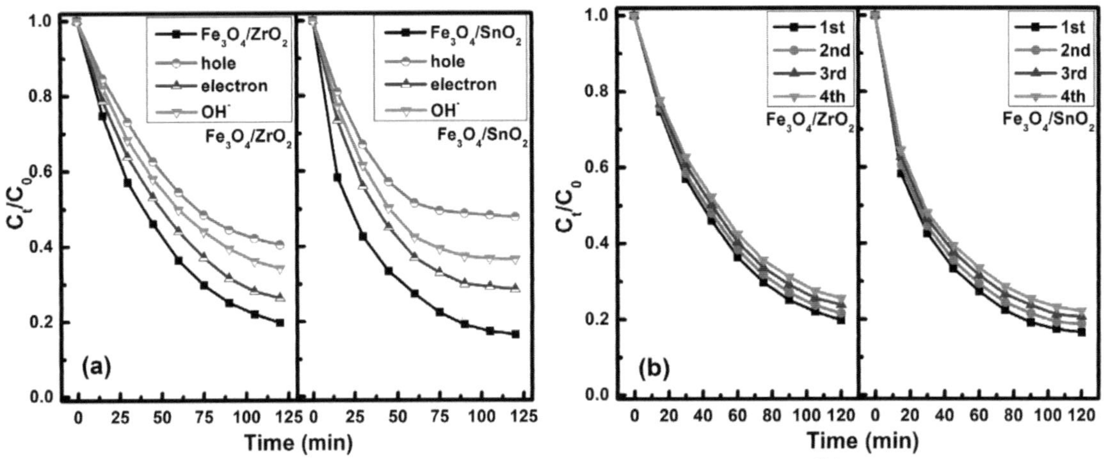

(a) (b)

Figure 4. (a) Effect of scavenger, (b) Reusability of Fe_3O_4/ZrO_2 and Fe_3O_4/SnO_2.

hysteresis of the Fe_3O_4 nanoparticle and both nanocomposites shows ferromagnetic behavior at room temperature with magnetization saturation (M_s) of 82, 62, and 59 emu g^{-1} for Fe_3O_4, Fe_3O_4/SnO_2, and Fe_3O_4/ZrO_2, respectively. The magnetization saturation of Fe_3O_4/ZrO_2 and Fe_3O_4/SnO_2 nanocomposites is smaller than that of the Fe_3O_4 nanoparticles. This is due to the addition of non-magnetic materials ZrO_2 and SnO_2 nanoparticles, which decreases the total magnet moment in the sample. If it is compared with the magnetic characteristics of Fe_3O_4/TiO_2 nanocomposite from previous research [20], as shown in figure 2, it can be seen that the magnetization saturation of Fe_3O_4/TiO_2 nanocomposite is higher than that of both Fe_3O_4/ZrO_2 and Fe_3O_4/SnO_2 nanocomposites.

3.3. Sonocatalytic activity

The results of sonocatalytic activity for all samples are shown as degradation rates and %degradation of methylene blue. The degradation rate of methylene blue in the sonocatalytic process was calculated using the kinetics equation $\ln(C_t/C_0) = -kt$, where k is the pseudo-first-order rate constant, C_0 is the initial concentration of MB, and C_t represents MB concentration at time t. The degradation efficiency of the samples was calculated using the equation %Degradation = 100 × (1 - C_t/C_0). Figure 3a shows the degradation rate of methylene blue through the sonocatalytic process. The result obtained shows that ZrO_2 nanoparticles and SnO_2 show a slower degradation rate than do Fe_3O_4/ZrO_2 and Fe_3O_4/SnO_2 nanocomposites; the rate constants and maximum degradation are shown in table 2. The sonocatalytic activity test was also done for TiO_2 nanoparticles and Fe_3O_4/TiO_2, as shown in figure 3a. The Fe_3O_4/ZrO_2 and Fe_3O_4/SnO_2 nanocomposites show better results for MB degradation capacity than do TiO_2 nanoparticles and the Fe_3O_4/TiO_2 nanocomposite. This is because the energy band gap of TiO_2 (~3.2 eV) is smaller than those of SnO_2 (~3.8eV) and ZrO_2 (~5eV), so the lifetime of generated electron–hole pairs increases and the recombination rate of electrons and holes decreases [21]. The sonocatalytic activity capacity of all nanocomposite samples compared to nanoparticles could be caused by the increased efficiency of separating electron–hole pairs on the nanocomposite [22]. The order of sonocatalytic activity efficiency based on test results is $Fe_3O_4/SnO_2 > Fe_3O_4/ZrO_2 > SnO_2 > Fe_3O_4/TiO_2 > ZrO_2 > TiO_2$.

3.4. *Effect of catalyst dosage*

Figure 3b shows the effect of catalyst dosage addition, Fe_3O_4/ZrO_2 and Fe_3O_4/SnO_2 nanocomposites from 0.1 g/L to 0.4 g/L, on the efficiency of degradation of methylene blue through the sonocatalytic process. As seen in figure 3b, after 120 min, the maximum degradation for Fe_3O_4/ZrO_2 with dosages of 0.1 g/L, 0.2 g/L, 0.3 g/L, and 0.4 g/L are 68.25%, 76.19%, 80.29%, and 72.19%, respectively, while for Fe_3O_4/SnO_2 with dosages of 0.1 g/L, 0.2 g/L, 0.3 g/L, and 0.4 g/L the maximum degradations are 72.19%, 80.04%, 83.5%, and 75.83%, respectively. The maximum degradation for both samples occurred with the addition of a dosage of 0.3 g/L. The catalyst dosage was also varied for the Fe_3O_4/TiO_2 nanocomposite (figure 3b), which resulted in maximum degradations for dosages of 0.1 g/L, 0.2 g/L, 0.3 g/L, and 0.4 g/L, respectively, of 61.4%, 73.77%, 76.75%, and 69.72%. The maximum degradation for Fe_3O_4/TiO_2 nanocomposite was similarly produced at a catalytic dosage of 0.3 g/L. The amount of nanocomposite added influences two things: the number of active sites on the sample and the amount of methylene blue adsorbed by the sample [23]. The dosage of the added catalyst will also increase the number of active sites. The results shown in figure 3b indicate that the optimum catalyst dosage for the nanocomposite sample is 0.3 g/L.

3.5. *Effect of scavangers*

To determine the reactive species most involved in MB degradation in the three samples, scavengers were added to the solutions. Figure 4a shows the results of MB degradation by sonocatalysis with the addition of scavengers. The scavengers serve to bind the reactive species so that they cannot take part in the sonocatalytic process. The obtained results show that the addition of hole scavengers resulted in the greatest decrease in degradation capacity. This indicates that holes are the most influential reactive species for the three samples.

3.6. *Reusability*

Cycling testing was carried out to analyze catalyst stability. The cycling test was done by reusing the catalyst four times for the sonocatalytic process. After each test, the sample was separated from its solutions using an external magnet as shown in figure 4b and then reused as the catalyst in the next experiment. Figure 4b shows the results of the reusability tests for the three samples. The three samples all showed good cycling capacity, with decreases of only about 6% after three reuses. This proves that sonocatalytic activity for the three nanocomposite samples is stable.

4. Conclusions

The addition of the magnetic material Fe_3O_4 to ZrO_2 and SnO_2 catalysts has been successfully carried out using the sol-gel method. The addition of Fe_3O_4 has been shown to increase the capacity of sonocatalysis to degrade methylene blue. Using magnetic material also aids the separation of catalyst material from the test solution, allowing reuse of the catalysts.

References

[1] Khodami Z and Nazamzadeh-Ejhieh A 2015 *J. Mol. Catal. A-Chem.* **409** 59-68
[2] Min Y, Zhang K, Chen Y and Zhang Y 2012 *Ultrason. Sonochem.* **19** 883-9
[3] Eskandarloo H, Badiei A, Behnajady M A and Ziarani G M 2015 *Ultrason. Sonochem.* **26** 281-92
[4] Li S, Wei C, Wang J, Zhang L and Li Y 2014 *Mat. Sci. Semicon. Proc.* **26** 438-47
[5] Zhai Y, Li Y, Wang J, Wang J, Yin L, Kong Y, Han G and Fan P 2013 *J. Mol. Catal. A-Chem.* **366** 282-7

[6] Khataee A, Vahid B, Saadi S and Joo S W 2016 *Ultrason. Sonochem.* **29** 146-55
[7] Song L, Chen C and Zhang S 2011 *Ultrason. Sonochem.* **18** 713-7
[8] Zhou M, Yang H, Xian T, Yang Y and Zhang Y 2015 *Chinese J. Catal.* **36** 1987-94
[9] Liu H, Liu T, Dong X, Lv Y and Zhu Z 2014 *Mater. Lett.* **126** 36-8
[10] Wang J, Sun W, Zhang Z, Xing Z, Xu R, Li R, Li Y and Zhang X 2008 *Ultrason. Sonochem.* **15** 301-7
[11] Guo G Y and Chen Y L 2005 *J. Solid State Chem.* **178** 1675-82
[12] Tahmasebpour M, Babaluo A A and Aghjeh M K R 2008 *J. Eur. Ceram. Soc.* **28** 773-8
[13] Zhu H Y, Xiao L, Jiang R, Zeng G M and Liu L 2011 *Chem. Eng. J.* **172** 746-53
[14] Majeed M I, Lu Q, Yan W, Li Z, Hussain I, Tahir M N, Tremel W and Tan B 2013 *J. Mater. Chem.* B **1** 2874-84
[15] Gao Y, Wei Z, Li F, Yang Z M, Chen Y M, Zrinyi M and Osada Y 2014 *Green Chem.* **16** 1255-61
[16] Ma J, Guo S, Guo X and Ge H 2015 *Appl. Surf. Sci.* **353** 1117-25
[17] Taufik A, Susanto I K and Saleh R 2015 *Mater. Sci. Forum* **827** 37-42
[18] Li Y, Guo Y, Tan R, Cui P, Li Y and Song W 2009 *Mater. Lett.* **63** 2085-8
[19] Jalaludin S, Arifin S A and Saleh R 2015 *Mater. Sci. Forum* **827** 13-8
[20] Arifin S A, Jalaludin S, Djaja N F and Saleh R 2015 *Adv. Mater. Res.* **1112** 221-7
[21] Pouretedal H R, Tofangsazi Z and Keshavarz M H 2012 *J. Alloy. Compd.* **513** 359-64
[22] Mageshari K, Nataraj D, Pal T, Sathyamoorthy R and Park J 2015 *J. Alloy. Compd.* **625** 362-70
[23] Chiang Y J and Lin C C 2013 *Powder Technol.* **246** 137-43

In silico mutation analysis of non-structural protein-5 (NS5) dengue virus

R D Puspitasari and U S F Tambunan

Bioinformatics Research Group, Department of Chemistry, Faculty of Mathematics and Natural Sciences Universitas Indonesia, Kampus UI Depok, Depok 16424, Indonesia

Corresponding author's e-mail: usman@ui.ac.id

Abstract. Dengue fever is a world disease. It is endemic in more than 100 countries. Information about the effect of mutations in the virus is important in drug design and development. In this research, we studied the effect of mutation on NS5 dengue virus. NS5 is the large protein containing 67% amino acid similarity in DENV 1-4 and has multifunctional enzymatic activities. Dengue virus is an RNA virus that has very high mutation frequency with an average of 100 times higher than DNA mutations, and the accumulation of mutations will be possible to generate the new serotype. In this study, we report that mutation occurs in NS5 of DENV serotype 3, glutamine mutates into methionine at position 10 and threonine mutates into isoleucine at position 55. These residues are part of the domain named *S-Adenosyl-L-Methionine-Dependent Methyltransferase* (IPR029063).

Keywords: NS5, DENV3, mutation, glutamine, threonine

1. Introduction

Until now, dengue fever remains a global health problem that endemic in more than 100 countries [1-2]. In 2015, an increasing number of cases were reported in Brazil and neighboring countries. An estimated 500,000 people have severe dengue fever, most of them were children, and around 25,000 were dead [3-4]. Dengue virus is a *flaviridae* virus transmitted by the *Aedes aegypti* and *Ae. albopictus* mosquito. Dengue virus has five serotypes (DENV1-5) [5-7]. DENV-1 and DENV-4 are often associated with primary infection, while DENV-2 and DENV-3 are twice more likely to infect and causing dengue hemorrhagic fever (DHF) to human as the secondary infection [8-9]. DENV-5 has been found in Sarawak Malaysia in 2007. The sample was taken from a farmer and has been officially announced in 2013 [6]. The discovery of the new serotypes possibly delays the development of vaccines and antiviral therapy, which have been done before [7].

Dengue virus is a ssRNA positive-strand, packed in nucleocapsid and enveloped in a lipid bilayer-glycoprotein. A genomic RNA consists of a single open reading frame (ORF) which encoded into 10 proteins. The three structural proteins (Capsid, pre-Membrane and Envelope) and seven non-structural proteins (NS1, NS2A, NS2B, NS3, NS4A, NS4B and NS5) [10]. Dengue virus is an RNA virus that has very high mutation frequency with an average of 100 times higher than the genomic DNA mutations. The accumulation of mutations is a continuous process, together with the possibility of intramolecular recombination for their simultaneous infection with a different serotype that could cause the appearance of a new serotype of dengue virus [11].

Mutation can affect the protein structure and function. It is located in the protein-coding region of the genome. The information about structural and functional effects of protein can contribute to further drug design research [12]. In this study, we used non-structural protein 5(NS5) dengue virus as a protein target. The non-structural protein 5 (NS5) has been reported as the best characterized DENV non-structural protein. NS5 is a large protein contains 67% amino acid similarity with DENV 1-4 and has multifunctional protein with several enzymatic activities [13]. The purpose of this study was to evaluate the effects of mutation on NS5 Dengue virus.

2. Methods

The NS5 amino acid sequence of dengue virus was downloaded in GenBank at NCBI (http://www.ncbi.nlm.nih.gov/). The 3D structure was downloaded from RCSB PDB (http://www.rcsb.org), and we used GIRAF [14] server to get the best protein model.

Amino acid sequence of dengue serotype was aligned using online Basic Local Aligment Search Tool (BLAST) for proteins at NCBI (http://blast.ncbi.nlm.nih.gov/), Multiple Sequences Alignment were conducted by using Clustal Omega at EMBL- EBI (http://www.ebi.ac.uk/Tools/msa/clustalo/) to search the mutated region. In order to find the conserved region, we used Cunsurf [15] server (http://consurf.tau.ac.il/). We used HOPE server [12] (www.cmbi.ru.nl/hope/) to predict the effects of the mutation on NS5 DENV. MOE 2014.09 was used to predict the binding site position.

3. Results and discussion

NS5 is the protein that plays an important role in the dengue virus replication [13]. Information about the effects of mutations on the NS5 is important in drug design and development. Comparison and classification of protein structure are essential to understand protein function. Due to the problem in computational capabilities and the increasing amount of data structures, it is not possible to compare all the existing structures using standard methods. In this research, we used GIRAF [14] server to search the best protein model using a similarity ligand binding site. From this server, we obtained the highest similarity score of ligand binding sites, a protein with PDBID 3P8Z chain A. This protein is a model of NS5 methyltransferase dengue serotype 3. Additionally, through this method, we obtained the protein function prediction. The function prediction of NS5 methyltransferase is shown in table 1. Researchers have reported that NS5 methyltransferase is one of the most important parts of dengue virus replication [13]. Based on the result of function prediction, NS5 proved to be one of the most important parts of dengue virus replication. Some researchers have used NS5 methyltransferase to design dengue drugs [16-17].

The sequence of NS5 methyltransferase dengue virus serotype 3 was aligned using BLAST, and then an alignment of multiple sequences was conducted by Clustal Omega at EMBL-EBI. Based on the results of multiple sequence alignment (MSA) in figure 1, amino acid glutamine (Q) at position 10 was mutated into methionine (M) and threonine (T) at position 55 was mutated to isoleucine (I). Amino acids have specific characteristics which can play a role in protein folding, stability, complex interactions of proteins and protein function [18]. Mutations can cause changes in the secondary and tertiary structure

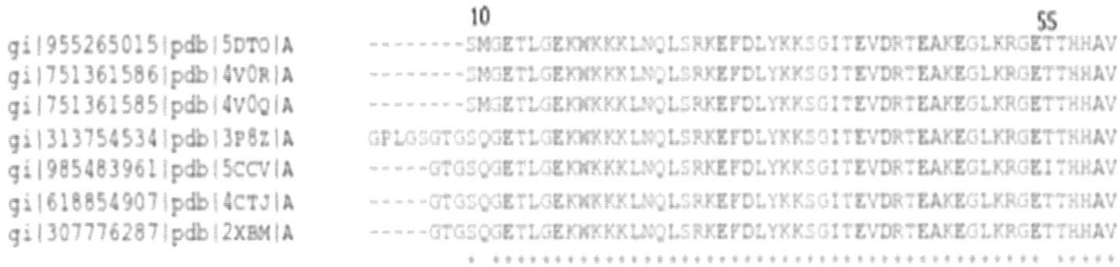

Figure 1. Multiple sequence alignment of NS5 methyltransferase DENV3 (http://www.ebi.ac.uk/Tools/msa/clustalo/)

Table 1. Molecular and biological function of non-structural protein 5 (NS5)

Chain	GOid	Contents
A	0005524	ATP binding
A	0008026	ATP-dependent helicase
A	0003725	Double-strand RNA binding
A	0046872	Metal ion binding
A	0004482	mRNA (guanine-N7-)-methyltransferase activity
A	0004483	mRNA (nucleoside-2'-O-)-methyltransferase activity
A	0003724	RNA Helicase activity
A	0003968	RNA-directed RNA polymerase activity
A	0004252	Serine-type endopeptidase activity
A	0070008	Serine-type exopeptidase activity
A	0005198	Structural molecule activity
A	0075512	Calthrin-mediated endocytosis of virus by host
A	0039654	Fusion of virus membrane with host endosome
A	0039520	Induction by virus of host autophagy
A	0006355	Regulation of transcription, DNA-templated
A	0039564	Supression by virus of host STAT2 activity
A	0039574	Supression by virus of host TYK2 activity
A	0039502	Supression by virus of host type I interferon-mediated signaling pathway
A	0006351	Transcription, DNA-templated
A	0039694	Viral genome replication
A	0019062	Virion attachment to host cell

of proteins. Changes in the secondary and tertiary structures affect the protein function and folding. The change of secondary and tertiary structure can affect the molecular docking simulation during drug design process.

The sequence was analyzed by Consurf server. From this server we obtained the variable region as well as the conserved region, which were marked using different colors on each amino acid residue. The amino acids glutamine and threonine are located in the variable regions (figure 2). At this position, the amino acids can be mutated.

Mutation analysis was done using HOPE server [12]. The structure of glutamine and methionine have different sizes (figure 3). The methionine residue has a smaller size than the glutamine residue. These residues are part of the domain named *S-Adenosyl-L-Methionine-Dependent Methyltransferase* (IPR029063). Although a structure was found, the protein might be still disordered in its native state under certain conditions. The protein is predicted to be partially disordered (61%) and the residue of interest is located in a disordered region.This residue is part of a domain named *S-Adenosyl-L-Methionine-Dependent Methyltransferase* (IPR029063). The other residue is part of a domain named mRNA Cap 0/1 Methyltransferase (IPR026490). This domain is annotated with the following Gene-Ontology (GO) terms to indicate its function as mRNA (nucleoside-2-O-)-Methyltransferase Activity (GO:0004483) and mRNA (guanine-N7-)-Methyltransferase Activity (GO:0004482).

The mutated residue is located in a domain that important for the protein activity. It is possible that this interaction is important for the correct function of the protein. The mutation can affect this interaction and protein function.

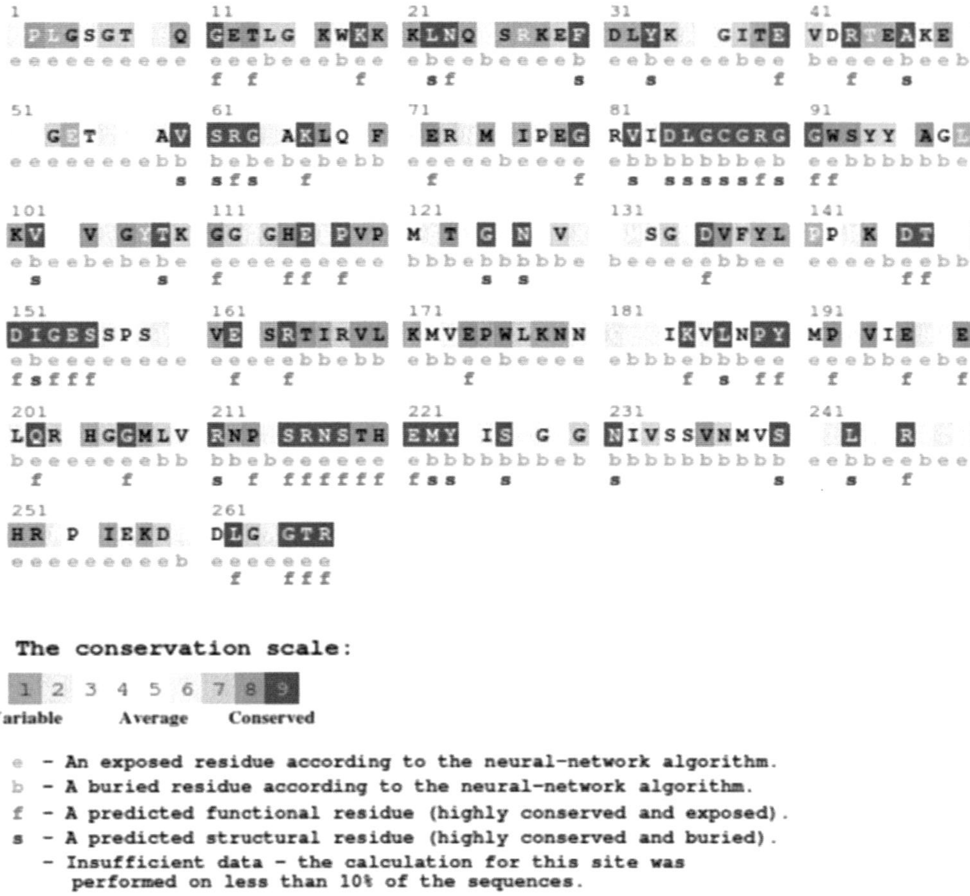

Figure 2. The conserved region of NS5 methyltransferase DENV3 (http://consurf.tau.ac.il/).

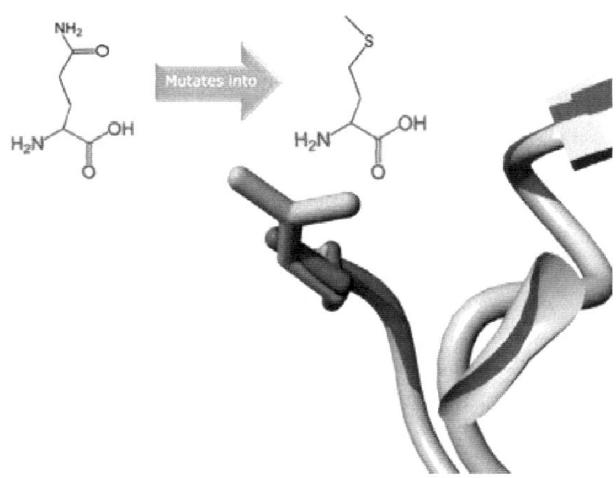

Figure 3. The structure of glutamin (green) mutates into methionine (red).

The structure of threonine and isoleucine have different sizes (figure 4). Isoleucine residue has a larger size and more hydrophobic than threonine. The mutation is located within a domain, annotated in

UniProt as mRNA cap 0-1 NS5-type MT. This residue is part of a domain named *S-Adenosyl-L-Methionine-Dependent Methyltransferase* (IPR029063). It is located on the surface of a domain with an unknown function. However, contact with other domain can affect protein function.

The residue is predicted (by KMAD) to be a phosphorylation site. Mutation into isoleucine disturbs the phosphorylation modification, because only serine, threonine and tyrosine can be phosphorylated.

In the process of in silico drug design and discovery, it is important to know the position of the ligand binding site. Mutations that occur in the ligand binding site would affect the interaction of the ligand to the receptor. Based on the result of the ligand binding site visualization, mutation does not occur in the ligand binding site, probably because it does not disturb the interaction of ligand to the receptor during drug design process (table 2).

4. Conclusions

The results of mutation analysis using HOPE server provide information about the possibility that the mutation would affect the structure and function of proteins. The mutation occurs in NS5 of DENV serotype 3, glutamine mutates into methionine at position 10 and threonine mutates into isoleucine at position 55. These residues are part of the domain named *S-Adenosyl-L-Methionine-Dependent Methyltransferase* (IPR029063). The amino acid residues Gln10 and Thr55 are not available in a binding

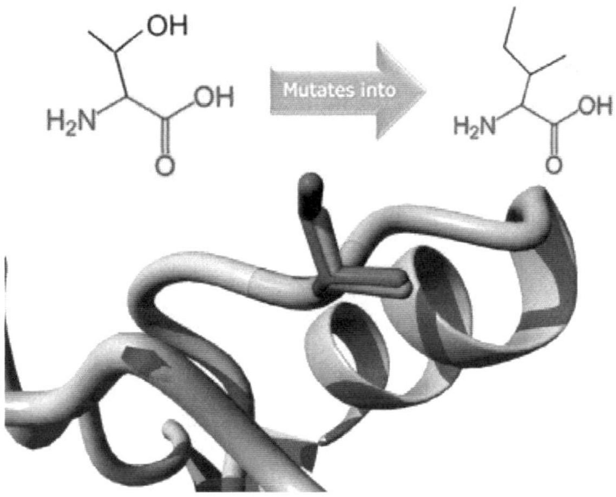

Figure 4. The structure of threonine (green) mutates into isoleucine (red)

Table 2. Amino acid residue ligand binding side

PDB ID 3P8Z	PDB ID 4V0Q	PDB ID 5CCV
Val 132	Val 132	Val 132
His 110	His 110	His 110
Asp 146	Asp 131	Asp 131
Gly 58	Gly 58	Gly 58
Ser 56	Ser 56	Ser 56
Gly 85	Gly 86	Cys 82
Gly 86	Cys 82	Asp 146
Cys 82	Asp 146	Glu 111
	Lys 105	

site pocket of proteins, so it might not affect the interaction of the ligand to the receptor in drug design. The result of this mutation will be used for further research design in dengue fever drug discovery.

Acknowledgements
The authors would like to thank Erwin Prasetya Toepak, who has proofread this manuscript. This research was funded by PITTA Grant Universitas Indonesia with contract No. 2049/UN2.R12/HKP.05.00/2016.

References

[1] Idrees S and Ashfaq U A 2014 *Asian Pac. J. Trop. Biomed.* **7** 513-6
[2] Mahmood S and Ashfaq U A 2015 *Sciforum Electronic Conference Series* **1** b032
[3] Mustafa M S, Bansal A S and Rastogi V 2011 *Med. J. Armed Forces India* **67** 192-3
[4] World Health Organization. Fact Sheet 117: Dengue and severe dengue [Internet]. Fact Sheet. World Health Organization; updated July 2016. p. 1-4. Available from: http://www.who.int/mediacentre/factsheets/fs117/en/
[5] Bartenschlager R and Miller S 2008 *Future Microbiol.* **3** 155-65
[6] Normile D 2013 *Science* **342** 415
[7] Mustafa M S, Rasotgi V, Jain S and Gupta V 2015 *Med. J. Armed Forces India* **71** 67-70
[8] Fried J R, Gibbons R V, Kalayanarooj S, Thomas S J, Srikiatkhachorn A, Yoon I K, Jarman R G, Green S, Rothman A L and Cummings D A 2010 *PLoS Negl. Trop. Dis.* **4** 1-6
[9] Garcia R A, Chismark E and Eggert J 2015 *J. Nurse Pract.* **11** 34-40
[10] Herrero L J *et al.* 2013 *Pharmacol. Ther.* **137** 266-82
[11] Monath T P 1994 *Proc. Natl. Acad. Sci. U S A* **91** 2395-400
[12] Venselaar H, Te Beek T A, Kuipers R K, Hekkelman M L and Vriend G 2010 *BMC Bioinformatics* **11** 548
[13] Noble C G and Shi P Y 2012 *Antiviral Res.* **96** 115-26
[14] Kinjo A R and Nakamura H 2012 *Biophysics (Nagoya-shi)* **18** 79-94
[15] Ashkenazy H, Erez E, Martz E, Pupko T and Ben-Tal N 2010 *Nucleic Acids Res.* **38** (Web Server Issue) W529-33
[16] Tambunan U S, Zahroh H, Utomo B B and Parikesit A A 2014 *Bioinformation* **10** 23-7
[17] Podvinec M, Lim S P, Schmidt T, Scarsi M, Wen D, Sonntag L S, Sanschagrin P, Shenkin P S, and Schwede T 2010 *J. Med. Chem.* **53** 1483-95
[18] Teng S, Madej T, Panchenko A and Alexov E 2009 *Biophys. J.* **96** 2178-88

Ligand 4'-(4-carboxyphenyl)-2,2':6',2''-terpyridine grafted with TiO_2 – complexation with the ions metal transition and application as fluorosensor for Na^+ ion

B Malau and A Zulys

Department of Chemistry, Faculty of Mathematics and Natural Sciences Universitas Indonesia, Kampus UI Depok, Depok 16424, Indonesia

Corresponding author's e-mail: zulys@ui.ac.id

Abstract. Ligand 4'-(4-Carboxyphenyl)-2,2':6',2"-Terpyridine (Hcptpy) has been synthesized using reaction of aldol condensation. The result was white-yellowish precipitate with yield of 62%. The Ligand hcptpy was grafted with TiO_2 by mass ratio 0.15 : 1 (wt/wt). The Hcptpy-TiO_2 was reacted in complexation reaction with transition metals (Fe, Mn, Ni) by chelating method, with the best mole chelating ratio of 0.15 : 1 (mole/mole). The ligand Hcptpy, Hcptpy-TiO_2 and complex assembly M-Hcptpy-TiO_2 were characterized by CHN Analyzer, HNMR, FT-IR, UV-VIS, and UV-DRS. The application of this research was to apply it as fluorosensor for sodium metal ion in sodium chloride. The study was conducted by spectrofluorometer. Types of fluoroscent for sodium metal ion detection are Chelation Quenching Effect (on-off signal) and the Chelation Enhanced Fluoroscence effect (off-on signal). The type of on-off signal causes a quenching in fluoroscence intensity, while the off-on signal causes an increase in fluoroscence intensity. The result showed that Hcptpy-TiO_2, Ni-Hcptpy-TiO_2 were the off-on fluoroscence type with the ability to respond the presence of sodium ion at concentration 3×10^{-10} M and 3×10^{-9} M, respectively. While Mn-Hcptpy-TiO_2 and Fe-Hcptpy-TiO_2 were the on-off signal type with the ability to respond the presence of sodium metal ion at concentration 3×10^{-8} M and 3×10^{-8}M, respectively. The process of sodium metal ions detection was carried out by substitution reaction of metal ion in complex with sodium metal ions.

Keywords: 4'-(4-carboxyphenyl)-2,2':6',2"-terpyridine (Hcptpy), semiconductor TiO_2, complex compound, fluorosensor, transition metals, complex assembly M-hcptpy-TiO_2

1. Introduction

The effective fluorescence sensor for metal ions is a system that interact with the presence signal of metals ions in solution by the changes of fluorescence properties, such as wavelength or emission intensity that different from free metals ions sensor. It consists of one or more fluorophore connected with active coordination species through spacer [1]. Fluorescent chemosensor usually can be found in organic compounds that has π-conjugate system [2]. It generates a better fluoresensor in terms of selectivity and practical applications for sensing method [3]. The fluorescence probe based on complex compounds has advantages in sensing method, such as longer lifetime of fluorescence and also wider stokes shift than organic fluorophore [4]. Fluorescence probe for metal ions sensor could be an organic molecule that can interact with a metal ion, either active or inactive [5]. Li *et al.* [6] has synthesized ligand 4'-(4-carboxyphenyl)-3,2':6',3''–terpyridine (3-Hcptpy) and complex compound of 3-hcptpy with zinc metal and the result showed a good intensity of fluorescence. Griffith *et al.* [7] reported that terpyridine has fluorescence properties in gel instead of sol form to detect sodium metal ions. The formed gel can give respond to the presence of sodium ions up to 98%-99.2%. Yang *et al.* [8] reported that the luminescence properties of complex compound Zn-Hcptpy have photofluorescence properties in solid state at room temperature condition.

The assembly of electron donor and acceptor plays a role in solar cell conversion. The synthesis of this assembly complex compound consists of TiO_2 nanoparticle as an electron acceptor, and the organic linker to absorb the visible light and also complex of Mn as an electron donor [9-11]. Alec C. Durrell *et al.* [12] reported that the effect of linker and anchoring to stabilize the assembly tpy-L-Anchor-TiO_2 as highly turn-on fluorescence type, for an example the assembly of Mn(II)-L-TiO_2 with sensitivity detection up to nanomolar detection. This research was to synthesize and characterize the hcptpy grafting-TiO_2 (Hcptpy-TiO_2) and M-Hcptpy-TiO_2 to obtain the role of TiO_2, diamagnetic metals and paramagnetic metals into complex assembly for electron transfer process and the fluorescence intensity shift, and to study the assembly ligand, hcptpy grafting TiO_2 and M-hcptpy-TiO_2 as fluorescent chemosensor for Na^+metal ion.

2. Materials and methods

2.1. Materials

The preparation of ligand hcptpy was performed according to the literature with some modifications [6]. Synthesis of Hcptpy-TiO_2 and M-Hcptpy-TiO_2 were prepared by modification of solvent (*wendu ding*). All other chemicals used were obtained from commercial source without further purification, such Merck and Sigma Aldrich. TruSpec LECO CHN Elemental analyzer was applied to determine the percentage of C, H, and N. H^1-NMR spectra were recorded at room temperature using JNM ECA 500 MHz. Infrared spectra were obtained from KBr pellet on Shimadzu IR 8400-S in the range 400-4000 cm^{-1}. Emission spectra were recorded by UV-Vis fluorescent (Shimadzu UV-1601, Shimadzu UV-2450) and Spectrofluorometer (Hitachi F-2000).

2.2. Synthesis of ligand 2-Hcptpy

Preparation: 2-acetylpyridine (1.21 g, 10 mmol) was added to the stirring solution of 4-carboxybenzaldehyde (0.75 g, 5 mmol), KOH (0.8 g, 14 mmol) and NH_4OH (20 mL) in 150 mL ethanol. The reaction mixture was stirred at room temperature whilst open to air for 48 h, until the colloid suspension is formed, indicated by the change of brown to yellow color. After that, the solution was adjusted to slightly acidic (pH 5) with the addition of HCl (aq., conc.). The light yellow precipitate product was filtered, recrystallized with a mixture of ethanol and methanol, and dried in vacuum to collect the colorless crystals. The results was 1.31 g (62%). Analysis for $C_{22}H_{15}N_3O_2$ (353.11): C, 74.78; H, 4.28; N, 11.89 showed the result as follows C, 74.37; H, 5.39; N, 11.96%. IR (KBr, cm^{-1}): 3445(s), 1715(s), 1631(m), 1490(w), 1400(m), 1255(w), 1118(m), 870(w), 769(w), 702(w), 617(w). H^1 NMR (400 MHz, DMSO-d6).

2.3. Synthesis of Hcptpy-TiO_2 and M-Hcptpy-TiO_2

Hcptpy-TiO_2 was prepared by stirring degussa P25 TiO_2 (1 g, 12.5mmol) with mass variation of ligand Hcptpy [0.1 g (0.28 mmol), 0.15 g (0.42 mmol), 0.2 g (0.56 mmol), and 0.3 g (0.84 mmol)] in 75 mL of DCM in reflux at temperature 40°C for 6 h. The suspension was cooled and filtered, and the white solid was washed with DCM then dried in vacuum. The best mole ratio of Hcptpy-TiO_2 used for complexation with the transition metals ($MnSO_4.H_2O$, $FeCl_2.4H_2O$, $Ni(NO_3)_2.3H_2O$) were 1:1, 1:0.3, and 1:0.15, respectively. All the transition metal was dissolved with 20mL of mQ water then added with Hcptpy-TiO_2 in 20 mL of DCM and ethanol mixture (1:1), next stirred for 6 h at room temperature. The surface of complex assembly was denoted as M-Hcptpy-TiO_2. The solid suspension was filtered and washed with the mQ water and dried in vacuum.

2.4. Characterization of Hcptpy-TiO_2 and M-Hcptpy-TiO_2
The samples were measured by UV-Vis spectroscopy with solvent of diethylether: ethanol (1:1) which showed the best mass and mole ratio to be used in fluorescence studies. To characterize UV-DRS measurement, the samples were collected as dry material and measured with $BaSO_4$ pellet. The data of UV-DRS showed the energy gap of the complex assembly, while the data of FT-IR spectra showed the binding of carboxylate linker on Deggusa P25 TiO_2, and also the interaction of transition metal ions with the ligand terpyridine.

2.5. Fluorescence studies
The excited and emission band of each sample were measured by Fluorescence Spectrophotometer Hitachi F-2700. The emission band of sample appeared at wavelength 470 nm, while excited spectra was recorded in the range 220-600 nm.

3. Results and discussion

3.1. Synthesis of ligand Hcptpy
Synthesis of Hcptpy with aldol condensation reaction method of 2-acetyl pyridine and 4-carboxybenzaldehide in ethanol and methanol medium (figure 1). The potassium hydroxide was added as catalysis base to play the role in formation of enolate, while ammonium hydroxides was added to close the chain ring in terpyridine, thus 4-(4'-Carboxyphenyl)-2,2':6',2''-terpyridine compound was obtained.

3.2. Synthesis of Hcptpy-TiO_2
The Grafting process occurred because of the carbonyl group in ligand that act as an anchor forced to bind with TiO_2 through covalent and covalent coordination bonds (figure 2). The best ratio of mass for grafting was 1 : 0.15 (wt:wt). Spectra FTIR in figure 3a of shows a decrease of vibration intensity of C=O at 1700 cm^{-1} and 2500 cm^{-1} while spectrum UV-VIS in figure 3b shows an adsorption transition orbital of $\sigma \rightarrow \sigma^*$ at wavelength 290 nm. The synthesis reaction of Hcptpy-TiO_2 occurred by the formation of bidentate bonds of carboxylic to Ti. The formed Hcptpy-TiO_2 bond is able to enhance the probability of electrons moving from HOMO to LUMO when the ligand adsorbs UV light.

The UV-DRS spectra is shown in figure 4a and the resulted plot in figure 4b showed that energy gap after grafting with TiO_2 increased from 2.89 eV to 3.21 eV. These changes occurred due to interaction between the p-orbitals from anchored to the d-orbital of the free TiO_2. The interaction was mutually reinforced to generate an increase in energy gap.

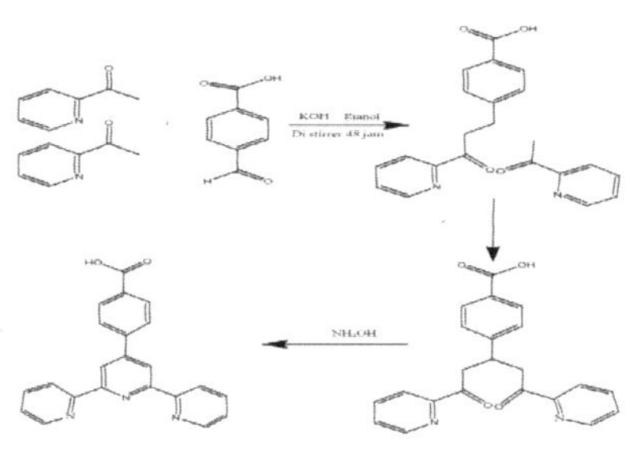

Figure 1. Scheme synthesis of Hcptpy

IOP Conf. Series: Materials Science and Engineering **188** (2017) 012044 doi:10.1088/1757-899X/188/1/012044

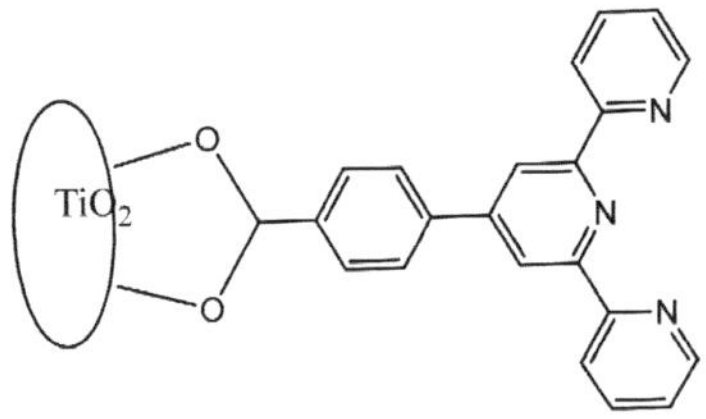

Figure 2. Scheme Hcptpy grafting TiO_2

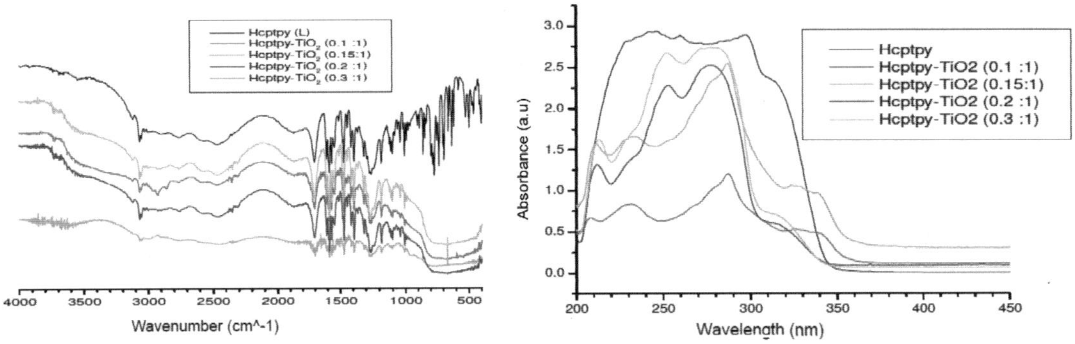

(a) (b)

Figure 3. (a) Spectra FTIR and (b) UV-VIS absorbance spectra of Hcptpy-TiO_2

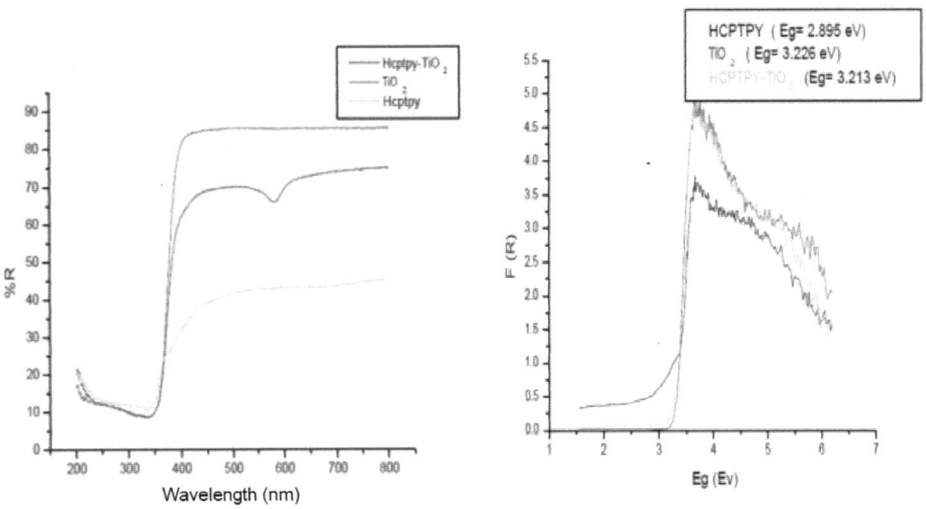

(a) (b)

Figure 4. (a) UV-VIS DRS spectra and (b) energy gap of Hcptpy-TiO_2 (1 : 0.15)

3.3. Synthesis M-Hcptpy-TiO_2

The formation of complex between terpyridine ligand and transition metal through chelating method. The formation of complex assembly through photoinducted electron transfer, which the metal ion acts as electrons donor and the ligand terpyridine as adsorber of UV light, while the carboxylic group as an anchor in transferring electron to TiO_2. The TiO_2 acts as electron acceptor which also adsorbs the UV light (figure 5). The best mole ratio of complex formation was 1 : 0.15. Figure 6 shows the results of

FT-IR measurement, the ratio showed a low interaction bonding between metal and terpyridine at 1260 cm^{-1}, and low site interaction bonding between carbonyl group and TiO_2 at 2500 cm^{-1}.

Figure 7 shows an electronic transition in d-d orbitals that affecting in a decrease of molar efficiency and wavelength shifting towards hypochromatic shift. The DRS spectra of complex assembly M-Hcptpy-TiO_2 is shown in figure 8a. The decrease of the energy gap (figure 8b) of complex assembly M-Hcptpy-TiO_2 caused by the chelation of metals, which had a weak bonding that caused decrease of gap energy.

3.4. Studies of complex assembly (Mn, Fe, Ni) M-Hcptpy-TiO_2

Photoinduced Electrons Transfer (PET) [1] in figure 9 occurred when the chromophore adsorbs the UV light, then the electron will be excited from HOMO to LUMO, therefore no electrons in the ground state. Electron in metals (Mn, Fe, Ni) will inject an electrons to HOMO in ligand terpyridine, and the electrons move to be anchored by energy and forces of scattering electrons. The electron injection process from an anchor to TiO_2 will occur in interfacial electron system to conduct in TiO_2 band.

When the electrons are scattering fast enough, the recombination of electrons occurs. The undesired recombination will occur. Photoinduced electrons transfer is activation process that involves an internal oxidation-reduction reaction between the excited state of fluorophore and another species that able to donate or accept an electron. The properties of excited state of the species are different from the ground

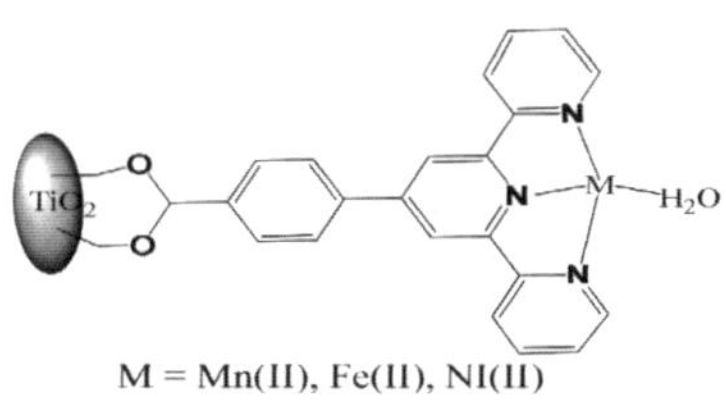

Figure 5. Functionalization M-Hcptpy-TiO_2

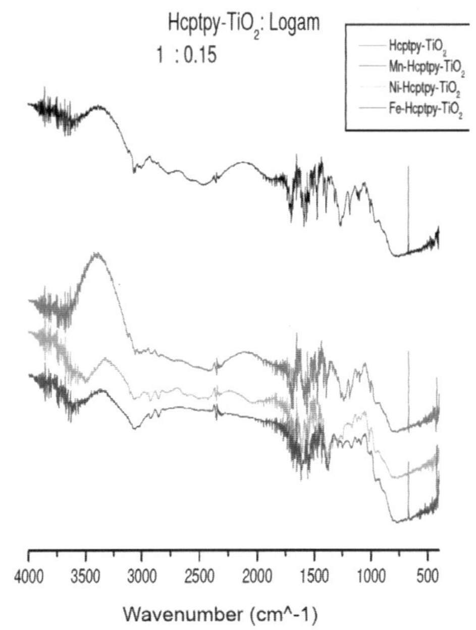

Figure 6. FTIR Spectra of M- Hcptpy-TiO_2

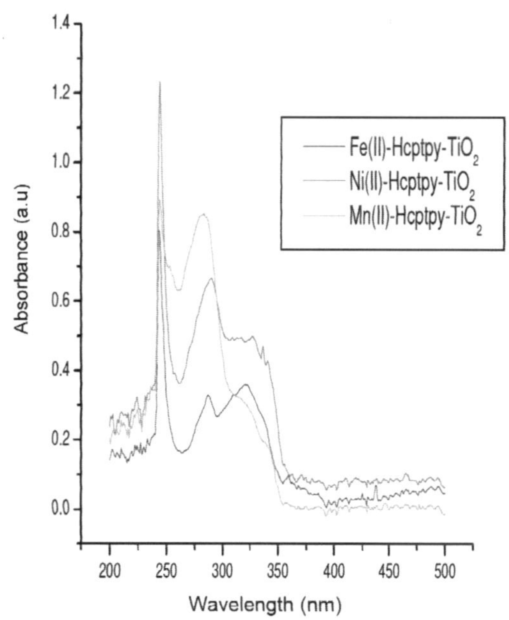

Figure 7. UV-VIS absorbance of M- Hcptpy-TiO_2

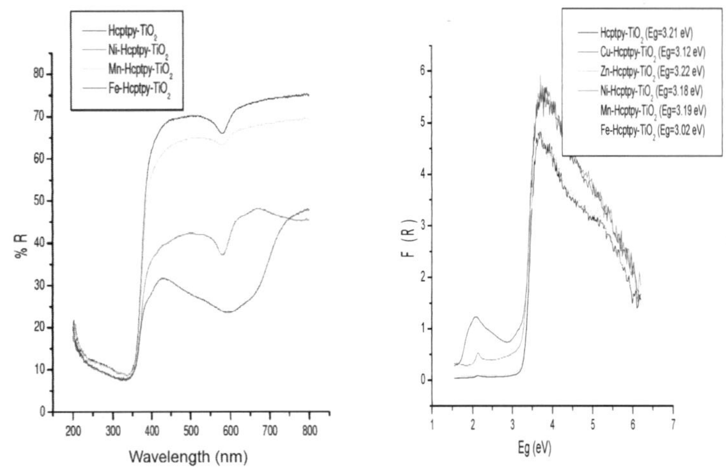

Figure 8. (a) UV-VIS DRS spectra and (b) energy gap of M- Hcptpy-TiO_2 (1 : 0.3) mol.

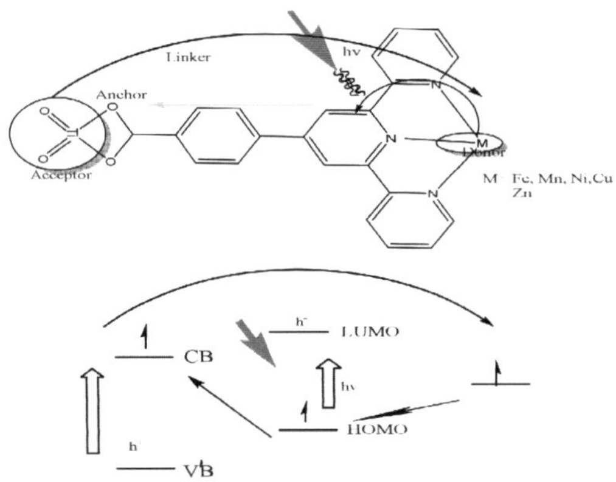

Figure 9. PET of M-Hcptpy-TiO_2 [1]

state due to the energy content. The excited state of reduction-oxidation reaction is stronger than the ground state. The PET process takes place in coordinating atom which has lone pair of electrons. The precense of ion metal coordination atom causes lower energy of lone pair electron. The PET process in fluorescent process did not cause any spectroscopic shift in the emission band upon a complexation. Its'process is strongly depended on the polarity of solvent which effecting the oxidation of lone pair electrons. As the consequence, the PET plays a role in quenching effect. The process of fluorescent occurs more quickly in polar solvent.

3.5. Fluorescence studies

Quenching effect in fluorescent happened when terpyridine adsorbs some UV light. The excitation process of chromophore ligand occurred from HOMO to LUMO. The electrons in LUMO has a short relaxation then continue emit to conduction band of paramagnetic metals. The result was magnetic resonance that triple state condition. An enhancement effect in fluorescent happened when the terpyridine adsorbs some UV light. The excitation process of chromophore ligand occurred from HOMO to LUMO. The electrons in LUMO has a short relaxation then continued emission to conduction band of diamagnetic metals. The results was magnetic resonance with singlet state condition.

IOP Conf. Series: Materials Science and Engineering **188** (2017) 012044 doi:10.1088/1757-899X/188/1/012044

3.6. Addition of Na^+ ion to Hcptpy-TiO_2 and Ni-Hcptpy-TiO_2

Figure 10a show the increase of fluorescent emission intensity. The Hcptpy-TiO_2 (3 x 10^{-6} M), change from 157.2 a.u to 970 a.u (by addition of Na^+ 10^{-4} M) 83.8% and 788.9 a.u (by addition of Na^+ 10^{-6} M) 80%, at the wavelength 246.5 nm. It indicated that Na^+ was able to interact with the N-atoms in the ligand. The posiblity interaction was chelation process with off-on (turn-on) fluorescent type. While figure 10b shows the limited respond of Hcptpy-TiO_2 to detect the presence of Na^+ ions up to 10^{-10} M by showing the increase of fluorescent intensity up to 62.6%. In figure 11b, the Ni-Hcptpy-TiO_2 (3 x 10^{-9} M) change from 53.48 a.u to 1088 a.u (by addition of Na^+ 10^{-4} M) 95.1% to 475.1 a.u (by addition of Na^+ 10^{-6} M) 88.7%, at wavelength 246.5 nm. The possible interaction of Ni^{2+} metal ion was substitution reaction in complex with the Na^+ ions and the type of fluorescent was the off-on (turn-on) type. Figure 11b shows the limited response of Ni-Hcptpy-TiO_2 to detect the presence of Na^+ ions up to 10^{-9}M with the decrease of fluorescent intensity at 46%.

3.7. Influence of Na^+ ions addition in Mn-Hcptpy-TiO_2 and Fe-Hcptpy-TiO_2 assembly

Figure 12a shows the decrease of fluorescent emission intensity of Mn-Hcptpy-TiO_2 (3 x 10^{-6} M) change from 32.28 a.u to 30.32 a.u (by addition of Na^+ 10^{-4} M) 6.07% to 26.59 a.u (by addition of Na^+ 10^{-6} M) 17.6%, at 246.5 nm. It indicated that Na^+ was able to interact with the N-atoms in the ligand. Figure 13a

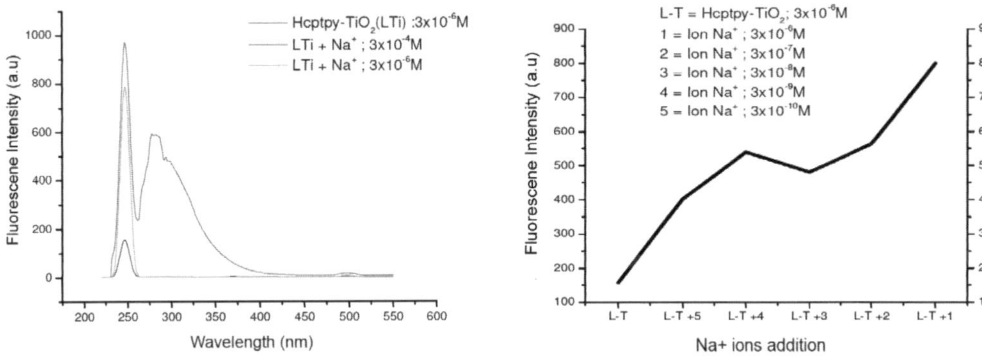

(a) (b)

Figure 10. Fluorescent emission intensity of Hcptpy-TiO_2 concentrate with the addition of Na^+ and (b) limited response of of Hcptpy-TiO_2 to the presence of Na^+ ions

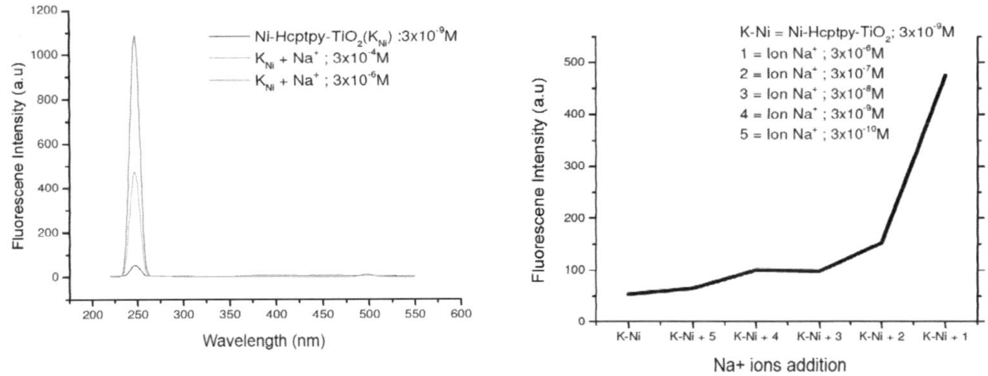

(a) (b)

Figure 11. Fluorescent emission intensity of Ni-Hcptpy-TiO_2 concentrate with the addition of Na^+ and (b) limited response of of Ni-Hcptpy-TiO_2 to the presence of Na^+ ions

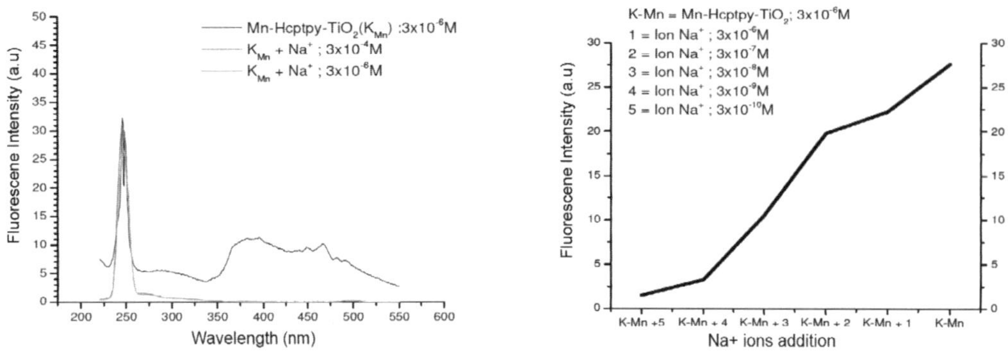

(a) (b)

Figure 12. Fluorescent emission intensity of Mn-Hcptpy-TiO_2 concentrate with the addition of Na^+ and (b) limited response of of Mn-Hcptpy-TiO_2 to the presence of Na^+ ions

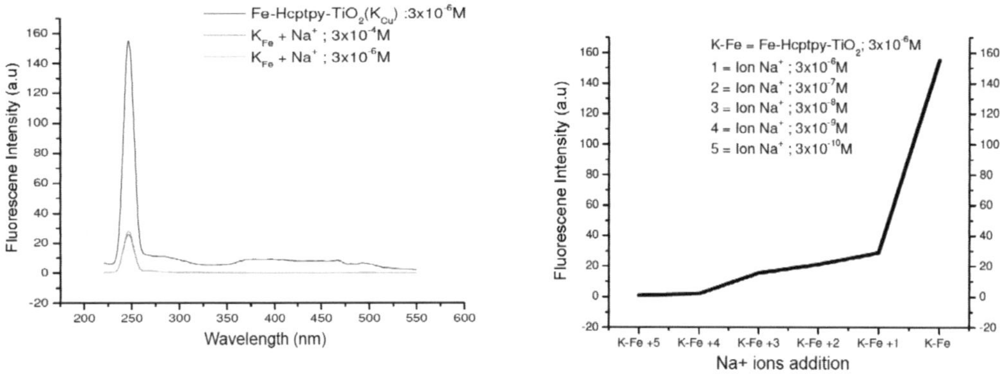

(a) (b)

Figure 13. Fluorescent emission intensity of Fe-Hcptpy-TiO_2 concentrate with the addition of Na^+ and (b) limited response of of Fe-Hcptpy-TiO_2 to the presence of Na^+ ions

shows the decrease in fluorescent emission intensity of Fe-Hcptpy-TiO_2 (3 x 10^{-6} M) changed from 155.1 a.u to 25.88 a.u (by addition of Na^+ 10^{-4} M) 83.3% to 28.5 a.u (by addition of Na^+ 10^{-6} M) 81.6%, at the wavelength 246.5 nm. The possible interaction was the substitution process of Mn^{2+} and Fe^{2+} in complex with Na^+ ions, with on-off (turn-off) fluorescent type. Figure 12b shows the limited response of Mn-Hcptpy-TiO_2 to detect the presence of Na^+ ions up to 10^{-8} M, with the decrease of fluorescent intensity at 67%, while Fe-Hcptpy-TiO_2 detected the presence of Na^+ ions up to 10^{-8} M, with the decrease of fluorescent intensity at 90% (figure 13b).

4. Conclusions

The characteristics of complex assembly through PET process, the transition metal ions act as electron donor while ligand terpyridine as the UV light adsorber with the anchoring carboxylic group as linker to inject an electron from ligand to TiO_2. The TiO_2 act as electron acceptor, also to follow and assist the ligand to adsorb some UV light. The transfer process of an electron from transition metal to ligand is in accordance with charge transfer. The studies of fluorescent in assembly compounds have 2 types. The first type is *Chelation Quenching Effect (decreased intensity fluorescent) (on-off signal)* showed by Mn-Hcptpy-TiO_2 and Fe-Hcptpy-TiO_2 (paramagnetic metals) and the second type is *Chelation Enhanced effect (off-on signal)* showed by Hcptpy-TiO_2 and Ni-Hcptpy-TiO_2 which enhance the fluorescent intensity (diamagnetic metals). Generally all the complex assembly can detect Na^+ ion in some concentrations.

Acknowledgements
We thanks to Indonesia Endowment Fund for Education-Ministry of Finance, for financial support.

References

[1] Mauro F, Vieri F, Luca G and Mauro M 2012 *Journal of Coordination Chemistry Reviews* **256** 170-92
[2] Fegley M E A, Pinnock S S, Malele C N and Jones Jr W E 2012 *Inorganica Chimica Acta* **381** 78-84
[3] Lakowicz J R 2006 *Principles of Fluorescence Spectroscopy 3rd Ed* (New York: Springer Verlag)
[4] Afrida M 2013 *Sintesis Senyawa Kompleks [ZnII(4'-(2-thienyl)- 2,2':6',2"-terpyridine)(NO3)2] dan Aplikasinya sebagai Fluorosensor Logam Berat* BSc projects (Depok: Departemen Kimia, FMIPA, Universitas Indonesia)
[5] Rurack K 2001 *Spectrochimica Acta A Mol. Biomol. Spectrosc.* **57** 2161-95
[6] Li N, Zhu Q E, Hu H M, Guo H L, Xie J, Wang F, Dong F X, Yang M L and Xue G L 2013 *Polyhedron* **49** 207-15
[7] Griffith A, Bandy T J, Light M and Stulz E 2013 *Chem. Commun.* **49** 731-33
[8] Yang J, Hui R X and Zhang M B 2012 *J. Solid State Chemistry* **196** 398-403
[9] McNamara W R, Snoeberger R C, Li G, Schleicher J M, Cady C W, Poyatos M, Schmuttenmaer C A, Crabtree R H, Brudvig G W and Batista V S 2008 *J. American Chemistry Society* **130** 14329-38
[10] McNamara W R, Snoeberger R C, Li G, Richter C, Allen L J, Milot R L, Schmuttenmaer C A, Crabtree R H, Brudvig G W and Batista V S 2009 *Energi Environ. Sci.* **2** 1173-75
[11] McNamara W R, Milot R L, Song H E, Snoeberger R C, Batista V S, Schmuttenmaer C A, Brudvig G W and Crabtree R H 2010 *Energi Environ. Sci.* **3** 917-23
[12] Durrell A C *et al.* 2014 *J. Catalysis* **310** 37-44

Improving the sonocatalytic performance of good crystallinity ZrO_2 nanocomposite through graphene addition

Y Kristianto[1,2], **A Taufik**[1,2] **and R Saleh**[1,2]

[1]Department of Physics, Faculty of Mathematics and Natural Sciences Universitas Indonesia, Kampus UI Depok, Depok 16424, Indonesia
[2]Integrated Laboratory of Energy and Environment, Faculty of Mathematics and Natural Sciences Universitas Indonesia, Kampus UI Depok, Depok 16424, Indonesia

Corresponding author's e-mail: rosari.saleh@gmail.com

Abstract. In this study, the catalytic performance of the ZrO_2 nanoparticles is enhanced by combining ultrasonic radiation with graphene material addition. The structural, morphological and surface properties of sol-gel synthesized ZrO_2 nanoparticles with three different annealing temperatures (350°C, 550°C, and 750°C) were investigated using various characterization methods, including X-Ray Diffraction, Transmission Electron Microscope measurements and Brunauer-Emmett-Teller measurements. The result show that if the annealing process does not take place, ZrO_2 nanoparticles have an amorphous structure. The formation of the crystalline structures of the tetragonal phase of the nanoparticle begins at an annealing temperature of 550°C. The highest sonocatalytic performance of nanoparticles is achieved at an annealing temperature of 550°C. The ZrO_2/graphene nanocomposites were produced via co-precipitation methods at all annealing temperatures. The sonocatalytic activities in the presence of ZrO_2/graphene nanocomposite indicate an enhancement, compared with ZrO_2 nanoparticles. ZrO_2 with and without graphene composites shows a level of stability that would allow reuse. A reasonable catalysis mechanism of ZrO_2 with and without graphene composites is proposed.

Keywords: ZrO_2, sonocatalytic, graphene, and annealing

1. Introduction

At present, the sonocatalytic process has received great attention as a useful and promising method for mineralizing organic pollutants, such as synthetic dyes, in aqueous media. In this process, water molecules are used to produce hydroxyl radicals – the primary oxidant in water purification treatment. Hydroxyl radicals are very reactive and non-selective oxidants, which are capable of decolorizing and mineralizing dyes to CO_2 and H_2O, as well as other simple inorganic products [1]. Heterogeneous sonocatalysis through ultrasonic treatment on a metal oxide semiconductor surface is an advanced oxidation process [2].

Among the various semiconductors, zirconium dioxide (ZrO_2) is an attractive candidate for degradation of dyes due to its interesting physical and chemical properties, low cost, high stability, and environmentally friendly status [3]. However, the degradation efficiency of ZrO_2 nanoparticles is not very high because of the high rate of recombination in electron-hole pairs. Therefore, the development of a new, efficient catalyst is still a major challenge.

Recently, a great deal of attention has been paid to the use of graphene materials in the catalysis field due to its large surface area [4], high rate of electronic charge transport [5], and excellent mechanical

IOP Conf. Series: Materials Science and Engineering **188** (2017) 012045 doi:10.1088/1757-899X/188/1/012045

and chemical properties [6, 7], which make it becomes a kind of attractive adsorbent in wastewater treatment. Therefore, efforts to combine graphene with metal-oxide semiconductors have become central to the production of catalysts with high performance in the decomposition of dyes. The purpose of combining graphene with metal-oxide semiconductors is to improve degradation efficiency by increasing the charge-separation process, reducing the rate of the recombination of electron-hole pairs, and increasing the surface area of the catalyst [8].

In this study, the prepared samples were characterized via various techniques, including X-ray Diffraction (XRD), Transmission Electron Microscope (TEM) measurement, and Brunauer-Emmett-Teller (BET) measurement. The catalytic performance of ZrO_2 nanoparticles was enhanced by combining ultrasonic radiation with graphene material addition. Methylene blue (MB) dye was selected as a model contaminant in wastewater because MB is widely used in the dyeing and printing of textiles, and it is reasonably chemically stable and non-biodegradable. The effects of annealing temperatures, reusability, and role of reactive agents on the catalytic reaction were investigated and the results are discussed below.

2. Experimental

2.1 Materials

All the chemical reagents were of an analytical grade and were used without further purification. Zirconium(IV) chloride ($ZrCl_4$) and sodium hydroxide (NaOH) were purchased from Merck. Graphene was purchased from Angstron Materials.

2.2 Preparation of ZrO_2-X (with X = dry, 350, 550, or 750)

ZrO_2-X nanoparticles were synthesized using the sol-gel method. Briefly, $ZrCl_4$ and NaOH were dissolved separately in distilled water through magnetic stirring. Then, the $ZrCl_4$ solution was added to the NaOH solution slowly, followed by magnetic stirring for 3 h, which allowed a homogenous solution to be achieved. The temperature of solution was maintained at 80°C during the stirring process. The mixed solution was then centrifuged to separate out the products that were precipitated from the solution. The precipitates were rinsed with distilled water and ethanol several times to remove impurities. The product was heated at 120°C for 12 h. It was labeled as ZrO_2-dry. To raise the crystallinity up in the product, the ZrO_2-dry nanoparticles were calcined at various temperatures: 350°C, 550°C, and 750°C. Calcination took place in the air for 5 h, at a heating rate of 10°C min^{-1}. The results were labeled ZrO_2-350, ZrO_2-550, and ZrO_2-750, respectively.

2.3 Preparation of ZrO_2-X/graphene (with X = dry, 350, 550, or 750)

ZrO_2-X/graphene nanocomposites were synthesized using the co-precipitation method. Briefly, the graphene powder was dispersed in distilled water and an ethanol solution via ultrasonic treatment for 2 h, which was followed by the addition of the prepared ZrO_2-X nanoparticles into a graphene solution. The mixed solution was stirred for 1 h in order to achieve a homogeneous suspension. The suspension was then heated at 120°C for 3 h to effect deposition on to the graphene sheets. The product was collected through centrifugation and was dried in a vacuum at 70°C for 12 h in order for ZrO_2-X/graphene nanocomposites to be obtained.

2.4 Characterization

XRD was performed using a Rigaku Miniflex 600 diffractometer with Cu K-α radiation (λ =1.5406 Å) in order to obtain the crystallized samples. Tecnai G2 supertwin transmission electron microscopy (TEM, accelerating = 200 kV) was used to observe particle size and the morphology of samples. The specific surface area of samples was determined using the BET method.

2.5 Sonocatalytic activity

MB was used as a model organic pollutant. The sonocatalytic activity of samples was investigated via the degradation of an MB aqueous solution with ultrasonic irradiations. Catalysts (30 mg) were dissolved in MB solution (20 mg/L). The solution kept stirred continuously in the dark condition for 30 min to ensure that an equilibrium was reached between the adsorption and desorption of MB molecules on the catalyst surface. The solution was placed in an ultrasonic bath with a fixed frequency and power of 40 kHz and 150 W, respectively. Subsequently, the solution was irradiated at regular intervals for 2 h. Every 15 min, a small amount of the water sample was collected through centrifugation so that the MB concentration could be measured with a UV-vis spectrophotometer.

2.6 Scavenger experiment. In order to determine the species involved in sonocatalytic activities, different radical scavengers were added to the solutions of MB. Ammonium oxalate, sodium sulfate, and tert-butyl alcohol (TBA) were used as scavengers for holes, electrons, and hydroxyl radical, respectively. The same measurements as described above were used.

3. Results and discussion

The XRD patterns of the as-prepared samples are shown in figure 1. The XRD spectra of tetragonal phase of ZrO_2-550 have four well resolved peaks at 2θ = 30.18°, 35.03°, 50.36°, and 59.94° which are indexed as reflection from [111], [200], [220], and [311] plane, respectively. However, the sample ZrO_2-750 has two additional diffraction peaks at 2θ = 28.13° and 31.51° which attributed to the existence of monoclinic phase of ZrO_2, indicating that sample annealing at 750°C has mixture phases of tetragonal and monoclinic. Unlike ZrO_2-550 and ZrO_2-750, the peaks which are reflected by ZrO_2-350 has no sharp peak. It indicates that with annealing at 350°C the crystallized has not formed properly yet. Meanwhile, both of ZrO_2-dry and graphene has broad peak which indicated them are amorphous. For samples ZrO_2-X/graphene, the addition peak from graphene cannot be observed because amorphous structure and low content of graphene on ZrO_2-X/graphene that is 5 wt.% of graphene. Furthermore, no characteristic peaks from other crystalline impurities were detected by XRD, suggesting that the as-prepared samples have good purity.

The average crystallite size for ZrO_2-550 and ZrO_2-750 is calculated from Debye Scherrer's formula (equation (1)):

$$D_{hkl} = \frac{0.89\lambda}{\beta cos\theta} \quad (1)$$

where D_{hkl} is the crystallite size along the (hkl) direction, λ is the wavelength of the X-ray, β is the full-

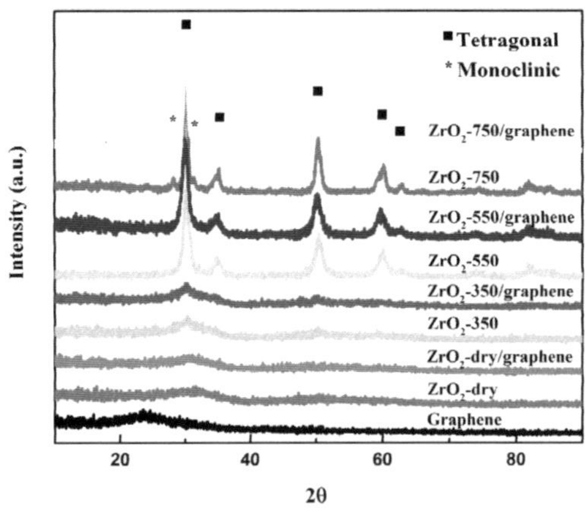

Figure 1. XRD patterns of the as-prepared samples and graphene.

Table 1. Specific surface area of samples.

Samples	Specific Surface Area (m^2/g)
ZrO_2-350	286.822
ZrO_2-550	160.875
ZrO_2-750	39.069
ZrO_2-550/graphene	176.306

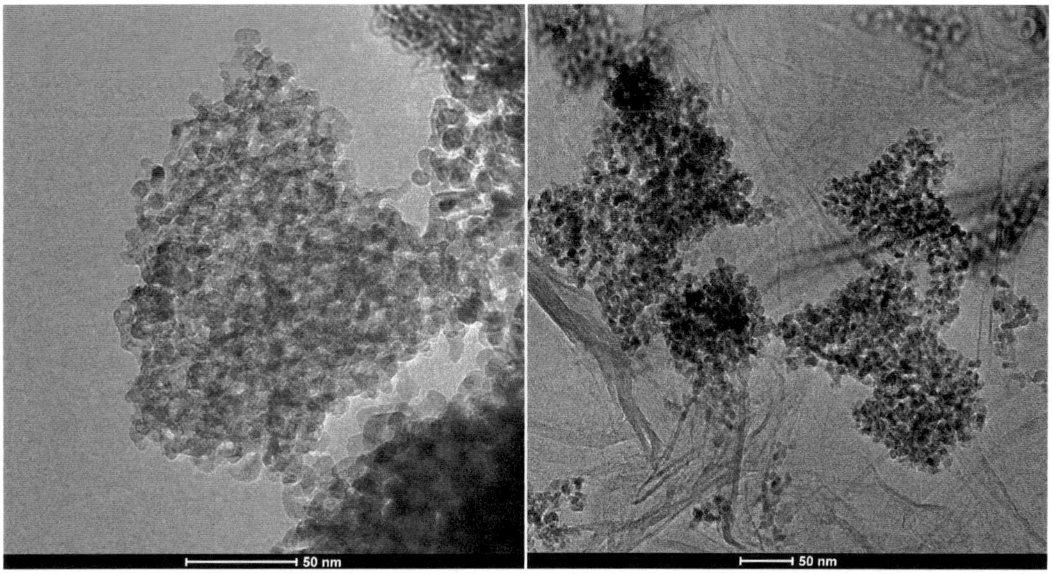

(a) (b)

Figure 2. TEM images of the as-prepared samples, (a) ZrO_2-550 and (b) ZrO_2-550/graphene.

width half maximum of the most intense diffraction peak, and θ is the diffraction angle. The average crystallite size to be approximately 13 and 24 nm for ZrO_2-550 and ZrO_2-750, respectively. The crystallite size is related to the contact area. Reducing the crystallite-size mean increases the contact area and improves the dispersion of the samples in the solution, thus promoting interaction between photons, the catalyst, and pollutants [9-11]. Briefly, samples with a small crystallite size exhibit good catalytic performance. In other words, ZrO_2-550 should exhibit a better catalytic performance than ZrO_2-750.

The specific surface area of all the samples in the annealing process is given in table 1. It can be seen that the surface area of ZrO_2 nanoparticles decreases as the temperature increases. However, the incorporation of graphene materials enhances the surface area significantly.

Figure 2 shows the TEM images of the pure ZrO_2-550 nanoparticles and the ZrO_2-550 nanocomposites with graphene addition. As is clear, there are agglomeration particles with a topography that is nearly spherical, indicating the presence of ZrO_2. The agglomeration probably occurred because of the high surface energy of ZrO_2 nanoparticles [12]. The aggregated structure of the ZrO_2 nanoparticles suggests a reduction in the total active surface area of the catalyst, which would have reduced catalytic performance. This might be one reason why the degradation efficiency of ZrO_2 nanoparticles was not very high.

As described earlier, the XRD measurements did not indicate the existence of graphene in ZrO_2-X/ graphene nanocomposites because of its amorphous structure and low graphene content. The TEM

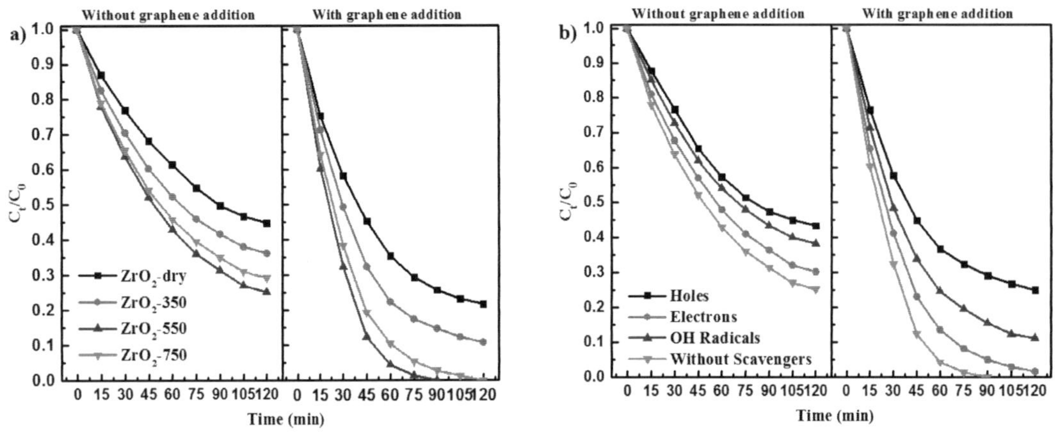

(a) (b)

Figure 3. (a) Sonocatalytic activity degradation; (b) scavenger experiment for ZrO_2-550/graphene.

images, however, show the presence of graphene clearly. Its existence was confirmed by the presence of a layer-like structure, as seen in figure 2b. This indicates that the ZrO_2-X/graphene nanocomposites were synthesized successfully through the co-precipitation method.

In order to investigate the degradation kinetics of MB using the samples, the pseudo-first-order reaction model was used. The following equation was utilized as equation (2) bellow:

$$\frac{C_t}{C_o} = e^{-K_{app}t} \tag{2}$$

where C_t is the concentration at any time, t, C_o is the initial concentration, K_{app} is the rate constant. As figure 3a shows clearly, the degradation efficiency is increased significantly by graphene addition to ZrO2-X nanoparticles. This is because of the role of graphene in increasing the charge-separation process that occurs between the MB molecules and the catalyst, which reduces the rate of the recombination of electron-hole pairs, as well as increasing the surface area of the catalyst [8]. With graphene addition, the degradation efficiency increased from 55.2% to 77.4%, 63.5% to 89.4%, 70.6% to 100%, and 75% to 100% for ZrO_2-dry, ZrO_2-350, ZrO_2-550, and ZrO_2-750, respectively. In spite of the graphene addition case, the degradation efficiency was enhanced as the annealing temperature increased. The sample that was annealed at 550°C had the best activity, when compared with the other samples. The decrease in the degradation reaction for ZrO_2-750 could be related to the decreasing catalyst surface area at higher temperatures, caused by aggregation and crystal growth [9-11, 13].

To ascertain the main reactive species responsible for the degradation of MB in the samples, a series of scavengers were employed. Figure 3b demonstrates the degradation of MB in the presence of selected scavengers. It can be seen that the addition of selected scavengers inhibits the sonocatalytic removal of MB in an aqueous solution. As a result, the role of the reactive species can be expressed as follows: hole > $OH^{\bullet}$ > electron. Therefore, it can be concluded that holes are the main reactive species.

Stability and reusability are important issues when assessing the practical and economic implications of catalysts. In order to evaluate stability and reusability, the same catalyst was used repeatedly over four cycles. At the end of each cycle, the catalyst was separated from the MB solution via centrifuge, allowing it to be reused under the same conditions with each MB solution (as shown in figure 4). The reuse of the catalyst did not lead to any significant changes in degradation efficiency where the sonocatalytic activity was concerned, which emphasizes the good level of stability of the catalysts that were used.

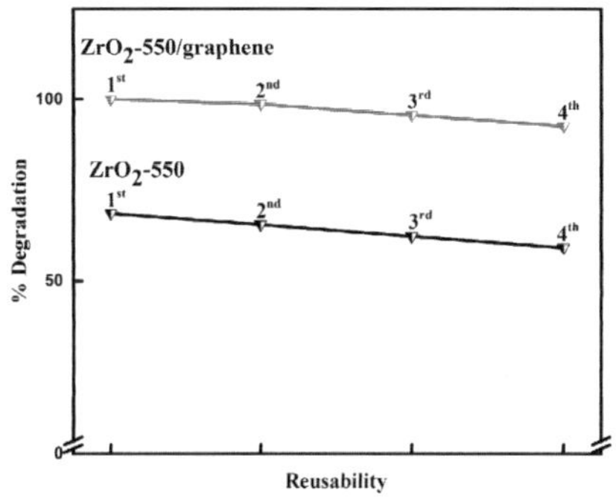

Figure 4. Reusability of ZrO_2-550 and ZrO_2-550/graphene.

The sonocatalytic mechanism of pure ZrO_2 has been studied previously [14]. In the presence of graphene, the rate of the recombination electron-hole pairs might be inhibited due to the trapping of electrons by graphene layers. This is caused by the properties of graphene, which is a good acceptor of electrons [15].

4. Conclusions

In this study, ZrO_2-X nanoparticles were synthesized successfully using the sol-gel method, and ZrO_2-X/graphene nanocomposites were synthesized successfully using the co-precipitation method. The results were confirmed via XRD and TEM measurements. The XRD measurements showed that ZrO_2 nanoparticles have an amorphous structure when the annealing process has not taken place. The crystalline structures of the tetragonal phase of the nanoparticle begin to be formed at an annealing temperature of 550°C, and at 750°C, the monoclinic phase of ZrO_2 begins. The BET results indicated that the surface area of samples decreases as the temperature increases. Adding graphene to ZrO_2 significantly improved degradation efficiency. The sample that was annealed at 550°C demonstrated the best activity levels, when compared with the other samples. The scavenger experiment showed that the reactive species operates via the following sequence: hole > $OH^{\bullet}$ > electron. In the sample, there was also a good level of stability in terms of reuse over four cycles.

References

[1] Ahmad M, Ahmed E, Hong Z L, Ahmed W, Elhissi A and Khalid N R 2014 *Ultrason. Sonochem.* **21** 761-73
[2] Shimizu N, Ogino C, Dadjour M F and Murata T 2007 *Ultrason. Sonochem.* **14** 184-90
[3] Poungchan G, Ksapabutr B and Panapoy M 2016 *Mater. Design* **89** 137-45
[4] Stoller M D, Park S, Zhu Y, An J H and Ruoff R S 2008 *Nano Lett.* **8** 3498-502
[5] Bolotin K I, Sikes K J, Jiang Z, Klima M, Funderberg G, Hone J, Kim P and Stormer H L 2008 *Solid State Commun.* **146** 351-5
[6] Balandin A A, Ghosh S, Bao W, Calizo I, Teweldebrhan D, Miao F and Lau C N 2008 *Nano Lett.* **8** 902-7
[7] Zhu Y, Murali S, Cai W, Li X, Suk J W, Potts J R and Ruoff R S 2010 *Adv. Mater.* **22** 3906-24
[8] Han L, Wang P and Dong S 2012 *Nanoscale* **4** 5814-25
[9] Fallet M, Permpoon S, Deschanvres J L and Langlet M 2006 *J. Mater. Sci.* **41** 2915-27
[10] Singh A, Dutta D P, Ballal A, Tyagi A K and Fulekar M H 2014 *Mater. Res. Bull.* **51** 447-54
[11] Maria A J, Yeung K L, Lee C Y, Yue P L and Chan C K 2000 *J. Catal.* **192** 185-96
[12] Liu Z L, Wang H B, Lu Q H, Du G H, Peng L, Du Y Q, Zhang S M and Yao K L 2004 *J. Magn. Magn. Mater.* **283** 258-62
[13] Kim S Y, Lim T H, Chang T S and Shin C H 2007 *Catal. Lett.* **117** 112-8
[14] Salavati H, Tavakkoli N and Hosseinpoor M 2012 *Ultrason. Sonochem.* **19** 546-53
[15] Zhang W, Dong F, Xiong T and Zhang Q 2014 *Ceram. Int.* **40** 9003-8

Preparation of glucose sensors using gold nanoparticles modified diamond electrode

Fachrurrazie, T A Ivandini and W Wibowo

Department of Chemistry, Faculty of Mathematics and Natural Sciences Universitas Indonesia, Kampus UI Depok, Depok 16424, Indonesia

Corresponding author's e-mail: ivandini.tri@sci.ui.ac.id

Abstract. A glucose sensor was successfully developed by immobilizing glucose oxidase (GOx) at boron-doped diamond (BDD) electrodes. Prior to GOx immobilization, the BDD was modified with gold nanoparticles (AuNPs). To immobilize AuNPs, the gold surface was modified to nitrogen termination. The characterization of the electrode surface was performed using an X-ray photoelectron spectroscopy and a scanning electron microscope, while the electrochemical properties of the enzyme electrode were characterized using cyclic voltammetry. Cyclic voltammograms of the prepared electrode for D-glucose in phosphate buffer solution pH 7 showed a new reduction peak at +0.16 V. The currents of the peak were linear in the concentration range of 0.1 M to 0.9 M, indicated that the GOx-AuNP-BDD can be applied for electrochemical glucose detection.

Keywords: Biosensors, glucose oxidase, gold nanoparticles, boron-doped diamond

1. Introduction

The basic for constructing a biosensor is to prepare the enzyme modified electrode. The enzyme needs to be immobilized on the electrode surface to obtain the electrochemical reaction immediately after enzyme bioactivity. However, by using enzyme, the accessibility of electron transferred between redox center and the surface of the electrode are limited by 3D protein structure of the enzyme.

On the other hand, boron-doped diamond (BDD) electrode offers special characteristics that required by electrochemical applications [1]. However, some challenges need to be resolved before it can be used as biosensors. To name a few, BDD has an inert surface, therefore modification on the surface needs to be performed.

In this research, glucose oxidase (GOx) was used as the biosensing/biocatalyst for the application in a glucose sensor. The redox active sites of this enzyme, i.e. flavin adenine dinucleotides (FADs) were located at the core center protein, thus limits its direct electron transfer to the electrode surface. It is known that Au offers good responses in electron transfer with enzyme hence increasing the surface area of Au in the system is important [2]. Accordingly, the surface of BDD was modified by gold nanoparticles (AuNPs) to achieve a more efficient electron transfer between the enzyme and BDD. Previously, our group has successfully developed AuNPs-modified BDD (AuNPs-BDD) for biochemical oxygen demand (BOD) sensing with yeasts as the biocatalysts [3-5]. Although the response of AuNPs-BDD electrode is smaller than that of bulk gold electrodes, the signal was generally clearer due to the smaller background current. In this work, the AuNPs-BDD was modified by GOx. This prepared electrode (GOx-AuNPs-BDD) was then investigated for a preliminary study of glucose sensors. The results showed good responses with linear peak currents to glucose concentrations.

2. Materials and methods

2.1. Materials
Glucose oxidase (GOx) from Aspergillus niger, D-glucose, $HAuCl_4$, $NaBH_4$, sodium citrate, and NH_4OH were supplied from Sigma or Sigma-Aldrich. Pt spiral was purchased from Nilaco Japan, while BDD was prepared in Keio University, Japan.

2.2. Methods
AuNPs was synthesized by mixing 10 mL of 0.5 mM M $HAuCl_4$ with 5 mL of 0.1 M sodium citrate solution followed by a vigorous stirring at 25°C. After the addition of sodium citrate solution, the color of gold solution became transparent. After 1 h stirring, 0.5 mL fresh solution of $NaBH_4$ (0.1 M) was added dropwise, while stirring continuously for another hour. The appearance of pink color indicated the formation of AuNPs.

Prior to modify the BDD electrodes with AuNPs, the electrodes were immersed in the concentrated NH_4OH solution and irradiated under a UV source (λ = 254 nm) for 6 h. Then, the electrodes were soaked in colloidal AuNPs for 20 min continued with drying at 60°C. XPS was used for the characterization process.

In order to immobilize GOx at AuNPs-BDD, 100 μL GOx solution (10 mg mL^{-1}) was poured to the surface of AuNPs-BDD and dried at 4°C using silica gel as water absorbent [6]. The modified electrodes were maintained at 4°C in dry condition when it is not in used. The electrode was then characterized by SEM.

The electrochemical analysis was performed in PBS pH 7 in the present of D-glucose to study the oxidation of glucose at the prepared electrodes.

3. Results and discussions
The color of $HAuCl_4$ solution changed from yellow to ruby red after the reduction reaction with $NaBH_4$ and sodium citrate, indicating the formation of AuNPs. Characterization of AuNPs by using UV-Vis spectroscopy shows a maximum absorption wavelength of AuNPs at 515 nm (figure 1a), while TEM image shows the main particle size around 30 nm (figure 1b). However, some particle sizes under 5 nm were also observed (figure 1). The nanoparticles size was not really homogenenous, therefore, it was expected that the immobilization will cover the surface of BDD well, with the small size filled the space between the big size of nanoparticles.

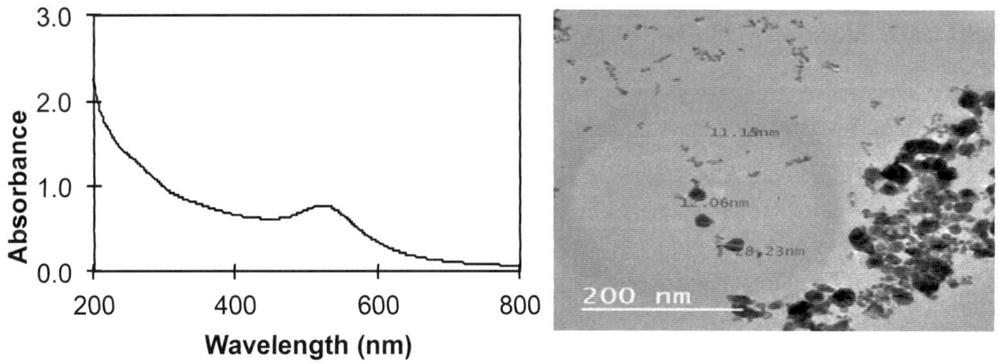

(a) (b)

Figure 1. (a) visible absorption spectra and (b) TEM image of AuNPs.

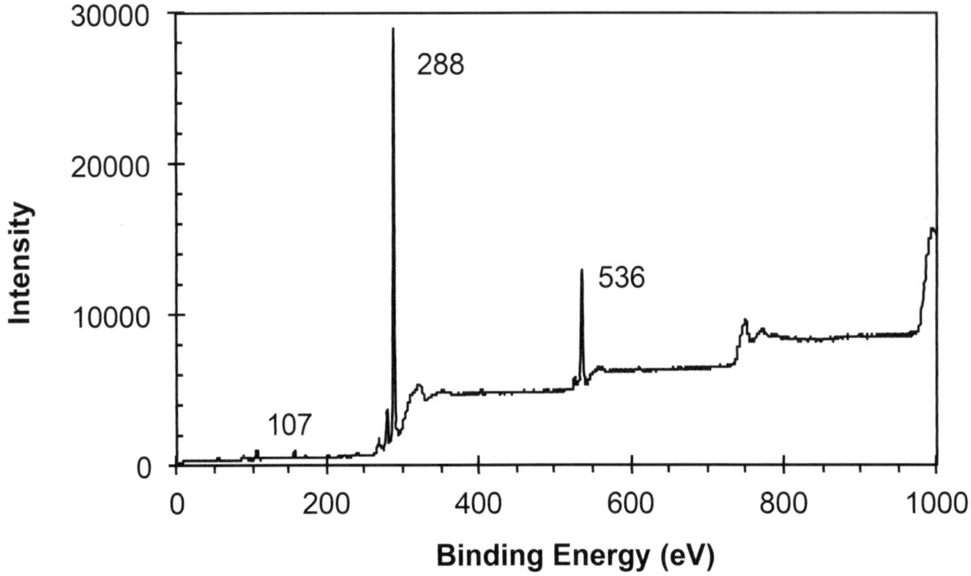

Figure 2. XPS Spectra of the GOx-AuNPs-BDD. The inset shows the magnification of gold binding energy at 107 eV and 109 eV.

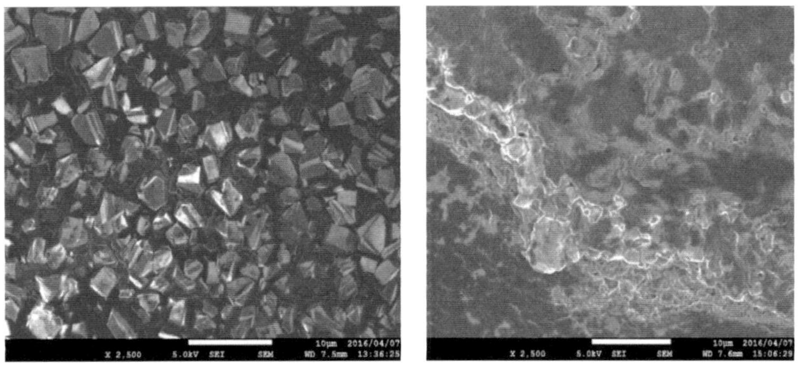

(a) (b)

Figure 3. (a) SEM Images of AuNPs-BDD and (b) GOx-AuNPs-BDD.

BDD surface is normally partly terminated by oxygen as the result of oxidation process in the air [3, 4]. In this case, the BDD surface needs to be modified to nitrogen termination by photochemistry reaction to exchange oxygen terminal with amine functional group (-NH_2) at the surface. Tian *et al.* [7] reported that self-assembly AuNPs-BDD can be formed at nitrogen-terminated BDD. It was proposed that covalent bonds between AuNPs and -NH_2 stabilize the position of AuNPs at the carbon surface [8] hence it was expected that a better stability of GOx at the surface of BDD can be generated.

XPS spectrum of GOx-modified AuNPs-BDD (figure 2) shows the diamond peak (C 1_s) at binding energy of 288 eV and a peak corresponds to C-OH at 536 eV with an O:C ratio of 0.54. In addition, a couple of peaks at 107 eV and 109 eV were also observed (Inset of figure 2), indicating the existence of Au at the surface of BDD.

Furthermore, surface characterization of the GOx-modified BDD in comparison with BDD using SEM shows that the surface of BDD has been covered by a film, indicating that the GOx visible surface has been successfully immobilized at the surface of AuNP-BDD (figure 3).

Figure 4(a) shows cyclic voltammograms (CVs) in 0.1 M PBS pH 7 at BDD (dashed line) and AuNPs-BDD (solid line). Two reduction peaks at +1.10 V and +0.46 V were observed. These peaks are typical oxidation-reduction peaks of gold, indicating that AuNP was successfully deposited at BDD surface. These peaks were disappeared after immobilizing GOx at AuNPs-BDD surface (Figure 4b).

In the presence of glucose, CVs in figure 5 (color lines) show a reduction peak at +0.16 V at GOx-AuNP-BDD, which is typical reduction peak of H_2O_2, indicating that the GOx could be successfully deposited at AuNP-BDD electrodes. It is well known that the presence of GOx oxidizes glucose to be H_2O_2, which is easier to be reduced by gold electrodes. Various concentrations of glucose from 0.1 to

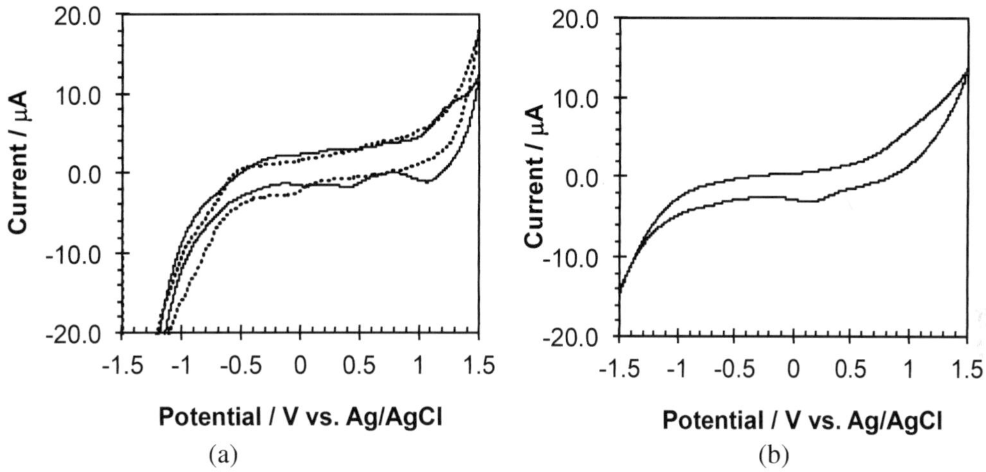

Figure 4. Cyclic voltammograms of 0.1 M PBS pH 7 at BDD (dashed line) and AuNPs-BDD (solid line) in comparison with (b) that at GOx-AuNPs-BDD. The scan rate was 100 mV/s.

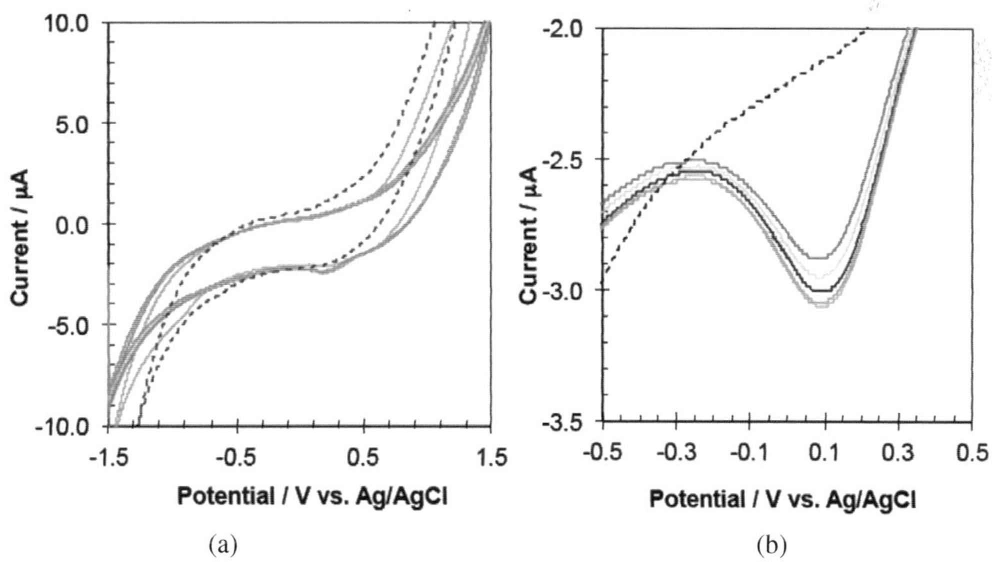

Figure 5. Cyclic voltammograms of 0.1 M PBS pH 7 in the absence (dashed line) and in presence of various concentrations of D-Glucose (0.1 M to 0.9 M, colored lines) at GOx-AuNPs-BDD. Figure (b) shows the magnification of the figure (a). The scan rate was 100 mV/s.

IOP Conf. Series: Materials Science and Engineering **188** (2017) 012046 doi:10.1088/1757-899X/188/1/012046

0.9 mM decreased this peak as shown in the right figure. The currents of this peak were linear with R2 of 0.99, indicating that the electrode can be applied as glucose sensors.

4. Conclusions

Glucose oxidase-modified AuNPs-BDD electrodes have been successfully prepared using self-assembly deposition of AuNPs with the size about 3 to 28 nm at N-terminated BDD. Furthermore, the prepared electrode can generate a reduction peak at +0.16 V in 0.5 M glucose solution, indicating that the electrode can be applied as glucose sensors.

References

[1] Ivandini T A and Einaga Y 2017 *Chem. Commun.* **53** 1338-47
[2] Holland J T, Lau C, Brozik S, Atanassov P and Banta S 2011 *J. Am. Chem. Soc.* **133** 19262-5
[3] Ivandini T A, Saepudin E, Wardah H, Harmesa Dewangga N and Einaga Y 2012 *Anal. Chem.* **84** 9825-32
[4] Ivandini T A, Wijaya L, Gunlazuardi J and Einaga Y 2012 *Makara J. Sci.* **16** 9-14
[5] Ivandini T A, Wicaksono W P, Saepudin E, Rismetov B and Einaga Y 2015 *Talanta* **134** 136-43
[6] Zhang Y, Li Y, Wu W, Jiang Y and Hu B 2014 *Biosensors Bioelectron.* **60** 271-6
[7] Tian R H, Rao T N, Einaga Y and Zhi J F 2006 *Chem. Mater.* **18** 939-45
[8] Folkers J P, Laibinis P E and Whitesides G M 1992 *Langmuir* **8** 1330-41

Natural dye sensitizer from cassava (*Manihot utilissima*) leaves extract and its adsorption onto TiO_2 photo-anode

Nurlela, R Wibowo and J Gunlazuardi

Department of Chemistry, Faculty of Mathematics and Natural Sciences Universitas Indonesia, Kampus UI Depok, Depok 16424, Indonesia

Corresponding author's e-mail: jarnuzi@ui.ac.id

Abstract. Interaction between TiO_2 and dyes sensitizer have been studied. The chlorophyll presents in the crude leave extract (CLE-dye) from cassava (*Manihot utilissima*) was immobilized on to the photo-anode, consists of TiO_2 supported by fluor doped Tin oxide (SnO_2-F) Glass. The TiO_2 was prepared by Rapid Breakdown Anodization (RBA) method then immobilized on to glass coated by SnO_2-F using doctor blade technique, to give CLE-dye/TiO_2/SnO_2-F/Glass photo-anode. The prepared photo-anode was characterized by UV-Vis-DRS, FTIR, XRD, SEM, electrochemical and spectro-electrochemical systems. In this study, the HOMO (highest occupied molecular orbital) and LUMO (lowest unoccupied molecular orbital) energy level of the CLE-dye were empirically determined by cyclic voltammetry method, while spectro-electro-chemistry method was used to determine the coefficient of degradation and formation of the dyes, and diffusion coefficient of the hole recombination as well. Good anchoring between TiO_2 with dye extracts (CLE-dye) can be seen from value of dye LUMO energy level (-4.26 eV), which is approaching the conduction band of TiO2 (-4.3 eV). The coefficient of degradation and formation of the CLE-dye showed a quasi reversible and diffusion coefficient hole recombination values were small, indicated that it is quite suitable as a sensitizer in a dyes sensitized solar cell.

Keywords: TiO_2, chlorophyll dyes, crude leave extract, cassava leaves, spectroelectrochemistry

1. Introduction

In addition to various types of solar cells, a solar cell based on dyes sensitized TiO_2 photo anode has been successfully developed by Gratzel [1]. This dyes sensitized solar cell (DSSC) is a type of solar cell that utilizes dye or pigment as harvesting photons from sun rays [1]. This type of solar cell is generally made up from various components such as photo anode support (metal oxide semiconductor coated on to a conductive oxide glass), photo sensitizer (dye), electrolyte and counter electrode. Optimization and compatibility of each components, especially the spectral response of photo anode and proper interaction between semiconductors and dyes [2] are very important in the dyes sensitized solar cell systems.

The dyes served for harvesting photons to provide an excited electron in the LUMO (lowest unoccupied molecular orbital) and positive hole in the HOMO (highest occupied molecular orbital) levels of the dyes. The excited electron will be transferred to the semiconductor and further transported to the counter electrode, *via* external circuit, then involved in redox couple reaction of the electrolyte before refilled the positive hole of the dyes. Among various dyes, the natural dyes are the interesting candidate to construct the DSSC, due to the availability of abundant resources of natural dyes, non-toxic property, environmentally friendly and readily available and affordable. In addition to these reasons, the selection of natural dyes also takes advantage of the natural behavior of the natural pigments, such as chlorophyll that can do a photochemistry naturally.

There are a lot of reports on the utilizing natural dyes as a sensitizer in the DSSC, such as chlorophyll, anthocyanin, flavonoid and carotenoids which are derived from leaves, flowers, fruits, seeds, etc. [3]. According to the reports, so far, the highest efficiency value was achieved by betanin pigment (3%), while the chlorophyll extracted from papaya leaves only give 0.28% efficiency. Compared to synthetic dyes, in which nearly 20% efficiency can be obtained, the natural dyes have much lower efficiency value. This might be due to a weak bond between the natural dyes with the metal oxide (TiO_2) and low charge transfer spectrum of natural dyes on the boundary solar ray spectra [4]. The lack of available bonding site between the dyes and TiO_2 molecules may hinder the electrons flow from the excited dye molecules to the TiO_2 films [5]. In addition, the timely imbalance in the refilling of electron in the HOMO level of the natural dyes may cause the degradation of the dyes, resulting a short life time of the sensitizer [6].

Based on the above conditions, the present work was designed to study the interaction between TiO_2 semiconductor with the sensitizer (natural dyes). The natural dye used in the present study was chlorophyll crude extract derived from cassava leave (*Manihot utilissima*). The TiO_2 was coated on to Fluor doped Tin Oxide (FTO) by Doctor Blade Method, in which it was immersed in the chlorophyll crude extract to obtain the sensitized TiO_2 photo anode. The photo anode was then subjected to a spectro-electro-chemical study to gain the insight of the energy level values (LUMO level position) and the stability feature of the dyes.

2. Materials and methods

2.1 Fabrication and characterizations of TiO_2

TiO_2 was synthesized by using rapid breakdown anodization (RBA) method. Ti foil (4 x 1 cm) was washed using acetone, ethanol and aquadest for 30 min by ultrasonication, respectively. Ti foil served as anode and Pt foil as cathode. Perchloric acid (0.15 M) was used as electrolyte and 15 V bias potential was applied during anodization. Anodization process was conducted until all Ti foil was submerged in the electrolyte decays perfectly. The obtained TiO_2 precipitate was then centrifuged and dried in oven over night, then calcinated for 3 h at 450°C. The prepared TiO_2 was characterized by UV-Vis Reflectance (UV-DRS), XRD and SEM-EDX, to get information on its band gap value, crystallinity, and morphology feature, respectively.

2.2 Extraction of natural dye

The cassava leaves (*Manihot utilissima*) were obtained from Cipeucang village, Pandeglang, Banten. The leaves were washed and dried under sun light. The dried leaves was then crushed to obtain the intended fine powder. Approximately 20 gram of the obtained fine powder of cassava leaves was macerated in the 200 mL ethanol for 3 hours with the aid of sonication then it was kept for another 3 hours, and followed by filtration with a filter paper (Whatman no.1). The aqueous filtrate was further extracted using hexane, to obtain two fractions, which are ethanol and hexane fractions, respectively. These two fractions were concentrated using rotary evaporator (40°C for 4 hours). The concentrated extract fractions were stored in dark bottle, ready to be used for the next step of the research. The obtained extracts was then characterized by FTIR to get an insight of the available functional groups and UV-Visible spectrometer to determine absorption spectra in visible light.

2.3 Preparation of photo-anode

FTO (fluorine-doped tin oxide, SnO_2-F) film was prepared on a glass slide (50x75x1 mm). Before coating, the glass plate was first cleaned in a detergent solution in an ultrasonic bath for 15 minutes,

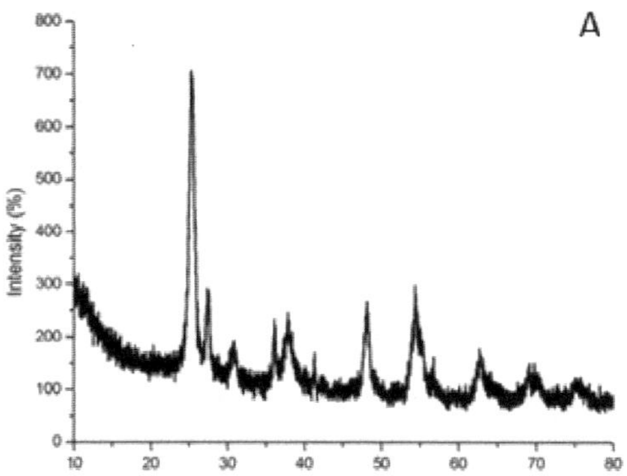

Figure 1a. XRD pattern of prepared TiO_2 by the RBA method.

Figure 1b. SEM image of prepared TiO_2 by the RBA method.

rinsed with mix of ethanol and acetone, followed by water, and then dried. The SnO_2-F precursor solution was homogenously sprayed on to the glass surface, followed by heat treatment, according to Supriyono *et al.* method [7], to obtain the proper FTO (good transparency and good conductivity). The self prepared TiO_2 was then deposited on the FTO conductive glass by doctor-blade technique in order to obtain a TiO_2 thin film. The TiO_2 film was preheated at 200°C for 10 min and then sintered at 450°C for 45 min. After cooling to 80°C, the TiO_2 electrode was immersed in a natural dye solution for 24 h. The fresh prepared photo-anode was then characterized by UV-Vis, UV-Vis-DRS, FTIR, spectro-electrochemical measurement to determine HOMO-LUMO energy level, degradation and formation coefficient and diffusion hole percolation coefficient.

3. Results and discussion

3.1. Fabrication and characterization of TiO_2

The self prepared TiO_2 by a Rapid Break-down Anodization (RBA) method was characterized by using XRD and SEM to see the phase of crystallite and morphology of TiO_2, respectively. Figure 1a shows XRD pattern of the prepared TiO_2, in which the characteristic observed the peaks of 2θ at 25°, 38°, 48°, and 54° indicated the existence of anatase crystalline phase. Further elaboration by using Scherer equation revealed the existence of particle with crystallite size of approximately 5.23 nm [8, 9]. While figure 1b shows morphology feature (SEM image) that indicated a nanotubular structure of the as-grown powder with a nanotube diameter of approximately 64 nm.

3.2. Characterization of cassava leave extract UV-Vis spectra

The fresh cassava leave extract (CLE) dye, TiO_2/SnO_2-F/Glass, and CLE-dye/TiO_2/SnO_2-F/Glass were characterized by UV-Vis spectrometry. Figure 2a shows the absorption spectra of CLE dye solution in the wavelength range of 350-800 nm. The CLE dye mainly consists of chlorophyll mixture, thus unsurprisingly have absorption peaks at wavelength of 410 and 665 nm. The observed absorption peaks are well matched with the absorption data characteristic of chlorophyll dye [10]. Chlorophyll absorbs strongly in the blue and red region but poorly in green region of light spectrum [11]. The excitation energy value corresponds to 410 nm absorption is equal to 3.03 eV photon energy, according to equation as follow,

$$E = \frac{h.c}{\lambda}$$

where the Planck constant, h, (6.63×10^{-34} Js), light velocity, c, (3×10^{8} m/s), and 1J equal to $6.24150934 \times 10^{18}$ eV). While figure 2b shows spectra pattern of CLE) dye, TiO_2/SnO_2-F/Glass, and

CLE-dye/TiO_2/SnO_2-F/Glass. It can be observed that almost no absorption peak of TiO_2/SnO_2-F/Glass is in the visible region. On the other hand, a typical chlorophyll absorption peaks are clearly observed in the CLE-dye/TiO_2/SnO_2-F/Glass and CLE dye. Furthermore an existence of red shift in absorption peak of pure dyes and TiO_2/Dyes was observed, that is from 665 nm of pure dyes to 670 nm of TiO_2/Dyes respectively. The red shift can be attributed to the characteristic of chlorophyll bonded on to TiO_2 surface [6].

3.3. Fourier transform infrared characterization

Fourier Transform Infra Red spectrometry was conducted, in order to get an insight on the functional group information of the fresh cassava leave extract (CLE), TiO_2/SnO_2-F/Glass, and CLE-dye/TiO_2/SnO_2-F/Glass. Figure 3a shows a wide range of absorption peaks appear, e.g. typical peak at wavenumber range of 3300–3500 cm^{-1} due to the O–H stretching vibration of the fresh CLE, a sharp peak at 2900–3000 cm^{-1} which corresponds to the C–H stretching vibration of chlorophyll. Other sharp peak at wave number 1640 cm^{-1} indicating the existence of-C=O group. While peak at 1040-1100 cm^{-1} and 1380 cm^{-1} are assigned as a C-C and C-N of a ring of chlorophyll domain. Figure 3b shows FTIR

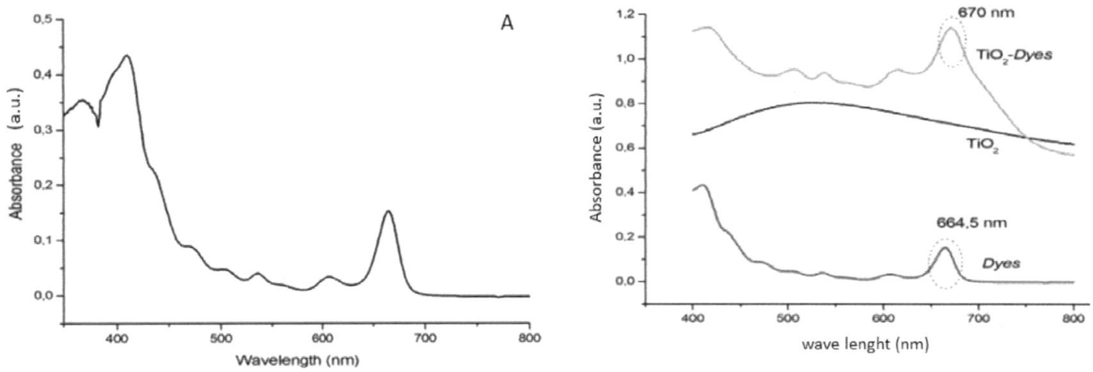

(a) (b)

Figure 2. Visible spectra of (a) the fresh cassava leave extract (CLE) dye, (b) CLE-dye, TiO_2/SnO_2-F/Glass, and CLE-dye/TiO_2/SnO_2-F/Glass.

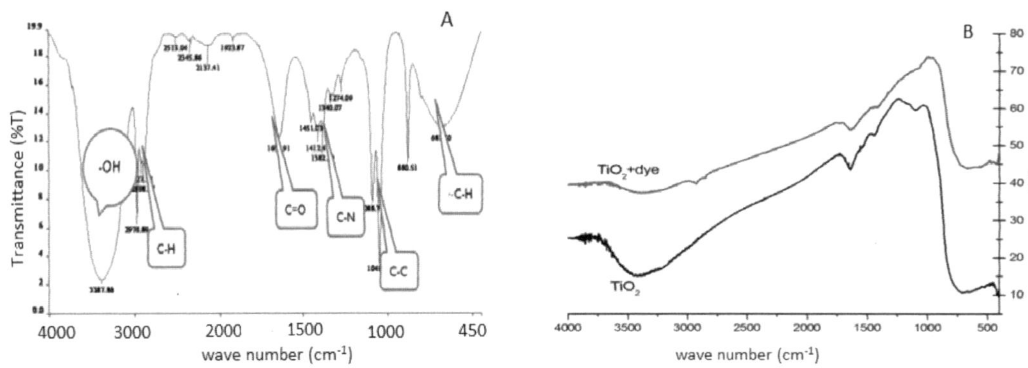

(a) (b)

Figure 3. FTIR spectra of (a) the fresh cassava leave extract (CLE) and (b) TiO_2/SnO_2-F/Glass (below), CLE-dye/TiO_2/SnO_2-F/Glass (above).

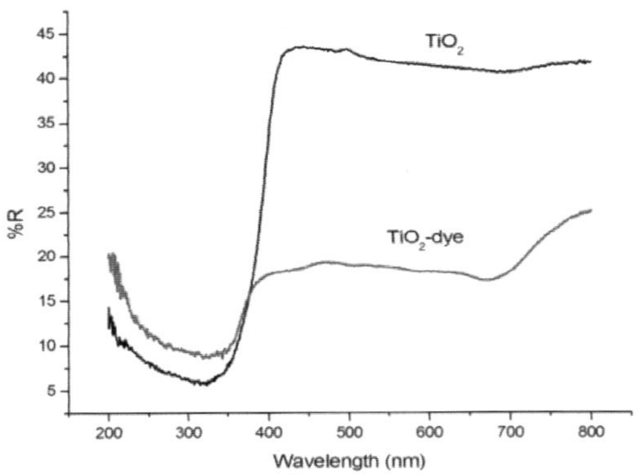

(a)

Figure 4. UV-Vis DRS spectra before and after dye loading.

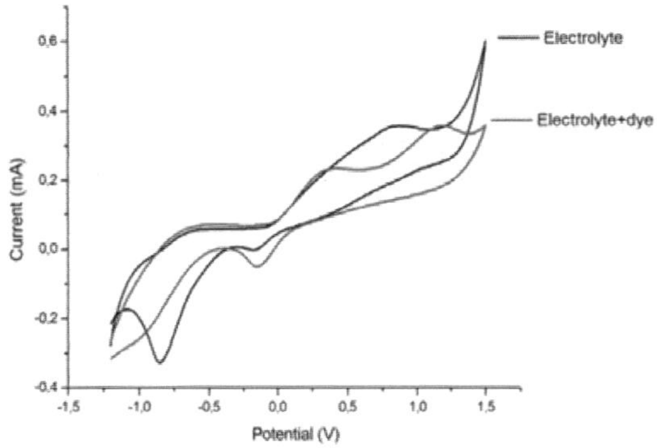

(b)

Figure 5. Cyclic voltamogram of TiO_2/SnO_2-F/Glass working electrode under electrolyte and electrolyte containing dyes.

spectra of TiO_2/SnO_2-F/Glass and CLE-dye/TiO_2/SnO_2-F/Glass, where a typical peak at 500-700 cm^{-1} due to –Ti-O-Ti- bound were observed in both samples. The disappearance of typical –OH bonding at 3300-3500 cm^{-1} in CLE-dye/TiO_2/SnO_2-F/Glass may suggest the –T-O-dye interaction, thus eliminate the –OH from titanyl (-Ti-OH) and hydroxyl (R-OH).

3.4 UV-Vis diffuse reflectant spectrometry characterization

The UV-Vis-DRS spectra provide information of the band gap value of TiO_2 and the visible response of the CLE-dye/TiO_2/SnO_2-F/Glass. It is clear that the typical UV-Vis-DRS spectra due to the first transition absorption corresponds to wavelength of about 398-410 nm were observed in both samples (figure 4), which indicates the energy gap value of 3.0-3.2 eV. Importantly, the spectrum of CLE-dye/TiO_2/SnO_2-F/Glass shows a significant absorption (inverse of reflectance) in the visible region. It is inferred the presence of CLE-dyeon TiO_2 surface.

3.5 Cyclic voltametry (CV) investigation

The CV investigations were conducted in an electrochemical cell comprise of TiO_2/SnO_2-F/Glass working electrode, Pt as counter electrode, and Ag/AgCl as reference electrode. The supporting electrolyte was solution of tetra butyl ammonium perchlorate (0.5 M) and ferrocene (0.002 M Fc/Fc^-) in acetonitrile. The solution of CLE-dye in acetonitrile was used to investigate CLE-dye/TiO_2/SnO_2-F/Glass. The investigation results are shown in figure 5, where definitive oxidation peaks are observed in the supporting electrolyte and in the electrolyte containing dye. Cyclic voltamogram of the system containing supporting electrolyte only shows an oxidation peak at (+) 0.75 V and two reduction peaks at (-) 0.176 V and (-) 0.83 V which are corresponded to the oxidation and reduction peak of ferrocene.

There are new peaks of oxidation and reduction after the CLE-dye was being added. These are oxidation peaks at (+) 1.17 V and reduction peak (-) 0.144 V, which can be attributed to the oxidation and reduction the CLE-dye. The bias potential values of oxidation and reduction peak of the CLE-dye can be correlated to HOMO and LUMO energy level of the dyes. Where the oxidation peak is corresponded to the HOMO and reduction peak is corresponded to the LUMO. The HOMO level acts as the ground state which is the highest energy molecular orbital (MO) containing at least one electron. Correspondingly, the LUMO is the lowest orbital energy level where electrons will excite. Referring to the energy diagram which is presented in figure 6, the energy level of the HOMO and LUMO of CLE-

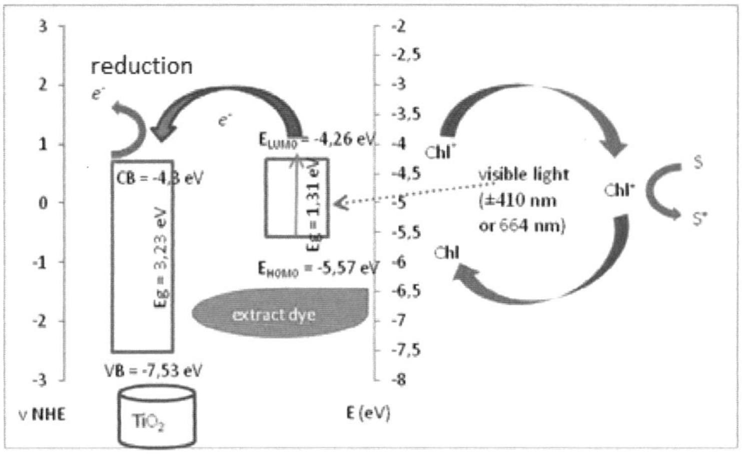

Figure 6. Illustration mechanism of chlorophyll-TiO_2 system by sunlight illumination (modified from Budhee *et al.* 2014 [2]).

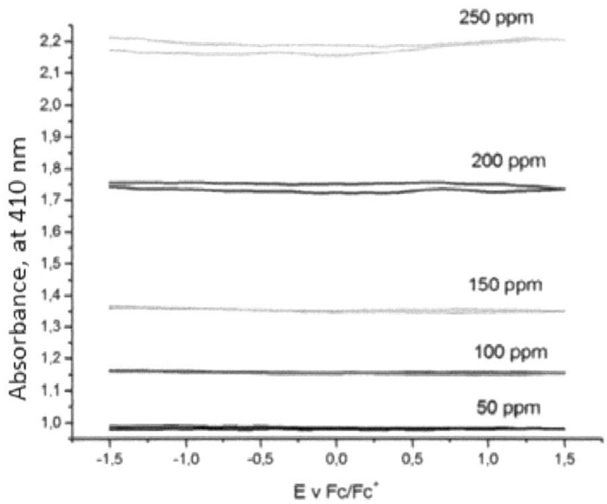

Figure 7. Absorbance at 410 nm values as a bias potential function at various CLE-dyes concentrations.

dye are -5.57 and -4.26 eV, respectively [12]. Thus, the CLE-dye LUMO level (-4.26 eV) is higher than the conduction band energy level of TiO_2, which may facilitates electron transfer to the conduction band.

3.6 Spectro-electro-chemical

To get an insight on the ratio of oxidation rate (degradation) and reduction rate (formation or regeneration) of the CLE-dyes and the hole percolation coefficient, a spectro-electrochemical investigation was conducted. The cyclic volta-absorptometry (CVA) which combines the electrochemical and UV-Visible spectrometer was applied in this study. CVA was measured at the highest absorbance value (410 nm, see figure 2a).The same wavelength was also used to investigate the changing of absorbance as bias potential function. The bias potential was given in the range from -1.5 V to +1.5 V, while the concentrations of CLE-dye were 50, 100, 150, 200, and 250 ppm.

Figure 7 shows the absorbance changes due to potential. When the bias potential shifted to the more positive voltage, the absorbance tends to be lower, while with the more negative voltage, the absorbance tends to be higher. This observed phenomena occurred because when the electrode was given a positive bias potential, energy level of electron from electrode became lower and at some point the electron from

IOP Conf. Series: Materials Science and Engineering **188** (2017) 012047 doi:10.1088/1757-899X/188/1/012047

solute in electrolyte is transferred to electrode. Electron leaping from solution to electrode can be attributed to the oxidation of CLE-dye to CLE-dyes⁺. On the contrary, when the electrode was given a negative bias potential, energy level of electron became higher and at some point electron from electrode is transferred to solute in electrolyte which causes leaping electron. That leaping electron is reduction process, where CLE-dyes⁺ (degradation) return back to CLE-dyes (formation). By that, we can determine $k_{degredation}$ and $k_{formation}$ of CLE-dyes from the change of absorbance versus time, which electrochemically a first order reaction.

$$\frac{-d[A]}{dt} = k[A] \quad (2)$$

$$ln[A] = -kt + ln[A]_0 \quad (3)$$

ln [A] is the concentration of CLE-dyes as y axis and time (*t*) as x axis. Slope value form that linear equation is the coefficient degradation and formation of the dye, as shown in figure 8.

Generally, it was observed that the degradation rate is smaller compared to the formation rate, as shown in table 1. It indicated that the recovery of the hole on HOMO level is fast enough to recover the oxidized dyes during operation. The reaction of CLE-dyes ↔ CLE-dyes $^{+}$+ e^{-} is quasi reversible, so the investigated CLE-dyes is suitable as a sensitizer.

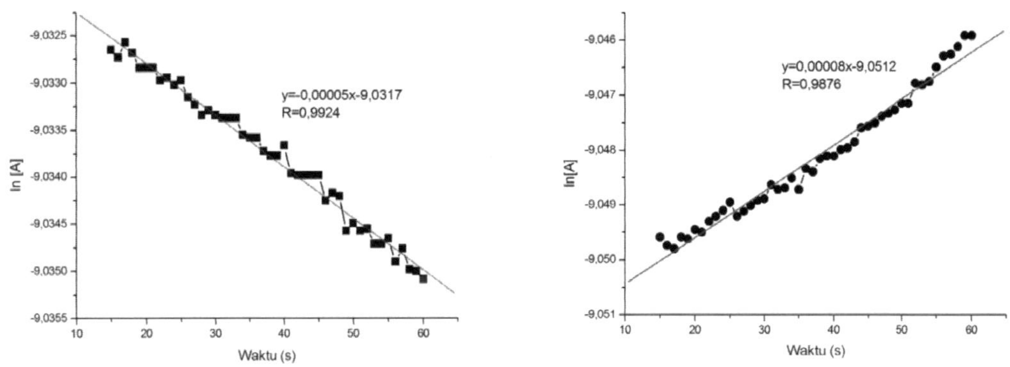

(a) (b)

Figure 8. (a) degradation and (b) formation coefficient value of dye.

Table 1. $k_{degradation}$ and $k_{formation}$ values at various CLE-dyes concentration

ppm (mg/L)	M (10^{-4})	$k_{degradation}$ (s^{-1})	$k_{formation}$ (s^{-1})
100	1.12	-0.00001	0.00003
150	1.69	-0.00005	0.00008
200	2.23	-0.00005	0.00008

Table 2. Hole diffusion coefficient of CLE-dye

ppm (mg/L)	M (10^{-4})	*Dapp* (cm^2/s)
100	1.12	$2.56x10^{-11}$
150	1.69	$1.21x10^{-10}$
200	2.23	$4.6x10^{-10}$

3.7 Diffusion hole coefficient

In this study, the dynamics of hole percolation (CLE-dyes$^+$) was investigated in a CLE-dye/TiO_2/SnO_2-F/Glass. The absorbance versus time data on the chronoabsorptometry mode of the reduction stage was used to determine the coefficient of hole percolation through the following equation [13],

$$D_{app} = \frac{\Delta OD^2 \; x \; \pi \; x \; d^2}{t \; x \; 2\Delta OD_f^2} \qquad (4)$$

where D $_{app}$: hole diffusion coefficient; ΔOD_f: change in absorbance; t: time; d: thick of the film (660 nm in this case). The resulting data of diffusion hole coefficient of dye is shown in table 2, where the value of hole diffusion coefficient is quite small, which means, the possibility of recombination is also small. Thus, the investigated CLE-dye has a potential for the sensitizer in a DSSC, since it will have a long life time.

4. Conclusions

The cassava leave extract dye (CLE-dye) coated on the TiO_2 layer resulted a good visible light response. Spectro-electrochemical study indicated that the LUMO energy level (-4.26 eV) of CLE-dye slight above the conduction of TiO_2 (-4.3 eV), makes it suitable for the sensitizer in the DSSC system. In addition, the CLE-dye has a $K_{formation}$ value higher than its $K_{degradatio}$ and has a quite low hole diffusion coefficient ($4.6x10^{-10}$- $2.56x10^{-11} cm^2/s$), which indicated a long life time, makes it suitable for a sensitizer in a DSSC.

Acknowledgements

Directorate of Higher Educatuion (DGHE) for the funding of the research, under PUPT scheme (JG).

References

[1] Gärtzel M 2003 *Journal of Photochemistry and Photobiology C: Photochemistry Reviews* **4** 145-53
[2] Buddee S, Wongnawa S, Sriprang P and Sriwong C 2014 *J. Nanopart. Res.* **16** 2336-42
[3] Ludin N A, Mahmoud A M A A, Mohamad A B, Kadhum A A H, Sopian K and Karim N S A 2014 *Renewable and Sustainable Energy Reviews* **31** 386-96
[4] Sengupta D, Mondal B and Mukherjee K 2015 *Spectrochimica Acta Part A: Molecular and Biomolecular Spectroscopy* **148** 85-92
[5] Torchani A, Saadaoui S, Gharbi R and Fathallah M 2015 *Current Applied Physics* **15** 307-12
[6] Suyitno S, Saputra T J, Supriyanto A and Arifin Z 2015 *Spectrochimica Acta Part A: Molecular and Biomolecular Spectroscopy* **148** 99-104
[7] Surahman H, Krisnandi Y K and Gunlazuardi J 2015 *Procedia Environmental Sciences* **28** 242-51
[8] Antony R P, Mathews T, Dagupta A, Dash S, Tyagi A K and Raj B 2011 *Journal of Solid State Chemistry* **184** 624-32
[9] Fahim N F and Sekino T 2009 *Chem. Mater* **21** 1967-79
[10] Al-Alwani M A M, Mohamad A B, Kadhum A A H and Ludin N A 2015 *Spectrochimica Acta Part A: Molecular and Biomolecular Spectroscopy* **138** 130-7
[11] Su Y H, Teoh L G, Lee J-H, Tu S-L and Hon M H **2009** *Journal of Nanoscience and Nanotechnology* **9** 960-4
[12] Zou Y, Tan S, Yuan Z and Yu Z 2005 *J. Mater. Sci.* **40** 3561-3
[13] Li X, Nazeeruddin M K, Thelakkat M, Barnes P R F, Vilara R and Durrant J R 2011 *Phys. Chem. Chem. Phys.* **13** 1575-84

IOP Conf. Series: Materials Science and Engineering **188** (2017) 012048 doi:10.1088/1757-899X/188/1/012048

Adsorption of β-sitosterol on molecularly imprinted polymer

N H Soekamto[1], St Fauziah[2], P Taba[2] and M B Amran[3]

[1]Organic Chemistry Laboratory, Universitas Hasanuddin, Jl. Perintis Kemerdekaan, Km 10, Tamalanrea, Makassar 90245, Indonesia
[2]Physical Chemistry Laboratory, Universitas Hasanuddin, Jl. Perintis Kemerdekaan, Km 10, Tamalanrea, Makassar 90245, Indonesia
[3]Analitical Chemistry Laboratory, Institute Technology of Bandung, Jl. Ganesa No.10, Bandung 40132, Indonesia

Corresponding author's e-mail: noek_1512@yahoo.com

Abstract. Molecularly Imprinted Polymer (MIP) has been synthesized using methacrylate acid (MAA) as a monomer with hydroxyl and carbonyl functional groups that can react with ethylene glycol dimethacrylate (EGDMA) as a cross-linking agent, and β-sitosterol as a template molecule. After the template was removed from the polymer, MIP_TFMAA was obtained. The MIP was used to adsorb β-sitosterol. The amount of β-sitosterol in solution after the adsorption was determined by HPLC. The results showed that the MIP was able to adsorb well the β-sitosterol at a pH 7 and the contact time of 90 min. The kinetic adsorption data obtained for β-sitosterol followed the pseudo-second-order model and consistent with the model of Feundlich isothermal with the adsorption capacity of 1.05 mg/g. The MIP was selective on β-sitosterol because it was able to adsorb β-sitosterol better than cholesterol.

1. Introduction

Molecularly imprinted polymers (MIPs) are polymers synthesized using a template molecule to create cavities after removing the template [1]. Main materials that have to be considered in synthesis of MIP are monomers, template molecules, crosslinking agents (cross-linkers, porogen solvents and initiators. Selection of monomer is very important to produce the appropriate interaction with the template molecule [2] and to determine the stability of the polymer formed during the polymerization process [3]. Methacrylic acid (MAA) can be used as monomer in the synthesis of MIP because it is acidic and it can interact with the template molecules through acid-base interactions and hydrogen bonding. Several studies showed that the MIPs synthesized using the MAA monomer and the β-sitosterol template have high affinity and selectivity against β-sitosterol [3, 4]. Cross-linkers are required to control the polymer matrices, to stabilize binding sites, and to maintain mechanical stability of polymers [3]. They can affect approximately 70% to 98% of the MIP final result [4] and also affect the adsorption selectivity [5]. Among the cross-linkers, EGDMA is widely used because its functional groups can easily interact with functional groups of monomers to produce a polymer with a high degree of rigidity [4]. Porogen solvents must have the ability to dissolve the template molecule, initiator, monomer and cross-linking agents to form the cavity [6]. They must have a relative low polarity to reduce interference when interactions occur between the template molecules and monomers to produce the high selectivity MIP [5]. In the polymerization process, the role of initiator is very important in the formation of free radicals by heating (thermolysis) and radiation (photolysis) [7]. One initiator that mainly used in the radical polymerization is 2-2'-azobis isobutyronitrile (AIBN) because it is available in the market and it can produce radicals at the low temperature.

In this study, a MIP was synthesized using MAA as a monomer, β-sitosterol as a template molecule, EDGMA as a cross-linker, AIBN as an initiator, and toluene as the porogen solvent. The use of β-sitosterol as a template molecule was conducted to find a material that is selective to the molecule that frequently interfere in the isolation and purification process of natural product compounds. The MIP produced was evaluated through the optimization test and determination of adsorption capacity before being used in the solid phase extraction (SPE). Parameters that influence the optimization i.e. time, pH and concentration were studied.

2. Materials and methods

2.1. Materials
The materials used in this study were β-sitosterol 97% (Sigma-Aldrich), methacrylic acid 99% (MAA) (Sigma_Aldrich) as a monomer, ethyleneglycol dimethacrylate (EGDMA) (Sigma-Aldrich) as a crosslinking agent (cross-linker), 2,2'-azobisisobutyronitrile (AIBN) as initiator, toluene as porogen, aquabidest, methanol (HPLC grade), tetrahydrofuran grade (THF), acetic acid industrial grade as the washing solvent, NaOH, HCl, and nitrogen gas.

The equipments used included glassware, analytical balance, shaker, water bath, sonication apparatus, oven, mini peristaltic pump (Gilson), mini glass column (length 3 cm), bottles, micropipette 10, 100, 1000 mL (Eppendorff), pH meter, 41 Whatman paper No. 41, aluminum foil, and paper towels. High Pressure Liquid Chromatography (HPLC) Agilent 1260 infinity using Cronus RP column type E18 C with the column length and diameter of 12.5 cm and 0.4 cm, Ultra violet-visible Spectrophotometer (UV-Vis) Agilent 8453, and Shimadzu.

2.2. Synthesis of MIP_MAA-co-EGDMA
A total of 50 mg β-sitosterol, 2 mL monomer MAA, and 3 mL EGDMA were dissolved with 5 mL toluene in a round bottom flask. The mixture was stirred slowly, and then added with 1.5 mL of AIBN. The solution was sonicated and purged with nitrogen. The polymerization was conducted by adjusting the temperature to 55°C in a water bath for 24 hours. The polymer formed was crushed, sieved, and washed with THF tetrahydrofuran, methanol: acetic acid (90:10% v / v) mixture, and methanol to release the β-sitosterol and porogen solvent [1]. Polymer or MIP was dried in an oven at a temperature of 50°C.

2.3. The pH influence on MIP_MAA-co-EGDMA adsorption
A total of 20 mg of MIP was put in four vial bottle then added with 3 mL of 10 ppm β-sitosterol that has been adjusted to the specified pH into each vial. The pH of solution was set at the pH of 4, 5, 6, and 7. The solution was mixed at room temperature for 6 hours. The solution was filtered and then β-sitosterol concentrations in the filtrate was analyzed by UV at a wavelength of 202 nm.

2.4. The effect of contact time influence on adsorption of MIP_MAA-co-EGDMA
A total of 20 mg MIP_MAA-co-EGDMA was put into five vials, 3 mL of 10 ppm β-sitosterol that has been set according to the optimum pH was added to the vial. The mixture was stirred with a shaker for 30, 60, 120, 180, 240 minutes at room temperature and filtered. The concentration of β-sitosterol in the filtrate was analyzed by the UV spectrophotometer at a wavelength of 202 nm.

2.5. The effect of concentration influence on adsorption of MIP_MAA-co-EGDMA
To determine the amount of β-sitosterol adsorbed by MIP, 3 mL of β-sitosterol with various concentrations of 2, 4, 6, 8, and 10 ppm at the optimum pH was put in a vial containing 20 mg of MIP_MAA-co-EGDMA. The mixture was then shaken at room temperature and at the optimum time. The solution was then filtered and the concentration of β-sitosterol in the filtrate was analyzed by the UV spectrophotometer at a wavelength of 202 nm. The adsorption capacity was determined by using Langmuir and Freundlich isotherms.

3. Results and discussion

3.1. The synthesis result of MIP_MAA-co-EGDMA

The polymer of MIP_MAA-co-EGDMA synthesized was obtained as white solid with the weight of 2.1352 g. The MIP was washed using THF, methanol:acetic acid mixture, and methanol, successively. The filtrate was tested with a solution of Liebermann Burchard and it was shown that β-sitosterol has been removed from the polymer.

3.2. The pH influence on adsorption of MIP_MAA-co-EGDMA.

The effect of pH on adsorption of β-sitosterol on MIP_MAA-co-EGDMA can be seen in figure 1. Figure 1 shows that the amount of β-sitosterol adsorbed on MIP_MAA-co-EGDMA gradually increases with the increasing pH. It is clear that the optimum pH in the range of studied pH for the adsorption of β-sitosterol was 7. The adsorption of MIP β-sitosterol at the low pH was very low. Yusof *et al.* [8] reported that at pH 3-5, a compound containing a particular functional group will be protonated so that it can affect the interaction with the active site of the MIP. Conversely, the increase of pH causes the reducing concentration of H^+ ions in the solution, resulting in the decrease of protonated functional groups on MIP and β-sitosterol, generating the formation of hydrogen bonds between -OH on β-sitosterol with functional groups on the active side of MIP. Thus, the best adsorption of β-sitosterol on MIP_MAA-co-EGDMA was achieved at the pH 7 with the adsorbed amount of 1.90 mg/g. The obtained data showed that the MIP's ability to adsorb a target molecule was strongly influenced by the pH. They also reported that adsorption of 2,4-dichlorophenol on MIP_acrylamide-co-EDGMA using 2,4-dichlorophenol as a template molecule decreased at the pH of 8.

3.3. Determination of the optimum time

The effect of time on the adsorption of β-sitosterol on MIP is given in figure 2. The amount of β-sitosterol adsorbed increases with the increasing contact time until achieves the maximum adsorbed amount at an equilibrium step. Further increase of time slightly decreases the amount adsorbed. The optimum time of MIP_MAA-co-EGDMA to adsorb β-sitosterol was 120 minutes and the amount of β-sitosterol adsorbed was 2.08 mg/g. The kinetic study of adsorption was evaluated using pseudo-first order and pseudo-second order equations [9]. Figure 3 shows the curve of log (q_e-q_t) vs time for the former order and the curve of t/qt vs t for latter order.

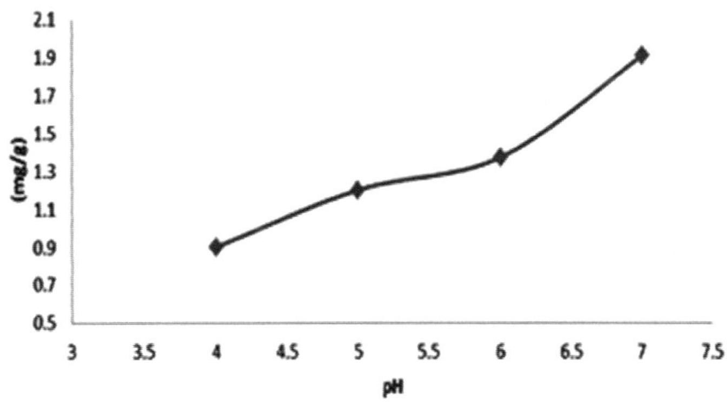

Figure 1. The amount of β-sitosterol adsorbed on MIP_MAA-co-EGDMA as a function of pH

The parameter suitability of adsorption kinetic model of first pseudo order and second pseudo order determined by linear regression correlation coefficient (R^2) of the curve obtained. The kinetic data for the adsorption of β-sitosterol are shown in table 1.

3.4. Determination of adsorption capacity of β-sitosterol on MIP

The adsorption isotherms of β-sitosterol on MIP_MAA-co-EGDMA are given in figure 4. The amount of β-sitosterol adsorbed increases with the increase of initial concentration of β-sitosterol solution. From figure 4, it is clear that the amount of β-sitosterol adsorbed continues to increase at the measured concentration. Therefore, Langmuir and Freundlich models were utilized to find the capacity of adsorption. Table 2 summarizes the constants of Langmuir and Freundlich for the adsorption of

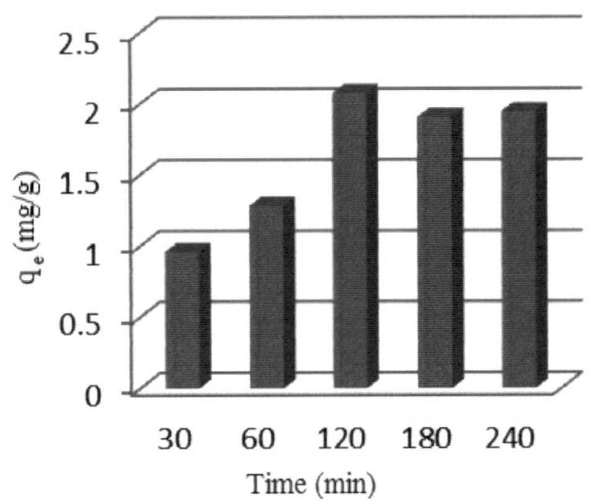

Figure 2. The influence of time on the β-sitosterol adsorption on MIP_MAA-co-EGDMA

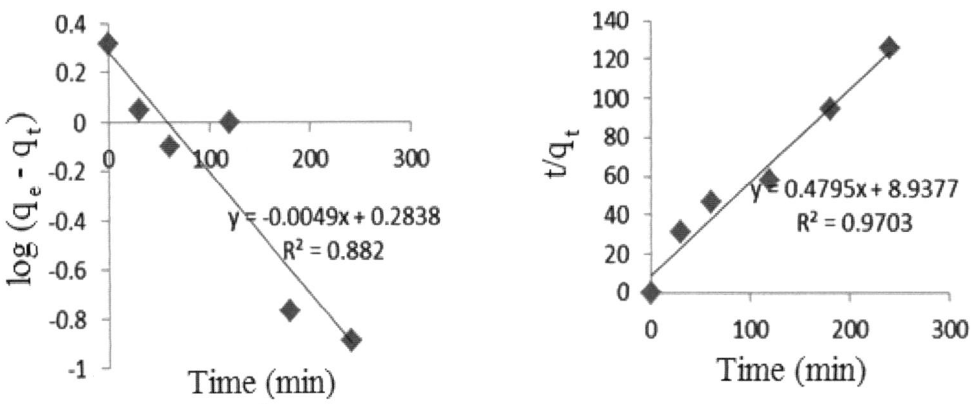

(a) (b)

Figure 3. (a) Kinetic curve of first pseudo order and (b) Kinetic curve of second pseudo order for β-sitosterol adsorption on MIP_MAA-co-EGDMA

Table 1. Kinetic data for adsorption of β-sitosterol by MIP

q_e Calculation		q_e Experiment	k_1	k_2	R^2 1st order	R^2 2nd order
1st order	2nd order					
1.92	2.09	2.08	0.0092	0.0257	0.882	0.970

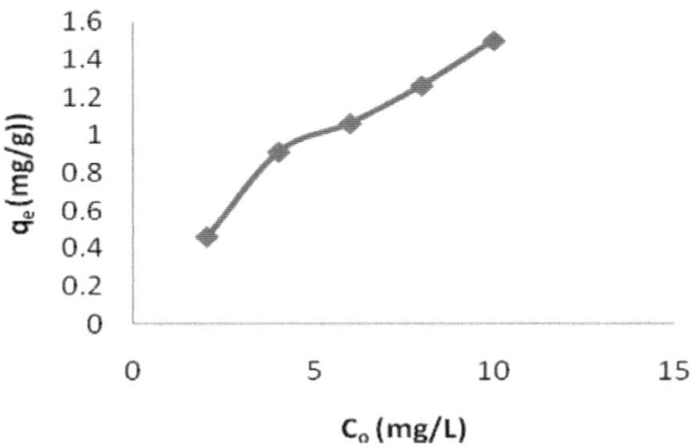

Figure 4. Adsorption isotherm of β-sitosterol on MIP_MAA-co_EGDMA

Table 2. Langmuir and Freundlich constant for the adsorption of β-sitosterol on MIP_MAA-co-EGDMA

Langmuir model			Freundlich model		
Q_o	b	R^2	k	n	R^2
2.03	1.07	0.966	1.05	1.89	0.967

β-sitosterol. Based on table 2, it can be stated that the adsorption of β-sitosterol on MIP_MAA-co-EGDMA followed both models with the adsorption capacity 2.03 and 1.05 mg/g for Langmuir and Freundlich models, respectively.

4. Conclusions

From the results above, it can be concluded that white solid of MIP_MAA-co-EGDMA has been synthesized. The obtained MIP was adsorbed well the β-sitosterol at pH 7. In addition, the optimum time for the adsorption of β-sitosterol was 120 min. Furthermore, the adsorption of β-sitosterol on MIP_MAA-co_EGDMA fitted the kinetic model of pseudo-second order with the rate constant of 0.0257 g $min^{-1}mg^{-1}$. The adsorption of β-sitosterol on MIP_MAA-co_EGDMA followed the Langmuir and Freundlich isotherm with the adsorption capacity of 2.03 and 1.05 mg/g for Langmuir and Freundlich models, respectively.

Acknowledgements

We thank the Directorate of Higher Education-Indonesia for the financial support and Universitas Hasanuddin for all facilities.

References

[1] Zhu T, Yoon C and Row K H 2011 *Chin. J. Chem.* **29** 1246-50
[2] Kim H, Spivak D A 2003 *J. Am. Chem. Soc.* **125** 11269-75
[3] Beltran A, Borrull F, Cormack P A G and Marcé R M 2010 *Trend in Analytical Chemistry* **29** 1363-75
[4] Mayes A G and Whitcombe M J 2005 *Adv. Drug Deliv. Rev.* **57** 1742-78
[5] Yan H Y and Row K H 2006 *Int. J. Mol. Sci.* **7** 155-78
[6] Holland N 2008 *An Investigation into the Role of Morphology on the Performance of Molecularly Imprinted Polymers* PhD thesis (Waterford: Waterford Institute of Technology Ireland)
[7] Walsh R 2010 *Development and Characterization of Molecularly Imprinted Suspension Polymers* PhD thesis (Waterford: Waterford Institute of Technology Ireland)
[8] Yusof N A, Rahman S K, Hussein M Z and Ibrahim N A 2013 *Polymers* **5** 1215-28
[9] Esmaeilian A, Mazdeh A M, Ghaforian H and Liaghat A M 2015 *Desalination and Water Treatment* **53** 2235-48

IOP Conf. Series: Materials Science and Engineering **188** (2017) 012049 doi:10.1088/1757-899X/188/1/012049

Better latent heat and specific heat of stearic acid with magnetite/graphene nanocomposite addition for thermal storage application

R Andiarto[1,2], **M K Nuryadin**[1,2], **A Taufik**[1,2] **and R Saleh**[1,2]

[1]Department of Physics, Faculty of Mathematics and Natural Sciences Universitas Indonesia, Kampus UI Depok, Depok 16424, Indonesia
[2]Integrated Laboratory of Energy and Environment, Faculty of Mathematics and Natural Sciences Universitas Indonesia, Kampus UI Depok, Depok 16424, Indonesia

Corresponding author's e-mail: rosari.saleh@gmail.com

Abstract. In our previous study, the addition of Magnetite (Fe_3O_4) into Stearic acid (Sa) as an organic phase change material (PCM) shows an enhancement in the latent heat for thermal energy storage applications. The latent heat of the PCM can also be increased by adding graphene material. Therefore, in this research, the thermal properties of Sa have been studied by the sonication method for several different concentrations of Fe_3O_4/Graphene nanocomposite additions. The structural properties of all of the samples were observed by X-Ray diffraction (XRD). Melting-solidifying behavior and specific heat value were measured by differential scanning calorimetry (DSC). The thermal degradation process of all samples was investigated by thermogravimetric analysis (TGA). Based on the DSC results, the presence of Fe_3O_4/Graphene in the Sa enhances the latent heat up to 20%. The specific heat value of the mixture was also found to be increased as the concentration of Fe_3O_4/Graphene to Sa increased. The TGA results show a lowered thermal degradation process of the Sa by the addition of the Fe_3O_4/Graphene which indicates a higher thermal stability of the mixture. In conclusion, the results demonstrate that the addition of Fe_3O_4/Graphene to Sa improves both the sensible heat and the latent heat of the mixture which are very important for thermal energy storage applications

Keywords: Phase change materials, Fe_3O_4, graphene, latent heat

1. Introduction

Thermal energy storage (TES) based on phase change material (PCM) utilization has resulted an advancement in thermal energy applications [1-2]. The use of PCM for TES applications also has a great potential in many fields such as solar cell efficiency, heating and cooling building conservation systems, and thermal regulating textile materials [3-5]. From the various types of PCM, Stearic acid (Sa) is considered to be the most recommended organic PCM for TES applications due to its advantageous features, such as: non-toxic, non-corrosive, and applicable phase change temperature [6-7]. However, the using of stearic acid as PCMs materials still has a limitation due to low thermal conductivity of stearic acid. many researchers are trying to find ways to enhance the ability of PCMs in order to store and release heat so it can be used in large scale applications [8]. One of the most promising methods for enhancing the ability of a PCM to store and release heat is by adding nanoparticles to the PCM [9].

In our previous study, the addition of the magnetite metal oxide material (Fe_3O_4) into the Sa resulted an enhancement in the latent heat value of the Sa [10]. Those results are supported by another study in which Fe_3O_4 nanoparticles were added to improve the ability of the organic PCM [11]. Moreover, the addition of carbon-based nanomaterials into the PCM also show an increment in the latent heat value of the PCM [12]. Based on above explanations, combining Fe_3O_4 and graphene materials is become a promising method for enhance the PCM performance. Therefore, in this study, Fe_3O_4/Graphene nanocomposite was synthesized using the hydrothermal method and then mixed with Sa in the various concentrations of 0.3, 0.5, 0.8, 1, and 5% wt (weight transfer) to form a Sa/Fe_3O_4/Graphene mixture. Thermal investigation was then carried out in order to observe the thermal properties of the samples to determine their effectiveness for use in TES applications.

2. Experimental setup

In this study, Stearic acid ($C_{17}H_{35}CO_2H$) was used as a phase change material. Iron (II) sulfate heptahydrate ($FeSO_4.7H_2O$), sodium hydroxide (NaOH), and ethylene glycol were used to synthesize Fe_3O_4 nanoparticles. The synthesis process followed the same method as our previous study [13]. All of the reagents were purchased from Merck and used without further purification. The synthesized Fe_3O_4 was then combined with Graphene (Angstron Materials) to form a Fe_3O_4/Graphene nanocomposite using the hydrothermal method with 5 wt.% Graphene added to the Fe_3O_4. In brief, the synthesis process of the Fe_3O_4/Graphene nanocomposite was started by sonicating Graphene with distilled water for 2 hours. The resulting solution was then mixed with Fe_3O_4 nanoparticles and then stirred with a magnetic stirrer. The mixture that formed was then heated at 70°C for 3 hours. The Fe_3O_4/Graphene nanocomposite which formed was then centrifuged and heated at 70°C for 12 hours in vacuum conditions. The combination of the Stearic acid (Sa) with the Fe_3O_4/Graphene nanocomposite was performed by dispersing the desired amount of the Fe_3O_4/Graphene (0.3, 0.5, 0.8, 1, and 5% wt) into the Sa in an ultrasonic bath for 2 hours.

The structural properties of the Fe_3O_4/Graphene nanocomposite and Sa/Fe_3O_4/Graphene mixtures were characterized by X-Ray diffraction (XRD). Thermogravimetric analysis (TGA) was also performed in order to prove the presence of Graphene in the Fe_3O_4/Graphene nanocomposite. The thermal degradation process of the samples was also observed through TGA while the differential thermogravimetric analysis (DTGA) was calculated based on the TGA data. For the thermal properties of the Sa/Fe_3O_4/Graphene samples, differential scanning calorimetry (DSC) was used in order to observe the phase transformation process, the latent heat value, and the specific heat value (C_p).

3. Results and discussion

The presence of Graphene in the Fe_3O_4/Graphene nanocomposite was proved by the TGA results as shown in figure 1a. The TGA measurements were investigated in order to determine the thermal degradation process of the nanocomposite. As can be seen in figure 1a, the degradation process was started at 400°C which indicates the combustion effect from the Graphene [14]. Moreover, the percentage loss residue at the end of measurement at 1000°C shows the composition of Graphene in the Fe_3O_4/Graphene nanocomposite which is about 5% wt. This result confirms the presence of Graphene in the Fe_3O_4/Graphene nanocomposite.

The XRD results of the Fe_3O_4/Graphene nanocomposite, along with the Sa and Sa/Fe_3O_4/Graphene 5% wt, are shown in figure 1b. Based on the XRD spectrum of Fe_3O_4/Graphene, there are peaks that can be seen at the $2\theta \approx 30.2°$, 35.5°, 53.4°, and 57.0° which represent the (220), (311), (511), and (440) planes of the inverse cubic spinel structure of Fe_3O_4. There is no sign of the Graphene peak pattern in the Fe_3O_4/Graphene XRD measurement which is due to Graphene's amorphous properties [15]. For the Sa spectrum shown in figure 1b, there are two readily indexed peaks at $2\theta \approx 21.3°$ and 24.7° which show the regular crystallization pattern of the Sa. The XRD measurement of the Sa/Fe_3O_4/Graphene shows the combination between the crystallinity peak patterns of Fe_3O_4/Graphene and Sa which confirms the presence of the desired Fe_3O_4/Graphene crystallinity in the Sa.The melting and solidifying

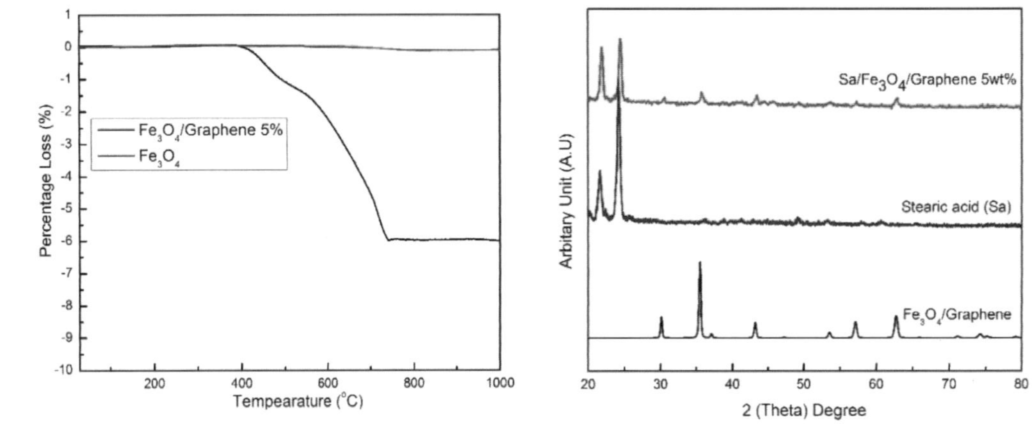

(a) (b)

Figure 1. (a) TGA results for Fe_3O_4 nanoparticle and Fe_3O_4/Graphene nanocomposite, and (b) XRD spectrum for Fe_3O_4/Graphene, Sa and Sa/Fe_3O_4/Graphene 5 wt.%.

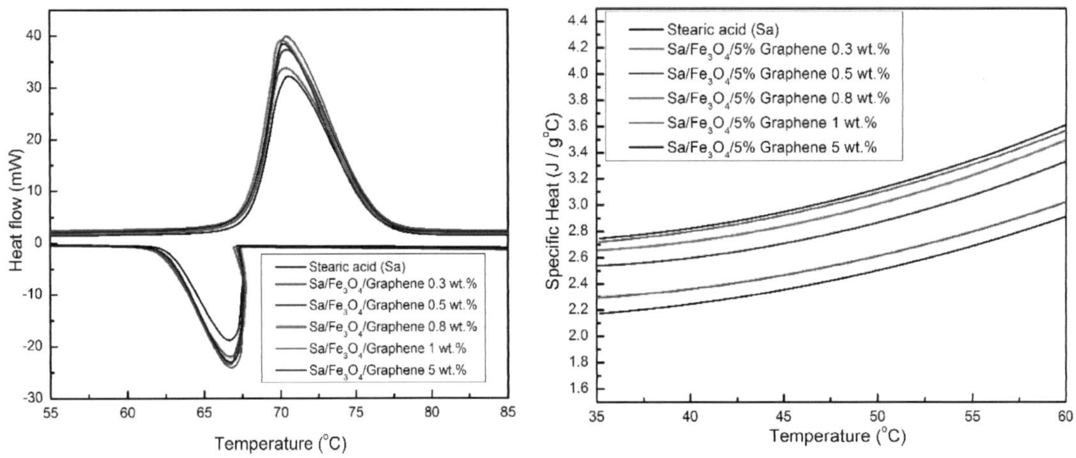

(a) (b)

Figure 2. (a) DSC curve of Sa/Fe_3O_4/Graphene in all variation, and (b) fitting curve of Sa/Fe_3O_4/Graphene specific heat in all variation.

process of the Sa/Fe_3O_4/Graphene mixtures was measured and analyzed through DSC. The phase change process of the PCM indicates the main characteristic of the PCM as an application for latent heat thermal storage (LHTS) [16]. The resulting curve of DSC measurements is shown in figure 2a. Based on the results, it can be seen that all of the samples performed the one-step process of endothermic and exothermic under melting and solidifying process. Qualitatively, the total area under the curve shows how much heat is involved in the phase change process. As can be seen in figure 2a, the area under the DSC curve of the Sa/Fe_3O_4/Graphene shows a larger curve than the Sa alone. This signifies the enhancement of the heat involved during the store and release process in the phase change process [11]. The resulting DSC curves shown in figure 2a also provide information about the melting point temperatures and solidifying point temperatures alongside the latent heat values which are calculated through the integration of the area under the DSC curve. All of the data from the DSC curve are tabula--ted in table 1. Based on the data, it can be concluded that the addition of the Fe_3O_4/Graphene nanocomposite to the Sa did not have any favorable effect on the melting and solidifying temperature behavior of the Sa. However, the latent heat value calculation for the Sa was found to be enhanced by the presence of Fe_3O_4/Graphene. The latent heat value of the Sa was found to increase as the concentration of Fe_3O_4/Graphene increased. The maximum enhancement was found at the 1% wt concentration of the Fe_3O_4/Graphene to Sa ratio. This result shows that the addition of Fe_3O_4/Graphene to Sa enhances the capability of the Sa to store and release heat. The enhancement of the Sa's latent heat value due to the presence of Fe_3O_4/Graphene can be explained by the intermolecular interaction between the Sa and the Fe_3O_4/Graphene [12].

The specific heat (C_p) calculations were also analyzed using the corresponding DSC curves in order to observe the sensible heat ability for the sensible heat storage application [17]. The specific heat calculations from 35°C to 65°C are shown in figure 2b. Based on the results, it can be determined that as the concentration of the Fe_3O_4/Graphene in the Sa increased, the specific heat value of the Sa/Fe_3O_4/Graphene mixture also increased. The enhancement of the specific heat can also be seen from the average heat value which is shown in table 2. Based on the average value, the maximum enhancement of specific heat was found at the concentration of 5% wt of Fe_3O_4/Graphene.

The thermal degradation process of the Sa/Fe_3O_4/Graphene in all variations was observed through TGA. The results are shown in figure 3a. The onset and percentage loss data based on the corresponding TGA curves are tabulated in table 2. Based on those results, it can be observed that the

Table 1. Solid-liquid, liquid-solid temperature data and latent heat value for melting and solidifying process of Sa/Fe_3O_4/Graphene in all variation.

Sample	Solid-liquid T_{peak}	liquid-solid T_{onset}	Melting ΔHm (J/g)	solidifying ΔHs (J/g)
Stearic acid (Sa)	70.5	67	177.6	174.1
Sa/Fe_3O_4/Graphene 0.3%	70.2	67.3	199.7	196.3
Sa/Fe_3O_4/Graphene 0.5%	70.4	67.4	210.3	206.8
Sa/Fe_3O_4/Graphene 0.8%	70.1	67.2	214.5	212
Sa/Fe_3O_4/Graphene 1%	70.2	67.3	216.6	213.2
Sa/Fe_3O_4/Graphene 5%	70.2	67.3	199	194.5

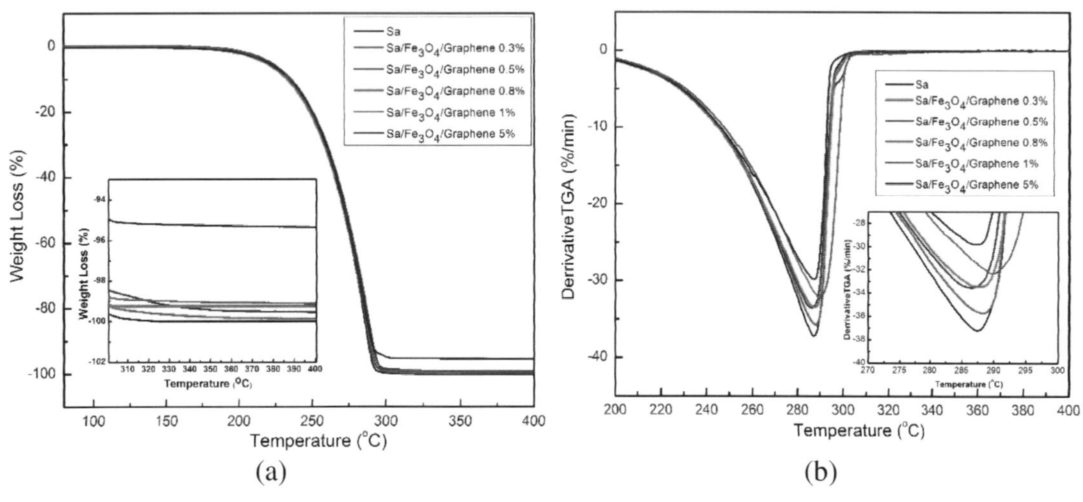

Figure 3. TGA curve (a) and DTGA curve (b) for the Sa/Fe_3O_4/Graphene in all variation.

Table 2. Average Specific heat (C_p) value calculation of Sa/Fe_3O_4/Graphene in all variation.

Samples	Average C_p (J/g°C)	Onset (°C)	Percentage weight lost (%)
Stearic acid (Sa)	2.48	251.6	100
Sa/Fe_3O_4/Graphene 0.3%	2.59	254	99.6
Sa/Fe_3O_4/Graphene 0.5%	2.84	253.3	99.59
Sa/Fe_3O_4/Graphene 0.8%	2.97	253.6	99.19
Sa/Fe_3O_4/Graphene 1%	3.03	255.5	99.08
Sa/Fe_3O_4/Graphene 5%	3.06	254.1	94.91

Sa/Fe_3O_4/Graphene mixture in all variations has a strong thermal stability up to 200 °C. The degradation process was found to suffer from gradual weight loss at 250°C which was due to the evaporation process of the Sa. Furthermore, the percentage loss data shows that the Fe_3O_4/Graphene was in the appropriate composition of variation.

Using the TGA results, differential thermogravimetric analysis was performed to calculate how fast the degradation process occurred due to temperature for each sample. The DTGA results are shown in figure 3b. The lowest peak indicates the faster degradation process of the Sa. As can be seen in the DTGA curve, the addition of Fe_3O_4/Graphene into the Sa can slow the degradation process. Some research has previously stated that the slower degradation process may result in an incremental increase in the thermal stability of the PCM [18].

4. Conclusions

Fe_3O_4/Graphene has been successfully synthesized through the sol-gel method and used as an additive material into Sa. The DSC results show that the latent heat value of the Sa increases due to the addition of the Fe_3O_4/Graphene with the maximum enhancement at 1% wt of Fe_3O_4/Graphene. Furthermore, the increments can be found in the specific heat results with the maximum enhancement at 5% wt of Fe_3O_4/Graphene. The enhancement of both latent heat and specific heat also shows that the Sa/Fe_3O_4/Graphene mixture is suitable for latent heat and sensible heat storage. Moreover, the thermal stability of the Sa was found to be enhanced due to the presence of the Fe_3O_4/Graphene as seen by the slower degradation process.

References

[1] Regin A F, Solnaki S C and Saini J S 2008 *Renew. Sustainable Energy Rev.* **12** 2438-58
[2] Sharma A, Tyagi V V, Chen C R and Buddhi D 2009 *Renew. Sustainable Energy Rev.* **13** 318-45
[3] Tyagi V V and Buddhi D 2007 *Renew. Sustainable Energy Rev.* **11** 1146-66
[4] Khodadadi J M, Fan L and Babaei H 2013 *Renew. Sustainable Energy Rev.* **24** 418-44
[5] Mondal S 2008 *Appl. Therm. Eng.* **28** 1536-50
[6] Zalba B, Marín J M, Cabeza L F and Mehling H 2003 *Appl. Therm. Eng.* **23** 251-83
[7] Cao L, Tang Y and Fang G 2015 *Energy* **80** 98-103
[8] Feldman D, Shapiro M M, Banu D and Fuks C J 1989 *Sol. Energ. Mat. Sol.* **18** 201-16
[9] Kibria M A, Anisur M R, Mahfuz M H, Saidur R and Metselaar H S C 2015 *Energy Convers. Manage.* **95** 69-89
[10] Andiarto R, Nuryadin M K and Saleh R 2016 *J. Phys. Conf. Ser.* **710** 012020
[11] Sahan N and Paksoy H O 2014 *Sol. Energ. Mat. Sol.* **126** 56-61
[12] Shaikh S, Lafdi K and Hallinan K 2008 *J. Appl. Phys.* **103** 094302
[13] Arifin S A, Jalaludin S and Saleh R 2015 *Adv. Mat. Res.* **1123** 264-9
[14] Loryuenyong V, Totepvimarn K, Eimburanapravat P, Boonchompoo W and Buasri A 2013 *Adv. Mater. Sci. Eng.* **2013** 923403
[15] Ferrari A C 2007 *Solid State Commun.* **143** 47-57

[16] Sharma R K, Ganesan P, Tyagi V V, Metselaar H S C and Sandaran S C 2016 *Appl. Therm. Eng.* **99** 1254-62
[17] Chieruzzi M, Cerritelli G F, Miliozzi A and Kenny J M 2013 *Nanoscale Res. Lett.* **8** 448
[18] Tahan S, Mehrali M, Mehrali M, Meurah T, Mahlia I, Simon H and Metselaar C 2013 *Energy* **61** 664-72

Electrosynthesis of vanillin from isoeugenol using platinum electrode

H Mubarok, Hilyatudini, E Saepudin and T A Ivandini

Department of Chemistry, Faculty of Mathematics and Natural Sciences Universitas Indonesia, Kampus UI Depok, Depok 16424, Indonesia

Corresponding author's e-mail: ivandini.tri@sci.ui.ac.id

Abstract. Vanillin was synthesized from isoeugenol through electrochemical method in one compartment cell using platinum electrode. Cyclic voltammetry in 0.1 M TBAP in methanol and acetonitrile indicated the first oxidation potential at +0.21 and +0.16 V (vs. Ag/AgCl), respectively. Isoeugenolis was proposed to undergo the oxidation accompanied by oxidative cleavage of alkene bond into aldehyde. Accordingly, the synthesis of vanillin was conducted using chronoamperometry technique. The electrosynthesis result was analyzed by HPLC and GC/MS. The optimum condition of the oxidation potential, solvent ratio, time of electrolysis and amount of water was investigated.

Keywords: electrosynthesis, vanillin, isoeugenol, oxidative cleavage

1. Introduction

Vanillin is one of main compound in vanilla. It is widely used as food and beverages flavoring, pharmaceutical substance, and fragrance [1]. The demand of vanillin was reported 15000 tons/year, where only 50 tons was produced naturally through extraction from vanilla beans. Natural vanillin was obtained from vanilla (*Vanila planifolia* or *Vanilafragnans*) through long process and needs a high cost, so its price ranges from \$2000/kg to \$4000/kg. Due to this reason, natural vanillin produces less than 1% from the market demand [2]. The demand of vanillin is catered by many industries producing synthetic vanillin, for example through catechol route which needs consecutive oxidation phase and the use of petroleum based phenol as the main source [3]. An alternative method is needed, especially using biomass derivative compound like isoeugenol, which is an essential oil compound, or lignin derivative compound.

Alternative methods of isoeugenol conversion into vanillin was reported, including synthesis using microorganism [4], cobalt porphyrin as heterogeneous catalyst through oxidative cleavage [5] and electrochemical synthesis from lignin compounds [6]. Synthesis using electrochemical method was the easiest way since the method minimizes the pollution, hazard, and toxicity risk [7]. In this work, electro oxidation of isoeugenol to form vanillin was performed using platinum electrode based on the report regarding electrooxidation of isoeugenol in methanol producing a fraction of vanillin [8]. The report indicated that by optimizing the potential and the reagent composition, a high percentage of vanillin can be produced by the method. Accordingly, the oxidation reaction was performed with different organic solvents, i.e. acetonitrile and methanol. The results showed that methanol is more suitable than acetonitrile to be applied as the solvent for isoeugenol conversion to vanillin. Furthermore, it was also found that water composition in the solvent is important to achieve a high percent yield of vanillin.

2. Materials and methods

2.1. Materials
Materials used in this work are mixture of cis/trans of isoeugenol (Wako), Vanillin (Wako), Tetrabutyl Ammonium Perchlorate (TBAP, Tokyo Chemical Industries), Methanol (Kanto Chemical Industries), Acetonitrile (Sigma-Aldrich), and Ultra High Purity Water (Merck Millipore).

2.2. Electrochemical study of isoeugenol and electrosynthesis of vanillin
Electrochemical study (CH Instrument, Inc.) was conducted using cyclic voltammetry in a self-fabricated one-cell compartment with three electrodes system. Pt foil (dia. 0.58 cm) was used as the working electrode, Pt spiral as the counter electrode, and Ag/AgCl system (ALS Co., Ltd) as the reference electrode. The electrochemical equipment was cleaned using sonicator for 5 min by water and isopropanol. Then, 2.5 mL of 0.1 M isoeugenol and 0.1 M TBAP in methanol or acetonitrile solvent and 1 mL of water were added into the compartment.

Electrosynthesis method was conducted using the same electrochemical cell based on the cyclic voltammetry results at +0.1 and +0.2 V for 30 and 60 min under 500 rpm stirring. Various volumes of water were used for the reagent.

2.3. Analysis of electrosynthesis product using HPLC and GC/MS
Electrosynthesis product was analyzed using HPLC (LC-20AD Prominence, Shimadzu) with C-18 column (YMC-Triart C-18). A binary gradient pump system with 0.1% formic acid (25%) and acetonitrile (75%) was used as mobile phase with UV detector at 280 nm. GC/MS (Agilent Technologies) was controlled by an increasing temperature from 70°C with 20°C/min rate to maximum temperature of 325°C.

3. Results and discussion

3.1. Electrochemical behavior of isoeugenol at Pt electrode
Cyclic voltammograms (CVs) of isoeugenol at Pt electrode in 0.1 M TBAP in methanol and acetonitrile are shown in figure 1a and 1b, respectively. An oxidation potential at +0.21 V can be observed in methanol solvent, whereas it was at +0.16 V in acetonitrile. The difference of polarity from solvent was responsible for the shift of the oxidation potential. Since acetonitrile is more polar than methanol, the faster mass transfer with respect of metallic electrode surface can lead to the lower oxidation potential. The mass transfer at this condition is assumed to be under diffusion controlled only. An investigation with various scan rates was found to provide linear current responses to the square of the scan rates (data not shown).

3.2. Analysis of product using HPLC and GC/MS
Table 1 displays the analysis products using HPLC and GCMS. Optimum condition at Pt electrode was obtained at potential +0.41 V in methanol solvent for 30 min with percent conversion up to 4.11% (162.74 mg/L). Although the cyclic voltammogram shows the onset of isoeugenol oxidation at +0.2 V (figure 1a). Table 2 shows that a higher potential or higher energy than +0.2 V was required to convert isoeugenol to vanillin. However, it was also observed that when the longer time applied to the system, the conversion percentage will be decreased because vanillin can be further oxidized to produce the dimerization. In acetonitrile solvent, the conversion of vanillin was less than in methanol. The possible reason is due to the different polarity between methanol and acetonitrile. Electrochemical reaction of isoeugenol needs water to produce hydroxyl radicals. On the contrary, isoeugenol is not completely

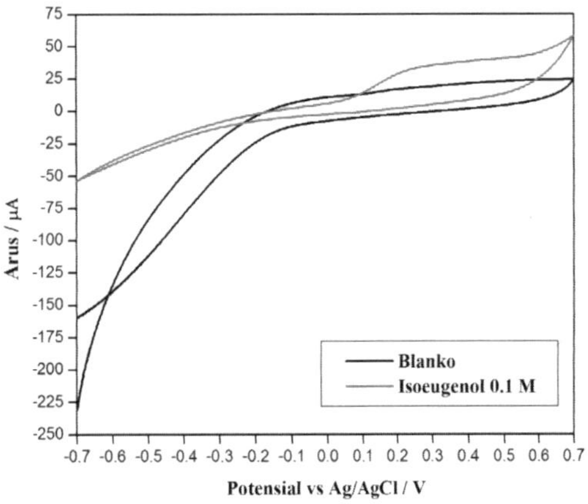

(a)

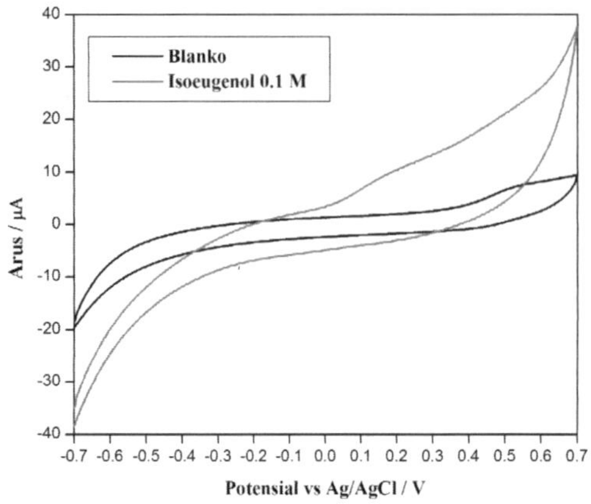

(b)

Figure 1. Cyclic voltammograms of 0.1 M isoeugenolin 0.1 M TBAP at Pt electrode in (a) methanol and (b) acetonitrile solvent.

Table 1. HPLC analysis of the electrosynthesis products

Solvent	Potential (V)	Time(minute)	Isoeugenol reacted (%)	Vanillin conversion (%)	Amount of vanillin (mg/L)
Methanol	0.21	30	31.82	0.47	12.01
		60	35.85	0.32	9.12
	0.31	30	33.47	0.48	12.91
		60	35.45	0.33	9.25
	0.41	30	49.51	4.11	162.74
		60	53.14	3.30	140.39
Acetonitril	0.16	30	22.02	0.44	7.73
		60	48.66	0.15	5.89
	0.26	30	40.00	0.25	7.91
		60	66.11	0.10	5.11
	0.36	30	57.63	0.11	5.07
		60	67.80	0.10	5.03

dissolved in polar solvent. Therefore, the reaction in polar solvent will be limited. Since acetonitrile is more polar than methanol, it is reasonable if the conversion of isoeugenol in acetonitrile was lower than that in methanol.

The vanillin conversion decreases as the water volume increases from 0 mL to 0.5 mL, it might be due to the hydroxyl radicals generated by water oxidation to form other products. The optimum addition of water was 1 mL. The vanillin product decreases at the water content of 1.5 mL, probably due to the change of isoeugenol solubility as it was reported that the electrochemically oxidative cleavage is strongly influenced by water content [9]. High water content also causes the formation of more intermediate species [10]. Moreover, the vanillin which has formed can react with residual water and lead to further oxidation.

Table 3 shows that compound 1 to 5 can be formed from oxidation of alkene group with respect of water, hydroxyl radical, or methoxy radical, while compound 6 can be formed with 2 radicals from isoeugenol which lead to dimerization product, i.e. licarin A. Based on the structure, it was clearly shown that compounds 1 to 5 can be formed through the backbone radicals of isoeugenol, while the compound no. 6 was formed due to the dimerization of the isoeugenol backbone. The result confirmed to the previous report by Sumi *et al.* [8].

Table 2. Variation of water addition.

System	Water Volume (mL)	Isoeugenol Reacted (%)	Vanillin Conversion (%)	Amount of Vanillin (mg/L)
Pt-methanol-0.41 V-30 min	0	18.8	1.10	16.56
	0.5	42.43	0.53	18.13
	1	49.51	4.11	162.74
	1.5	54.35	0.40	17.21

Table 3. Compound formed at vanillin electrosynthesis characterized by GC/MS.

No	Structure	Name	Fragmentation (m/z)
1	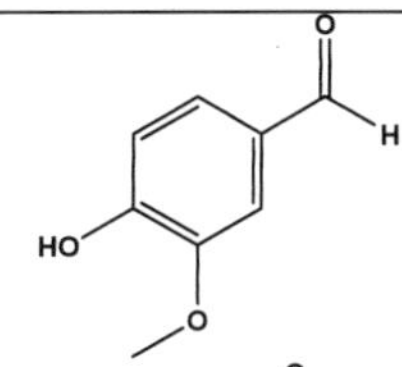	Vanillin (4-hydroxy-3-methoxybenzaldehyde)	151, 109, 81, 53, 29
2	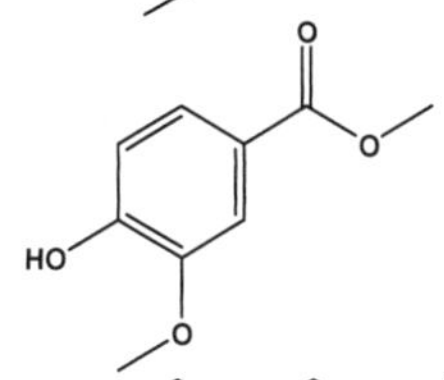	Methyl vanillate (methyl 4-hydroxy-3-methoxybenzoate)	182, 151, 123, 80, 52
3	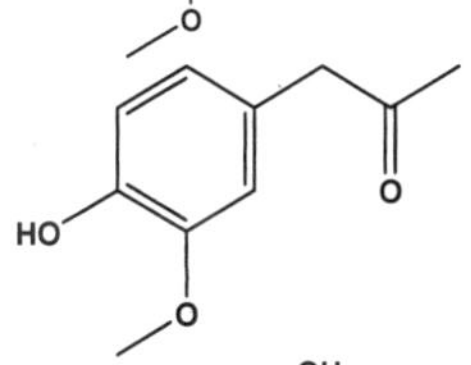	1-(4-hydroxy-3-methoxyphenyl) acetone	180, 137, 94, 43
4	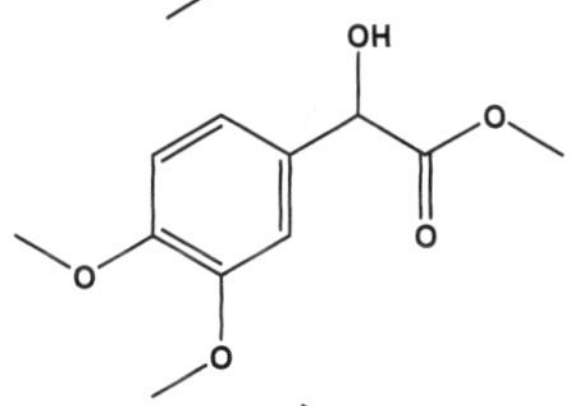	Methyl (3,4-dimethoxyphenyl) hydroxy acetate	226, 195, 167, 151, 137, 109
5	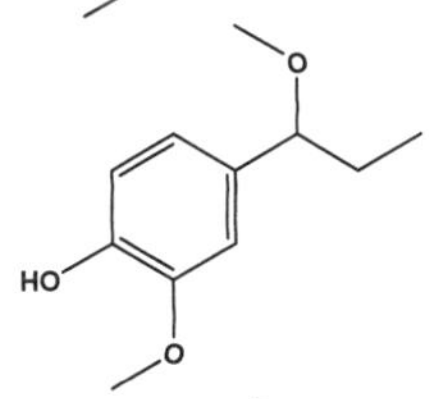	2-methoxy-4-(1'-methoxypropyl) phenol	196, 167, 151, 137, 57
6	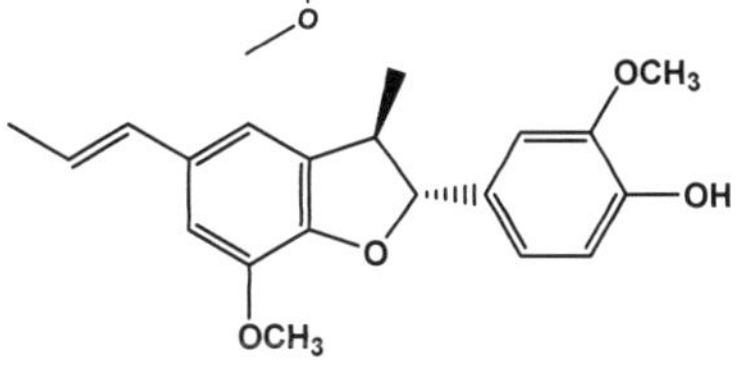	Licarin A – 98%	326, 202

3.3. Proposed mechanism of reaction

Isoeugenol reaction into vanillin is one example of oxidative cleavage from C=C bond into C-C followed by the cleavage into aldehyde or carboxylic group. This reaction takes place at electrode surface with rearrangement in the medium. There are two proposed reaction mechanisms in this oxidative cleavage, which are oxidation at the vinyl (figure 2) [9] alkene and the phenol groups (figure 3) [11].

The first step of the oxidation at the alkene group is the formation of radical cation, when one electron is taken from phi bond at the vinyl group. Then, the nucleophilic attack of water occurs, followed by loss of proton to form neutral benzylic radical intermediate. The second step is the oxidation followed by the second nucleophilic attack to form diol intermediate. After diol formation, the oxidation of one OH from the diol group forms radical cation, followed by the cleavage of C-C sigma bond and the loss of proton. The final step is the rearrangement to form aldehyde group. Oxidation through this mechanism needs two equivalents of water and four electrons as well as loss of four protons.

The oxidation at phenol group starts from the formation of radicals, which is able to resonate (figure 4) [12]. As the consequence, the loss of electrons from phenol groups occurs. After the formation of the fourth resonance intermediate, the next step is the attack of hydroxyl radicals formed through water oxidation at the electrode surface. Then, the arrangement of OH group will form the epoxide by losing 1 proton. In the presence of water and proton, the epoxide ring can be opened to form diol. The next step is the cleavage of C-C sigma bond and formation of aldehyde as in the first proposed reaction mechanism.

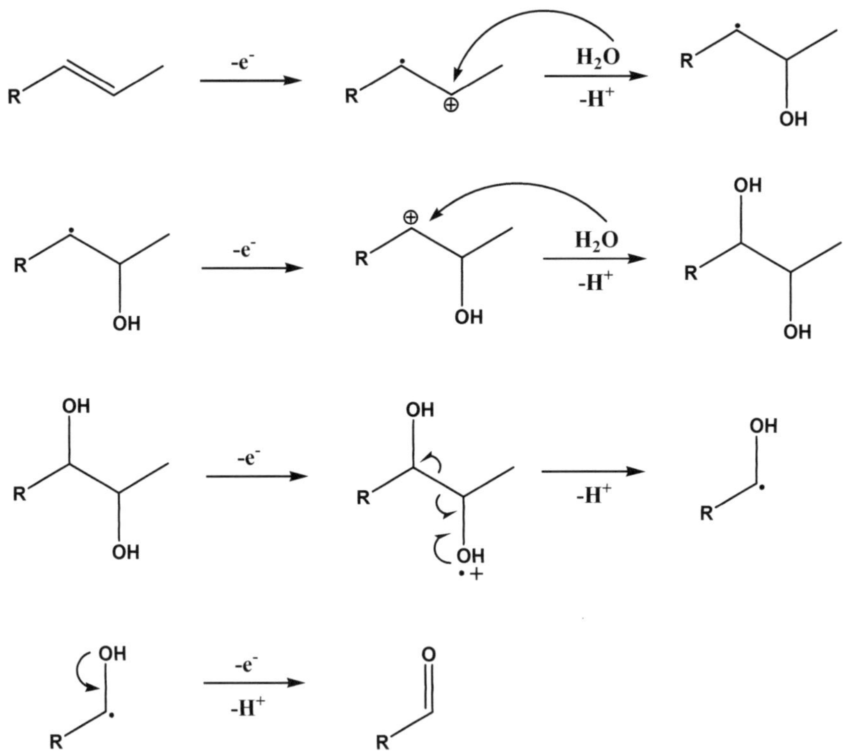

Figure 2. Mechanism of vanillin electrosynthesis through vinyl group.

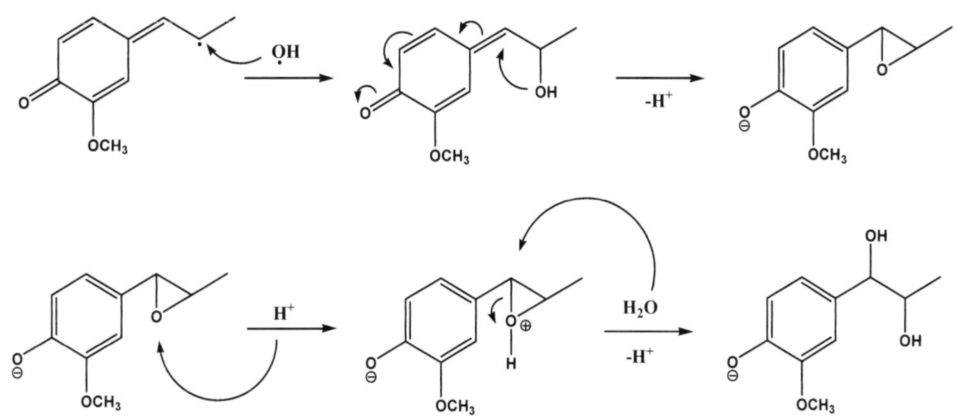

Figure 3. Mechanism of vanillin electrosynthesis via phenol group.

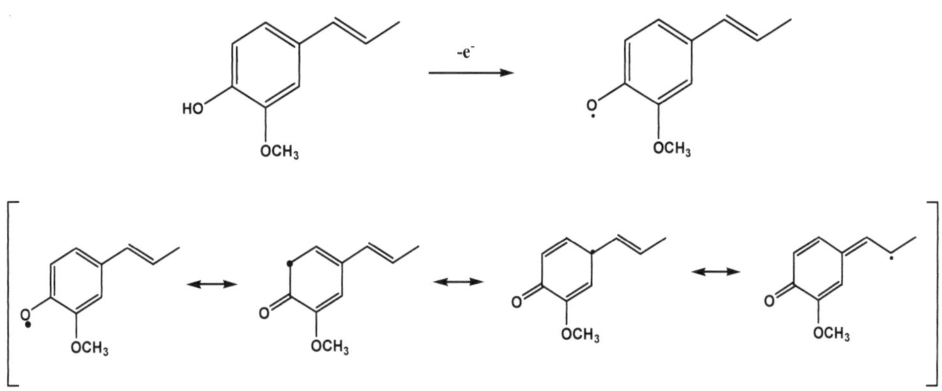

Figure 4. The resonance of radical isoeugenol.

4. Conclusions

Electrosynthesis of vanillin from eugenol has been successfully performed using Pt electrode. Optimum condition of the electrosynthesis were obtained at a potential of 0.41 V for 30 min in methanol solvent. Influence of the feedstock water was found to affect the oxidation process with 1 mL of water as the optimum condition. A conversion up to 4.11% (162.74 mg/L) could be achieved. The reaction mechanism was proposed through the oxidation of the C=C bond and the phenol group.

References

[1] Walton N J, Mayer M J and Narbad A 2003 *Phytochemistry* **63(5)** 505-15
[2] Priefert H, Rabenhorst J and Steinbuchel A 2001 *Appl.Microbiol. Biotechnol.* **56** 296-314
[3] Metivier P 2001 *Method for preparing a 4-Hydroxybenzaldehyde and derivatives* United States Patent 6184421 B1 February 6
[4] Rana R, Mathur A, Jain C K, Sharma S K and Mathur G 2013 *Intl. J. Biotechnol. Bioeng. Res.* **4** 227-34
[5] Adilina I B, Hara T, Ichikuni N and Shimazu S 2012 *J. Molecular Catal. A: Chemical* **361-362** 72-9
[6] Reichert E, Wintringer R, Volmer D A and Hempelmann R 2012 *Physical Chemistry Chemical Physics* **14(15)** 5214
[7] Shibano S, Ivandini T A, Terashima C, Nakata K and Einaga Y 2014 *Chem. Lett.* **43** 1292-3
[8] Sumi T, Saitoh T, Natsui K, Yamamoto T, Atobe M, Einaga Y and Nishiyama S 2012 *Angew. Chem. Int. Ed.* **51** 1-5

[9] Wu X, Davis A P and Fry A J 2007 *Organic Letters* **9(26)** 5633-6
[10] Lee J H Q, Tay B K, Ganguly R and Webster R D 2015 *Electrochim. Acta* **184** 392-402
[11] Smith J G 2011 *Organic chemistry* (New York: McGraw Hill Education) p 323, pp 343-6
[12] Smith J G 2011 *Organic chemistry* (New York: McGraw Hill Education) pp 575-6

The copolimeryzation synthesis and swelling capacity of cellulose-poly superabsorbent (acrylic acid-co-acrylamide) based on rice straw

Helmiyati, A Fitriyani and F Meyanti

Department of Chemistry, Faculty of Mathematics and Natural Sciences Universitas Indonesia, Kampus UI Depok, Depok 16424, Indonesia

Corresponding author's e-mail: helmi-yt@sci.ui.ac.id

Abstract. A superabsorbent has been synthesized by copolymerization of rice straw cellulose as the back bone with the composition of 0.724 mol/L acrylamide and 1.429 mol/L acrylic acid as the monomers, 2.32 mmol/L N, N'-methylene-bis-acrylamide as the crosslinker, and 7.94 mmol/L potassium persulfate as the initiator. The rendement of cellulose obtained from rice straw isolation is 33.55% with the size of 34.06 nm nanocrystalline cellulose, obtained from XRD diffraction pattern. The copolymerization results in the spectrum characterization of Cellulose-Poly superabsorbent (AA-co-AM) with FTIR shows OH stretching vibration, NH and C=O stretching of monomer acrylic acid and acrylamide at wave number about 3343 cm^{-1} and 1600 cm^{-1}. The surface morphology analyzed with SEM shows the superabsorbent has rough surface morphology compared to acrylic acid-acrylamide copolymer. The results of grafting efficiency increases with the increasing amount of the reacted monomer. The characterization of result shows that the grafting process of acrylic acid-acrylamide on cellulose has been formed. The swelling capacity of superabsorbent in water is 691.18 g/g, and 765.58 g/g in urea. This result is quite satisfactory and can be applied for slow release superabsorbent.

1. Introduction

In recent years, research has been done to study the superabsorbent and its application in agricultural. Usually, the synthesis of superabsorbent is based on acrylic acid and acrylamide monomer. However the materials are expensive and can be hazardous to the environment [1]. Therefore, superabsorbent polymers which have biodegradable properties started to be developed. The synthesis of superabsorbent from waste wheat straw has been done, using acrylic acid and acrylamide as monomer, dimethyl-diallyl-ammonium chloride as initiator, and N, N'-metilenbisakrilamida (MBA) as crosslinking agent [1]. Utilization of mulberry twig waste as materials for the superabsorbent has been studied [2]. Ma *et al.* [3] has successfully synthesized the superabsorbent with the addition of inorganic materials. The addition of inorganic substances will create a higher thermal stability superabsorbent. Cellulose polymer is one of the most abundant, renewable, and biodegradable materials among the natural polymers. Cellulose can be converted into nanocellulose. Compared to other ordinary cellulose, nanocellulose has a smaller diameter which makes it has many unique properties, such as high mechanical strength and large surface area [4]. The cellulose can be isolated from rice straw, which contains 43-49% cellulose, 25-28% hemicellulose, lignin 15.8%, 12.5% silica, resin, gum, protein, and mineral compounds [5]. In this research, the cellulose is isolated from rice straw, and will be used to synthesize the superabsorbent.

IOP Conf. Series: Materials Science and Engineering **188** (2017) 012051 doi:10.1088/1757-899X/188/1/012051

2. Material and methods

2.1. Materials
Rice straw was obtained from Bukittinggi, Indonesia. Acrylic acid and acrylamide (Nippon Shokubai) were used as monomers. Toluene (Merck Co) and ethanol (Merck Co) were used to extract the cellulose from rice straw. Hydrogen peroxide (Merck Co) and potassium hydroxide (Merck Co) were used as the impurity removal from rice straw cellulose. Potassium persulfate (Merck Co) was used as initiator, and N,N′-methylenebisacrylamide (Sigma) as crosslinker.

2.2. Isolation of cellulose and synthesis of superabsorbent
The isolation of rice straw cellulose was performed according to the reported method by Fan *et al.* [6, 7]. The rice straw was washed with warm water at temperature 40°C to remove impurities and solutes then dried at 60°C. The dried and cleaned rice straw was then grinded into powder. 10 grams of the powder was extracted with a mixture of toluene: ethanol at a ratio of 2 : 1 for 72 hours with maceration method while stirring with a stirrer to eliminate the extracted substances, wax, pigment, and oil. Then the straw powder was filtered and dried at 50°C for 24 hours. Removing the hemicellulose and lignin was done by adding 350 mL KOH and H_2O_2, and then stirred at room temperature. The precipitate was washed to neutral pH and then dried in an oven at 50°C until completely dry and the weight constant was obtained. The dissolution process of rice straw cellulose and synthesis of superabsorbent was performed as reported by Cay *et al.* [8] and Helmiyati *et al.* [9].

3. Results and discussion

3.1. Isolation of cellulose from rice straw
The average of cellulose rendements can be seen in table 1. From table 1, it can be seen the rendement of cellulose obtained after isolation is 33.55%. The rendement of cellulose obtained in this work is smaller compared to Fan et al with the rendement of 58.0% [6], and Jiang *et al.* [10] with the rendement of 35.06%. This difference was caused by rice straw source.

3.2. Diffraction pattern of rice straw cellulose by XRD
The diffraction patterns of rice straw cellulose is shown in figure 1. In figure 1 it can be seen the peaks of the cellulose are obtained at about 22.4° with a sharp peak. The peak at 22.4° called intensity of crystalline, and the peak at 18.25° called intensity of amorphous. The crystallinity degree of cellulose is 72.753%, means cellulose is semi crystalline and the calculation by Scherrer's law obtained the size of crystal cellulose of 34.06 nm which indicated the nanocrystal size.

3.3. FTIR spectra of superabsorbent
Figure 2 shows the FTIR spectra of rice husks, acrylate acid-acrylamide copolymer (AA-AM copolymer) and Cellulose-Poly (AA-co-AM). Figure 2c shows the infrared absorptions of Cellulose-Poly superabsorbent (AA-co-AM). The main absorption band shows stretching vibration of OH and NH from monomer at 3343 cm^{-1}, while at 1600 $cm-^{1}$ indicates the C=O stretching vibration of the carbonyl

Table 1. Rendement of cellulose

Weigth of rice straw (g)	Weight of rendement (g)	Rendement of cellulose (%)
10.00	3.10	31.01
10.01	3.49	34.84
10.00	3.17	31.69
10.01	3.67	36.64
Average of cellulose rendement (%)		33.55

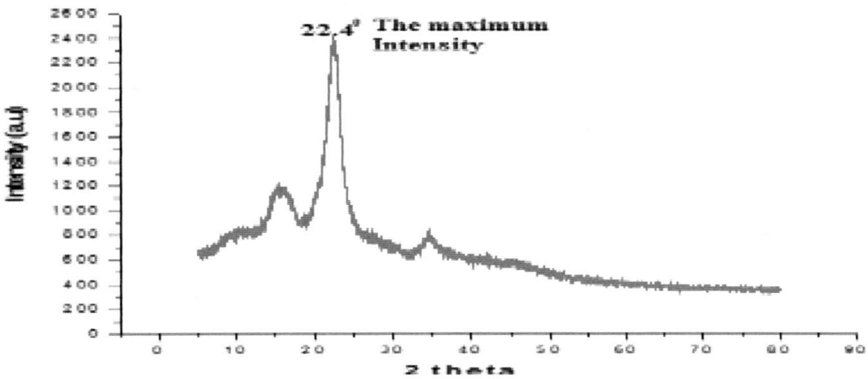

Figure 1. XRD diffraction pattern of rice straw cellulose

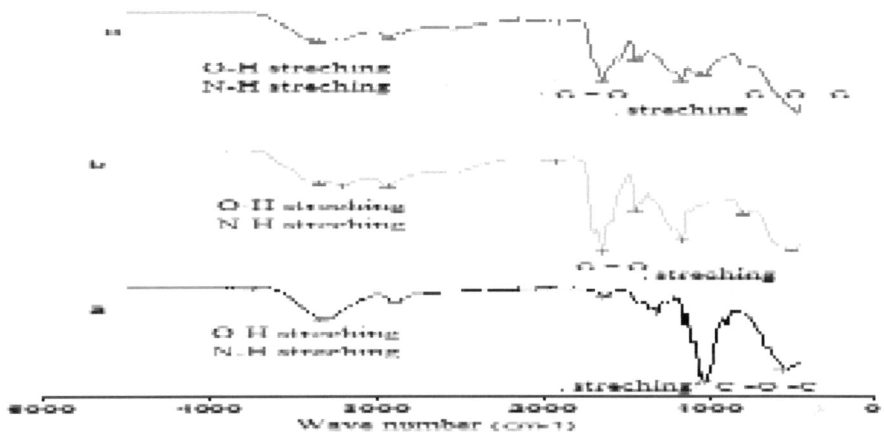

Figure 2. Spectra of FTIR (a) rice straw (b) copolymer AA-AM, and (c) cellulose-poly (AA-co-AM)

group of acrylic acid and acrylamide. Overall, it was found that the spectrum of the superabsorbent is a combination from spectrum straw cellulose (figure 2a) and copolymers (AA-AM) (figure 2b). It indicates that theacrylic acid and acrylamide has been successfully grafted on the cellulose.

3.4. Surface morphology of superabsorbent

Figure 3a shows the surface morphology of Cellulose-Poly superabsorbent copolymerization (AA co AA) in porous form, more uniform and rough compared to figure 3b. This is due to grafting process of acrylic acid and acrylamide monomer to cellulose, so the superabsorbent network was formed more tightly and the pores can be utilized to absorb water and fertilizer when carried out the swelling process. Figure 3b shows a more smooth surface morphology of acrylic acid-acrylamide copolymer without cellulose. This is similar to the results conducted previously [10].

3.5. Swelling capacity

The swelling characteristics of the superabsorbent were measured by gravimetric analysis [11]. Superabsorbent was synthesized with different conditions to get different swelling capacity. The results of the analysis with orthogonal show the sequence of swelling influence factors in water and urea are MBA>AM>AA>KPS. MBA crosslinking agent is the most important factor, followed by the monomer AM, and AA, then the KPS initiator. Based on the experimental results, the optimum combination for the synthesis of Cellulose-Poly superabsorbent copolymerization (AA co AA) are [AA] with the

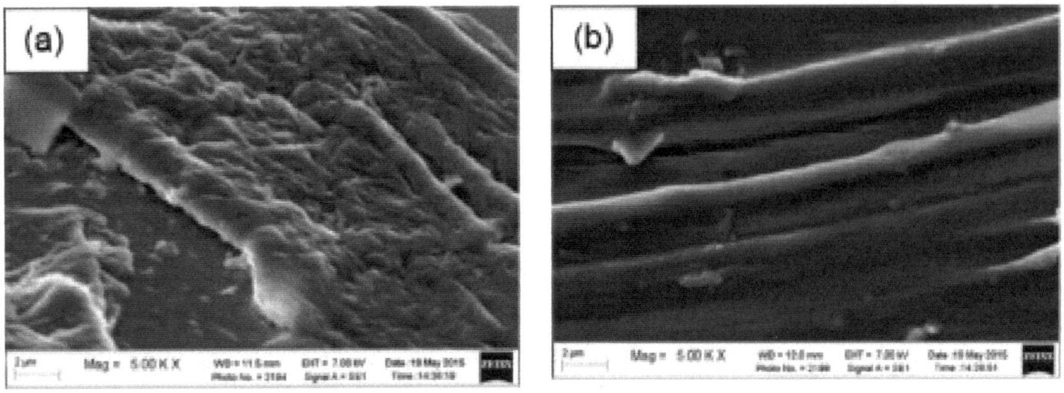

(a) (b)

Figure 3. SEM micrograph of (a) Cellulose-Poly Superabsorbent (AA-co-AM) and (b) Copolymer AA-AM

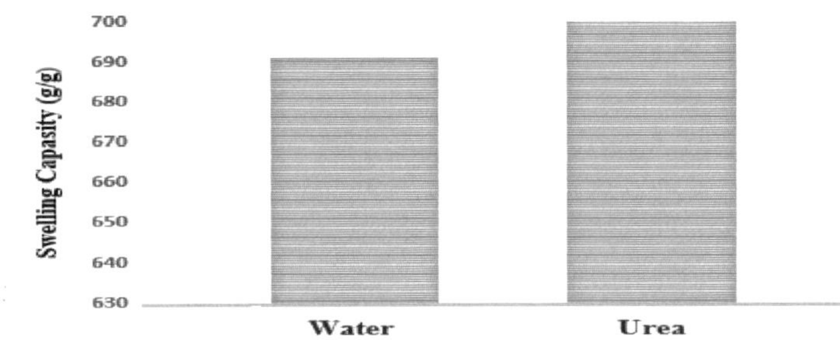

Figure 4. Swelling capacity cellulose poly (AA co AA) superabsorbent on the water and urea

concentration of 1.429 mol/L, followed by 0.724 mol/L [AM], 2.32 mmol/L [MBA], and 7.94 mmol/l [KPS]. The synthesized superabsorbent has swelling capacity in the water and urea of 691.18 g/g and 765.58 g / g, respectively, as can be seen in figure 4.

4. Conclusions

The cellulose has been isolated from rice straw and can be separated from the lignin and hemicellulose with the average rendement of 33.55%, and the size of cellulose crystal at 34.06 nm, which indicated nanocrystal size. Cellulose isolation was used as the backbone in the synthesis of superabsorbent. The result of grafting was characterized by FTIR and SEM. The optimum condition of the Cellulose-Poly superabsorbent synthesis (AA co AA) were 1.429 mol/L acrylic acid, 0.724 mol/L acrylamide, 2.32 mmol/L MBA crosslinking agent,and 7.94 mmol/L KPS initiator, with the capacity of swelling in water and ureawere 691.18 g/g and 765.58 g/g, respectively.

Acknowledgements

The authors would like to thank Universitas Indonesia for funding this research through PITTA Grant Universitas Indonesia with contract No: 2045/UN2.R12/HKP.05.05/2016.

References

[1] Li Q, Ma Z, Yue Q, Gao B, Li W and Xu X 2012 *Bioresour. Technol.* **118** 204-9
[2] Zhang Y, Liang X, Yang X, Liu H and Yao J 2014 *ACS Sustainable Chem. Eng.* **2** 1871-8
[3] Ma G, Yang Q, Ran F, Dong Z and Lei Z 2015 *Appl. Clay Sci.* **118** 21-8
[4] Wei H, Rodriguez K, Renneckar S and Vikesland P J 2014 *Environmental Science: Nano* **1** 302-16

[5] Sahu M, Samal R and Sahoo P K 2015 *Asian Journal of Chemistry* **27** 3377-82
[6] Fan G, Wang M, Liao C, Fang T, Li J and Zhou R 2013 *Carbohydrate Polymers* **94** 71-6
[7] Helmiyati and Kurniawan S 2015 *Journal of Chemical and Pharmaceutical Research* **7** 878-85
[8] Cai J *et al.* 2008 *Macromolecules* **41** 9345-51
[9] Helmiyati, Saefumillah A and Yulianti W 2014 *Asian Journal of Chemistry* **26** 7337-42
[10] Jiang M, Zhao M, Zhou Z, Huang T, Chen X and Wang Y 2011 *Industrial Crops and Products* **33** 734-8
[11] Anirudhan T S, Tharun A R and Rejeena S R 2011 *Ind. Eng. Chem. Res.* **50** 1866-74

Capillary electrophoresis of adenosine phosphates using boron-doped diamond electrodes

B D Firmansyah, T A Ivandini and J Gunlazuardi

Department of Chemistry, Faculty of Mathematics and Natural Sciences Universitas Indonesia, Kampus UI Depok, Depok 16424, Indonesia

Corresponding author's e-mail: ivandini.tri@sci.ui.ac.id

Abstract. A capillary electrophoresis coupled with electrochemical detection using boron-doped diamond electrode was developed for simultaneous detection of adenosine phosphates, i.e. adenosine monophosphate (AMP), adenosine diphosphate (ADP), and adenosine triphosphate (ATP). In phosphate buffer solution pH 7, these three adenosine phosphates have similar oxidation potentials at around +0.9 V (vs. Ag/AgCl), which indicated that the oxidation occurred at the same moiety. Capillary electrophoresis, which was then performed using fused silica capillary (dia. 0.05 mm) at an applied potential of 10 KV can separate ATP, ADP and AMP with the retention times of 848 s, 1202 s, and 1439 s, respectively. Linear calibration curves with the limits of detection of 0.59 μM, 0.56 μM and 1.78 μM, respectively, can be achieved, suggested that capillary electrophoresis with electrochemical detector is promising for simultaneous detection of adenosine phosphates.

Keywords: capillary electrophoresis, adenosine phosphates, electrochemical detection, boron-doped diamond

1. Introduction

Adenosine phosphate is a nucleotide composed of pentose, adenine bases (purine), and phosphates. Adenosine phosphates have a various number of phosphates, that are adenosine monophosphate (AMP), adenosine diphosphate (ADP), and adenosine triphosphate (ATP). Among these three types of adenosine phosphates, only ATP is used as a medical treatment for amyotrophia disease, cerebral hemorrhage and hepatitis [1]. Accordingly, a sensitive method to measure ATP in the mixture of those adenosine phosphates is necessary. Generally, HPLC is employed to determine the adenosine phosphates. However, HPLC is not a cheap instrument. A capillary electrophoresis method, which is performed based on the movement of the charged molecules through a fluid under an electric field influence, can be used as an alternative for the separation process of adenosine phosphates. Then, by combining the electrophoresis with a detector, the separated adenosine phosphates can be quantified.

The detectors commonly used in capillary electrophoresis are UV-visible and laser induced fluorescence [2]. UV-visible has the disadvantage in the detection limit, while the laser induced fluorescence is an expensive instrument. Meanwhile, boron-doped diamond (BDD) is established as an alternative of solid electrodes due to its superior properties among other electrodes, including its wide potential window, low background current, and high surface stability [3-5]. BDD have been investigated for electrochemical detection of adenosine phosphates [1], showing similar oxidation

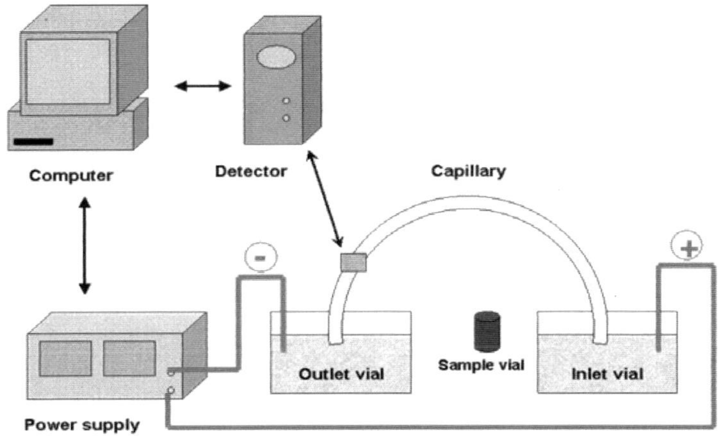

Figure 1. Scheme of the electrophoresis cell.

potentials at around +0.9 V for those three adenosine phosphates. In this work, BDD was employed as the detector for capillary electrophoresis of AMP, ADP, and ATP. Succesfull separation with good detection limits could be achieved, suggested that the method was promising for the real applications.

2. Materials and methods

Prior to use, the BDD film was cleaned subsequently by ultrasonication in 1-propanol and water for 15 min each. Cyclic voltammetry was performed in one-container electrochemical cell to study the electrochemical behaviour of adenosine phosphates in phosphate buffer solution (PBS).

The capillary electrophoresis cell was self-fabricated and modified from the design of Shin *et al.* [2]. This cell was composed of a couple of platinum electrodes immersed in two separate containers. The electrodes were connected to a power supply to provide negative and positive poles, while between the containers, a silica fused capillary tube (i.d. 0.05 mm) was placed. A BDD film was arranged at the end of the capillary tube in the negative pole container and connected to a potensiostat with a Pt wire as the counter electrode and an Ag/Ag system as the reference electrode. The scheme of capillary electrophoresis cell is displayed in figure 1.

3. Results and discussion

Cyclic voltammograms (CVs) of 0.1 M PBS pH 7 in the absence and in the presence of AMP (figure 2a) in comparison with those of ADP and ATP (figure 2b and figure 2c, respectively) at BDD electrode show an oxidation peak of all types of adenosine phosphates at the potential around +0.90 to +0.95 V (vs. Ag/AgCl). It can be seen that the more phosphate groups contained in the adenosine phosphates, the higher oxidation potential was required. However, the oxidation potential was observed in the narrow range between +0.90 to +0.95 V, suggested that the oxidation occurred at the same moieties. Adenine, the nitrogenous bases with electroactive properties, is the most possible moiety to undergo the oxidation reaction [1]. Furthermore, there was no reduction peak observed in the potential range between -1.0 to +1.5 V indicated that the oxidation reaction was irreversible.

Influence of the pH was also studied. Linear relationships were observed in the pH range between pH 2 and pH 8 for all adenosine phosphates with a slope of around -0.59 mV/pH (data not shown) suggested that the reaction involving the same numbers of proton and electron. Furthermore, linear calibration curves showed good linearity of the adenosine phosphates in the concentration range of 50 μm to 500 μm, indicated that the electrode can be applied for detection of all adenosine phosphates.

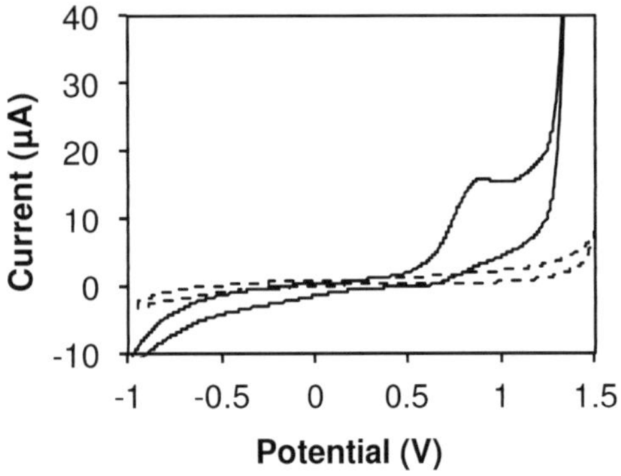

(a)

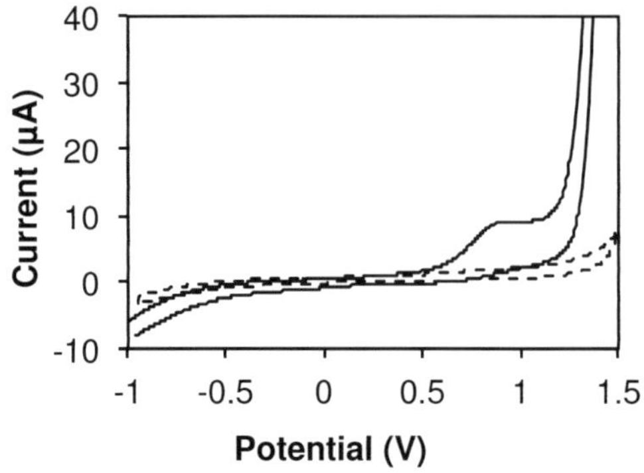

(b)

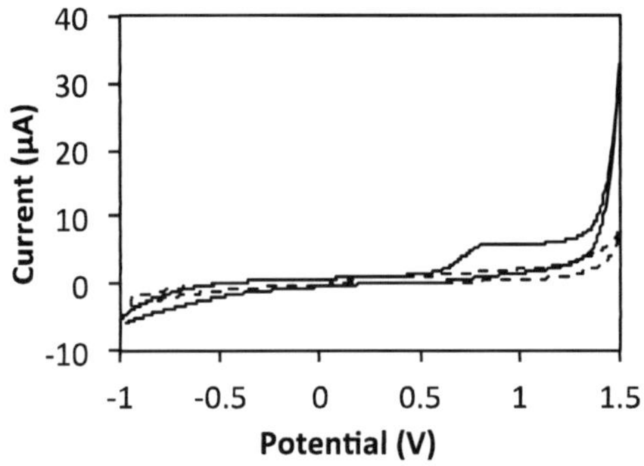

(c)

Figure 2. Cyclic voltammetry of 0.1 M phosphate buffer solution in the absence (dashed line) and in the presence (solid line) of 0.05 mM (a) AMP, (b) ADP, and (c) ATP in pH 7 at BDD electrodes.

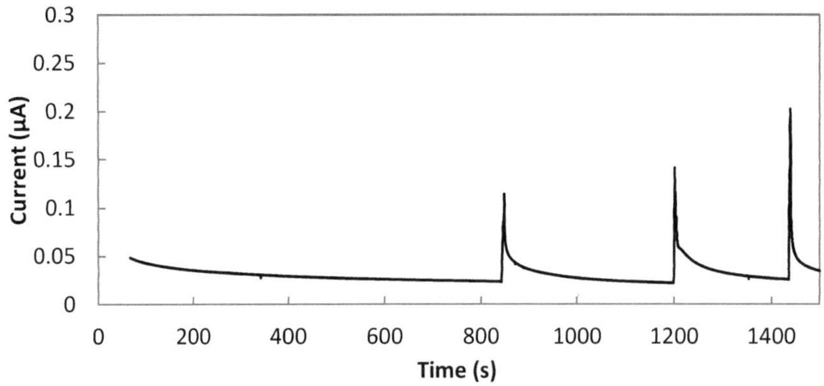

Figure 3. Typical amperogram of the mixture of 0.05 mM AMP, 0.05 mM ADP, and 0.05 mM ATP in 0.1 M phosphate buffer solution pH 7 after separating process using capillary electrophoresis method. Separating voltage was 10 KV. BDD film was used as the electrochemical detector at an applied potential of +0.95 V.

Capillary electrophoresis of the adenosine phosphates was performed at the separating potential of 10 KV. At the end of capillary tube in the negative pole container, BDD film was placed.

Amperommetry technique at an applied potential of +0.95 V was applied and all types of adenosine phosphate was added into the sample container with PBS pH 7 was used as the mobile phase. Figure 3 shows the amperogram of a mixture solution of 0.05 mM ATP, 0.05 mM ADP, and 0.05 mM AMP in 01 M PBS pH 7. The peaks of ATP, ADP and AMP were observed at the retention times of 848 s, 1202 s, and 1439 s, respectively. ATP appeared as the first peak due to its higher charge, followed by ATP and AMP. Further, investigation of the mixture with various concentrations of adenosine phosphates can achieve linear calibration curves with the limits of detection of 0.59 μM, 0.56 μM and 1.78 μM, respectively. The results suggested that electrophoresis with electrochemical detector is promising for the simultaneous detection of adenosine phosphates.

4. Conclusions

BDD electrode was successfully applied in the electrochemical detection using a capillary electrophoresis method for mixtures solution of ATP, ADP, and AMP. The peaks of ATP, ADP and AMP were observed at the retention times of 848 s, 1202 s, and 1439 s, respectively, with the limits of detection of 0.59 μM, 0.56 μM and 1.78 μM.

Acknowledgements

This work was funded by Hibah Kompetensi Publikasi Internasional Terindeks Universitas Indonesia, Contract No. 2016/UN2.R12/HKP.05.00/2016.

References

[1] Asai K, Ivandini T A, Falah M M and Einaga Y 2016 *Electroanalysis* 28 177-82
[2] Shin D C, Sarada B V, Tryk D A, Fujishima A and Wang J 2003 *Anal. Chem.* **75** 530-4
[3] Ishii Y, Ivandini T A, Murata K and Einaga Y 2013 *Anal. Chem.* **85** 4284-8
[4] Suzuki A, Ivandini T A, Kamiya A, Nomura S, Yamanuki M, Matsumoto K, Fujishima A and Einaga Y 2007 *Sensors Actuat. B: Chemical* **120** 500-7
[5] Macpherson J V 2015 *Phys. Chem. Chem. Phys.* **17**, 2935-49

IOP Conf. Series: Materials Science and Engineering **188** (2017) 012053 doi:10.1088/1757-899X/188/1/012053

Ulva fasciata-mediated preparation of zinc oxide nanocrystalline for one-pot multicomponent synthesis of 6-amino-3-methyl-4-phenyl-2,4-dihydropyrano[2,3-c]pyrazole-5-carbonitrile

A H Cahyana[1], Nurhayati [1,2] , B S B Utomo[2] and B Ardiansah[1]

[1] Department of Chemistry, Faculty of Mathematics and Natural Sciences Universitas Indonesia, Kampus UI Depok, Depok 16424, Indonesia
[2] Research and Development Center for Marine and Fisheries Products Competitiveness and Biotechnology, Ministry of Marine Affairs and Fisheries Republic of Indonesia, Jl. KS Tubun, Petamburan VI, Jakarta, 10260, Indonesia

Corresponding author's e-mail: herry@ui.ac.id

Abstract. ZnO has been successfully prepared by a simple co-precipitation method using sodium hydroxide, zinc acetate as precursor and *Ulva fasciata* polysaccharides as capping agent. Characterization using FTIR, XRD and SEM-EDS indicated the successful formation of ZnO nanocrystalline with crystallite size of 27.01 nm. In order to promote ZnO as a green catalyst, its application for one-pot multicomponent pyranopyrazole synthesis has been performed. Several reaction conditions have been investigated, and the highest yield (98%) was observed in the use of 10% mole of ZnO at room temperature for 90 min reaction time in aqueous medium to produce 6-amino-3-methyl-4-phenyl-2,4-dihydropyrano[2,3-c]pyrazole-5-carbonitrile.

1. Introduction

The study based on nanotechnology is growing rapidly, especially in the field of catalysts [1]. The principle of nanotechnology is to manipulate the properties and performance of materials in such a way that it becomes more effective, efficient and powerful. Nanoparticles have a very small particle size ranging from 1 to100 nm. Through minimizing the size, nanoparticles will have unique physical and chemical characteristics, which are very different than the materials with larger particle sizes [1]. One of synthetic method of nanoparticles is biosynthesis. Biosynthesis of nanoparticles is performed using media from biological materials such as microorganisms, enzymes, or plant extracts. Biosynthesis has advantages over other methods, such as chemical and physical methods, which require a high cost, long time process, and produce a large amount of waste from the chemicals used. Thus, it can be an alternative and environment-friendly way to generate nanoparticles [1].

Recently, several studies have shown that seaweed can be used for the synthesis of silver and gold nanoparticles [2-4]. Azizi *et al.* (2014) [5] reported the use of brown seaweed in the biosynthesis of ZnO nanoparticles. The synthesized nanoparticles are known to have anti-bacterial activity, anti-fungal, wound dressings, photochemical and catalytic agent, so it can be applied in various fields, such as biomedical, pharmaceutical, and cosmetics [2]. One of the many synthesized nanoparticles is zinc oxide. This material has a number of potential applications in the fields of electronics and sensors due to its optical, acoustic and electrical properties. In addition, ZnO has several advantages including stable chemical structure, non-toxic, cheap, highly abundant in nature, and can be used as an additive in various materials [5].

Ulva fasciata is a type of non-economic seaweed that has not been used commercially. Consumers of this seaweed are still limited to the fishermen families or coastal communities, namely as seaweed chips. Green seaweed contains polysaccharides in bulk, especially cell wall polysaccharides such as *ulvan*. It is a water-soluble sulfate polysaccharides found in *Ulva* sp. [6]. The presence of sulfate groups in the polysaccharide is likely to become an active force in the formed nanoparticles.

Pyranopyrazole is widely used in biomedical and pharmaceutical applications. Its compounds have antibacterial, anti-inflammatory, insecticidal, antitumor and anticancer activity [7]. Synthesis of pyranopyrazole derivatives can be done through multicomponent reactions [7-9]. The advantages of using multicomponent reactions are short time, simple, and easy to purify. Pagore and co-workers reported that ammonium chloride can be utilized as an effective catalyst in the synthesis of pyranopyrazole with the use of a microwave [7]. In the other research, Vasuki & Kumaravel (2008) [10] have used piperidine as a catalyst. Although the reported methods are effective, but the methods have some disadvantages, such as the use of microwave is not suitable when used for large-scale synthesis. In addition, the use of toxic chemicals such as piperidine can also harm the environment. Therefore, in this research, we will investigate the performance of synthesized zinc oxide as solid catalyst for one-pot multicomponent pyranopyrazole synthesis.

2. Materials and methods

2.1. Materials
Green seaweed, *Ulva fasciata,* was obtained from Binuangeun, Banten. Seaweed was stored in plastic container and placed inside the cooling box, cooled with ice cubes and then brought to Research and Development Centre for Marine and Fisheries Product Competitiveness and Biotechnology, Jakarta. Furthermore seaweed was washed with fresh water and dried. The dried seaweed was milled into flour. All the chemicals required were purchased from Merck and used without further purification. Phase purity and particle size were determined by X-ray diffraction (XRD) analysis and recorded by diffractometer MiniFlex. The morphology and elemental composition of the sample were characterized by using Scanning Electron Microscope tandem Energy Dispersive X-Ray Spectroscopy (SEM-EDS) JEOL JCM-6000. The distribution and particle size were measured using Particle Size Analyser (PSA) Beckman Coulter DelsaMax Pro. Infrared spectra were recorded on PerkinElmer Spectrum One FTIR spectrophotometer using KBr pellets. ^{13}C-NMR and ^{1}H-NMR spectra were recorded using DMSO-d_6 solvent on JEOL 400 MHz with TMS as an internal standard. Mass spectra were scanned on LCMS-IT-TOF Shimadzu.

2.2. Methods
2.2.1. Extraction of water soluble polysaccharide. Dry *U. fasciata* powder (5 g) was extracted in ultrapure water (1:15 w/v) for 3 h at 80°C. The extract was filtered and centrifuged at 8,000 rpm for 10 min. The hot water soluble polysaccharide was precipitated by adding ethanol 95% (v/v). The precipitated polysaccharide was filtered, washed with ethanol, and dried for 12 h at 40°C in an oven [2].

2.2.2. Preparation of ZnO nanocrystalline. Synthesis of ZnO nanocrystalline refers to a method performed by Pandimurugan and Thambidurai (2014) [11]. Polysaccharide (0.5 g) was mixed with 25 mL of 25% zinc acetate solution. The mixture was heated at 70°C for 2 h while stirring on magnetic stirrer. Subsequently, the aqueous NaOH solution (25% b/b) was slowly added to the mixture until the mixture reached pH 10. The precipitate was separated with centrifuge at 8,000 rpm for 10 min. Furthermore, the precipitate was dried at 105°C for 20 h.

2.2.3. Catalytic activity test of ZnO on pyranopyrazole synthesis. Benzaldehyde (1 mmol), ethyl acetoacetate (1 mmol), hydrazine hydrate (1.5 mmol), malononitrile (1 mmol), various amount of ZnO (2.5, 5, 7.5 and 10 % mol catalyst) and ultrapure water were mixed in a 250 beaker glass. The suspension was stirred using a magnetic stirrer at various temperatures and times (reaction time: 30, 60, 90, 120, and 180 min; temperature: 30, 70 and 100°C). Once the reaction was completed, the reaction product was cooled, filtered and washed with hot ethanol (5 mL) to separate the product from the catalyst.

Ethanol from the filtrate was allowed to evaporate at room temperature in order to obtain the product. The product was purified by recrystallization using warm ethanol. The crystal formed was separated from the solution and characterized.

3. Results and discussion

3.1. Catalyst characterization

One type of green seaweed, *Ulva fasciata* is still not widely known and commercially exploited, so it can be regarded as a non-economic seaweed. Seaweed is rich in polysaccharide consisting of *ulvan*, cellulose, xyloglucan and glucurunan [12]. Based on this research, it was known that the yield of hot water soluble polysaccharide from *U. fasciata* is equal to 17.32 ± 0.09% of the dry weight of the seaweed. Siddhanta *et al.* (2001) [13] reported that water-soluble polysaccharides were divided into two categories, soluble in cold water and hot water with a yield value of 1.2 to 6.5% and from 5.2 to 16.9%, respectively.

As a sulfate polysaccharide, the sulfate content of *U. fasciata* polysaccharides was 15.05 ± 0.15 %. These polysaccharides has higher sulfate content than *Ulva armoricana* (10.3 to 13.8 %) and *Ulva rotundata* (9.2 to 12.5 %) [14-15], but lower than *Ulva conglobata* (23.04 to 35.2 %) [16]. With the sulfate polysaccharide, Zn ions would be covered by the sulfate groups in the polysaccharide to prevent the formation of the large-sized particles. This is consistent with the research of Mahdavi *et al.* (2013) [17], who reported that the polysaccharide acts as a capping agent in the production of metal oxide nanoparticles

The FTIR spectra of *U. fasciata* (a), polysaccharides (b), and ZnO (c) can be seen in figure 1. Figure 1 shows that the algae and polysaccharide has a similar spectrum. It means that the main component of algae is polysaccharides. The characteristic frequencies of *U. fasciata* were observed at 3429 cm^{-1} for vibration mode of O-H group, 2924 cm^{-1} due to asymmetric C-H. While the peaks at 1648 and 1427 cm^{-1} due to the characteristics of the symmetric stretching vibration of C=O. Whereas the peak at 1031 cm^{-1} shows guluronic acid groups [11]. Sulfate band is appeared at 850cm^{-1}(C-O-S). In ZnO sample, wavenumbers below 450 cm^{-1} is attributed to Zn-O stretching vibration. This is consistent with the study results of Pandimurugan & Thambidurai (2014) [11], who stated that the Zn-O band is observed at 423 cm^{-1}.

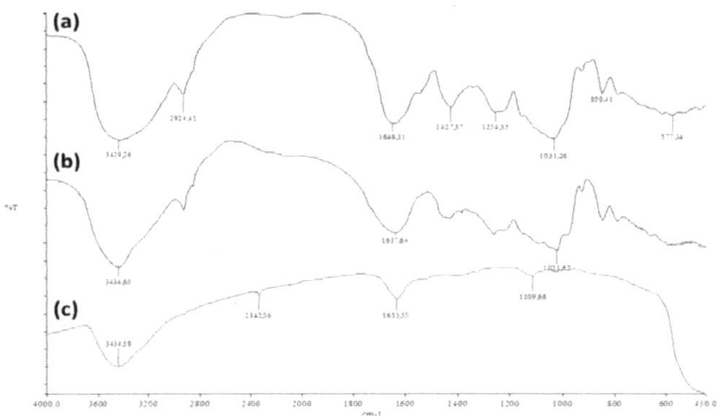

Figure 1. IR spectra of (a) algae flour, (b) polysaccharide and (c) ZnO.

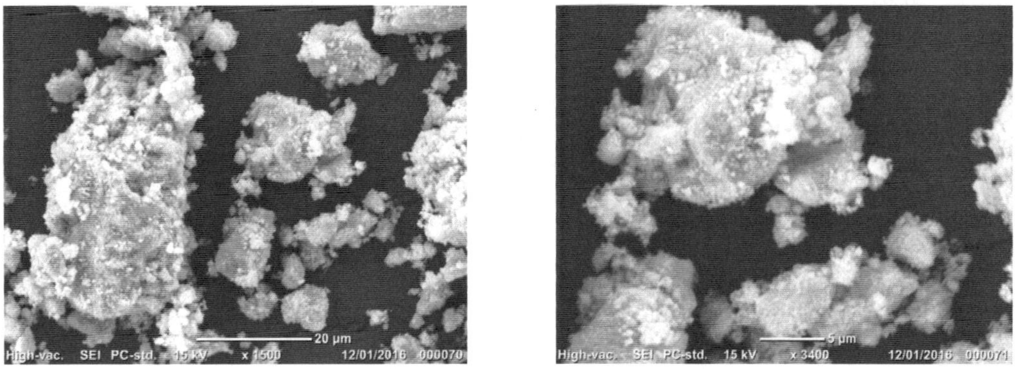

(a) (b)

Figure 2. SEM analysis of ZnO with magnification (a) 1500 × and (b) 3400 ×.

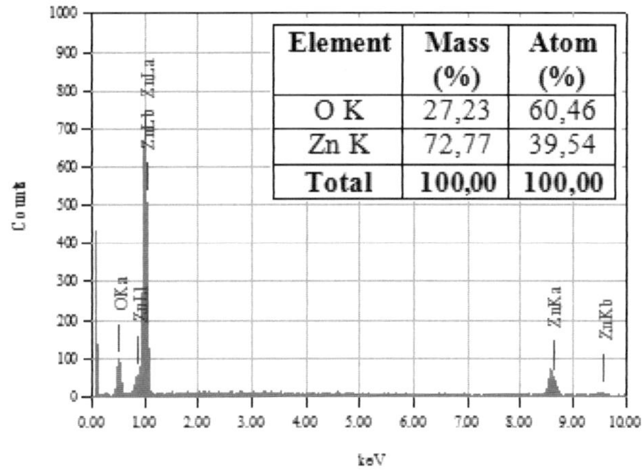

Figure 3. The result of EDS analysis of ZnO.

SEM-EDS analysis was conducted to determine the surface morphology and elemental composition of a solid sample. Figure 2 shows the results of the SEM analysis with 2 magnifications, 1500 and 3400 times magnification. Measurement with smaller magnification (1500 times) was conducted to determine the morphology of the whole sample. SEM analysis showed the shaped of particles like blobs, although there were also the small particles. Meanwhile, further analysis with EDS (figure 3) shows that ZnO metal oxide is formed, with Zn atoms and O at 39.54% by 60.56%, respectively.

XRD analysis was conducted to determine the structure and phase of ZnO samples. In figure 4, the XRD diffractogram of the ZnO has a dominant 2θ at 36.3216; 31.8360; and 34.4914. The crystallite size of ZnO was calculated using Debye-Scherrer equation ($D = K.\lambda/\beta.\cos\theta$), based on the most dominant peak intensity in the XRD analysis, where K = factor Crystals (0.9 for a sphere); λ = wavelength used (0.154 nm Cu-Kα); β = the value of FWHM (full width at half maximum) and θ = diffraction angle Brag. From the calculations, the average crystallite size for ZnO is 27.01 nm.

3.2. Synthesis of pyranopyrazole using ZnO

Synthesis of pyranopyrazole was performed using 4 components: ethyl acetoacetate, hydrazine hydrate, malononitrile and benzaldehyde. To investigate the effect of catalyst amount, zinc oxide was varied from 2.5-10 mol%. The reaction was performed for 1.5 hours at 70°C with a solvent of 10 mL ultrapure

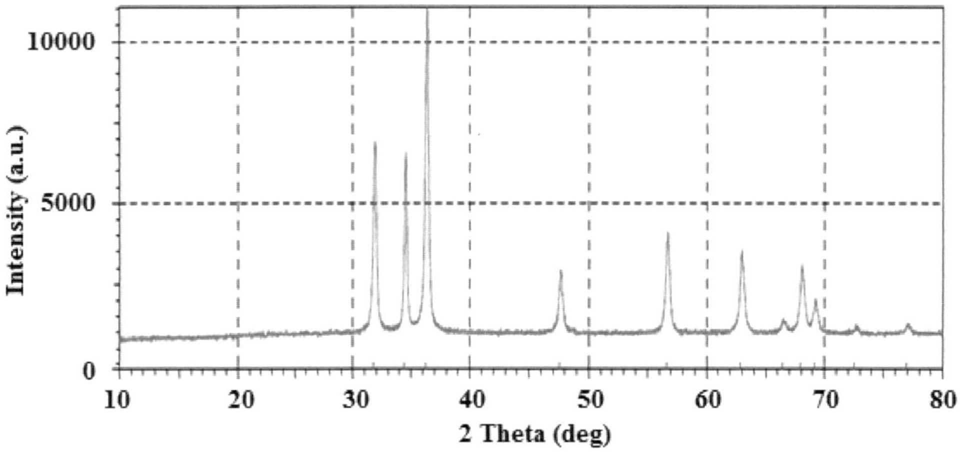

Figure 4. XRD patterns of ZnO nanocrystalline.

Table 1. Several optimizations in pyranopyrazole synthesis*

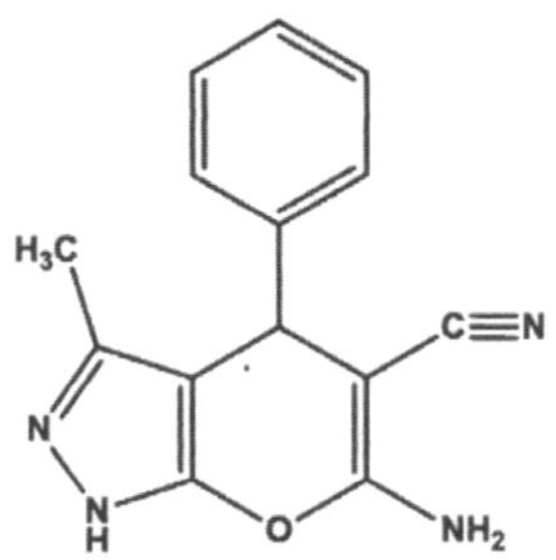

Temperature (°C)	Catalyst (% mol)	Time (min)	Yield (%)
30	10	90	98
70	10	90	88
100	10	90	73
70	7.5	90	65
70	5	90	45
70	2.5	90	41
70	10	30	67
70	10	60	75
70	10	120	86
70	10	180	85

*Reaction conditions: benzaldehyde (1 mmol), ethyl acetoacetate (1 mmol), hydrazine hydrate (1.5 mmol) and malononitrile (1mmol).

water. Table 1 shows that the amount of 10 mol% catalyst gives the highest yield, so this amount of catalyst is used as a reference for subsequent synthesis. The second observed variable is the effect of reaction time (30, 60, 90, 120 and 180 min) on the yield of product. The reaction was performed at 70°C with the use of 10 mol% catalyst and 10 mL water solvent. The results showed that the highest yield was obtained at 90 min after the reaction. The longer the reaction time, the yield of the product will be slowly decreased. The third variable is the effect of the reaction temperature (30, 70 and 100°C) to the yield of product. The reaction was conducted with the optimum amount of catalyst (10% mol), in 10 mL of ultrapure water for 90 min. The experimental results showed that the optimum yield (98%) was obtained at 30°C, when the temperature was increased to 70°C, the yield of product decreased to 88%. Increasing the temperature may cause the degradation of the desired product.

Stability of ZnO as a solid catalyst was determined experimentally under the optimum conditions. In each reaction cycle, the catalyst which has been separated from the reaction mixture was washed with ethanol, then dried at 50°C under vacuum to remove the residual solvent, and then reused for the next cycle. The catalyst can be reused up to five times without significant loss of catalytic activity. The result of recoverability showed that the catalyst can be well-recovered (figure 5). It is an advantage for the large scale and sustainable synthesis of pyranopyrazole using ZnO nanocrystalline.

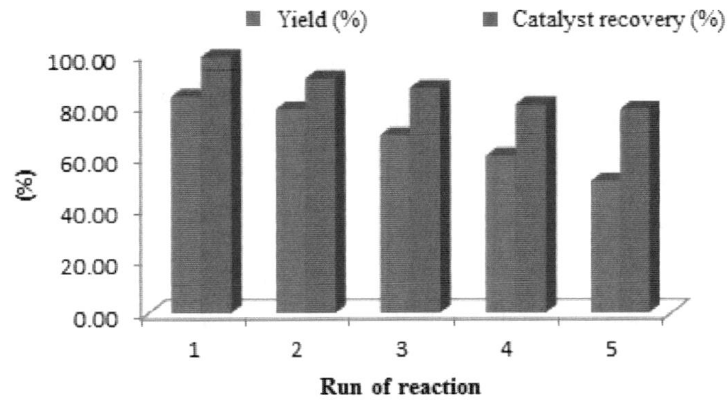

Figure 5. Activity and recoverability of ZnO catalyst.

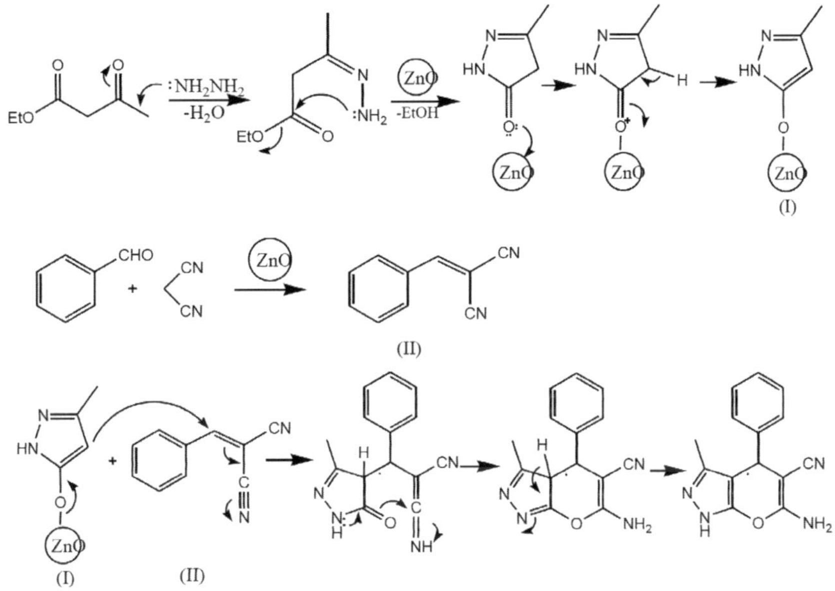

Figure 6. Proposed and plausible mechanism for pyranopyrazole synthesis.

The product was successfully characterized as 6-amino-3-methyl-4-phenyl-2,4-dihydropyrano [2,3-c] pyrazole-5-carbonitrile. Yellowish crystals; IR absorption FT-IR) υmax (cm^{-1}): 3372 and 3310 (NH_2), 2192 (C≡N), 1649 (C=C), 1276 (C-C, pyrazole), 1184 (C-O), and 1073 (N-N). ^{1}H-NMR (400 MHz, DMSO-d_6), δ H (ppm): 12.11 (*s*, 1H, -NH); 7.3-7.1(*m*, -ArH); 6.86 (*s*, 2H, -NH_2); 4.58 (*s*, 1H, -CH-); 1.06 (*s*, 1H, -CH_3). ^{13}C-NMR, δ C (ppm): 160.95; 154.21; 144.52; 135.71; 128.55; 127.56; 126.85; 120.93; 97.73; 57.19; 36.27; 9.82.

In the synthesis of pyranopyrazole, ZnO can serve as Lewis acidic and base sites (figure 6). The Lewis acid site of ZnO (Zn^{2+}) will receive a pair of electrons on oxygen atom of the carbonyl group of the pyrazolone, which is the result of the reaction of ethyl acetoacetate with hydrazine hydrate. Meanwhile, the Lewis base of ZnO (O^{2-}) will enable malononitrile resulting in deprotonated C-H bond to form an active methylene group. The existence of an active methylene group will initiate the Knoevenagel condensation reaction between benzaldehyde and malononitrile to form arylidene malononitrile. Furthermore, Michael addition reaction occurs between pyrazolone and arylidene malononitrile followed by cyclization and tautomerization to form pyranopyrazole.

4. Conclusions

ZnO was successfully synthesized using *Ulva fasciata* as capping agent. The optimum conditions of the pyranopyrazole synthesis was obtained in aqueous solvent, 10 mol% of catalyst with reaction time of 90 min at 70°C, and the name ofthe pyranopyrazole compound is 6-amino-3-methyl-4-phenyl-2,4-dihydropyrano [2,3-c] pyrazole-5-carbonitrile.

Acknowledgements

This research was fully supported by Universitas Indonesia through PITTA Grant Universitas Indonesia with contract No: 2035/UN2.R12/HKP.05.00/2016.

References

[1] Pal S L, Jana U, Manna P K, Mohanta G P and Manavalan R 2011 *J. Appl Pharm. Sci.* **1** 228-34
[2] El-rafie H M, El-rafie M H and Zahran M K 2009 *Carb. Polym.* **96** 403-10
[3] Dhanalakshmi P K, Azeez R, Rekha R, Poonkodi S and Nallamuthu T 2012 *Phykos* **42** 39-45
[4] Kayalvizhi K, Asmathunisha N, Subramanian V and Kathiresan K 2014 *J. Med. Plants Stud.* **2** 32-7
[5] Azizi S, Ahmad M B, Namvar F and Mohamad R 2014 *Mat. Lett.* **116** 275-7
[6] Kraan S 2012 *Algal Polysaccharides, Novel Applications and Outlook. In Carbohydrates-Comprehensive Studies on Glycobiology and Glycotechnology,* Chang C F ed. (Vienna: InTech) 459-532
[7] Pagore V P, Rupnar B D, Tekale S U and Pawar R P 2015 *Der Pharma Chemica* **7** 312-7
[8] Ebrahimi J, Mohammadi A, Pakjoo V, Bahramzade E and Habibi A 2012 *J. Chem. Sci.* **124** 1013-7
[9] Pawar P B, Jadhav S D, Patil B M, Shejwal R V and Patil S 2014 *Arch. Appl. Sci. Res.* **6** 150-8
[10] Vasuki G and Kumaravel K 2008 *Tetrahedron Lett.* **49** 5636
[11] Pandimurugan R and Thambidurai S 2014 *J. Appl. Polym. Sci.* **131** 40948
[12] Lahaye M and Robic A 2007 *Biomacromolecules* **8** 1765-74
[13] Siddhanta A K, Goswami A M, Ramavat B K, Mody K H and Mairh O P 2001 *Indian J. Marine Sci.* **30** 166-72
[14] Robic A, Sassi J F and Lahaye M 2008 *Carb. Polym.* **74** 344-52
[15] Robic A, Bertrand D, Sassi J F, Lerat Y and Lahaye M 2009 *J. Appl. Phycol.* **21** 451-6
[16] Peña-Rodríguez A, Mawhinney T P, Ricque-Marie D and Cruz-Suárez LE 2011 *Food Chem* **129** 491-8
[17] Mahdavi M, Namvar F, Ahmad M B and Mohamad R 2013 *Molecules* **18** 5954-64

Enhancement of photocatalytic activities of perovskite $LaFeO_3$ composite by incorporating nanographene platelets

N Afifah[1,2] and R Saleh[1,2]

[1]Department of Physics, Faculty of Mathematics and Natural Sciences Universitas Indonesia, Kampus UI Depok, Depok 16424, Indonesia
[2]Integrated Laboratory of Energy and Environment, Faculty of Mathematics and Natural Sciences Universitas Indonesia, Kampus UI Depok, Depok 16424, Indonesia

Corresponding author's e-mail: rosari.saleh@gmail.com

Abstract. $LaFeO_3$/NGP composites with several concentrations of NGP by weight percent were prepared using co-precipitation methods. The ferromagnetic behavior of $LaFeO_3$/NGP composites was characterized by vibrating sample magnetometer (VSM), while the structural properties were investigated using X-ray diffraction (XRD), Fourier transform infrared (FT-IR) spectroscopy, and Brunauer–Emmett–Teller (BET) surface area analysis. The $LaFeO_3$ nanoparticles with and without NGP show orthorhombic structure and ferromagnetic behavior. The samples were used to degrade methylene blue using visible light irradiation. The obtained results revealed that the photocatalytic performance of $LaFeO_3$/NGP composites increased with increasing NGP from 3 wt.% up to 5 wt.%. The main factors influencing the photocatalytic activity were studied and are discussed herein.

1. Introduction

The carbon-based material graphene has attracted much attention because of its large specific surface area, flexible structure, excellent mobility of charge carriers, and good electrical and thermal conductivity [1-3]. Graphene, with its unique properties, could serve as a support material and dispersion medium for catalysts, which could inhibit the aggregation of catalyst while it is dissolving into aqueous organic dye [4-5]. The photocatalytic activity of graphene as a supporting material has been reported by several studies. Li *et al.* (2013) [6] and Morales-Torres *et al.* (2013) [2] reported the photocatalytic degradation of organic pollutants in wastewater by P25-graphene composite. Their results showed that the photocatalytic activity of P25-graphene composite is better than that of pure P25. Xiang *et al.* (2012) [7] and Tu *et al.* (2009) [8] also reported the photocatalytic activity of a graphene/TiO_2 composite. The role of graphene in their composite is an electron transport channel, which could inhibit electron–hole recombination and enhance catalytic performance.

Graphene could support lanthanum ferrite ($LaFeO_3$) nanoparticles as well as TiO_2. $LaFeO_3$ is an ABO_3 perovskite type-oxide that is widely used as a catalyst due to its unique optoelectronic properties and narrow optical band gap, making it suitable for application as a visible light photocatalyst [9, 10]. The photocatalytic activity of $LaFeO_3$ nanoparticles was reported in our previous work [11]. However, in our previous studies, $LaFeO_3$ nanoparticles could not completely degrade organic pollutants in the photocatalytic process. Therefore, in this research we combined nanographene platelets (NGP) with $LaFeO_3$ in composite form to increase the photocatalytic performance. $LaFeO_3$/NGP with various weight percent (wt.%) of NGP were synthesized using the co-precipitation method. The photocatalytic activity of all samples was evaluated under visible light irradiation. Methylene blue (MB) was used as a model organic pollutant. The stability of the catalyst and the main active species playing an important role in the photocatalytic process were also studied.

2. Experimental details

Catalyst preparation: $LaCl_3 \cdot 7H_2O$, $FeCl_2 \cdot 4H_2O$, and sodium hydroxide (NaOH) were purchased from Merck (Kenilworth, NJ, USA). The chemical reagents were used to synthesis $LaFeO_3$ nanoparticle based on our previous work [11]. Nanographene platelets (NGP), purchased from Angstron Materials (Dayton, OH, USA), were then used to synthesis $LaFeO_3$/NGP composites using the co-precipitation method. The weight ratio of NGP to $LaFeO_3$ varied at 3, 5, and 10 wt.%. Briefly, NGP was dissolved into 80 mL distilled water and 40 mL ethanol through ultrasonic treatment for 2 h. Then, $LaFeO_3$ nanoparticles were added to the NGP solution, followed by stirring for 2 h. The suspension was then centrifuged and dried at 70°C for 12 h.

Characterization: The samples were characterized by X-ray diffraction (XRD), Fourier transform infrared (FT-IR), vibrating sample magnetometer (VSM), and Brunauer–Emmett–Teller (BET) surface area analysis spectroscopy.

Photocatalytic activity: The photocatalytic activity was determined by observing the degradation of MB under visible light irradiation. First, 0.4 g/L catalyst was dissolved into 100 mL of 20 mL/g MB. The resulting solution was stirred in the dark for 30 min to allow it to reach absorption–desorption equilibrium. Then, the suspension was irradiated using a 40 W Xe lamp as the visible light source for 2 h. The degradation of MB was monitored using a Hitachi (Tokyo, Japan) UH5300 UV-visible spectrophotometer. The degradation of MB was plotted as C_t/C_0, where C_t is the concentration of MB at each measurement interval during irradiation and C_0 is the initial concentration of MB after reaching adsorption–desorption equilibrium. The stability of the catalyst was also checked in this research by reusing the catalyst for four times cycle processes. The main active species involved in the catalytic degradation of MB were studied using sodium sulfate, di-ammonium oxalate and tert-butyl alcohol, which are recognized as scavengers for electrons, holes, and hydroxyl radical species, respectively.

3. Results and discussion

The structural properties of all samples were characterized using XRD. Figure 1(a) shows the XRD pattern of the $LaFeO_3$/NGP composites with different wt.% NGP. The XRD pattern of $LaFeO_3$ nanoparticles is also shown as a comparison. The diffraction peak at 2θ of 26.4° confirmed the graphite structure of NGP. The obtained diffraction peak of NGP increased with increasing wt.% NGP. The orthorhombic structure of $LaFeO_3$ could also be found in the XRD pattern of all composites. No other phases or impurities were found, indicating that the desired synthesized sample was obtained. Along with XRD measurements, the crystal size value could also be calculated using Scherrer's formula [12]. The crystal size and lattice parameters of all samples are summarized in table 1.

The incorporation of NGP into the composite was confirmed by FT-IR measurements. Figure 1(b) shows the FT-IR spectra of the $LaFeO_3$/NGP composites with various wt.% NGP. The FT-IR spectrum of pure NGP is also shown for comparison. The strong absorption at wavenumber 555 cm^{-1} is the Fe–O stretching vibration of octahedral FeO_6, which was characteristic of the material forming the perovskite $LaFeO_3$ [12-14], while the C=O stretching vibration at 1460-1750 cm^{-1} confirmed the presence of NGP in the sample [15-16]. The O–H stretching vibration was also found at wavenumber 3440 cm^{-1} [16].

Magnetic saturation of the $LaFeO_3$ nanoparticles and $LaFeO_3$/NGP composites with various wt.% NGP can also be seen in figure 1(c). All samples showed ferromagnetic behavior at room temperature. The magnetic saturation of the composite increased as the wt.% of NGP increased up to 5 wt.%. The magnetic saturation of $LaFeO_3$/NGP composites is summarized in table 1.

The photocatalytic activity of the $LaFeO_3$/NGP composites with various wt.% NGP is shown in figure 2(a). In comparison, the photocatalytic activity of $LaFeO_3$ nanoparticles is also shown. The results showed that the degradation of MB increased with the addition of NGP up to 5 wt.% and then decreased with the addition of 10 wt.% NGP. The degradation rate of MB was also calculated using the pseudo-first-order kinetic rate equation [17]. The obtained result (shown in figure 2(b)) revealed that $LaFeO_3$

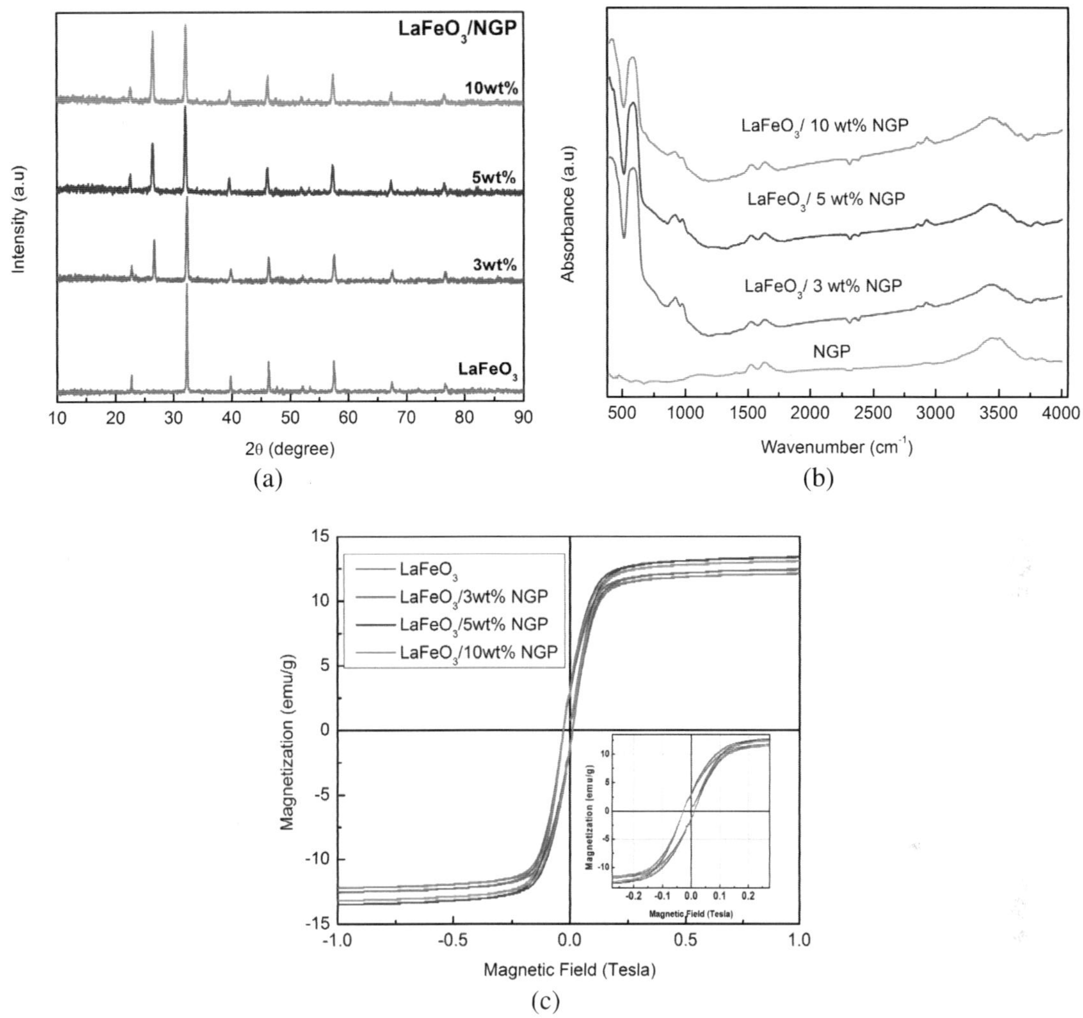

Figure 1. (a) XRD pattern; (b) FTIR spectra; and (c) VSM spectra of $LaFeO_3$/NGP composites with different wt.% NGP.

Table 1. The lattice parameter, grain size <D>, magnetic saturation (M-S) and surface area of $LaFeO_3$/NGP composites.

Sample	Lattice Parameter			<D> (nm)	M-S (emu/g)	Surface area (eV)
	a (Å)	b (Å)	c (Å)			
$LaFeO_3$	5.5436	7.8456	5.5528	58	12.08	-
$LaFeO_3$/3 wt.% NGP	5.5326	7.8376	5.5592	28	12.41	43
$LaFeO_3$/5 wt.% NGP	5.5632	7.8687	5.5727	27	13.38	76
$LaFeO_3$/10 wt.% NGP	5.5703	7.8782	5.5743	28	13.06	90

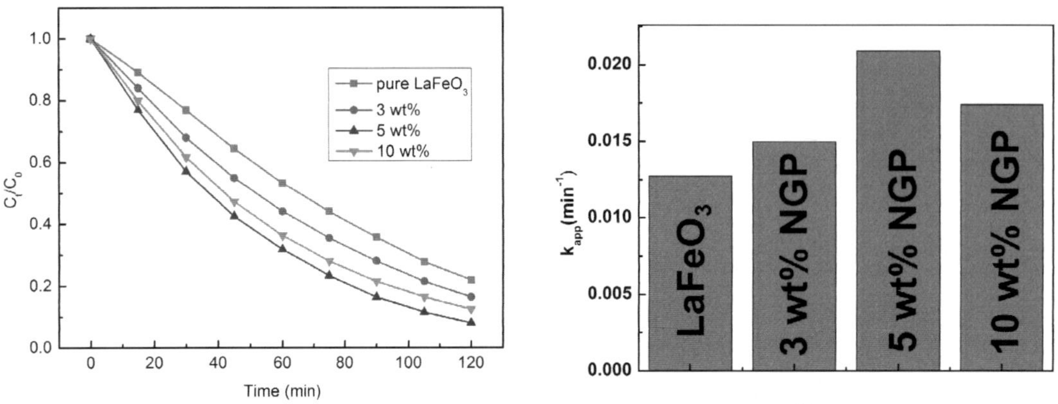

(a) (b)

Figure 2. (a) The photocatalytic activity of $LaFeO_3$/NGP composites with various wt.% NGP; and (b) The degradation rate of MB using $LaFeO_3$/NGP composites with various wt.% NGP.

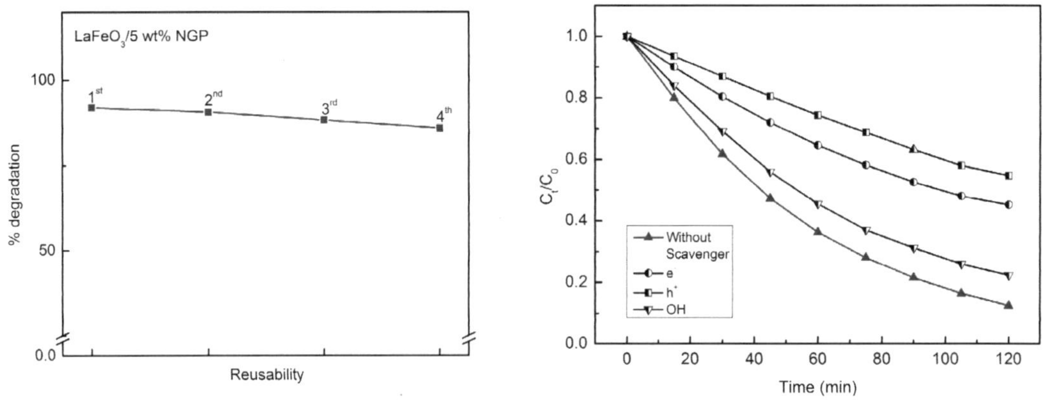

(a) (b)

Figure 3. (a) The stability of catalyst using $LaFeO_3$/5 wt.% NGP; and (b) The effect of scavengers on photocatalytic degradation of MB using $LaFeO_3$/5 wt.% NGP.

with added NGP shows more rapid degradation of MB than $LaFeO_3$ without NGP. It has been suggested that the photocatalytic efficiency depends on several factors, one of which is surface area. A large surface area may increase the opportunity for contact between the catalyst and light, thereby increasing the rate at which the oxidation process happens [18]. The surface area of the $LaFeO_3$/NGP composites is summarized in table 1. The results show that increasing the NGP content increased the surface area of the catalyst. However, the highest photocatalytic activity was observed with 5 wt.% added NGP. This indicates that the photocatalytic degradation efficiency of the composite was more influenced by the capacity of the electron transport layer of NGP, which could inhibit electron–hole recombination and enhance photocatalytic performance [18-20].

After determining the catalyst with the maximum degradation rate, the stability of the catalyst was also observed. MB solutions were degraded using the same catalyst four times under the same treatment conditions. The results of reusing the $LaFeO_3$/NGP composite (5 wt.%) as a catalyst can be seen in figure 3(a). After four repetitions, the catalyst did not show a significant decrease in activity, indicating that the catalyst has good stability.

To study the main active species in the process of MB degradation using the $LaFeO_3$/NGP composite (5 wt.%), the experiment was repeated with the addition of active species like electrons, holes, and hydroxyl radicals. $Na_2S_2O_8$ and di-ammonium oxalate were used as electron and hole scavengers, while tert-butyl alcohol was used as a hydroxyl radical scavenger. The influence of these scavengers on the photocatalytic activity of $LaFeO_3$/NGP composites (5 wt.%) is shown in figure 3(b). The presence of scavengers decreased the degradation rate of MB; the largest decrease was obtained with di-ammonium oxalate. This result indicates that the holes pay the most important role in the photocatalytic activity.

4. Conclusions

The photocatalytic activity of $LaFeO_3$/NGP composites with various wt.% NGP has been investigated under visible light irradiation. The composite with incorporated NGP shows better photocatalytic performance than the $LaFeO_3$ nanoparticles. The $LaFeO_3$/NGP composites also showed stability as they were able to degrade MB through four photocatalytic cycles without a decrease in efficiency. The role of NGP in composites could inhibit electron–hole recombination and enhance catalytic performance.

References

[1] Aleksandrzak M, Adamski P, Kulkulka W, Zielinska B and Mijowska E 2015 *Appl. Surf. Sci.* **331** 193-9
[2] Morales-Torres S, Pastrana-Martínez L M, Figueiredo J L, Faria J L and Silva A M 2013 *Appl. Surf. Sci.* **275** 361-8
[3] Machado B F and Serp P 2012 *Catal. Sci. Technol.* **2** 54-75
[4] Zhou K, Zhu Y, Yang X, Jiang X and Li C 2011 *New J. Chem.* **35** 353-9
[5] Liang D, Cui C, Hu H, Wang Y, Xu S, Ying B, Li P, Lu B and Shen H 2014 *J. Alloy. Compd.* **582** 236-40
[6] Li J, Zhou S, Hong G and Chang C 2013 *Chem. Eng. J.* **219** 486-91
[7] Xiang Q, Yu J and Jaroniec M 2012 *Chem. Soc. Rev.* **41** 782-96
[8] Tu W, Zhou Y, Liu Q, Yan S, Bao S, Wang X, Xiao M and Zou Z 2009 *Adv. Funct. Mater.* **19** 894-904
[9] Grabowska E 2016 *Appl. Catal. B: Environ.* **186** 97-126
[10] Tang P, Tong Y, Chen H, Cao F and Pan G 2013 *Curr. Appl. Phys.* **13** 340-3
[11] Afifah N and Saleh R 2016 *J. Phys. Conf. Ser.* **710** 012030
[12] Abazari R and Sanati S 2013 *Superlattice. Microst.* **64** 148-57
[13] Noroozifar M, Khorasani-Motlagh M, Ekrami-Kakhki M S and Khaleghian-Moghadam R 2014 *J. Power Sources* **248** 130-9
[14] Kaiwen Z, Xuehang W, Wenwei W, Jun X, Siqi T and Sen L 2013 *Adv. Powder Technol.* **24** 359-63
[15] Fujii T, Matsusue I, Nakatsuka D, Nakanishi M and Takada M 2015 *Mater. Chem. Phys.* **129** 805-9
[16] Taufik A and Saleh R 2016 *AIP Conf. Proc.* **1725** 020089
[17] Saleh R and Djaja N F 2014 *Superlattice. Microst.* **74** 217-33
[18] Kim T W, Park M, Kim H Y and Park S J 2016 *J. Solid State Chem.* **239** 91-8
[19] Hu J, Men J, Ma J and Huang H 2014 *J. Rare Earth.* **32** 1126-34
[20] Hu J, Ma J, Wang L and Huang H 2014 *J. Alloy. Compd.* **583** 539-45

IOP Conf. Series: Materials Science and Engineering **188** (2017) 012055 doi:10.1088/1757-899X/188/1/012055

Theoretical exploration of optical response of Fe_3O_4-reduced graphene oxide nanoparticle system within dynamical mean-field theory

M A Majidi[1], R Kusumaatmadja[1], A D Fauzi[1], W Y Phan[1], A Taufik[1], R Saleh[1] and A. Rusydi[2,3]

[1]Department of Physics, Faculty of Mathematics and Natural Sciences Universitas Indonesia, Kampus UI Depok, Depok 16424, Indonesia

[2]NUSNNI-NanoCore, Department of Physics, Faculty of Science, National University of Singapore, Singapore 117576, Singapore

[3]Singapore Synchrotron Light Source, National University of Singapore, Singapore 117603, Singapore

Corresponding author's email: aziz.majidi@sci.ui.ac.id

Abstract. We theoretically investigate the optical conductivity and its related optical response of Fe_3O_4-reduced graphene oxide (rGO) nanoparticle system. Experimental data of magnetization of the Fe_3O_4-rGO nanoparticle system have shown that the saturation magnetization can be enhanced by controlling the rGO content with the maximum enhancement reached at the optimal rGO content of about 5 weight percentage. We hypothesize that the magnetization enhancement is due to spin-flipping of Fe ions at tetrahedral sites induced by oxygen vacancies at the Fe_3O_4 nanoparticle boundaries. These oxygen vacancies are formed due to adsorption of oxygen atoms by rGO flakes around the Fe_3O_4 nanoparticle. In this study, we aim to explore the implications of this effect to the optical response of the system as a function of the rGO content. Our model incorporates Hubbard-repulsive interactions between electrons occupying the e g orbitals of Fe^{3+} and Heisenberg-like interactions between electron spins and spins of Fe^{3+} ions. We treat the relevant interactions within mean-field and dynamical mean-field approximations. Our results are to be compared with the existing experimental reflectance data of Fe_3O_4 nanoparticle system.

1. Introduction

Fe_3O_4, also known as magnetite, has long been investigated most notably for its unique magnetic characteristics and the related potential applications [1]. One of the research directions being pursued by many researchers is to enhance the saturation magnetization. Magnetization of bulk Fe_3O_4 is known to be limited due to its ferrimagnetic structure [2, 3]. Namely, the face-centered-cubic inverse spinel crystal structure of Fe_3O_4 makes two Fe ions at the octahedrally coordinated sites in each formula unit favor to align in the same direction while the remaining Fe ion at the tetrahedrally coordinated site favors to align oppositely to the other two. Recently, a magnetization measurement on a mixture of Fe_3O_4 nanoparticles and reduced graphene oxyde (rGO) flakes has shown that the mixture can have saturation magnetization higher than that without the presence of rGO, and it reaches a maximum magnetization enhancement at 5 wt% rGO content [4]. In another theoretical study we propose a hypothesis that the rGO flakes located nearby each Fe_3O_4 nanoparticle cluster adsorb some oxygen atoms from the surface of the Fe_3O_4 nanoparticle cluster during the synthesis of the sample, causing formation of oxygen vacancies on the surface of the cluster. According to a previous experimental study [5], which is suported by a theoretical study following it [6], oxygen vacancies in nano-Fe_3O_4 material can induce spin flippings of the Fe ions located at the tetrahedrally coordinated sites around the oxygen vacancies, which in turn enhances the magnetization of the nano-Fe_3O_4 material. Meanwhile, graphene is a thin layer of carbon in the form of two-dimensional honeycomb lattice. Researchers found that setting adatoms on graphene has been a promising breakthrough because it could tailor the electronic properties of graphene, such as transport, optical and magnetic properties [7]. The hypothesis that rGO tends to adsorb oxygens is motivated by an experimental study on graphene exposed with oxygen plasma [8], and a previous theoretical study on graphene with adatoms [9].

Here, we propose to explore implications of the magnetization enhancement of the Fe_3O_4-rGO nanoparticle system on the optical response of the system. We model the Fe_3O_4 system without and with oxygen vacancies as proposed in reference [6]. While, for the rGO, we use a model for graphene with adatoms as proposed in reference [9]. In addition, we also formulate how the total optical conductivity evolves a function of rGO content. Our calculation result of the optical response of pure Fe_3O_4 nanoparticle system surprisingly shows a very good agreement with the available experimental data. While, the calculation results for the mixture of Fe_3O_4 and rGO nanoparticles need to be verified with more experimental data.

2. Model and method

We use the model Hamiltonian proposed in reference [6] to describe the systems of Fe_3O_4 without and with oxygen vacancies, and the model of graphene with oxygen adatoms proposed in Ref. [9] to represent an rGO layer in an rGO flake. Here we assume that as soon as the nanoparticles are distributed in the sample at the end of the syntesis process there occurs oxygen adsorption from the surfaces of the Fe_3O_4 nanoparticle cluster by the rGO flakes located nearby the Fe_3O_4 nanoparticle cluster. This leads to formation of oxygen vacancies on the surfaces of the Fe_3O_4 nanoparticle clusters. It is then assumed that every unit cell of Fe_3O_4 on the surface of the Fe_3O_4 nanoparticle cluster carrying an oxygen vacancy undergoes a spin flipping that causes its net magnetic moment in the unit cell to enhance by a factor of 3 compared to that of the unit cell that does not have an oxygen vacancy. The overall system is then assumed to consist of the inner part of Fe_3O_4 nanoparticle clusters that have no oxygen vacancies, the surface part of Fe_3O_4 nanoparticle clusters that have oxygen vacancies, and the rGO flakes represented by graphene layers with oxygen adatoms. Figure 1 illustrates the model of the Fe_3O_4-rGO nanoparticle system we propose.

To explore the optical response of the system, we first calculate the retarded Green functions of each component of the system within the dynamical mean-field theory [10, 11], then using the obtained Green functions we calculate the corresponding optical conductivity. The optical conductivity of Fe_3O_4 nanoparticle clusters without oxygen vacancies, denoted as $\sigma^{O}_{1,F}(\omega)$, and with oxygen vacancies, denoted as $\sigma^{V}_{1,F}(\omega)$, are derived using the Kubo formula as implemented in reference [12], based on the model presented in reference [6]. While, the optical conductivity of the rGO, $\sigma_{1,G}(\omega)$, is represented by that of graphene with adatoms as presented in reference [6]. Note that the symbol σ_1 denotes the real part of the complex optical conductivity.

Based on the consideration explained above, we propose that the total optical conductivity of the Fe_3O_4-rGO nanoparticle system is an additive combination of all the components, that is

$$\sigma_1(x,\omega) = P_0(x)\sigma^{O}_{1,F}(\omega) + (1-P_0(x))\left[(1-\gamma)\sigma^{O}_{1,F}(\omega) + \gamma\sigma^{V}_{1,F}\right] + 2\beta(x)\sigma_{1,G}(\omega). \quad (1)$$

In equation (1), several variables are introduced as the following: x is the measure of weight percentage of rGO in the nanoparticle system; $P_0(x)$ is the probability of finding a space between two neighboring Fe_3O_4 nanoparticle clusters with no rGO flake at all; γ is the portion of the unit cells that have oxygen

Figure 1. Cartoon of our model for Fe_3O_4-rGO nanoparticle system. Blue chunks and black slabs represent Fe_3O_4 nanoparticle cluster and rGO flake, respectively. Left: Fe_3O_4 nanoparticle system without rGO flakes. Middle: Fe_3O_4 nanoparticle system with some rGO flakes. Right: Illustration of oxygen adsorption from Fe_3O_4 nanoparticle clusters by rGO flakes at the end of the synthesis of the sample.

vacancies in an Fe_3O_4 nanoparticle cluster; $\beta(x)$, which is a function of x, is the ratio of the total number of existing rGO layers to the total number of "rGO slots" in all spaces between two neighboring Fe_3O_4 nanoparticle clusters in the sample. Statistically, $P_0(x)$ depends on $\beta(x)$ in the following way

$$P_0(x) = (1-\beta(x))^n, \tag{2}$$

where n is the maximum number of rGO layers filling the space between two neighboring Fe_3O_4 nanoparticle clusters above which no emply spaces between two neighboring Fe_3O_4 nanoparticle clusters can be found. $\beta(x)$ itself depends on x in the following manner

$$\beta(x) = \frac{\alpha x}{2(1-x)}, \tag{3}$$

where α is the ratio between mass of an Fe_3O_4 nanoparticle cluster and mass of an rGO layer.

As we aim to compare our results with experimental data available in the form of reflectance, we need to convert our formula of optical conductivity to reflectance. This can be done as the following. First, having the real part of the complex optical conductivity, $\sigma_1(\omega)$ for the whole spectrum of photon energy, we can compute the corresponding imaginary part, $\sigma_2(\sigma)$, through the Kramers-Kronig relation

$$\sigma_2(\omega) = \frac{2\omega}{\pi}\int_0^\infty d\omega' \frac{\sigma_1(\omega')}{\omega'^2-\omega^2}. \tag{4}$$

Knowing the complex optical conductivity, $\sigma(\omega) = \sigma_1(\omega) + i\sigma_2(\omega)$, we can calculate the complex dielectric function through the relation

$$\varepsilon(\omega) = \frac{i}{\varepsilon_0\omega}\left(\sigma_1(\omega) + i\sigma_2(\omega)\right). \tag{5}$$

The complex dielectric function itself is related to the complex refractive index through

$$n(\omega) + i\kappa(\omega) = \sqrt{\varepsilon(\omega)}. \tag{6}$$

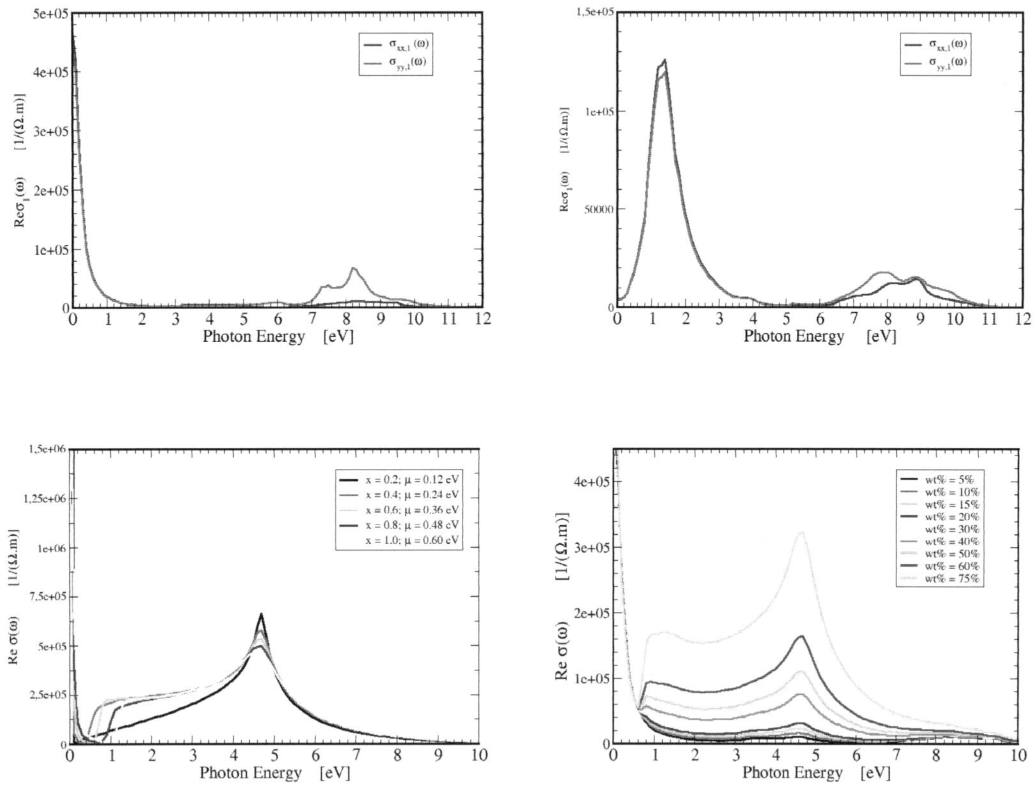

Figure 2. Calculation results of optical conductivity at a room temperature. Top-left: Fe_3O_4 without oxygen vacancies, top-right: Fe_3O_4 with oxygen vacancies; bottom-left: rGO wth various adatom concentrations and positions chemical potential; bottom-right: Fe_3O_4-rGO system at various wt% of rGO.

Finally, the reflectance, $R(\omega)$, is defined as

$$R(\omega) = |r(\omega)|^2 = \left|\frac{n(\omega)+i\kappa(\omega)-1}{n(\omega)+i\kappa(\omega)+1}\right|^2 = \left|\frac{\sqrt{\varepsilon(\omega)}-1}{\sqrt{\varepsilon(\omega)}+1}\right|^2. \tag{7}$$

3. Results

Comparing between tho top-left and the top-right of figure 2, especially in the low energy region, our calculation results of optical conductivity at a room temperature show that Fe_3O_4 without oxygen vacancies has a Drude peak indicating the metallic (more precisely half-metallic) character, whereas Fe_3O_4 with oxygen vacancies is an insulator or semiconductor. Meanwhile, the bottom-left figure shows the optical conductivity profile of graphene with adatoms at various values of adatom concentration (x) and chemical potential (μ). The bottom-right figure shows our prediction of optical conductivity profile of the Fe_3O_4-rGO nanoparticle system for various wt% values of rGO.

The left panel of figure 2 shows the comparison between the calculated reflectance of Fe_3O_4 system without oxygen vacancies and the experimental data of Fe_3O_4 nanoparticle system without rGO [4].

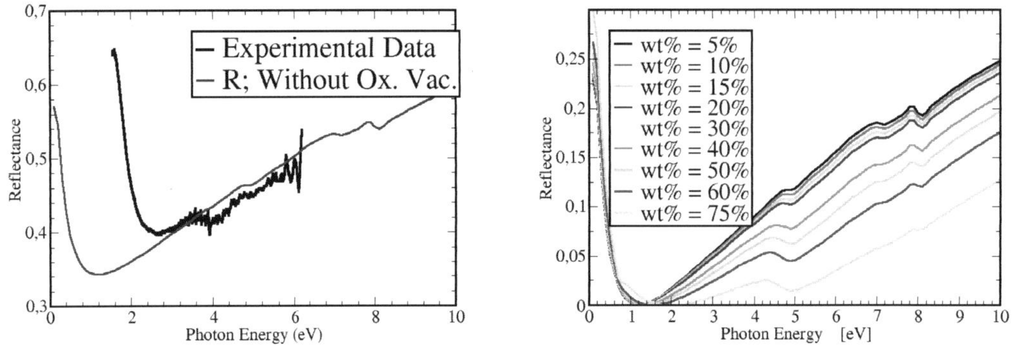

Figure 3. Left: Calculated reflectance of Fe_3O_4 without oxygen vacancies and the corresponding experimental data Fe_3O_4 nanoparticle system for comparison. Right: Calculated reflectance of Fe_3O_4-rGO nanoparticle system at various wt% values of rGO.

There we can see a very good similarity of the trend of the two curves. In particular, both curves agree very well even quantitatively in the region of 2.5-6 eV. We argue that this is a strong indication of the legitimacy and adequacy of our model. One little discrepancy appears in the width of the zero-frequency peak, for which our calculation result gives a too small width compared to that of the experimental data. However, this is not a very crucial discrepancy as we believe a further fine-tuning of our model parameters could remedy this problem. Based on the consideration just discussed, on the right panel of figure 2 we present our prediction of the reflectance of the Fe_3O_4-rGO nanoparticle system for various wt% values of rGO, that we hope to be able to compare with experimental data in the near future.

4. Conclusions

In conclusion, we have developed a model to predict the optical response of the Fe_3O_4-rGO nanoparticle system for various wt% of rGO, as an implication of the phenomenon of magnetization enhancement occuring in the system upon the addition of rGO flakes into the Fe_3O_4 nanoparticle system. The very good qualitative agreement of the calculated reflectance curve of the Fe_3O_4 system without oxygen vacancies may strongly indicate that the model of Fe_3O_4 system is adequate as a starting point, and hence the calculation results for the optical conductivity or the reflectance of the Fe_3O_4-rGO nanoparticle system are worthy of experimental verification.

Acknowledgement

We gratefully acknowledge the full funding support provided by Universitas Indonesia, Indonesia, for the completion of this project through PITTA Research Grant No. 2022/UN2.R12/HKP.05.00/2016.

References

[1] Gutfleisch O, Willard M A, Brück E, Chen C H, Sankar S G and Liu J P 2010 *Adv. Mater.* **23** 821–42
[2] Néel L 1948 *Ann. Phys.* **3** 137
[3] Smit J and Wijn H P J 1959 *Ferrites* (Eindhoven: Philips' Technical Library)
[4] Taufik A and Saleh R 2016 *AIP Conf. Proc.* **1725** 020089
[5] Herng T S *et al.* 2015 *Nano Res.* **8(9)** 2935–45
[6] Fauzi A D, Majidi M A and Rusydi A 2017 *J. Phys.: Condens. Matter* **29** 135802
[7] Neto A H, Kotov V N, Nilsson J, Pereira V M, Peres N M R and Uchoa B 2009 *Sol. State Comm.* **149** 1004

[8] Santoso I, Singh R S, Gogoi P K, Asmara T C, Wei D, Chen W, Wee A, Pereira V M and Rusydi A 2014 *Phys. Rev.* B **89** 0752134
[9] Majidi M A, Siregar S and Rusydi A 2014 *Phys. Rev.* B **90** 195442
[10] Georges A, Kotliar G, Krauth W and Rozenberg M 1996 *Rev. Mod. Phys* **68**
[11] N. Furukawa, J. Phys. Soc. Jpn. 63, 3214 (1994); in Proceedings of the International Conference on Physics of Manganites, Michigan State University, 1998, edited by Kaplan T A and Mahanti S D (Kluwer Academic/Plenum, New York, 1998), p. 1.
[12] Majidi M A, Su H, Feng Y P, Rübhausen M and Rusydi A 2011 *Phys. Rev.* B **84** 075136

Enhancement of the stability of silver nanoparticles synthesized using aqueous extract of *Diospyros discolor* Willd. leaves using polyvinyl alcohol

H K Ardani[1], C Imawan[1], W Handayani[2], D Djuhana[1], A Harmoko[1] and V Fauzia[1]

[1]Department of Physics, Faculty of Mathematics and Natural Sciences Universitas Indonesia, Kampus UI Depok, Depok 16424, Indonesia
[2]Department of Biology, Faculty of Mathematics and Natural Sciences Universitas Indonesia, Kampus UI Depok, Depok 16424, Indonesia

Corresponding author's e-mail: cuk.imawan@sci.ui.ac.id

Abstract. Biosynthesis of silver nanoparticles is recently attracting considerable attention because of it reduces the environmental impact and already used in numerous applications. However, the disadvantages such as easy aggregation and instability properties, prevent its' application. In this papers, biosynthesis of silver nanoparticles using aqueous extract of *Diospyros discolor* Willd. leaves have been prepared. The effect of biosynthesis variables, like ratio of reactants and reduction time on the particle size distribution, stability, and morphology of the silver nanoparticles were investigated. The resulted silver nanoparticles were characterized using UV spectroscopy, Transmission Electron Microscopy, and Particles Size Analyzer. Polyvinyl alcohol (PVA) was used to enhance the stability of the silver nanoparticles. Silver nanoparticles modification with 1% PVA concentration has produced a better characteristic of particle size distribution compared to the original silver nanoparticles, from highly polydisperse into moderately disperse. The results of the Zetta potential measurement also confirmed the increase stability of cluster distribution in the colloidal Ag/PVA compared to the original Ag.

1. Introduction

The nanoparticles research have given a lot of contribution in material engineering, which opening the opportunities for its' usage and applications in various fields, such as medical [1], pharmaceutical [2, 3], environmental [4, 5], and sensors [6, 7]. From a wide range of the applied nanoparticle materials, gold and silver nanoparticles are the most renowned [8, 9].

Nanoparticles have physical and chemical properties which are different from its bulk material. This uniqueness in character grants nanoparticles a lot of remarkable properties that it is used for various applications. On the other hand, nanoparticles are usually unstable since it easily agglomerates, forming particles which are bigger in size. Additionally, creating nanoparticles with homogeneous particle size distribution is a challenging task [10].

In this research, silver nanoparticles (silver NPs) biosynthesis using aqueous extract of *Diospyros discolor* Willd. leaves have been performed. Biosynthesis method is chosen as it is environmentally friendly [11]. The stable character of silver nanoparticles is aimed to be fixed by modifying it with PVA. Moreover, PVA influence on the resulted nanoparticles polydisperse characteristic improvement is also observed.

2. Materials and methods

Leaf extract of *D. discolor* was prepared by using the raw leaf material, previously dried in the oven at the temperature of 40°C for 24 hours. The leaf was then ground into powder, weighed 2 gram, and subsequently boiled for 10 minutes with 100 mL of aquabidest. The extract was then filtered with Whatman filter paper no 1.

$AgNO_3$ solution with the concentration of 1 mM was then prepared from $AgNO_3$ powder, obtained from Dhucefa Biochemies. Next, in order to synthesize nanoparticles, 10 mL of *D. discolor* leaf extract was mixed with 100 mL $AgNO_3$ solution and stirred using magnetic stirrer for 2 hours at the temperature of 24°C.

The influence of Polyvinyl Alcohol (PVA) modificator on the resulted silver nanoparticles was studied by adding PVA solvent in different volume ratio to the mixture of leaf extract and $AgNO_3$ solution. PVA solution was prepared by dissolving PVA powder into aquabidest. PVA solution concentration was then varied to 1%, 2%, 3%, 4%, and 5%. This solution was then heated at the temperature of 80°C for 5 hours. In order to synthesize PVA-modified silver nanoparticles, the ratio of leaf extract volume, $AgNO_3$, and PVA solution of 1:10:3 was applied, respectively. The mixture was then stirred using magnetic stirrer for 2 hours at the temperature of 24°C.

Characterization of the resulted silver nanoparticles was performed using UV-Vis spectrometer Thermo UV-vis 10 S Genesys, transmission electron microscopy TEM FEI Tecnai G2 Supertwin, and particle size analyzer (PSA) Delso Nano Beckman Coulter. The reduction of Ag^+ ion into silver was monitored by observing the color change and UV-visible spectrum of the solution orderly in the period of 2, 6, 24, 30, 36, 44 hours. Meanwhile, the PVA-modified silver nanoparticles formation was observed five times at 24, 30, 36, 42, 48 hours reaction time.

The granule size of the formed nanoparticles was then measured using PSA and TEM. This measurement was performed for original and PVA-modified silver nanoparticles samples. The achieved stability of the produced nanoparticles was observed by measuring Zeta potential, while silver nanoparticles molecular weight distribution was studied by calculating the Polydipersity Index.

3. Results and discussion

Figure 1 shows the photograph of the solutions' color and absorption spectrum result achieved from UV-visible spectroscopy measurement of silver nanoparticles biosynthesis by using aqueous extract of *D. discolor* leaves. The color of leaf extract and $AgNO_3$ solution changed after being mixed. The color of the solvent mixture becomes yellow, and the yellow color becomes darker until it turns tawny over the mixing process period. These changes of colors are in line with the absorption spectrum change from the result of UV-visible measurement in figure 2. The spectrum of AgNO3 and leaf extract solution does not have a peak of absorption around the wavelength of 400 nm. After the two solutions are mixed, an absorption peak appears in the wavelength of approximately 418 nm. As known, silver nanoparticles has value of surface plasmon resonance (SPR) around the wavelength of 400 nm to 500 nm [12]. A similar absorption peak in these areas has also been achieved by Nagarajan *et al.* [13]. From figure 1, it can also be observed that silver nanoparticles nucleation is influenced by time of reaction. The longer reaction time, the more quantity of silver nanoparticles are formed, as seen in the parallel condition between the gradual increase of absorbance value in UV-visible spectrum and the amount of reaction time passed in figure 2.

The result of UV-visible spectrum and PVA modification of silver nanoparticles is given in figure 3. Such observation was done for the reaction time of 24 h. The wavelength of absorption peaks and PVA-modified silver nanoparticles was not shifted; however, the absorbance value at the peak was changed. The change in absorbance value from the observed absorption peak was increased along with more addition of PVA concentration. This result proves that after nanoparticles are modified with PVA, there is no change in value of the surface plasmon resonance.

TEM image from the original and PVA-modified silver nanoparticles are displayed in figure 4. Both types of nanoparticles have spherical basic form. The particle size of the original silver nanoparticles is around 5 nm, while the size of silver nanoparticles modified with 1% PVA is approximately 20 nm. The electron diffraction measurement shows that the achieved silver nanoparticles possess good crystallinity.

IOP Conf. Series: Materials Science and Engineering **188** (2017) 012056 doi:10.1088/1757-899X/188/1/012056

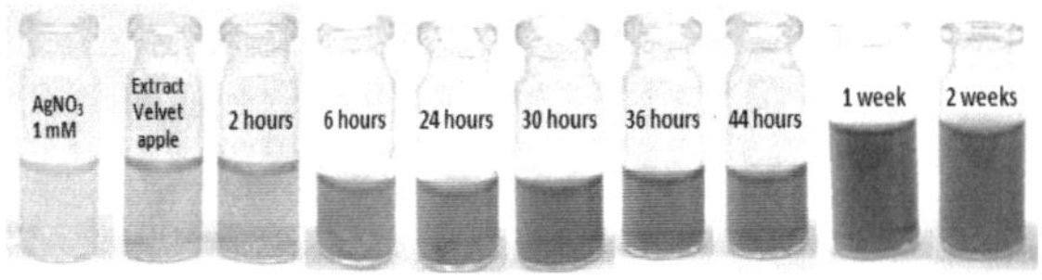

Figure 1. The difference of color between $AgNO_3$ and leaf extract solution. After reaction occurs, solution color changes over the time, from bright yellow to dark brown.

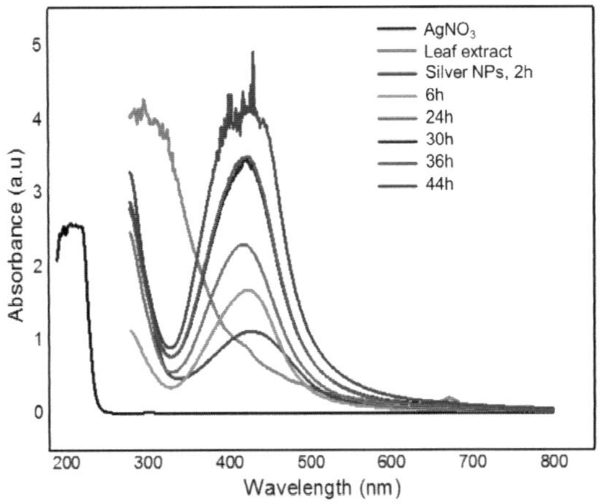

Figure 2. UV-visible recorded spectrum as the function of silver ion reduction time.

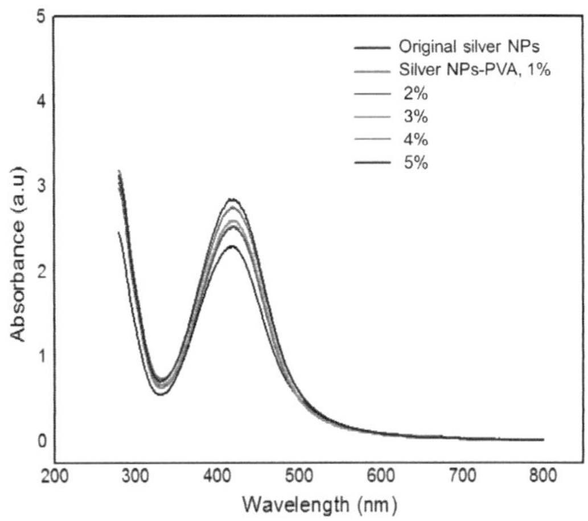

Figure 3. UV-visible Spectrum from silver nanoparticles modified with PVA which is recorded as the function of PVA volume ratio change. The observation was done for the reaction time of 24 h.

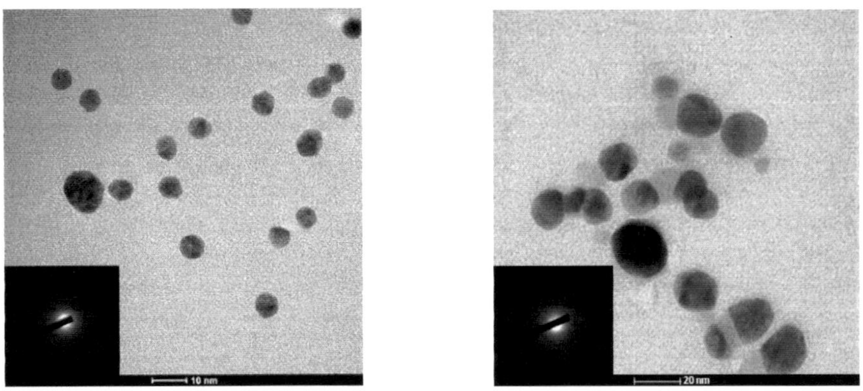

(a) (b)

Figure 4. TEM imaging for silver nanoparticles (a) original and (b) PVA(1%)-modified.

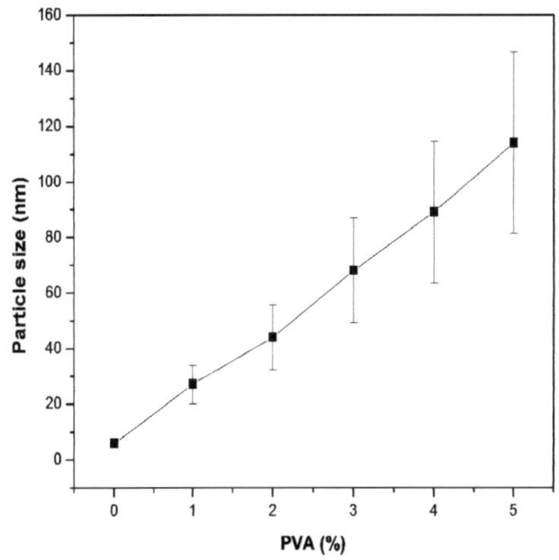

Figure 5. The measured particle size resulted from PSA for original and PVA-modified silver nanoparticles.

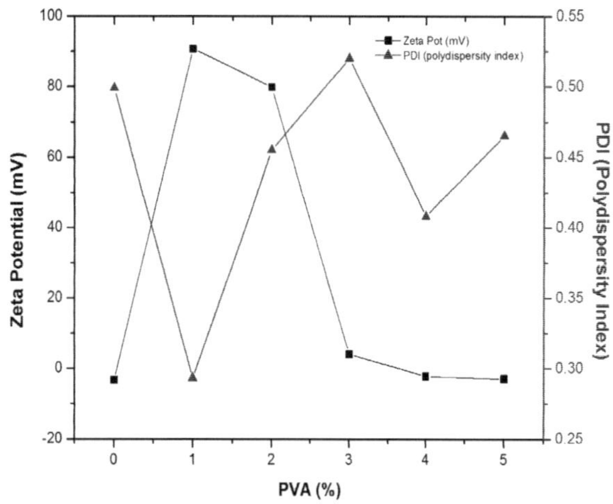

Figure 6. Zeta potential and polydispersity index from original and PVA-modified silver nanoparticles as a function of PVA concentration.

The result of the particle size observation through TEM image is sufficiently consistent with the particle size measurement using PSA. Based on PVA measurement, the original silver nanoparticles is approximately 6 nm in size and the 1% PVA-modified silver nanoparticles average size is approximately 27 nm.

The measurement result of average particle size using PSA for original silver nanoparticles and the nanoparticles which are modified with PVA at the concentration of 1% to 5% is shown in figure 5. After silver nanoparticles were modified with PVA, the nanoparticles size was increased. The higher PVA concentration resulted in a bigger nanoparticle size. This increase in nanoparticle size is apparently linear with the increment of addition of PVA concentration.

Figure 6 shows the measurement result of Zeta potential (ZP) and Polydispersity Index (PDI) of original and PVA-modified silver nanoparticles as the function of PVA concentrations. The Zeta potential value shows nanoparticles colloid stability, which is formed in the solution [14]. The Zeta potential value that ranges from ±0-10 mV shows a highly unstable colloid, with ZP value of ±10 - 20 mV, ±20 - 30 mV and > ± 30 mV shows relatively, moderately, and highly stable colloid in the respective order. Figure 5 shows the original silver nanoparticles has -2.4 mV ZP, which means the formed colloid system is highly unstable, and this conclusion is also confirmed with UV-Visible measurement result in figure 1. The figure shows that the formed silver nanoparticles is still changing, proven with great value change of absorbance, even after 44 hours measurement of the solution mixture.

After the silver nanoparticles were modified with 1% PVA, a drastic change of ZP value was observed to be 90.8 mV. When the PVA concentration was raised to 2%, the ZP value decreased to 79.8 mV. Both ZP values are far beyond the stable colloid border value, which is > ± 30 mV. PVA addition to the concentration of 3%, 4%, and 5% gives ZP value in the respective order of 4.1 mV, -2.1 mV, and -3.0 mV. These three ZP values show that the resulted silver nanoparticles colloids are highly unstable. Consequently, the 1% PVA concentration value for modifying silver nanoparticles biosynthesis using aqueous extract of *D. discolor* leaves have given the highest ZP value, resulting in highly stable colloid.

Colloid stability is one of the properties of nanoparticles. Another important characteristic is the particle size distribution in non particle system. Nanoparticle size distribution can be articulated through polydispersity index (PDI) value. A nanoparticle system with PDI value < 0.1 is considered as highly monodisperse, while PDI value > 0.4 and value in range of 0.1 – 0.4 are indicated that the system has highly polydisperse and moderately disperse distribution in the respective order [15]. The PDI measurement of original and PVA-modified silver nanoparticles with different concentrations is shown in figure 6. Original silver nanoparticles have PDI value of 0.50 that shows as highly polydisperse. After they were modified with 1% PVA, the PDI value sharply decreased to the value of 0.29 that shows a moderate dispersity. When PVA concentration was raised to 2%, 3%, 4% and 5%, PDI value once again increased, each of which has the value of 4.6, 5.2, 4.1, and 4.7 respectively. These PDI values give highly polydisperse characteristic. Thus, silver nanoparticles modification with 1% PVA concentration has produced a better characteristic of particle size distribution compared to the original silver nanoparticles.

4. Conclusions

Silver nanoparticles have been successfully prepared with biosynthesis method using aqueous extract of *D. discolor* leaves. The resulted silver nanoparticles are small in particle size, which is around 6 nm, and have highly polydisperse and highly unstable properties. After the modification with PVA was performed, the properties were improved. PVA concentration as high as 1% shows an improvement from highly polydisperse into moderately disperse with remarkable improvement in colloid stability, which becomes highly stable. Even so, this improvement in properties has a cost to the increase of nanoparticle size into 27 nm.

References

[1] Debjani N and Pratyusha B 2013 *Environ. Toxicology and Pharmacology* **36** 997-1014
[2] Yan H B, Zhang Y Q, Ma Y L and Zhou L X 2009 *J. Nanopart. Res.* **11** 1937-46
[3] Elechiguerra J L, Burt J L, Morones J R, Bragado A C, Gao X, Lara H H and Yacaman M J 2005 *Journal of Nanobiotechnology* **3** 6
[4] Mandal D, Bolander M E, Mukhopadhyay D, Sarkar G and Mukherjee P 2006 *Appl. Microbiol. Biotechnol.* **69** 485-92
[5] Prema P and Raju R 2009 *Biotechnology and Bioprocess Engineering* **14** 842-47
[6] Sachdev D, Kumar V, Maheshwari P H, Pasricha R, Deepthi and N Baghel 2016 *Sensors and Actuators B: Chemical* **228** 471-9
[7] Liang A, Liu Q, Wen G and Jiang Z 2012 *Trends in Analytical Chemistry* **37** 32-47
[8] Das S K and Marsili E 2010 *Rev. Environ. Sci. Biotechnol.* **9** 199-204
[9] Douglas F, Yanez R, Ros J, Marin S, Escosura-Muniz A, Alegret S and Merkoci A 2008 *J. Nanopart. Res.* **10** 97-106
[10] Mohanpuria P, Rana N K and Yadav S K 2008 *J. Nanopart. Res.* **10** 507-17
[11] Mittal A K, Chisti Y and Banerjee U C 2013 *Biotechnol. Adv.* **31** 346-56
[12] Djuhana D, Imawan C, Fauzia V, Harmoko A, Handayani W and Putra M H 2015 *KnE Engineering ICoSE Conference on Instrumentation, Environment and Renewable Energy* **2016**
[13] Nagarajan R and Hatton TA 2008 *Nanoparticles: Synthesis, Stabilization, Passivation, and Functionalization* (Washington, DC: American Chemical Society)
[14] Patel V R and Agrawal Y K 2011 *J. Adv. Pharm. Technol. Res.* **2** 81-7
[15] Bhattacharjee S 2016 *Journal of Controlled Release* **235** 337-51

Synthesis, characterization, and electrochemical behavior of Au@Pd core shell nanoparticles

W P Wicaksono[1] and T A Ivandini[2]

[1]Department of Chemistry, Faculty of Mathematics and Natural Sciences Universitas Islam Indonesia, Kampus Terpadu UII, Sleman, Yogyakarta, Indonesia
[2]Department of Chemistry, Faculty of Mathematics and Natural Sciences Universitas Indonesia, Kampus UI Depok, Depok 16424, Indonesia

Corresponding author's e-mail: ivandini.tri@sci.ui.ac.id

Abstract. Au@Pd core shell nanoparticles (Au@Pd CSNPs) were successfully synthesized using a seed-mediated growth method. Firstly, a pale pink gold seed solution was used to produce a pale purple gold nanoparticles (AuNPs) core solution. Then, three series of Pd shell thickness using 20µ, 100 µL, and 500 µL of $PdCl_2$ produced purple, brown, and deep brown of Au@Pd CSNPs respectively. A strong absorbance UV-Visible spectrum with peaks at 285 nm and 535 nm was identified for AuNPs formation. The disappearance of the peak at 535 nm was indicated the Au@Pd CSNPs formation. The electrochemical properties were examined in phosphate buffer pH 7 using cyclic voltammetry technique with boron-doped diamond (BDD) as working electrode showed a couple oxidation and reduction peak of gold at 0.67 V and at 0.33 V, respectively. The Au@Pd CNPs will be used for modification of BDD electrodes.

1. Introduction

A core shell structure bimetallic nanoparticles (BNPs) are a heterogeneous nanomaterial which have unique catalytic, magnetic, and optical properties, that are very dissimilar from their monometallic nanoparticles constituents [1, 2]. These superior properties are related to their electronic effect and or geometric effect [3]. Interaction between heterometallic (core and shell) surface atoms can modify the surface electronic structure of the BNPs [4]. Hereby, BNPs in the form core shell frequently perform a better catalytic activities compared to their monometallic nanoparticles mixtures or to alloyed counterparts [5, 6]. One of the most fascinate BNPs are gold-palladium core shell nanoparticles (Au@Pd CSNPs), which have been reported to be useful for electrocatalyst of selective oxidative of benzyl alcohol [7], oxygen reduction reaction [2], hydrogen production [8, 4], formic acid oxidation [9], and methanol electro-oxidation [10].

In this work, we reported the synthesis of Au@Pd CSNPs and its characterization using a UV-Visible spectroscopy and their electrochemical properties.

2. Materials and methods

2.1. Chemicals and reagents

Chloroauric acid ($HAuCl_4.3H_2O$), palladium (II) chloride ($PdCl_2$), Cetyltrimethylammonium bromide (CTAB), sodium borohydride ($NaBH_4$), Potassium hydrogen phosphate (K_2HPO_4), Potassium dihydrogen phosphate (KH_2PO_4), and ascorbic acid (AA) were purchased from Wako Pure Chemical Industries, ltd, while Hydrochloric acid (HCl) was supplied from Merck.

2.2. Preparation of gold core nanoparticles (AuNPs)

AuNPs were synthesized based on modified seed-mediated grow method reported by Sau *et al.* [11]. Briefly, a 10 mL aqueous solution consists of CTAB 75 mM and $HAuCl_4$ 0.25 mM was mixed. Then, a 0.6 mL of 10 mM $NaBH_4$ fresh-cold solution was quickly added with vigorous stirring to the mixture, followed by slow stirring at 35°C for 3 hour, generating a light brown solution. A volume of 2 mL of this solution was diluted to 50 mL of water to produce a seed solution. Then, 2 mL of the seed solution was immediately added to a 25 mL solution containing 0.1 mL 10 M of $HAuCl_4$, 2 mL of 0.2 M CTAB, and 1.5 mL of 0.1 M AA. The mixture was then stirred slowly at 308 K for 8 h until produced light purple of AuNPs. Then, the AuNPs were characterized using UV-Visible Spectrophotometer in a wavelength of 200-800 nm.

2.3. Preparation of gold-palladium core shell nanoparticles (Au@Pd CSNPs)

Au@Pd CSNPs were synthesized based on the method of Fan *et al.* [12]. A solution of 10 mM H_2PdCl_4 was prepared from 44.5 mg of $PdCl_2$ dissolved into 25 mL of 20 mM HCl in a boiling water bath. To control the Pd shell, a 20 mL of the prepared AuNPs was then added by 20 μL, 100 μL, and 500 μL of H_2PdCl_4, to produce one atomic layer, several layers, and nanocubes, respectively. Finally, several brown solutions of Au@Pd CSNPs were observed. Characterization was performed using UV-Visible Spectrophotometer in a wavelength 0f 200-800 nm.

2.4. Electrochemical study of Au@Pd CSNPs at Boron Doped Diamond (BDD) electrodes

A volume of 2 mL of Au@Pd CSNPs was diluted in 2 mL of 0.1 M phosphate buffer solution (PBS). Then, cyclic voltammetry technique was performed in a potential range of -1 to 1.75 V with a scan rate of 100 mV/s to evaluate the oxidation and reduction properties of Au@Pd CSNPs in compare to AuNPs.

3. Results and discussion

3.1. The formation of Au@Pd CSNPs

Figure 1a show the photograph of a light brown Au seed solution which was produced from the reduction of $HAuCl_4$ by $NaBH_4$ using CTAB as the stabilizing agent. CTAB is one of cationic surfactant that usually use as stabilizing and shape-directing agent in the particle synthesis [13]. A strong absorbance of UV-Visible spectrum at 285 nm and an intense Surface Plasmon resonance (SPR) band at 535 nm (figure 2a) were identified from the light purple, indicated that the formation of AuNPs with average diameter of 11 nm [1] (figure 2b). AA was used couple with CTAB to control the morphology and dimension of core AuNPs [11]. Figure 2 (d, e, f) show the photograph of Au@Pd CNPs with Pd shell thickness varying with 500, 100, and 20 μL, respectively. The disappearance of the peak at 535 nm and

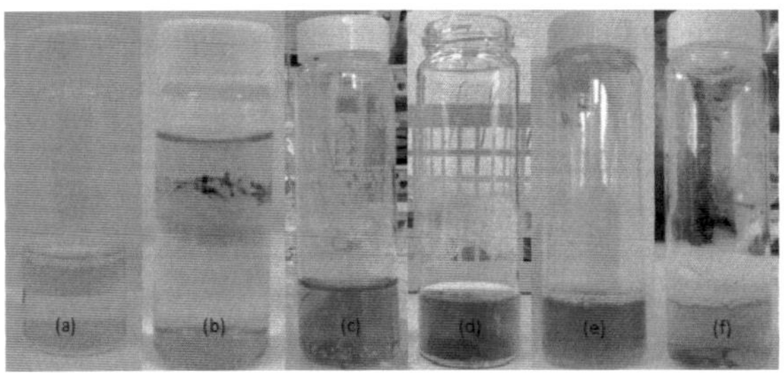

Figure 1. Photograps of (a) Au seed, (b) Core AuNPs, (c) H_2PdCl_4 solution, and Au@Pd CSNPs with (d) 500 μL of Pd shell, (e) 100 μL of Pd shell, (f) 20 μL of Pd shell

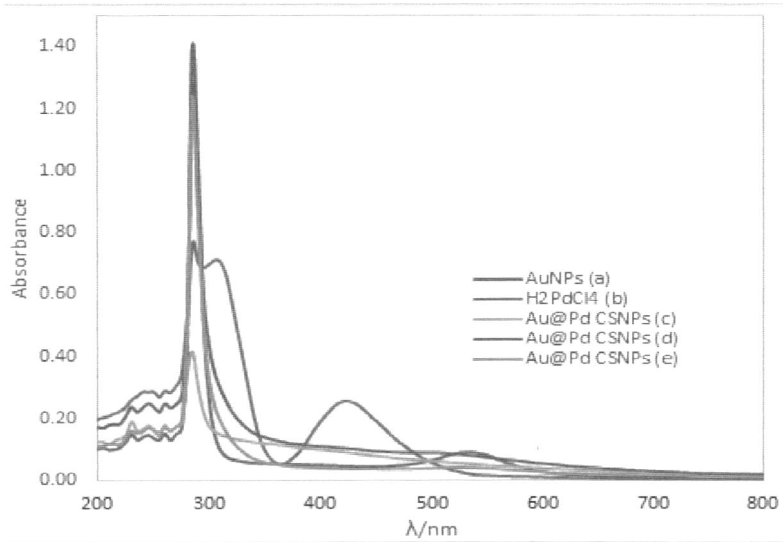

Figure 2. UV-Visible spectrum of (a) core AuNPs, (b) H_2PdCl_4, and Au@Pd CSNPs with (c) 500 μL of Pd shell, (d) 100 μL of Pd shell, (e) 20 μL of Pd shell

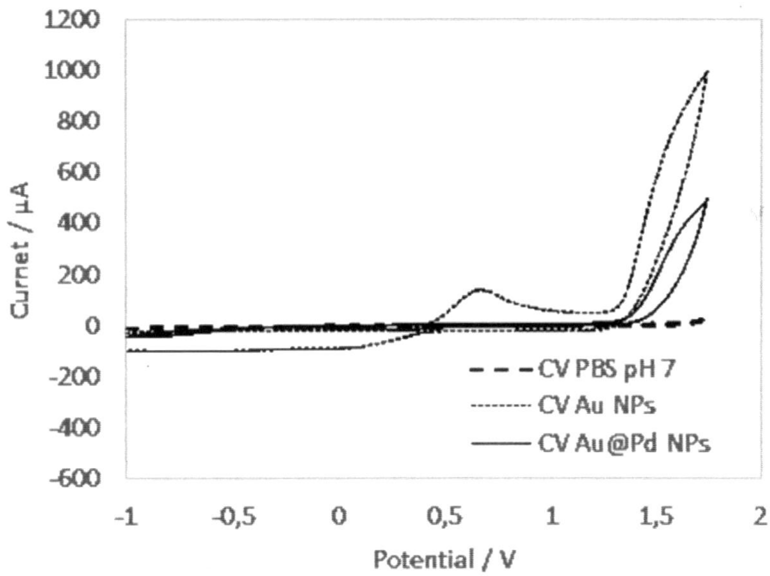

Figure 3. The electrochemical properties of Au@Pd CSNPs in 0.1 M PBS pH 7

420 nm were correspondent to the decreasing of core AuNPs and H_2PdCl_4 characteristics, respectively, and was indicated the Au@Pd CNPs formation [14] (figure 2b and 2c). The increasing of Pd shell thickness decreased of the peak absorbance at 535 nm, it means no SPR band was detected [14] (figure 2 c, 2d, 2e).

3.2. The electrochemical properties of Au@Pd CSNPs at boron doped diamond (BDD) electrode

Electrochemical behaviour of the Au@Pd CSNPs was examined in 0.1 M PBS pH 7 using cyclic voltammetry technique. Figure 3 shows a couple reduction and oxidation peaks of gold at +0.67 V and at +0.33 V, respectively, which were identified for both AuNPs and Au@Pd CSNPs. The strong oxidation peak of core AuNPs was identified as the oxidation of Au^0 to Au^{3+} [15], whereas the absence of the oxidation peak of Au@Pd CNPS indicated that the Pd shell may inhibit the core from the oxidation. However, a unique reduction peak can be exhibited in Au@Pd CNPs, which can be explained due to the effect of lattice strain produced by core shell structure nanoparticle which increased the reduction catalytic activity [2]. This behaviour suggested that Au@Pd CSNPs may be more sufficient as catalyst for reduction reaction.

4. Conclusions

Au@Pd CSNPs were successfully synthesized using a seed-mediated growth method. UV-Vis spectrum of Au@Pd CSNPs shown that core AuNPs have been successfully coated by Pd shell to form Au@Pd CSNPs. The electrochemical examination showed that a better reduction peak was obtained for Au@Pd CSNPs, whereas a better oxidation peak occurred at the AuNPs core.

References

[1] Ding Y, Fan F, Tian Z and Wang Z 2010 *J. Am. Chem. Soc.* **16** 12480-6
[2] Chen D, Li C, Liu H, Ye F and Yang J 2015 *Sci. Rep.* **5** 11949
[3] Sinfelt J H and Meitzner GD 1993 *Acc. Chem. Res.* **26** 1-6
[4] Tedsree K, Ling T, Jones S, Chan C W A, Yu K M K, Bagot P A J, Marquis E A, Smith G D W and Tsang S C E 2011 *Nat Nanotechnol.* **6** 302-7
[5] Alayoglu S, Nilekar A U, Mavrikakis M and Eichhorn B 2008 *Nat. Mater.* **7** 333-8
[6] Serpell C J, Cookson J, Ozkaya D and Beer PD 2011 *Nat. Chem.* **3** 478-83
[7] Silva T A G, Teixeira-Neto E, Lo´pez N and Rossi L M 2014 *Sci. Rep.* **4** 5766
[8] Su R *et al* 2014 *ACS Nano* **4** 3490-7
[9] Hsu C, Huang C, Hao Y and Liu F 2013 *Nanoscale Res. Lett.* **8** 113
[10] Li N, Tang S and Meng X 2014 *J. Mater. Sci. Technol.* **30** 1071-7
[11] Sau T K and Murphy C J 2004 *J. Am. Chem. Soc.* **126** 8648-9
[12] Fan F R, Liu D Y, Wu Y F, Duan S, Xie Z X, Jiang Z Y and Tian Z Q 2008 *J. Am. Chem. Soc.* **130** 6949-51
[13] Tebbe M, Kuttner C, Männel M, Fery A and Chanana M 2015 *ACS Appl. Mater. Interfaces* **7** 5984-91
[14] Liu F, Wechsler D and Zhang P 2008 *Chem. Phys. Lett* **461** 254-9
[15] Ivandini T A, Wicaksono W P, Saepudin E, Rismetov B and Einaga Y 2015 *Talanta* **134** 136-43

Crystallite size dependence on electrical properties of $LaFeO_3.0.1Fe_3O_4$ nanocomposite material

I Rhidwan, D Triyono and H Laysandra

Department of Physics, Faculty of Mathematics and Natural Sciences Universitas Indonesia, Kampus UI Depok, Depok 16424, Indonesia

Corresponding author's e-mail: djoko.triyono@sci.ui.ac.id.

Abstract. $LaFeO_3.0.1Fe_3O_4$ nanocomposite was prepared by sintering method. $LaFeO_3$ was synthesized by the sol-gel method and mixed with Fe_3O_4. The mixture was pressed and then sintered at 1300°C for 1 h resulting nanocomposite. After that, $LaFeO_3.0.1Fe_3O_4$ nanocomposite was annealed at temperatures of 1000°C, 1100°C, and 1200°C for 12 h. Structural analysis by using X-ray diffraction (XRD) confirm the double phase of $LaFeO_3$ and Fe_3O_4. The crystallite sizes calculated by using Scherrer's formula are in the range of 30-60 nm. By using impedance spectroscopy method, the electrical properties were evaluated as functions of frequency (1 kHz-1 MHz) and temperature (RT - 373 K). Electrical properties were represented in Nyquist plot and Bode plot as a function of temperature. Bode plot of Z'' vs f shows that relaxation peaks shift to the higher frequency. From the plot of ln τ vs 1/T, the activation energy was calculated.

1. Introduction

Perovskite oxide materials, especially the orthoferrites, are important class of materials due to the large variety of application in SOFC and chemical sensors [1]. They can be used in various applications because their stable structure, high thermal stability, high dielectric constant, low dielectric loss, and high performances [2]. Among these perovskite oxides, $LaFeO_3$ and doping in La and Fe-site of $LaFeO_3$ have been studied for application in oxide fuel cell cathode, oxygen sensors, etc. [1]. The structural and electrical properties of $LaFeO_3$ is very depend on method of preparation, grainsize, doping and composites, sintering temperature, etc. [3-4]. Study of dielectric properties in this material as a function of frequency and temperature with variation of compositions may provide valuable information to obtain good candidate material for application.

Nanocomposite are drawing attention to investigate the correlation between structural and electrical properties with composition and grain size. In this research, we want to investigate the correlation between crystallite size and electrical properties of $LaFeO_3$-Fe_3O_4. We have synthesized a homogeneous composites $LaFeO_3$ as a main phase with 0.1 at.% Fe_3O_4 as a second phase of this composite. We were trying to vary the crystallite size of $LaFeO_3$ phase by annealing treatment. We focused on the effect of annealing treatment on the structural and electrical properties of composites of $LaFeO_3.0.1Fe_3O_4$ materials.

2. Materials and methods

$LaFeO_3$ was synthesized by the sol-gel method using lanthanum oxide (La_2O_3), iron (III) nitrate nonahydrate ($Fe(NO_3)_3.9H_2O$) and citric acid as a precursor. The precursors were mixed into solution. The solution was stirred and heated at 120°C until gel was formed. Then, the gel was calcinated in

900°C for 6 h to form $LaFeO_3$ powder. $LaFeO_3$ powder were mixed with 0.1 at.% of Fe_3O_4 powder and grind to form $LaFeO_3.0.1Fe_3O_4$. Finally, the powder was pressed into pellet with diameter of 10 mm at pressure of 5kN and sintered at 1300°C for 1 h. Finally, the $LaFeO_3.0.1Fe_3O_4$ was annealed at 1000°C, 1100°C, and 1200°C for 12 h. The phase identification was checked by using X-Ray Diffractometer (XRD) with Cu Kα (λ =1.54060 Å) radiation source in the 2θ range of 15° to 90°. Electrical properties of $LaFeO_3.0.1Fe_3O_4$ were characterized by *RLC-Meter FLUKE-PM6306* at frequency range of 1kHz-1MHz with the temperature range of RT-373K.

3. Results and discussion

3.1. Structure analysis

Figure 1 shows the XRD pattern of $LaFeO_3.0.1Fe_3O_4$ sintered and annealed at 1000°C, 1100°C, and 1200°C. The XRD analysis data were collected at room temperature, with the 2θ range from 15° to 90° and Cu Kα radiation source. All samples exhibit two phases i.e. $LaFeO_3$ orthorhombic structure as a main phase and Fe_3O_4 cubic structure as a second phase. The crystallite size of $LaFeO_3.0.1Fe_3O_4$ is determined by Scherrer method using HighScore plus software as tabulated in Table 1. As shown in Table 1, the crystallite size of $LaFeO_3$ is increasing when annealed at 1000°C, then it starts to decrease when annealing temperature was increased. Decreasing in crystallite size might be due to the phase Fe_3O_4 that inhibit the grain growth of the main phase of $LaFeO_3$. This is similar to the results of research from Xia *et al.* [5] that showed the presence of a second phase that is segregated in grain boundary.

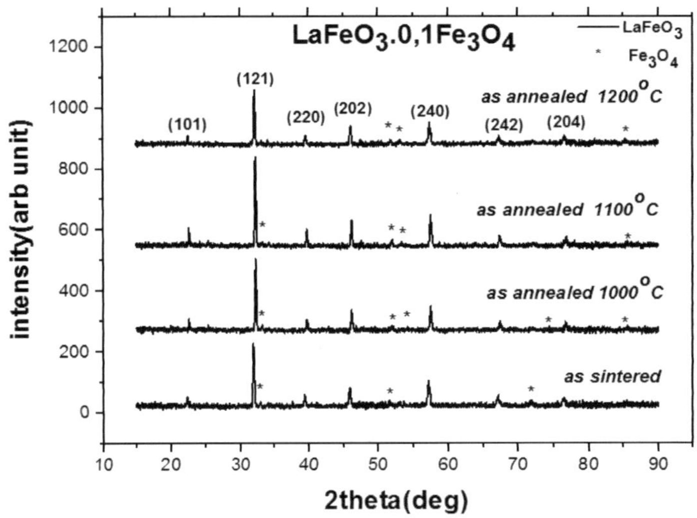

Figure 1. XRD curves of $LaFeO_3.0.1Fe_3O_4$ that sintered and annealed for 12 h.

Table 1. Crystallite size of $LaFeO_3.0.1Fe_3O_4$ that sintered and annealed for 12 h.

Sample	Crystallite size (nm)
As Sintered	38.1
As Annealed 1000°C	60.1
As Annealed 1100°C	58.2
As Annealed 1200°C	40.6

3.2. Electrical properties

Impedance spectroscopy method is commonly used to characterize electrical properties of ceramic material. The impedance data is representated by Nyquist plot and Bode plot [6, 7]. The variation of imaginary part and real part of impedance (Nyquist plot) for both sintered and annealled samples (1000°C, 1100°C, 1200°C) is given in figure 2. Nyquist plot show that the diameter of semicircle decreases with increasing temperature. The similar phenomenon was reported by Nasri *et al.* [8] on $LiFeP_2O_7$ ceramic. The high-frequency semicircle is caused by the grain contribution and at low frequencies is caused by grain boundary contribution that arising due to equivalent circuits of two parallel resistance-capacitance (RC) elements similar as reported by Triyono *et al.* [9] for $Sr_2(Fe,Ti)O_6$ compounds. From the figure 2, it can be seen that the radius of the semicircle at low frequencies is greater than the semicircle at high frequencies. This means the effect of grain boundaries to resistance is more effective than grains at temperature range for all samples.

Using Zsimpwin 7.0 software, the grain resistance (R_g) for all samples can be calculated. Table 2 shows the value of R_g at 323 K. It can be shown that the R_g is decreasing when the annealing temperature increases indicating that grain resistance in the system is affected by crystallite size.

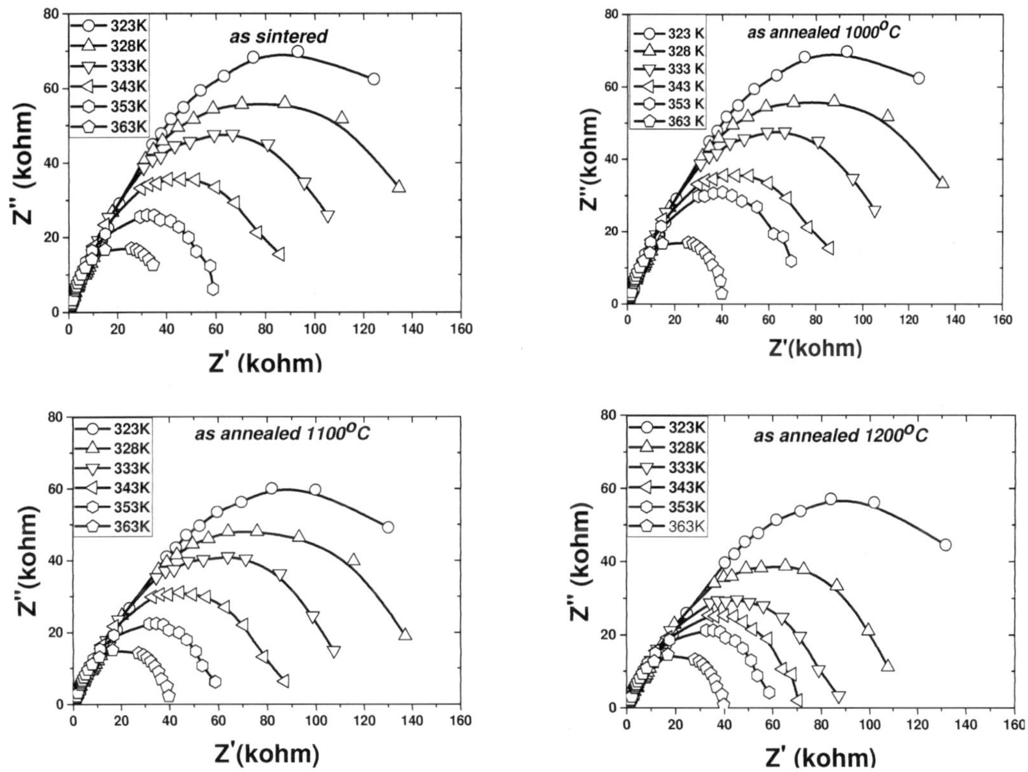

Figure 2. Nyquist plots of $LaFeO_3.0.1Fe_3O_4$ that sintered and annealed for 12 h.

Table 2. R_g of $LaFeO_3.0.1Fe_3O_4$ that sintered and annealed for 12 h at 323 K.

R_g (kOhm)	323 K
As Sintered	108.7
As Annealed 1000°C	129.9
As Annealed 1100°C	113.0
As Annealed 1200°C	112.9

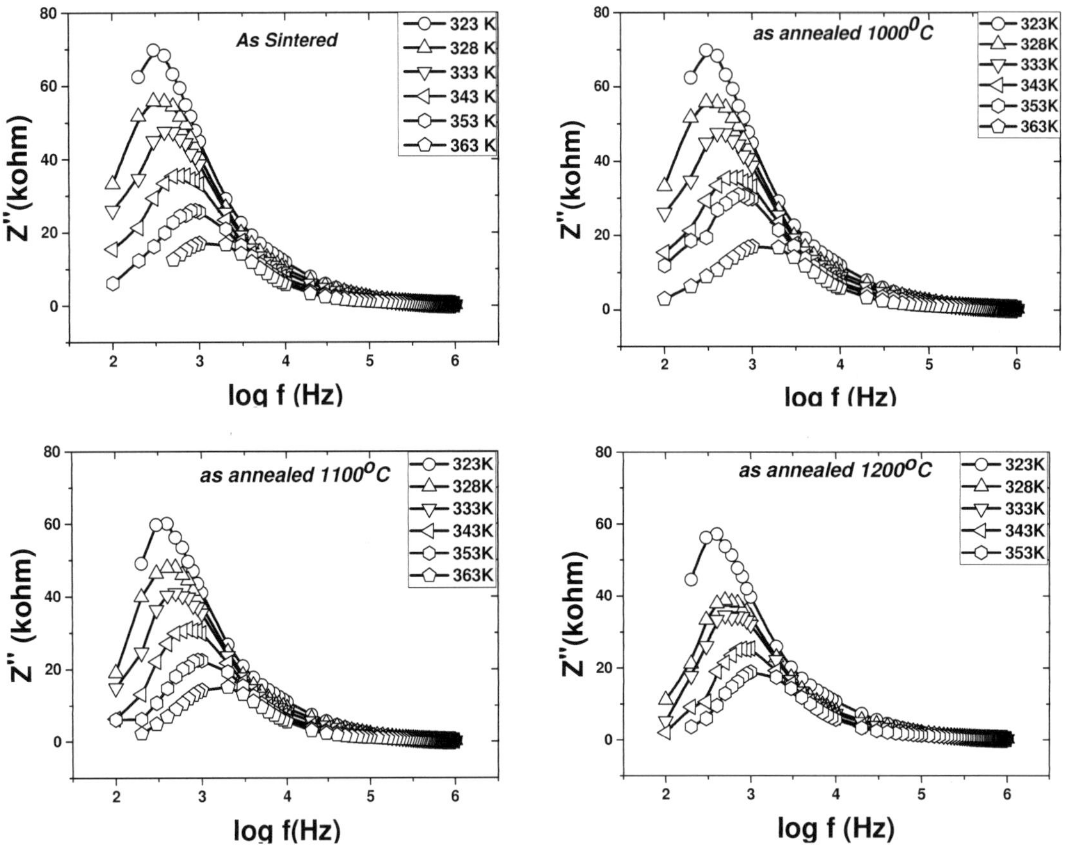

Figure 3. Variation of imaginary part of impedance with frequencies at different temperatures for $LaFeO_3.0.1Fe_3O_4$ that sintered and annealed for 12 h.

Figure 3 shows that imaginary part of impedance, *Z" (f)* at different temperatures for $LaFeO_3.0.1Fe_3O_4$ for both sintered and annealled samples (1000°C, 1100°C, 1200°C). *Z"* values increase with increasing of frequency in all temperature range, and exhibited a maximum value before starts decreasing rapidly. From figure 3, it can also be noted that the maximum value show relaxation peak. The *Z"* peak shifts towards higher frequency with the increase of temperature which indicate the relaxation phenomenon [8]. The activation energy for relaxation process was calculated from plot of ln τ vs 1/T following Arrhenius law. The observed relaxation times and activation energies [9] are shown in figure 4. It is observed that the value of τ decreases with rise of temperature. Activation energy indicates the type of charge carrier related to the relaxation process which is accepted that for n-type polaron Ea < 0.2 eV, whereas for p-type polaron Ea > 0.2 eV [10]. The activation energies for all samples have been calculated. They show that the relaxation process is accepted for n-type polaron. Activation energies for all samples are similar indicates that the crystallite size has no effect for type of charge carrier.

4. Conclusions

Nanocomposite $LaFeO_3.0.1Fe_3O_4$ has been successfully synthesized. All samples exhibit two phases i.e. $LaFeO_3$ orthorhombic structure as a main phase and Fe_3O_4 cubic structure as a second phase. The grain resistance is affected by crystallite size. The relaxation peak as shown in bode plot *Z"* vs *f* shifts

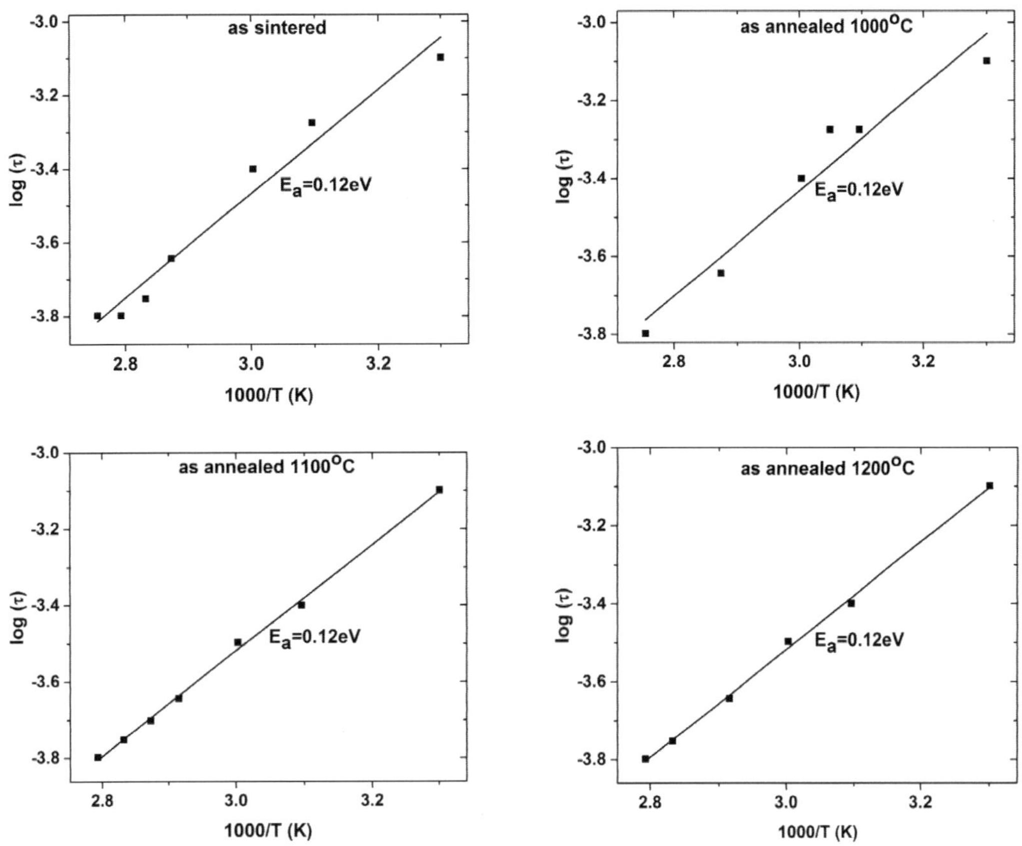

Figure 4. Plot between In τ vs 1/T for $LaFeO_3.0{,}1Fe_3O_4$ that sintered and annealed for 12 h.

to the higher frequency indicates relaxation phenomenon. From τ as a function of a temperature, the activation energy for all samples was calculated and its values indicate as n-type polaron.

Acknowledgements
The authors would like thanks to HIBAH PITTA (2007/UN2.R12/HKP.05.00/2016) from Universitas Indonesia for financial support.

References

[1] Benali A, Bejar M, Dhahri, Graça M F P and Costa L C 2015 *J. Alloys and Compounds* **653** 506-12
[2] Cui X, Li S and Zhu X 2014 *Mater. Lett.* **130** 267-70
[3] Shen M, Ge S and Cao W 2001 *Appl. Phys.* **34** 2935-8
[4] Waernhus I, Grande T and Wiik K 2005 *Solid State Ionic* **176** 2609-16
[5] Xia Z C, Yuan S L, Feng W, Zhang L J and Zhang G H 2003 *Solid State Comm.* **126** 567-71
[6] MacDonald J R 1987 *Impedance Spectroscopy Emphasising Solid Materials and Systems* (New York: Wiley)
[7] Sutar B C, Choudhary R N P, Das P R 2013 *J. Ceram. Int.* **40** 7791-8
[8] Nasri S, Megdiche M and Gargouri M 2016 *Ceram. Int.* **42** 943-51
[9] Triyono D and Laysandra H 2016 *AIP Conference Proceedings* **1729** 20-2
[10] Devi Chandrasekhar K, Mallesh S, Krishna Murthy J, Das A K and Venimadhav A 2014 *Physica* B **448** 304-11

IOP Conf. Series: Materials Science and Engineering **188** (2017) 012059 doi:10.1088/1757-899X/188/1/012059

Photodegradation study of TiO_2-organoclay modified acetate cellulose bioplastic

Y K Krisnandi, D G W K D Rasanji, S Z Luthfiyah, Zahara and R Sihombing

Department of Chemistry, Faculty of Mathematics and Natural Sciences Universitas Indonesia, Kampus UI Depok, Depok 16424, Indonesia

Corresponding author's e-mail: yuni.krisnandi@sci.ui.ac.id

Abstract. Biodegradable cellulose acetate films have been synthesized using TiO_2 modified organobentonite as nanofiller. The aim of titania addition is to add photocatalyst to the biocomposite, so it has self-photodegradation properties. Organobentonite was prepared from Tapanuli bentonite, previously purification and Na^+-exchanged, modified with cationic surfactant hexadceyltrimethylammonium bromide (HDTMABr). The composition of bioplastic cellulose acetate was 7 wt.% organobentonite and certain amount of TiO_2 (0 wt.%, 1 wt.%, 3 wt.%, 5 wt.%, 10 wt.%) of the total composite weight. Fabrication of nanocomposite film was carried out using acetone as solvent and through solvent casting method. FTIR analysis showed the intercalation with surfactant was successfully carried out, indicated by new absorption band at 2636 cm^{-1} and 2569 cm^{-1}. Nanocomposite application in photodegradation test was carried out under direct sunlight irradiation, UV light, and without irradiation for thirty days. It is found that the greater the amount of TiO_2 added into the composites, the more weight loss occurred due to photodegredation. Percent weight loss in the UV light irradiation are 4.02%, 13.45%, 18.66%, 22.35%, 27.86%, respectively for (TiO_2 0 wt.%, 1 wt.%, 3 wt.%, 5 wt.%, 10 wt.%). While for bioplastic irradiated by direct sunlight, the weight loss was 2.15%, 8.49%, 13.85%, 14.70%, 15.02%, respectively. In contrast, without irradiation, the weight reduction of bioplastic was insignificant. The results indicate that the addition of TiO_2 to the composition of bioplastic has given the ability of self-photodegradation to the composite.

Keywords: bioplastic, biodegradable, nanofiller, organobentonite, self-photo degradation, TiO_2

1. Introduction

The use of materials as fabrication of polymer has been increased since some last decades. Plastic becomes popular because it is low cost production and easy to be produced. It also has good tenacity and light of weight. Although it is very helpful in human daily activity, the exceed use of plastic become hazard waste for environment. Plastic wastes need many time to be decomposed naturally by land microbe. This waste quantity increases day by day significantly and become the biggest polluting agent for environment.

Many researches have focused to prevent and minimalize the harmful effect of plastic wastes. One method that can be suggested is by employing photocatalytic reaction. This method is known for eco-friendly technique and the result for degrading organic pollutant is significant. It has been reported that TiO_2 is an effective photocatalyst because this molecule is non-toxic, has ability to do photo electric conversion and oxidative capacity, low cost and stable [1].

This present work focused on fabrication of biodegradable cellulose acetate films modified by variation amount of TiO_2-organobentonite as nanofiller and then the nanocomposite application in photodegradation test carried out under direct sunlight radiation, UV light, and without irradiation.

Our previous work has successfully prepared biocomposite membrane from acetic cellulose using organobentonite made of hexadecyl trimethyl ammonium bromide (HDTMABr) as well as octadecyl trimethyl ammonium bromide (ODTMABr) surfactants [2, 3]. The bentonite initially has hydrophilic properties. Therefore, in order to be 'compatible' with organic compounds such as polymers, it needs to be modified so that it has organophylic properties [4]. Modification is often done by inserting a cationic surfactant into intergallery of bentonite to produce organobentonite or also named organoclay [5]. Other research showed that nano composite membrane can be synthesized from acetic cellulose–butyric cellulose using organoclay modificated with same surfactant [6].

In our work, the organoclay was then modified further with TiO_2 anatase in order to add photocatalytic properties to the composite. Generally, TiO_2 is added to the bentonite~~,~~ as a pillar in intergallery of clay [7]. Until now, titanium dioxide still dominate photocatalytic applications. Andhika [8] and Luthfiyah *et al.* [9] have synthesized nanocomposite using TiO_2-bentonite intercalated with hexadecyl trimethyl ammonium bromide (HDTMABr) as surfactant in order to increase nano composite cellulose acetic photoctalytic activity.

2. Materials and methods

2.1. Materials
Natural bentonite was supplied from Tapanuli, North Sumatera Utara. It was fractioned, purified and Na-exchanged before being used follow [10]. All other materials in this research were used without further purification, e.g. synthetic plastic, NaOH and NaCl (Merck), hexadecyl trimethyl ammonium bromide (HDTMA-Br), acetic cellulose p.a, acetic acid p.a and silver nitrate p.a. (Sigma Aldrich), and TiO_2 P25 Evonik.

2.2. Synthesis of organoclay
Follow previous procedure [9], Na-bentonite was dispersed in distilled water and mixed with hexadecyl trimethyl ammonium Bromide (HDTMA-Br) surfactant by 1 CEC. The suspension was stirred at 60°C for about 1 hour. After stirring completed, then suspension was sonicated at 150 ppm about 3 m and decanted. The precipitation was dried at 70°C.

2.3. Synthesis of TiO_2-organoclay
TiO_2 with variation of weight 1%, 3%, 5% and 10% and 7% organoclay from total composite was dispersed into 25 mL acetone. The suspension was stirred at temperature of 60°C for about 1 hour, follow by sonication [10].

2.4. Fabrication of TiO_2-organoclay/acetic cellulose composite
TiO_2-organoclay that has been dispersed into acetone was added into acetic cellulose with certain weight. The suspension then was stirred at room temperature for 1 h. After stirring was completed, the suspension was sonicated for 2 m. Then, suspension was poured into bio-plastic molder. Suspension was dried at 40°C. The Films that have been formed, were stored in poly propylene bag for further characterization.

2.5. Photo degradation test
The as-prepared of TiO_2-organoclay/ acetic cellulose was irradiated using UV light, direct sun light and without irradiation for 30 days. Each irradiation was performed 6 h a day for 5 days. UV light used in this research has wavelength range 280-320 nm and 8 W power. The distance between sample and UV light source was about 17 cm. Irradiation process took place at room temperature. Sample weighing was performed once in five days. The same treatment was carried out on the experiments with direct sun light (at 9.00 am until 15.00 pm) and without irradiation.

IOP Conf. Series: Materials Science and Engineering **188** (2017) 012059 doi:10.1088/1757-899X/188/1/012059

3. Results and discussion

3.1. Preparation of TiO_2-organoclay/acetic cellulose composite

An addition of organobentonite nanofiller will strengthen the acetic cellulose plastic but its colour become dark brown, as can be seen in figure 1a.

The gradual addition of TiO_2 to acetic cellulose (through the organobentonite nanofiller) somehow has whitened the composite (figure 1a). It was expected that the TiO_2 addition into acetic cellulose composite would give self-photodegradation activity to the composite.

The modification of bentonite with surfactants was aimed to change the nature of the properties of the aluminasilicate from hydrophilic to organophilic character. It is reported that this action is a strategic step to create a mineral that can absorb or react with organic pollutants [11]. The change of structure property can be observed in infrared spectra displayed in figure 2. The decrease of stretching band assigned for hydrogen bonded silanol group at 3600-3500 cm^{-1} and the bending peak for water molecules at 1620 cm^{-1} altogether with the presence of symmetric and asymmetric peaks of $–CH_n$- at 2850-2950 cm^{-1} are evidence of the transformation.

3.2. Photodegradation test on as-prepared composite

The photodegradation test was focused on three irradiation sources, i.e. direct sunlight, UV light and without irradiation.

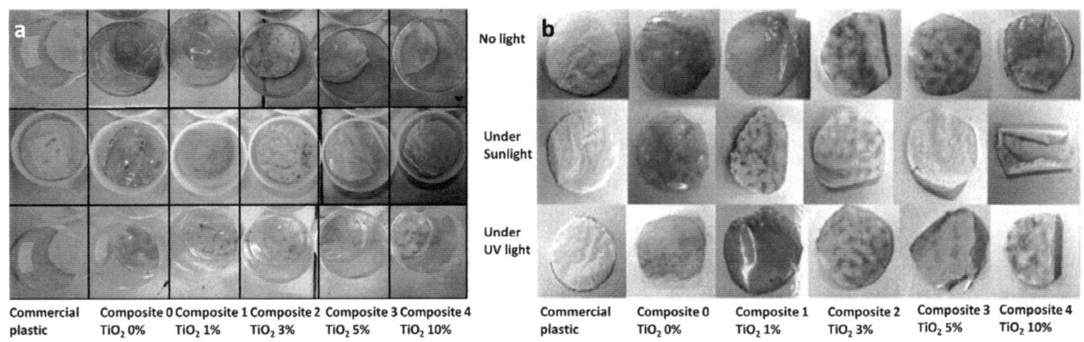

(a) (b)

Figure 1. Nanocomposites, (a) before irradiation, and (b) after irradiation.

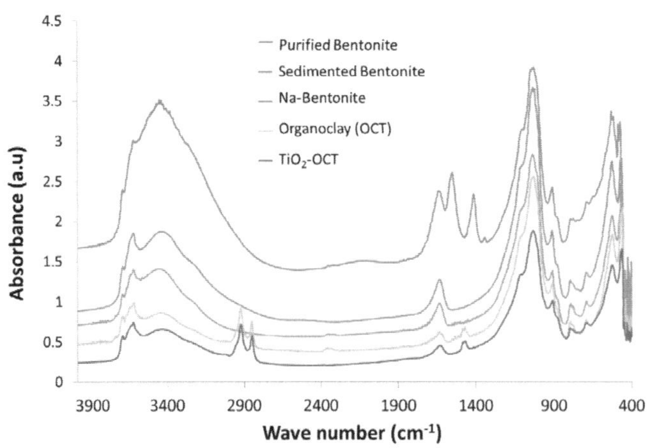

Figure 2. FTIR spectra of raw and modified bentonites.

Shorter UV light with wavelength less that 280 nm can degrade acetic cellulose through photo-chemical reaction. But, its ability is limited because there are not enough chromophore groups available to adsorb the UV light [11]. Self-photodegradation ability can be increased by adding a certain amount of TiO_2 to the composite. Photo-degradation activity of TiO_2 will damage the surface of the composite. Combination of photo process and biodegradation may increase total degradation process of composite. Besides, degradation rate is also affected by physical design of acetic cellulose and by the environmental condition [11].

The three different irradiation process were performed in order to know the effect of environmental conditions to the self-degradation of the composites. The results can be seen in figure 3, 4 and 5. Figure 3 shows that the as-prepared and TiO_2-added organobentonite-acetic cellulose composites retained their weights when being kept in the dark room without irradiation. While the commercial available plastic only lost insignificant weight (~1.4%) after 6 week. From figure 4 to 5, it can be seen that there are differences in pattern on lighting using direct sun light and UV light. Irradiation using UV light results more linear graphic and more weight loss than using direct sun light. This is caused by consistency of intensity used. The use of UV light is more stable and more consistent than direct sun light which is affected by environmental condition. Furthermore, direct sunlight has wavelength range 409-435 nm (UV A), which has energy lower than the energy of UV light used in this research (UV B, 280-320 nm).

The variation of weight of TiO_2 added into composite was then studied in order to know the decrement of composite weight. It is found that the greater the amount of TiO_2 added into the composites, the more weight loss occurred due to photodegradation. This is because there is no TiO_2 as photocatalyst agent. As comparison, the synthetic plastic was also tested. The result is decrement of weight of synthetic plastic is more than the acetic cellulose composites with and without TiO_2 except with direct sunlight as

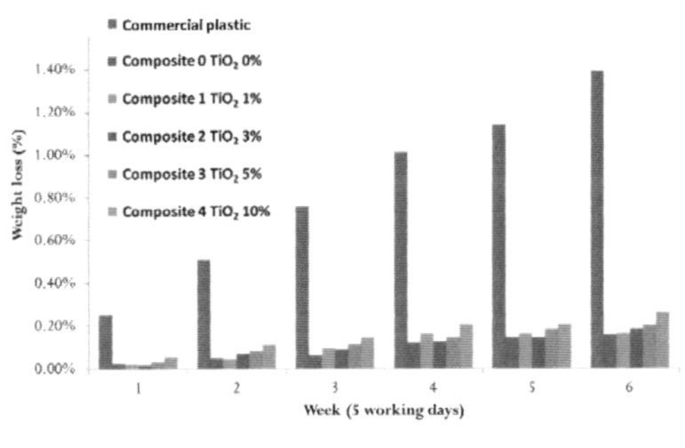

Figure 3. Nanocomposite without irradiation.

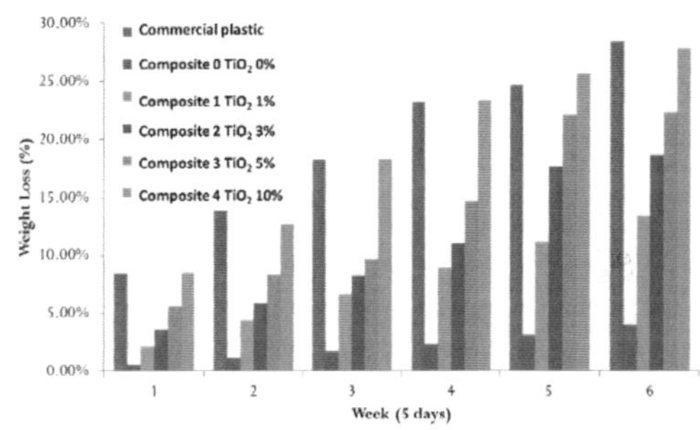

Figure 4. Nanocomposite irradiated using UV light.

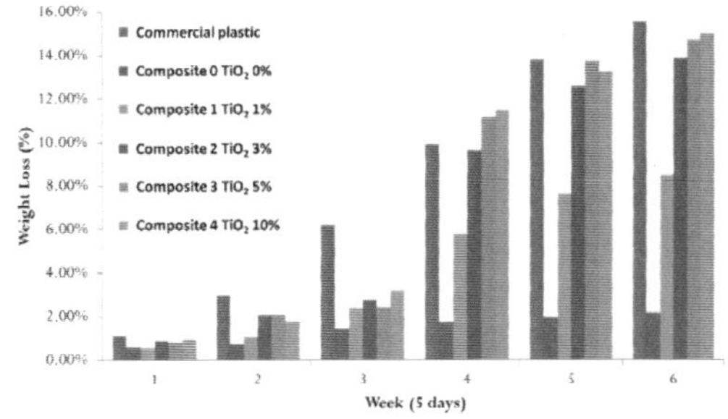

Figure 5. Nanocomposite irradiated using direct sunlight.

shown in figure 4 and 5. The evidence of weight loss and damage in the composites after irradiation for 30 days are shown in figure 1b. The plastics somehow look darker and more fragile after the treatment.

4. Conclusions

Increasing amount of TiO_2 photocatalyst agent into composite can increase the decrement of weight of composite with irradiation. Irradiation using direct sunlight is affected by light intensity. In contrast, without irradiation, the weight reduction of bioplastic was insignificant. The results indicate that the addition of TiO_2 to the composition of organobentonite modified acetic cellulose plastic has given the ability of self-photodegradation to the composite. ability of self-photodegradation to the composite.

Acknowledgements

This work was funded by PITTA Grant No. 1753/UN2.R12/PPM.00.00/2016 from Universitas Indonesia.

References

[1] Chen X and Mao S S 2007 *Chem. Rev.* **107** 2891-959
[2] Fitriyani A 2013 *Sintesis dan Karakterisasi Membran Biokomposit Terinterkalasi Surfaktan HDTMABr sebagai Nanofiller* BSc projects (Depok: Departemen Kimia, FMIPA Universitas Indonesia)
[3] Fitriah A 2013 *Sintesis dan Karakterisasi Membran Biokomposit Terinterkalasi Surfaktan DTMABr sebagai Nanofiller* BSc Projects (Depok: Departemen Kimia, FMIPA Universitas Indonesia)
[4] He H, Zhu J, Yuan P, Zhou Q, Ma Y and Frost R L 2008 *Proc. ICAM 2008 – Ninth Int. Congress for Applied Mineralogy* (Brisbane: AusIMM) 321-7
[5] Rodriguez F J 2012 *Polymer Degradation and Stability* **97** 1996-2001
[6] Chen D Du G, Zhu Q and Zhou F 2013 *J. Colloid and Interface Science* **409** 151-7
[7] Juwono A L, Sihombing R, Krisnandi Y K, Sutarno, Subawi H and Chitraningrum N 2010 *Int. J. Modern Physics* B **24** 148-56
[8] Andhika K 2014 *Sintesis Selulosa Asetat dengan TiO_2-Bentonit Terinterkalasi Surfaktan Heksadesiltrimetilamonium Bromida (HDTMABr) sebagai Nanofiller* BSc Projects (Depok: Departemen Kimia, FMIPA Universitas Indonesia)
[9] Luthfiyah S Z, Krisnandi Y K, Andhika K, Sihombing R and Fisli A 2016 *AIP Conf. Proc.* ***1729*** 161
[10] Bergaya F, Theng B K G and Lagaly G 2006 *Handbook of Clay Science* (Amsterdam: Elsevier)
[11] Puls J, Wilson S A and Holter D 2011 *J. Polym. Environ.* **19** 152-65

Enhancement of photoresponse to ultraviolet region by coupling perovskite $LaMnO_3$ with TiO_2 nanoparticles

N Afifah[1,2] and R Saleh[1,2]

[1]Department of Physics, Faculty of Mathematics and Natural Sciences Universitas Indonesia, Kampus UI Depok, Depok 16424, Indonesia
[2]Integrated Laboratory of Energy and Environment, Faculty of Mathematics and Natural Sciences Universitas Indonesia, Kampus UI Depok, Depok 16424, Indonesia

Corresponding author's e-mail: rosari.saleh@gmail.com

Abstract. Mixed oxides of rare-earth elements and 3d transition metal perovskite materials were studied for their catalytic application. We have previously reported the catalytic performance of perovskite $LaMnO_3$ nanoparticles under visible light and ultrasonic irradiation separately and simultaneously. However, $LaMnO_3$ perovskite's narrow optical gap can absorb only a small portion of the solar spectrum in the ultraviolet region, resulting in low catalytic performance. Coupling $LaMnO_3$ with other metal oxides with an optical gap in the ultraviolet light range offers an effective solution to this problem by improving catalytic performance. To this end, sol-gel was used to prepare $LaMnO_3/TiO_2$ nanocomposites with varying $LaMnO_3$ to TiO_2 molar ratios. The $LaMnO_3/TiO_2$ nanocomposites were characterized using X-ray diffraction (XRD), a vibrating samples magnetometer (VSM), and UV-visible diffuse reflectance spectroscopy (DRS). The anatase phase of TiO_2 and the orthorhombic structure of $LaMnO_3$ facilitate the coupling of $LaMnO_3$ with TiO_2, and the incorporation of TiO_2 nanoparticles tends to increase the optical gap of the nanocomposite. The samples were used to remove organic dyes by means of simultaneous light and ultrasonic irradiation. The results confirm that the photosonocatalytic performance of $LaMnO_3/TiO_2$ nanocomposites successfully extends to the ultraviolet light region, and the main factors influencing photosonocatalytic activity are studied and discussed.

1. Introduction

ABO_3-type perovskite (A: rare earth; B: transition metal) has been widely used as a catalyst to degrade organic pollutants contained in wastewater because of its high chemical stability, high electron mobility, non-toxicity, and environmental friendliness [1-2]. $LaMnO_3$ is one ABO_3-type perovskite that can be used in this way; its special physical and chemical characteristics and an optical band gap value corresponding to the range of visible light irradiation make it a strong candidate for use as a catalyst. The use of $LaMnO_3$ as catalyst is of interest because solar energy contains about 46% visible light [1, 3-5].

Catalytic performance could also be improved by combining TiO_2 and $LaMnO_3$ in nanocomposite form. Catalysts in nanocomposite form are known to inhibit electron-hole recombination, so improving catalytic performance [6-7]. Additionally, the optical gap of TiO_2 (~3.2 eV) would be expected to extend the response of nanocomposite light in the ultraviolet (UV) light range, which accounts for about 5% of solar energy. TiO_2 itself is used as a catalyst because it is high catalytic activity, non-toxicity, and low cost [8].

The catalytic activity of $LaMnO_3$ nanoparticles has been reported in our previous work [9]. The combination of visible light and ultrasonic (Vis+US) in photosonocatalytic activity achieved maximum performance in degrading methylene blue (MB) as organic pollutant model. In the present study, catalyst performance was improved by synthesizing $LaMnO_3/TiO_2$ nanocomposite using the sol-gel method. TiO_2's influence on the photosonocatalytic activity of $LaMnO_3/TiO_2$ nanocomposite was assessed using different irradiation sources: UV light and ultrasonic (UV+US) and visible light and ultrasonic (Vis+US). The main factors influencing photosonocatalytic activity are investigated and discussed below.

2. Experimental details

Chemicals: All chemicals were of analytical grade and were used without further purification. Reagents included $LaCl_3{\cdot}7H_2O$, $MnCl_2{\cdot}4H_2O$, titanium dioxide (TiO_2), and sodium hydroxide (NaOH), (Merck).

Catalyst preparation: $LaMnO_3$ nanoparticles were first prepared following our procedure in previous work [9]. $LaMnO_3/TiO_2$ with different molar ratios (1:0.1, 1:0.3, and 1:0.5) were synthesized using the sol-gel method [10], involving the following steps. First, 0.00125 mol TiO_2 was dissolved in 30 mL distilled water under the influence of a magnetic stirrer. Then, 0.00125 NaOH was dissolved in 65 mL distilled water and added to the mixture (solution A). Next, solution A was stirred and heated to a temperature of 80°C. Meanwhile, the prepared $LaMnO_3$ nanoparticle was dissolved into 30 mL a distilled water (solution B). Solution B was then mixed with solution A under the continuous influence of a magnet stirrer at 80°C for 2 hours. The final product was allowed to stand for one night at room temperature and then heated at 120°C for an hour in vacuum conditions.

Characterization: The crystal structure was obtained using X-ray diffraction (XRD) spectroscopy (Rigaku Miniflex 600) with monochromatic Cu-K_α radiation source (λ = 1.54060 Å), operating at 40 kV and 20 mA in the range 10-90°. The XRD pattern was then analyzed using the MAUD program and the Rietveld method for obtaining crystal structure and lattice parameters. Magnetic characteristics were measured using an Oxford Type 1.2 T vibrating sample magnetometer (VSM) (from 0 to ± 1 Tesla). Optical characteristics were measured using a UV-Vis diffuse reflectance spectrophotometer.

Photosonocatalytic activity: Methylene blue (MB) was chosen as a model organic pollutant. For present purposes, MB degradation was executed using $LaMnO_3/TiO_2$ nanocomposite as catalyst and a combination of light and ultrasonic irradiation, known as photosonocatalysis. This photosonocatalytic procedure was described in detail in our previous work [9], using a Hitachi UH5300 UV-visible spectrophotometer for measurement. Moreover, in this work, the photosonocatalytic were monitored using different irradiation sources: UV light and ultrasonic (UV+US) and visible light and ultrasonic (Vis+US). MB degradation percentage was measured as C_t/C_0, where C_t is the MB concentration at a specified time following irradiation and C_0 is the earlier concentration of MB after reaching adsorption–desorption equilibrium. To identify the main active species in MB degradation, chemical compounds such as di-ammonium oxalate, sodium sulfate, and tert-butyl alcohol were used as hole, electron, and hydroxyl radical scavengers. To check its stability, the catalyst was reused 4 times.

3. Results and discussion

The crystal structure of the sample was studied using XRD spectroscopy. Figure 1(a) shows the diffraction pattern of $LaMnO_3/TiO_2$ samples with different molar ratios, as well as the diffraction patterns of TiO_2 and $LaMnO_3$. The dominant diffraction pattern of the $LaMnO_3/TiO_2$ nanocomposite was that of orthorhombic $LaMnO_3$ [11-12]. The diffraction peak at 2θ 25.2° confirmed the presence of the tetragonal structure in the anatase phase of TiO_2 [7]. There was no impurity phase in the diffraction pattern of the $LaMnO_3/TiO_2$ nanocomposite. Table 1 summarizes lattice parameter values for the diffraction pattern of the sample obtained by the Rietveld refinement method and crystal size values calculated using Scherrer's formula [13].

VSM measurement was used to investigate the magnetic properties of samples. Figure 1(b) shows the magnetic saturation of $LaMnO_3/TiO_2$ nanocomposites with different molar ratios. All samples showed ferromagnetic behavior at room temperature. The results also confirm that magnetic saturation were influenced by the presence of TiO_2, and that magnetization decreased with increased TiO_2 molar ratio. This decrease in magnetization with increased TiO_2 molar ratio may be caused by the decrease of $LaMnO_3$ mass in the presence of TiO_2 as non-magnetic material. TiO_2 is also known to have diamagnetic characteristics, which may reduce dipolar interaction of the $LaMnO_3/TiO_2$ nanocomposite [10]. Magnetic saturation values for $LaMnO_3/TiO_2$ nanocomposite are summarized in table 1.

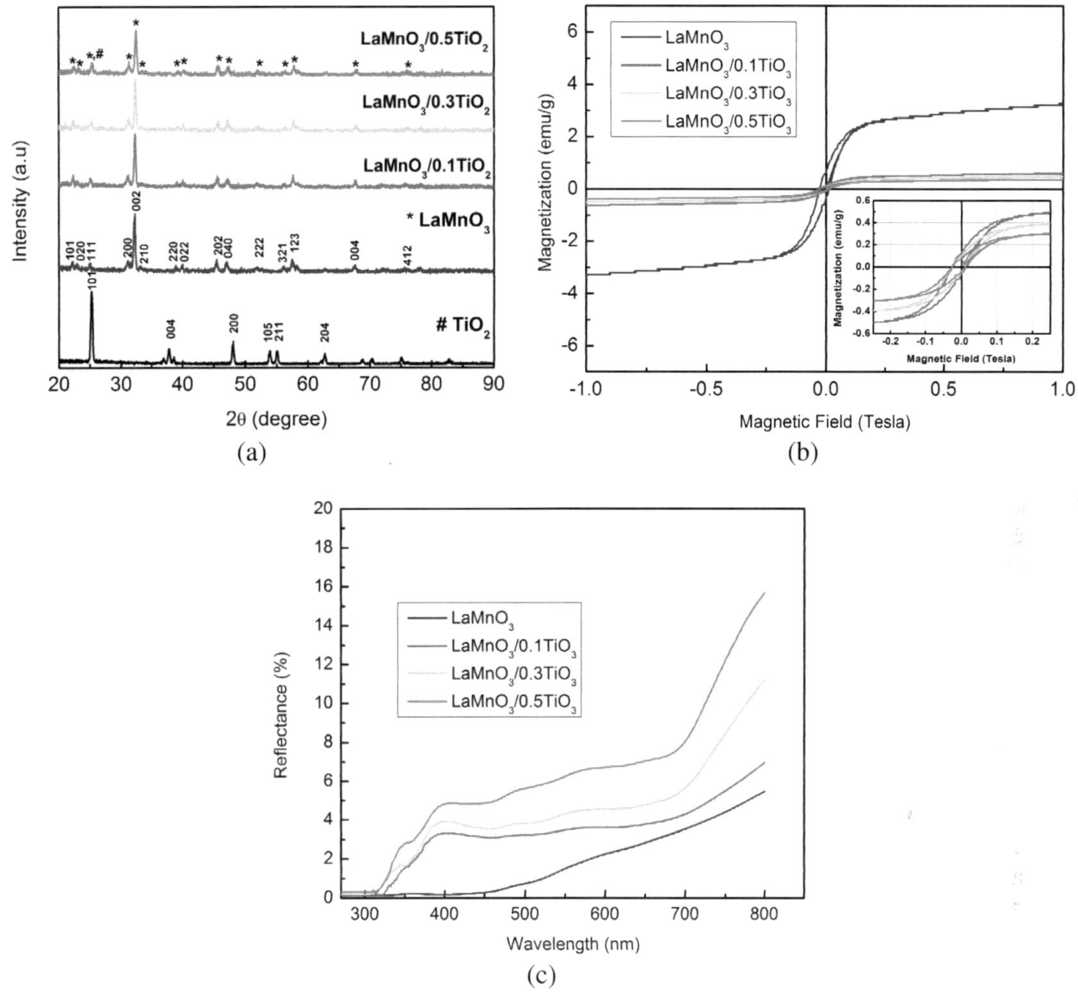

Figure 1. (a) XRD pattern; (b) VSM spectra; and (c) UV-Vis reflectance spectra of $LaMnO_3/TiO_2$ nanocomposites with different molar ratios.

Table 1. The lattice parameter, grain size <D>, magnetic saturation (M-S) and optical gap of $LaMnO_3/TiO_2$ nanocomposite.

Sample	Lattice Parameter $LaMnO_3$			Lattice Parameter TiO_2		<D> $LaMnO_3$ (nm)	<D> TiO_2 (nm)	M-S (emu/g)	Optical gap (eV)
	a (Å)	b (Å)	c (Å)	a=b (Å)	c (Å)				
$LaMnO_3$	5.7419	7.7215	5.5403	-	-	44	-	3.20	1.24
$LaMnO_3/0.1TiO_2$	5.7382	7.7187	5.5430	3.8215	9.4850	24	23	0.60	1.48
$LaMnO_3/0.3TiO_2$	5.7269	7.6995	5.5321	3.7824	9.5055	23	22	0.48	1.59
$LaMnO_3/0.5TiO_2$	5.7136	7.6787	5.5176	3.7685	9.4907	20	21	0.36	1.62

Optical characteristics of the samples were studied using UV-Vis diffuse reflectance spectroscopy (DRS). Figure 1(c) shows the UV-Vis reflectance spectra of $LaMnO_3$ and $LaMnO_3/TiO_2$ nanocomposite with variation of molar ratio. Extrapolating the curve using Kubelka-Munk analysis [14] provided the optical band gap value of each sample. These values are summarized in table 1. The results indicate that the adsorption capacity of $LaMnO_3/TiO_2$ nanocomposite is in the range of visible light radiation. Additionally, optical band gap value was found to increase with the molar ratio of TiO_2. The increase in optical band gap value would be expected to increase the response of $LaMnO_3/TiO_2$ nanocomposite light in the UV light radiation.

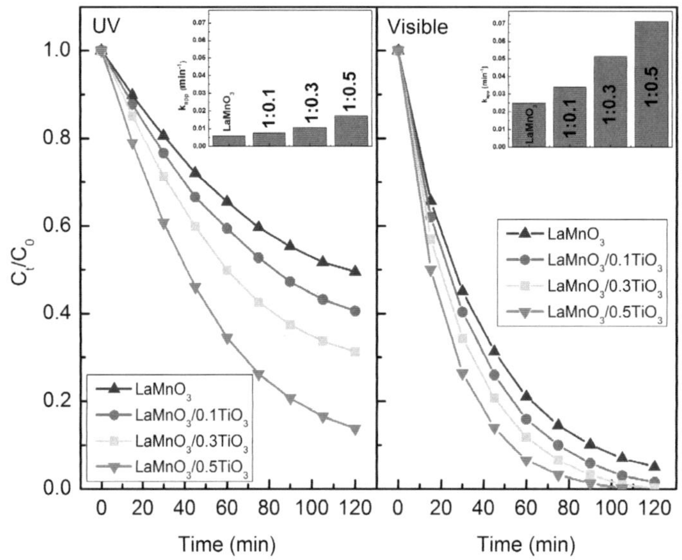

Figure 2. The photosonocatalytic activity of $LaMnO_3/TiO_2$ nanocomposite with different molar ratios of TiO_2 under UV (left) and Visible (right) light irradiation. The inset shows MB degradation rate using $LaMnO_3/TiO_2$ nanocomposite.

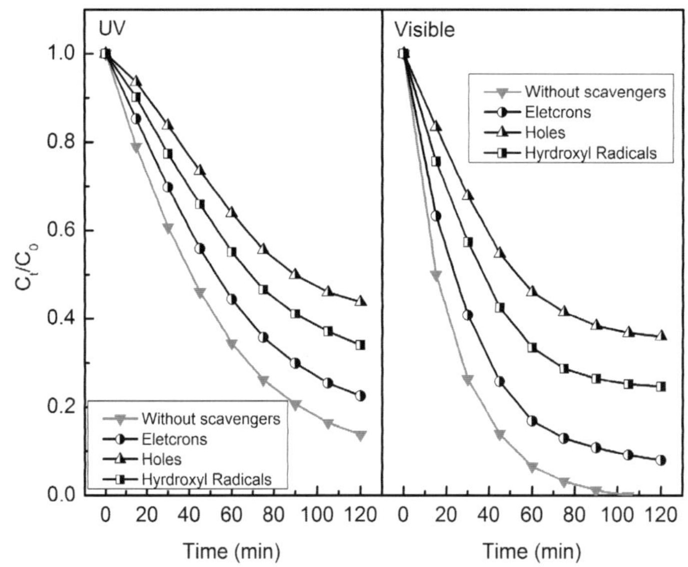

Figure 3. The effect of scavengers on photosonocatalytic degradation of MB using $LaMnO_3/0.5TiO_2$ nanocomposite.

The photosonocatalytic activity of $LaMnO_3/TiO_2$ nanocomposite was used to degrade MB as model organic pollutant under different light irradiation sources, combining UV light and ultrasonic (UV+US) and visible light and ultrasonic (Vis+US). Figure 2 shows the photosonocatalytic activity of $LaMnO_3/TiO_2$ nanocomposite with different molar ratios. For comparison, the photosonocatalytic activity of $LaMnO_3$ nanoparticle is also shown. The results demonstrate that the presence of TiO_2 in the nanocomposite also increases the photosonocatalytic activity of $LaMnO_3$ nanoparticles. Increased TiO_2 molar ratio increases the photosonocatalytic activity of $LaMnO_3/TiO_2$ nanocomposite in both UV+US and Vis+US. The highest degradation was obtained using $LaMnO_3/TiO_2$ nanocomposite with molar ratio 1:0.5. The inset in figure 2 shows MB degradation rate using $LaMnO_3/TiO_2$ nanocomposite, calculated by a pseudo-first order kinetics model [15], in which $\ln (C_t/C_0) = k_{app}\, t$. The results show that photosonocatalytic activity using Vis+US irradiation was faster than when using UV+US irradiation; this may be caused by the greater $LaMnO_3$ content as a catalyst of visible light. The optical gap values for the nanocomposite also confirmed its absorption capacity in the range of visible light irradiation.

In general, photosonocatalytic activity was influenced by catalyst interaction with ultrasonic and light waves. The increase of photosonocatalytic activity following combination of the two wave interactions was described in our previous work [9]. The influence of different light irradiation on catalyst interaction can be described as follows. Light waves can influence photosonocatalysis by forming electrons and holes. The electrons in conduction band interact with oxygen (O_2) molecules to form superoxide radicals ($\bullet O_2^-$), and holes in the valence band interact with water molecules to form hydroxyl radicals (•OH). Superoxide radicals and hydroxyl radicals are known as strong reductor and oxidator agents and can degrade the complex chemical bonds of organic pollutants. It follows that the time needed for electrons and holes to recombine is strongly influenced by photosonocatalytic activity [16]. In the nanocomposite, using visible light as irradiation source, electrons in the conduction band of $LaMnO_3$ did not return directly to the valence band but were first transferred to the conduction band of TiO_2. In this way, the recombination of electrons and holes can be inhibited, and photosonocatalytic activity is seen to increase by comparison with the performance of the $LaMnO_3$ nanoparticle alone. The opposite was found to occur when UV light was used as irradiation source. At that moment, electrons from the TiO_2 conduction band did not return directly to the valence band but were transferred to the conduction band of $LaMnO_3$, inhibiting the recombination of electrons and holes, and increasing photosonocatalytic activity.

The present study also sought to identify the most influential factors in photosonocatalytic activity by adding several kinds of different scavenger. The degradation of methylene blue due to the existence

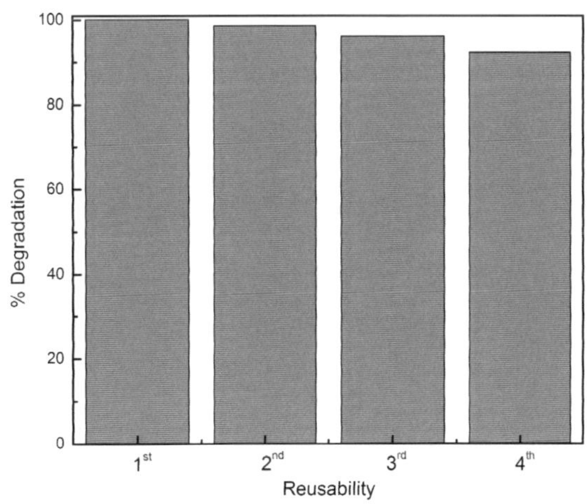

Figure 4. The reusability on the photosonocatalytic degradation of MB using $LaMnO_3/0.5TiO_2$ nanocomposite

of $Na_2S_2O_8$, di-ammonium oxalate, and tert-butyl alcohol as scavengers of electrons, holes, and hydroxyl radicals is shown in figure 3. The use of di-ammonium oxalate as a hole scavenger inhibited MB degradation rate, indicating that holes played the most important role in photosonocatalytic activity. Catalyst stability during degradation is shown in figure 4. Reused 4 times, 1:0.5 $LaMnO_3/TiO_2$ nanocomposite using Vis+US irradiation succeeded in degrading 92% of MB, indicating the good stability of $LaMnO_3/TiO_2$ nanocomposite as catalyst.

4. Conclusions

The use of $LaMnO_3/TiO_2$ nanocomposite's photosonocatalytic activity to degrade methylene blue was successful using both UV+US and Vis+US irradiation. The incorporation of TiO_2 on the nanocomposite increased the photosonocatalytic activity and successfully extended the response of catalyst to the ultraviolet light region. The use of $LaMnO_3/TiO_2$ in nanocomposite form plays an important role in the improvement of photosonocatalytic performance by reducing electron-hole recombination.

References

[1] Hu J, Ma J, Wang L, Huang H and Ma L 2014 *Powder Technol.* **254** 556-62
[2] Lotey G S and Verma N K 2014 *Mat. Sci. Semicon. Proc.* **21** 206-11
[3] Hu J, Ma J, Wang L and Huang H 2014 *Mater. Sci. Eng.* B **180** 46-53
[4] Hu J, Ma J, Wang L and Huang H 2014 *J. Alloy. Compd.* **583** 539-45
[5] Hu J, Men J, Ma J and Huang H 2014 *J. Rare Earth.* **32** 1126-34
[6] Tang P, Tong Y, Chen H, Cao F and Pan G 2013 *Curr. Appl. Phys.* **13** 340-3
[7] Gao K and Li S 2012 *Appl. Surf. Sci.* **285** 6460-4
[8] Grabowska E 2016 *Appl. Catal. B: Environ.* **186** 97-126
[9] Afifah N and Saleh R 2016 *Mater. Sci. Forum* **864** 99-105
[10] Arifin S A, Jalaludin S and Saleh R 2015 *Mater. Sci. Forum* **827** 49-55
[11] Galal A, Atta N F and Ali S M 2011 *Electrochim. Acta* **56** 5722-30
[12] Zhang C, Guo Y, Guo Y, Lu G, Boreave A, Retailleau L, Baylet A and Giroir-Fendler A 2014 *Appl. Catal. B: Environ.* **148-149** 490-8
[13] Abazari R and Sanati S 2013 *Superlattice. Microst.* **68** 148-57
[14] Köferstein R, Jäger L and Ebbinghaus S G 2013 *Solid State Ionics* **249-250** 1-5
[15] Saleh R and Djaja N F 2014 *Superlattice. Microst.* **74** 217-33
[16] Tijare S N, Joshi M V, Padole P S, Mangrulkar P A, Rayalu S S and Labhsetwar N K, 2012 *Int. J. Hydrogen Energ.* **37** 10451-6

IOP Conf. Series: Materials Science and Engineering **188** (2017) 012061 doi:10.1088/1757-899X/188/1/012061

Dynamical mean-field theoretical approach to explore the temperature-dependent magnetization in Ta-doped TiO_2

M A Majidi[1], A S Umar[1] and A Rusydi[2,3]

[1]Department of Physics, Faculty of Mathematics and Natural Sciences Universitas Indonesia, Kampus UI Depok, Depok 16424, Indonesia

[2]NUSNNI-NanoCore, Department of Physics, Faculty of Science, National University of Singapore, Singapore 117576, Singapore

[3]Singapore Synchrotron Light Source, National University of Singapore, Singapore 117603, Singapore

Corresponding author's email: aziz.majidi@sci.ui.ac.id

Abstract. TiO_2 has, in recent years, become a hot subject as it holds a promise for spintronic application. Recent experimental study on anatase $Ti_{1-x}Ta_xO_2$ ($x \sim 0.05$) thin films shows that the system changes from non-magnetic to ferromagnetic due to Ti vacancies that are formed when a small percentage of Ti atoms are substituted by Ta. Motivated by those results that reveal the ferromagnetic phase at room temperature, we conduct a theoretical study on the temperature-dependent magnetization and the Currie temperature of that system. We hypothesize that when several Ti vacancies are formed in the system, each of them induces a local magnetic moment, then such moments couple each other through Ruderman-Kittel-Kasuya-Yosida (RKKY) interaction, forming a ferromagnetic order. To study the temperature dependence of the magnetization and predict the Curie temperature, we construct a tight-binding based Hamiltonian for this system and use the method of dynamical mean-field theory to perform calculations for various temperatures. Our work is still preliminary. The model and method may need further improvement to be consistent with known existing facts. We present our preliminary results to show how the present model works.

1. Introduction

The phenomenon of room-temperature ferromagnetism (RTFM) in doped transition-metal oxides has renewed interest of a lot of researchers in the related area because of their potential applications for spintronic devices. Among the seemingly promising materials is titanium dioxide (TiO_2) [1, 2, 3, 4, 5, 6, 7, 8]. A recent experimental study on anatase Ta-doped TiO_2 ($Ti_{1-x}Ta_xO_2$) thin films shows that the system changes from non-magnetic to ferromagnetic at $x \sim 0.05$ [9]. The authors observed that some Ti sites are occupied by Ta atoms, but in addition to those, some Ti sites become vacant. They reported that these Ti vacancies (V_{Ti}) induce the formation of local magnetic moments. A recent theoretical study shows that the Ruderman-Kittel-Kasuya-Yosida (RKKY) exchange interaction is likely responsible to make the magnetic moments align to form the ferromagnetic order [10]. Although it has been confirmed that the ferromagnetism of this system exists at room temperature, there has not been any report on the measurement or prediction of the ferromagnetic Curie temperature. Motivated by this fascinating RTFM induced by cationic vacancies with yet a lot of more issues needing to explore further, we hereby present

IOP Conf. Series: Materials Science and Engineering **188** (2017) 012061 doi:10.1088/1757-899X/188/1/012061

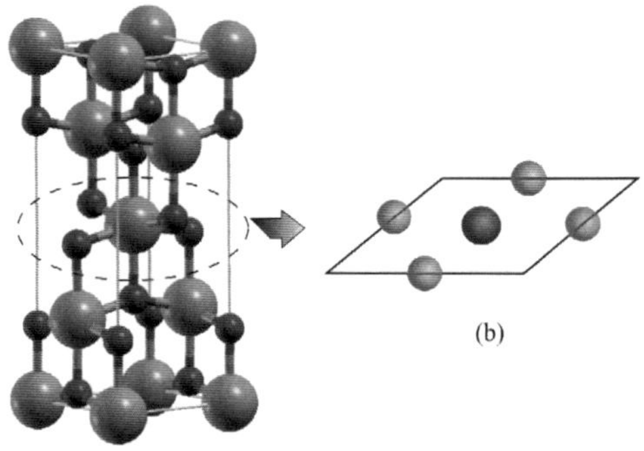

(a)

Figure 1. (a) Actual unit cell of anatase TiO_2 that consists of Ti (grey spheres) and oxygen atoms (red spheres). (b) Portion of the actual unit cell that taken to be the model unit cell.

a theoretical study on the temperature-dependent magnetization and the calculation of Curie temperature of the system.

TiO_2 is a naturally occurring oxide of titanium which comes in three types of crystal. The anatase TiO_2 is formed as a tetragonal with lattice parameters $a = b = 37.84$ nm and $c = 95.15$ nm [11]. Anatase crystal has a unit cell that consists of 12 atoms, that is, eight oxygen atoms and four titanium atoms. Each titanium atom bonds with six neighboring oxygen atoms. While, each oxygen atom bonds with three titanium atoms. The electron configuration for titanium is [Ar] $3d^2 4s^2$ and for the oxygen is [He] $2s^2 2p^4$. Another experiment shows that the pure anatase TiO_2 has an optical band gap of 3.32 eV, while the gap is slightly larger (3.4 eV) for the system doped with tantalum [9].

2. Model

The actual crystal structure of anatase TiO_2 is shown in figure 1(a) [12]. A combined experimental-theoretical study reveals the location of Ti vacancy in a unit cell [13]. The Ti vacancy is surrounded by four O atoms that are almost coplanar and form a square. Assuming that the most important physics leading to the ferromagnetism lies in this layer, we propose to simplify the crystal structure such that the model unit cell forms a square at the center of which is located the Ti atom surrounded by four O atoms as depicted in figure 1(b). Effectively this two-dimensional model unit cell contains only one Ti atom and two O atoms. In addtion to such a unit cell having a Ti atom at its center, we also consider another unit cell whose center is a vacant site. It is this unit cell with Ti vacancy to carry a magnetic moment. The net magnetic moment at each unit cell is assumed to come from four unpaired electrons each of which resides in an oxygen orbital and spends most of its time in the region close to the Ti vacancy site as illustrated in figure 2.

Referring to the model unit cells described in figures 1 and 2, we construct the set of basis atomic states for our model. Note that to distinguish between the two oxygen atoms in a unit cell we label them as O_A and O_B. Keeping two p orbitals (p_x and p_y) and one d orbital at the Ti site, two p orbitals (p_x and p_y) belonging to O_A, and another two p orbitals (p_x and p_y) belonging to O_B, we have seven atomic orbital states to use. Further, considering the two possible spin orientations, we have in total 14 basis

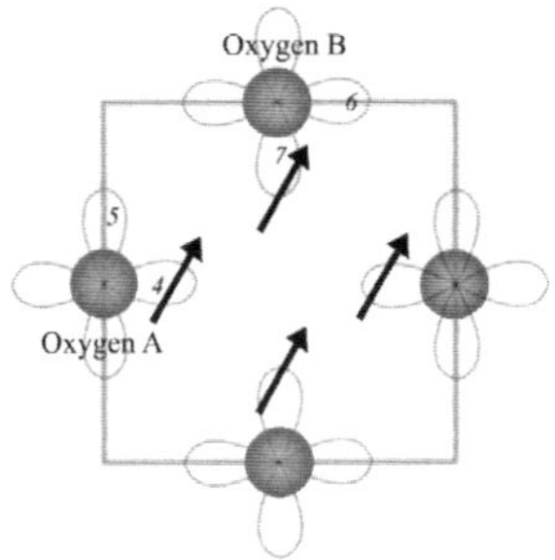

Figure 2. Model unit cell with a Ti vacancy. The blue arrows represent the unpaired electron spins that form the net magnetic moment in a unit cell.

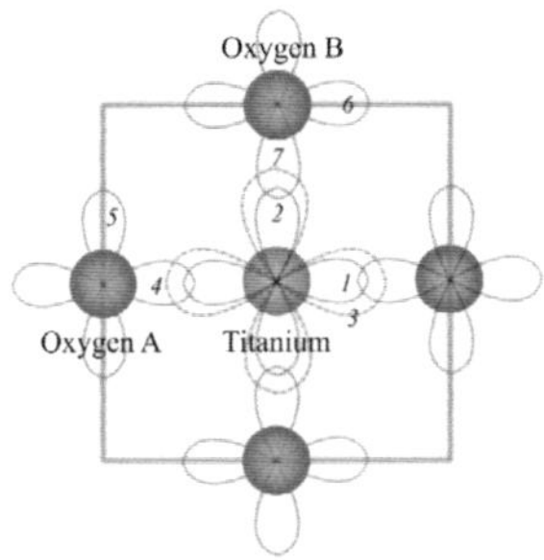

Figure 3. The seven basis states for one spin component in a unit cell.

states to construct the Hilbert space of our model that we order as below

$$\begin{array}{ll} 1:|\mathrm{Ti}_{p_x\uparrow}\rangle & 8:|\mathrm{Ti}_{p_x\downarrow}\rangle \\ 2:|\mathrm{Ti}_{p_y\uparrow}\rangle & 9:|\mathrm{Ti}_{p_y\downarrow}\rangle \\ 3:|\mathrm{Ti}_{d\uparrow}\rangle & 10:|\mathrm{Ti}_{d\downarrow}\rangle \\ 4:|\mathrm{O}_{A_{px}\uparrow}\rangle & 11:|\mathrm{O}_{A_{px}\downarrow}\rangle \\ 5:|\mathrm{O}_{A_{py}\uparrow}\rangle & 12:|\mathrm{O}_{A_{py}\downarrow}\rangle \\ 6:|\mathrm{O}_{B_{px}\uparrow}\rangle & 13:|\mathrm{O}_{B_{px}\downarrow}\rangle \\ 7:|\mathrm{O}_{B_{py}\uparrow}\rangle & 14:|\mathrm{O}_{B_{py}\downarrow}\rangle. \end{array} \tag{1}$$

The seven basis states for one spin component are shown in figure 3.

Now that we have defined our basis states, we construct our model Hamiltonian as the following

$$H = \sum_{\mathbf{k},\sigma} \hat{c}^{\dagger}_{\mathbf{k},\sigma}[H_0(\mathbf{k})]\hat{c}_{\mathbf{k},\sigma} + U \sum_{i\in \mathrm{Ti\ sites}} n_{di\uparrow}n_{di\downarrow} - J_H \sum_{j\in V_{\mathrm{Ti\ sites}}} \mathbf{S}_j.\mathbf{s}_j. \tag{2}$$

The first term is the kinetic part that is constructed within tight-binding approximation. Here $\hat{c}^{\dagger}_{\mathbf{k},\sigma}$ is a row vector consisting of the creation operators, while $\hat{c}_{\mathbf{k},\sigma}$ is its hermitian conjugate. $[H_0(\mathbf{k})]$ is a 14×14 matrix that contains two identical diagonal blocks $[H_0(\mathbf{k})]_{\uparrow\uparrow(\downarrow\downarrow)}$ of size 7×7 corresponding to each spin direction as follows

$$[H_{\rm O}(\mathbf{k})]_{\uparrow\uparrow(\downarrow\downarrow)} = \begin{bmatrix} \varepsilon_{\mathrm{Ti}_p} & 0 & 0 & A_x(\mathbf{k}) & 0 & 0 & 0 \\ 0 & \varepsilon_{\mathrm{Ti}_p} & 0 & 0 & 0 & 0 & A_y(\mathbf{k}) \\ 0 & 0 & \varepsilon_{\mathrm{Ti}_d} & B_x(\mathbf{k}) & 0 & 0 & B_y(\mathbf{k}) \\ A_x(\mathbf{k}) & 0 & B_x(\mathbf{k}) & \varepsilon_{\rm O} & 0 & 0 & 0 \\ 0 & 0 & 0 & 0 & \varepsilon_{\rm O}+C_y(\mathbf{k}) & D(\mathbf{k}) & 0 \\ 0 & 0 & 0 & 0 & D(\mathbf{k}) & \varepsilon_{\rm O}+C_x(\mathbf{k}) & 0 \\ 0 & A_y(\mathbf{k}) & B_y(\mathbf{k}) & 0 & 0 & 0 & \varepsilon_{\rm O} \end{bmatrix}. \quad (3)$$

The elements of $[H_0(\mathbf{k})]_{\uparrow\uparrow(\downarrow\downarrow)}$ reflects the on-site energy of each orbital and the hopping-energy dispersion between different orbitals, with the symbols in equation (3) standing for

$$\begin{aligned} A_x(\mathbf{k}) &= -2t_{\mathrm{Ti}_p\mathrm{O}}\cos(\frac{k_x a}{2}), \\ A_y(\mathbf{k}) &= -2t_{\mathrm{Ti}_p\mathrm{O}}\cos(\frac{k_y a}{2}), \\ B_x(\mathbf{k}) &= -2t_{\mathrm{Ti}_d\mathrm{O}}\cos(\frac{k_x a}{2}), \\ B_y(\mathbf{k}) &= -2t_{\mathrm{Ti}_d\mathrm{O}}\cos(\frac{k_y a}{2}), \\ C_x(\mathbf{k}) &= -2t_{\mathrm{OO}_{far}}\cos(k_x a), \\ C_y(\mathbf{k}) &= -2t_{\mathrm{OO}_{far}}\cos(k_y a), \\ D(\mathbf{k}) &= -2t_{\mathrm{OO}_{near}}\cos\left(\frac{(k_x-k_y)a\sqrt{2}}{2}\right) - 2t_{\mathrm{OO}_{near}}\cos\left(\frac{(k_x+k_y)a\sqrt{2}}{2}\right). \end{aligned} \quad (4)$$

The second term in equation (2), where $n_{di\sigma} = c^{\dagger}_{di\sigma}c_{di\sigma}$, is the Hubbard repulsive interactions for two electrons occupying the same Ti_d orbital. Whereas, the third term represents the Hund's coupling between the net spin $\mathbf{S}$ of the localized-unpaired electrons and spin $\mathbf{s}$ of an itinerant electron hopping into the Ti-vacancy site.

3. Method

To simplify our method, we treat the Hubbard term within mean-field approximation. This implies simply adding the on-site energy $\varepsilon_{\mathrm{Ti}_d}$ of the Ti_d orbital with spin σ, by $\langle n_{d(-\sigma)}\rangle U$, where $\langle n_{d(-\sigma)}\rangle$ is to be determined self-consistently. As for the magnetic interaction term, we treat it within dynamical mean-field theory (DMFT), such that we can track the temperature-dependent behavior of the system due to spin fluctuation. The general flow of DMFT algorithm for handling a problem with Hund's magnetic interactions is standard as prescribed by Furukawa [14]. Some examples of specific DMFT implementation of such a problem can be found, for instance, in references [15, 16, 17].

The uniqueness of the DMFT treatment in this work lies in the definition of the "local" self energy matrices for our problem. As part of DMFT algorithm, local self energy, usually defined in form of a matrix, is where interaction terms of the Hamiltonian occuring locally in a site or a unit cell is inserted. For our problem, Ta doping is believed to cause addition of conduction electrons into the undoped system, making the originally insulating system transform into an n-type semiconductor. The Fermi level then falls into the Ti_d band. The magnetic moments residing at the oxygen atoms around the Ti vacancies are assumed to interact mainly with itinerant electrons of the Ti_d band. This is our first hypothesis. Although Ta doping adds more electrons to the system, but the formation of magnetic moments around the Ti vacancies cause some itinerant electrons to become localized at the oxygen sites. This then causes the effective number of itinerant electrons to decrease, hence shifting the Fermi level to the Op band and

turn the system to be a p-type semiconductor. The magnetic moments around the Ti vacancies are also assumed to interact mainly with itinerant electrons of the Op band. This is our second hypothesis. When Ta doping is introduced into the system, 99% of the tantalum replaces titanium site [9]. However, the Ti substitution by Ta does not cause any magnetization in the system. Therefore, we do not need to define a specific local self energy for the unit cells where Ti sites are occupied by Ta. However, this second type of unit cell induces the appearance of the third type of unit cell where the Ti atom is missing but not replaced by Ta, i.e. forming Ti vacancies. It is in these unit cells the magnetic interactions occur and play an important role to lead the system to become ferromagnetic. Below we detail the local self-energy matrices for two assumptions:, that is, if the system is considered as n-type and p-type semiconductors.
a). If the system is considered as n-type semiconductor:

$$\left[\Sigma_{loc\uparrow}\right] = \begin{bmatrix} \Delta\varepsilon_{\mathrm{Ti}_p} & 0 & 0 & t_{\mathrm{Ti}_p\mathrm{O}} & 0 & 0 & 0 \\ 0 & \Delta\varepsilon_{\mathrm{Ti}_p} & 0 & 0 & 0 & 0 & t_{\mathrm{Ti}_p\mathrm{O}} \\ 0 & 0 & \Delta\varepsilon_{\mathrm{Ti}_d} - U\langle n_{\mathrm{Ti}_d\downarrow}\rangle - \frac{J_H S}{2}\cos\theta & t_{\mathrm{Ti}_d\mathrm{O}} & 0 & 0 & t_{\mathrm{Ti}_d\mathrm{O}} \\ t_{\mathrm{Ti}_p\mathrm{O}} & 0 & t_{\mathrm{Ti}_d\mathrm{O}} & -t_{\mathrm{OO}_{far}} & 0 & 0 & -t_{\mathrm{OO}_{far}} \\ 0 & 0 & 0 & 0 & 0 & 0 & 0 \\ 0 & 0 & 0 & 0 & 0 & 0 & 0 \\ 0 & t_{\mathrm{Ti}_p\mathrm{O}} & t_{\mathrm{Ti}_d\mathrm{O}} & -t_{\mathrm{OO}_{far}} & 0 & 0 & -t_{\mathrm{OO}_{far}} \end{bmatrix} \quad (5)$$

b). If the system is considered as p-type semiconductor:

$$\left[\Sigma_{loc\uparrow}\right] = \begin{bmatrix} \Delta\varepsilon_{\mathrm{Ti}_p} & 0 & 0 & t_{\mathrm{Ti}_p\mathrm{O}} & 0 & 0 & 0 \\ 0 & \Delta\varepsilon_{\mathrm{Ti}_p} & 0 & 0 & 0 & 0 & t_{\mathrm{Ti}_p\mathrm{O}} \\ 0 & 0 & \Delta\varepsilon_{\mathrm{Ti}_d} - U\langle n_{\mathrm{Ti}_d\downarrow}\rangle & t_{\mathrm{Ti}_d\mathrm{O}} & 0 & 0 & t_{\mathrm{Ti}_d\mathrm{O}} \\ t_{\mathrm{Ti}_p\mathrm{O}} & 0 & t_{\mathrm{Ti}_d\mathrm{O}} & -t_{\mathrm{OO}_{far}} & 0 & 0 & -t_{\mathrm{OO}_{far}} \\ 0 & 0 & 0 & 0 & -\frac{J_H S}{2}\cos\theta & 0 & 0 \\ 0 & 0 & 0 & 0 & 0 & -\frac{J_H S}{2}\cos\theta & 0 \\ 0 & t_{\mathrm{Ti}_p\mathrm{O}} & t_{\mathrm{Ti}_d\mathrm{O}} & -t_{\mathrm{OO}_{far}} & 0 & 0 & -t_{\mathrm{OO}_{far}} \end{bmatrix} \quad (6)$$

Note that, other than the terms proportional $J_H S$ appearing because of the magnetic interactions, the rest of the elements of the local self-energy reflects the differences in the on-site energies and the hopping parameters between unit cell having Ti at the center and that where Ti is missing.

4. Result and discussion

The results and discussion presented below are divided according to the two assumptions: TiO_2 as an n-type and as a p-type semiconductor. The physical parameter values are set based on typical values of tight-binding parameters for transition-metal oxides, and tuned to suit the actual band gap, and other known empirical data. They are tabulated in table 1.

4.1. n-type

From the density of states (DOS) shown in figure 4, we roughly see that the band is splitted into three parts. With the help of projected DOS shown in figure 5, we can see contribution of each orbital to the total DOS. The left red-orange band is contributed by Ti_{p_x} and Ti_{p_y} orbitals. The middle band is almost contributed by all orbitals, but the most pronounced peak is from O_{A_x} and O_{B_y} orbitals (blue-yellow) followed by O_{A_y} and O_{B_x} orbitals (cyan-black). The right part is dominated by Ti_d orbital (thick green). Here, with the right choice of parameters we successfully capture an n-type semiconductor as the chemical potential lies slightly inside the conduction band. However, we do not obtain an impurity band near the conduction band as this may be related to non-magnetic trend we obtain for the n-type from our calculations as shown in the magnetization curve in Figure 6. That our model for the n-type Ta:TiO_2 does not capture a ferromagnetic order, we consider that the model may still lack of important key points not incorporated in the calculations. Thus, the model needs further checking and improvement.

Table 1. List of physical parameters used in the calculation.

Physical Parameter	Symbol	Value
Lattice parameter	a	3.874 Å
V-Ti concentration	x	3%
Filling value for Ti_d	n_{Ti_d}	$\{n$: 0.025 ; p: 0.0$\}$
Total filling value	$n_{filling}$	$\{n$: 12.025 ; p: 11.94$\}$
Hund's exchange coupling constant	J_H	3.0 eV
Hubbard's on-site Coulomb repulsion constant	U	6.0 eV
On-site energy of O_A and O_B	ε_O	-5.0 eV
On-site energy of Ti_p	ε_{Ti_p}	-20.0 eV
On-site energy of Ti_d	ε_{Ti_d}	0.3 eV
Corection for Ti_p on-site energy	$\Delta\varepsilon_{Ti_p}$	10.0 eV
Corection for Ti_d on-site energy	$\Delta\varepsilon_{Ti_d}$	11.0 eV
Hopping parameter of Ti_p-O	t_{Ti_pO}	3.3 eV
Hopping parameter of Ti_d-O	t_{Ti_dO}	5.0 eV
Hopping parameter of near O-O	$t_{OO_{near}}$	1.4 eV
Hopping parameter of far O-O	$t_{OO_{far}}$	0.8 eV
Ion spin of oxygen	S	2.0

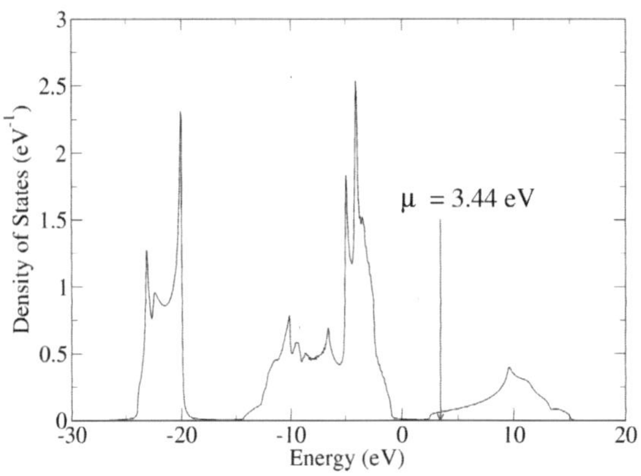

Figure 4. n-type Ta:TiO_2 density of states at 300 Kelvin. The band gap of $\sim$ 3.3 eV is obtained upon tuning the related parameters. The calculated chemical potential (μ) is 3.4475 eV. From the location of the chemical potential, the material is verified as an n-type semiconductor.

4.2. p-type

The DOS curve shown in figure 7 indicates that with the right choice of parameters we capture a p-type semiconductor. The chemical potential now lies at the impurity band (figure 7). In addition, as shown in figure 8, we obtain that the system can have a stable magnetization at zero external magnetic field. As we increase the temperature, the system spontaneous magnetization decreases, and drops to zero as the $T_C \sim$ 150 Kelvin.

It is interesting to notice that the p-type model does show a ferromagnetic order although with a $T_C \sim$ 150K which is much lower than a room temperature, while the n-type model does not. In response to these results, there are two possible implications. First, it may be that if one manages to experimentally

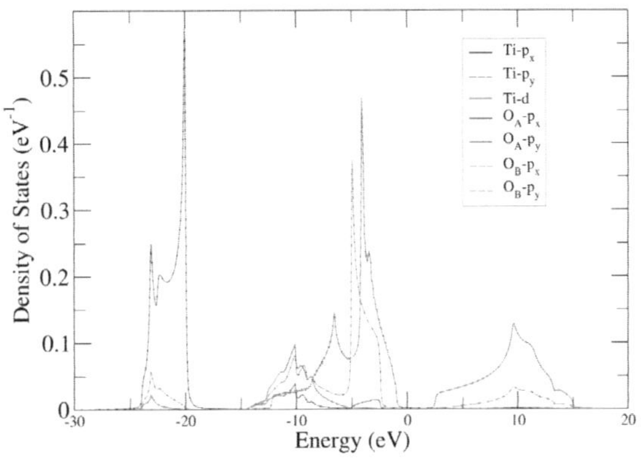

Figure 5. n-typed Ta:TiO_2 projected density of states at 300 Kelvin. The thick green curve is the Ti_d band. It is almost the only orbital that contributes electrons at the Fermi energy, μ.

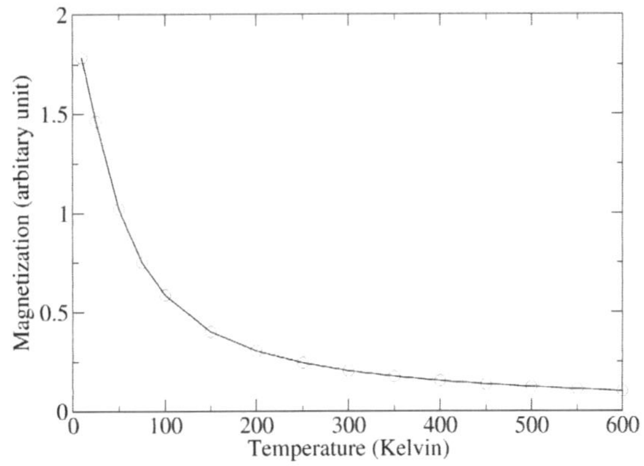

Figure 6. The spontaneous magnetization vs temperature for the modeled n-type with $\delta H = 10^{-4}$ eV. The exponentially decaying curve, signifies that the system is not ferromagnetically ordered.

dope TiO_2 with a transition metal element similar to Ta, while making sure that the semiconductor becomes p-type, it might lead the doped system to ferromagnetic although perhaps with a low T_C. Second, the both model for n-type and p-type may have to be re-inspected as the experimental data shows that Ta-doped TiO_2 is an n-type semiconductor and exhibits an RTFM.

5. Conclusion

We have developed models to study temperature-dependent magnetization of Ta-doped TiO_2 using the assumptions of both as an n-type and a p-type semiconductor. We demonstrate how the model and the method work to capture the corresponding types semiconductor, through the density of states calculations. However, our present n-type model does not lead to a ferromagnetic order, whereas our present p-type model does although with $T_C \sim 150$K much below a room temperature. Although the

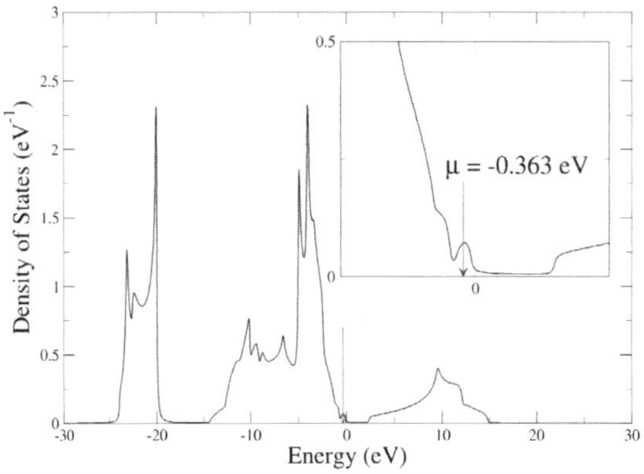

Figure 7. Density of states of p-type Ta:TiO_2. The chemical potential lies at the impurity band of the system.

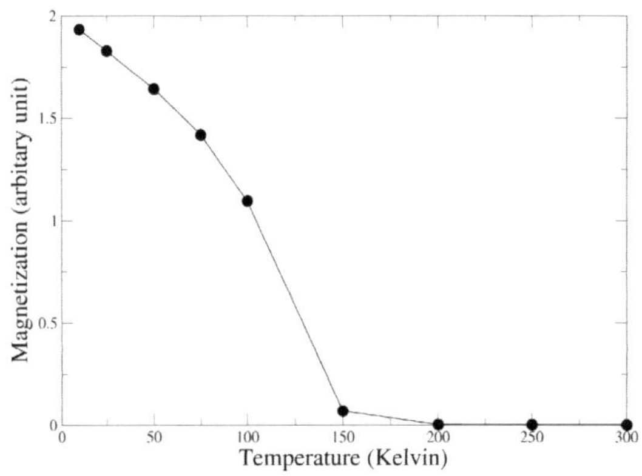

Figure 8. The spontaneous magnetization vs temperature for the modeled p-type with $\delta H = 0$. The curve does show a ferromagnetic trend, with $T_C \sim 150$ Kelvin.

result for the p-type model may be interpreted as there is a possibility to make a p-type ferromagnetic semiconductor from TiO_2 doped with some transition metal element, the fact that our present n-type model has not shown to capture to a ferromagnetic order, we consider that re-inspection and improvement of our present model and method is necessary. This work is still preliminary, hence all the conclusion presented in this paper are not yet final.

Acknowledgement
We gratefully acknowledge the full funding support to this project from Universitas Indonesia through PITTA Research Grant No. 2020/UN2.R12/HKP.05.00/2016.

References

[1] Matsumoto Y, Murakami M, Shono T, Hasegawa T, Fukumura T, Kawasaki M, Ahmet P, Chikyow T, Koshihara S and Koinuma H 2001 *Science* **291** 854-6
[2] Matsumoto Y, Takahashi R, Murakami M, Koida T, Fan X J, Hasegawa T, Fukumura T, Kawasaki M, Koshihara S and Koinuma H 2001 *Jpn. J. Appl. Phys.* **40** L1204-6
[3] Bapna K, Choudhary R J, Pandey S K, Phase D M, Sharma S K and Knobel M 2011 *Appl. Phys. Lett.* **99** 112502
[4] Nguyen H H, Sakai J, Prellier W, Hassini A, Ruyter A and Gervais F 2004 *Phys. Rev. B* **70** 195204
[5] Patel S K S, Kurian S and Gajbhiye N S 2012 *AIP Adv.* **2** 012107
[6] Nguyen H H, Sakai J and Hassini A 2004 *Appl. Phys. Lett.* **84** 2602
[7] Yildirim O, Cornelius S, Butterling M, Anwand W, Wagner A, Smekhova A, Fiedler J, Bottger R, Bahtz C and Potzger K 2015 *Appl. Phys. Lett.* **107** 242405
[8] Yang J Y, Han Y L, He L, Dou R F, Xiong C M and Nie J C 2012 *Appl. Phys. Lett.* **100** 202409
[9] Rusydi A *et al.* 2012 *Phil. Trans. R. Soc. A* **370** 4927-43
[10] Majidi M A, Bupu A and Fauzi A D 2016 *Physica B* http://dx.doi.org/10.1016/j.physb.2016.08.031
[11] Shang-Di M and Ching W Y 1995 *Phys. Rev. B* **51** 13023
[12] Cromer D T and Herrington K 1955 *J. Am. Chem. Soc.* **77** 4708
[13] Qi D C *et al.* 2013 *Phys. Rev. B* **87** 245201
[14] Furukawa N 1994 *J. Phys. Soc. Jpn.* **63** 3214; in Proceedings of the International Conference on Physics of Manganites, Michigan State University, 1998, edited by Kaplan T A and Mahanti S D (Kluwer Academic/Plenum, New York, 1998), p. 1.
[15] Aryanpour K, Moreno J, Jarrell M and Fishman R 2005 *Phys. Rev. B* **72** 045343
[16] Majidi M A, Jarrell M, Moreno J, Fishman R and Aryanpour K 2006 *Phys. Rev. B* **74** 115205
[17] Majidi M A, Siregar S and Rusydi A 2014 *Phys. Rev. B* **90** 195442

IOP Conf. Series: Materials Science and Engineering **188** (2017) 012062 doi:10.1088/1757-899X/188/1/012062

Structural and morphological characterization of TiO_2-SnO_2 thin film prepared by combining doctor-blade and sol-gel techniques

S R Adawiyah and Endarko

Physics Department, Institut Teknologi Sepuluh Nopember, Kampus ITS, Sukolilo, Surabaya 60111, East Java, Indonesia

Corresponding author's email: endarko@physics.its.ac.id

Abstract. The TiO_2-SnO_2 thin film has successfully been synthesized using the co-precipitation method and was coated on ITO (Indium Tin Oxide) substrate by doctor-blade technique. The structure and morphology of the film were investigated by XRD and SEM, respectively. The results showed that the film with SnO_2 has a stronger formation of anatase phase compared to TiO_2 film. The morphological study is also revealed that the TiO_2-SnO_2 film has a more porous nature and uniform particle aggregates, and the presence of SnO_2 has been confirmed with EDX spectra.

Keywords: co-precipitation, doctor blade, TiO_2-SnO_2 thin film, microstructure, nanoporous

1. Introduction

Titanium dioxide (TiO2) is one of the most selected and studied extensively as photoanodes materials in DSSCs [1-2]. TiO2 can be found in its three polymorphs in nature: anatase, brookite, and rutile. However, anatase is mostly used due to its excellent stability and photoactivity. As the main material of the photoanode, characteristics of TiO2 such as structure and morphology will certainly affect the whole characteristics of DSSC.

The photoanode in DSSC should possess stable and porous structure, enormous surface area, high molecules dye-loading capacity and solid structure of the surface film [2-4]. The physical and chemical properties of TiO_2 can be modified by incorporating metals and nonmetals element to tailor the properties of TiO_2 for enhancing the photovoltaic application especially in DSSCs [1]. Doping with SnO_2 can improve the stability of crystallinity of TiO_2 because it has porous nature, fast rate of electron transport, higher electric mobility, long-term stability under UV-illumination and excellent performance in an acidic condition [2, 5, 6].

Many methods have been employed to synthesize TiO_2 nanoparticles, such as Stöber [7], hydrothermal [8] and sol-gel [9]. Recently, the co-precipitation method has been extensively investigated to synthesized TiO_2 nanoparticles due to low processing temperature, ease of composition control, high purity and good chemical homogeneity [10]. In this work, we synthesized TiO_2 and TiO_2-SnO_2 nanocomposites by a co-precipitation method, then coated on ITO film. The structure and morphology of TiO_2-SnO_2 film were characterized and compared to TiO_2 film.

2. Materials and methods

2.1. TiO_2 synthesis

TiO_2 powders were synthesized by mixing 20 mL of Titanium (III) chloride ($TiCl_3$) with 100 mL of aquades and was stirred for 1 hour. To this mixture, NH_4OH solution was added dropwise until pH reached to 9. The resultant solution was stirred until resulting white precipitate. The precipitate was filtered and was then washed several times with distilled water, reaching a value of pH equal to 7. Removal process of residual organics and the stabilization of the materials were carried out by calcination for three hours at 450°C.

2.2. TiO_2-SnO_2 synthesis

SnO_2 nanopowders was purchased from Sigma Aldrich. SnO_2 powders were mixed with TiO_2 nanopowders. The powders mixture was grinded with a grinding bowl and then annealed at 450°C for 30 min. To the mixture aquades, ethylcellulose, terpineol and ethanol were added and kept stirring for 10 min. The mixed oxide pastes were obtained from 0.7 mL of acetic acid, 0.3 g of PEG, and 0.7 mL of Triton X-100. The TiO_2 pastes were prepared in the same way.

2.3. TiO_2-SnO_2 thin films

ITO glass substrates were purchased from Mianyang Prochema Commercial Co., Ltd., China. ITO with a size of 1 × 1 cm^2 were thoroughly rinsed with deionized water and anhydrous ethanol and dried on a hot plate. TiO_2 and TiO_2-SnO_2 pastes were deposited onto a conductive glass substrates-ITO glass using doctor blade technique. The films were heated at 450°C for one hour and cooled naturally to obtain a nanoporous film. The structure and morphology of the films were characterized by XRD and SEM-EDX.

3. Results and discussion

3.1. Structural study

The crystalline phase of photoanodes films was evaluated by XRD analyses, and the result is shown in figure 1. It can be seen that the films are polycrystalline, and the diffraction peaks observed around 26 and 49 degrees correspond to the (101) and (200) of the anatase phase of TiO_2 with the tetragonal crystal structure (PDF no. 00-078-2486, ICDD) [11]. The TiO_2 film exhibits a new diffraction peak (222) plane around 31 degree, which belongs to the ITO peak. These results agree with the analysis of the microstructure of pure TiO_2 film by Arunachalam *et al.* [5].

Furthermore, the diffraction peaks from rutile phase appear in the X-ray patterns due to the addition of the SnO_2 in the film. The (110), (101) and (211) planes at 2θ values 27, 34.1 and 53 degrees were observed as the characteristic peaks of SnO_2 in the doped TiO_2 film. The similar results have also been reported in the previous studies [11-13].

Nevertheless, the intensity of (101) plane is higher than the intensity of TiO_2 film because of the electronegativity, and the ionic radius of Sn^{4+} ions are larger than Ti^{4+}. It allowed easily for Sn^{4+} ions to replace and occupy the oxygen position in the TiO_2 lattice [3, 5] Thus, the spectrum shows the intensity of (101) plane is increased with the decrease ITO peak (222) plane, which indicated the better crystallinity than the TiO_2 film was obtained. This feature gives a more stable chemical bond and permits excellent interconnection and continuity between titania nanoparticle, which in turn enhances electron transfer efficiency in photoanode [3-4]. This phenomenon showed that the inclusion of SnO_2 in the TiO_2 may stabilize the anatase as the main and strongest phase [5, 13]. The smaller radius of Ti^{4+} (0.68 Å) as compared to Sn^{4+} (0.69 Å) [5] also made the crystallite size of doped TiO_2 film is bigger than the undoped one. The average crystallite size, which is calculated from XRD data using Rietveld method, is 34.2 and 10.3 nm for the doped TiO_2 and the undoped TiO_2 films respectively.

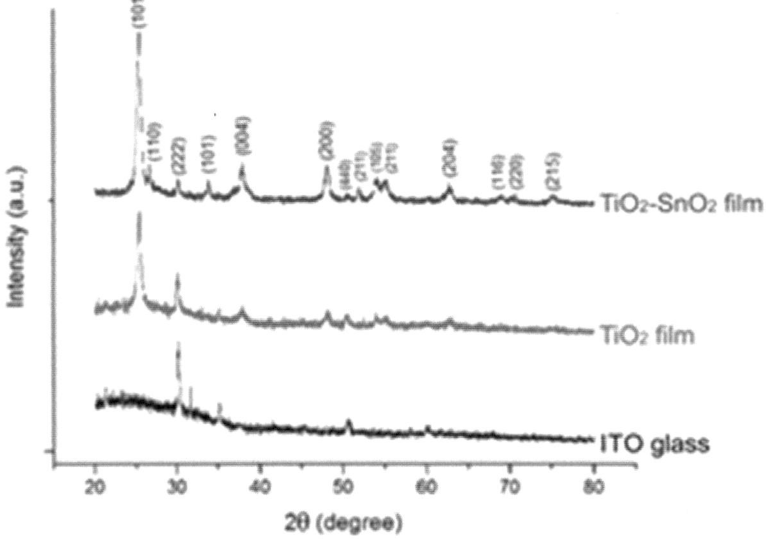

Figure 1. XRD spectra of ITO, TiO_2 and TiO_2-SnO_2 films.

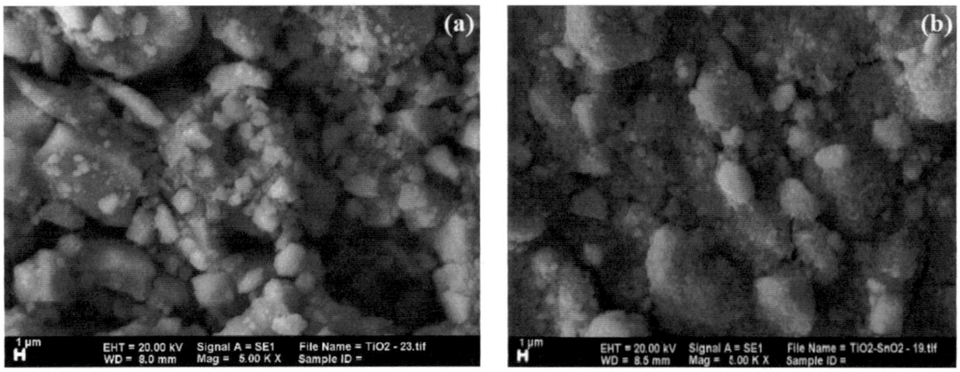

(a) (b)

Figure 2. SEM micrographs of (a) TiO_2 and (b) TiO_2-SnO_2 films

3.2 Morphological study

Figure 2 represents the top-view SEM images of the TiO_2-SnO_2 and TiO_2 films. These images confirm that the microstructure of both samples exhibits spherical shaped particle with irregular morphology due to the agglomeration of primary particles during the anneal treatment. It can be seen that smaller particles with an average diameter of 10-11 nm were observed for TiO_2 film and around 35 nm for SnO_2 doped TiO_2 film. The enlarging particle size of the doped TiO_2 film results in the larger surface area of the film [9], thus enabling a high dye loading capacity [14] as well as enhancing the photosensitivity to solar radiation [15]. The porous nature was observed in both films and this structure also plays a role in enhancing the surface area of the film photoanode [5, 9, 16]. A bit rough, large and intense inhomogeneous agglomerations were formed in the pure TiO_2 film. On the other hand, TiO_2-SnO_2 film exhibits smooth and rather well-distinguished uniform aggregates, although also there are few voids and cracks which may be due to losing of the binder during the annealing process [16]. These results indicate that the presence of SnO_2 can effectively suppress the grain growth of anatase compare with the pure TiO_2 [5].

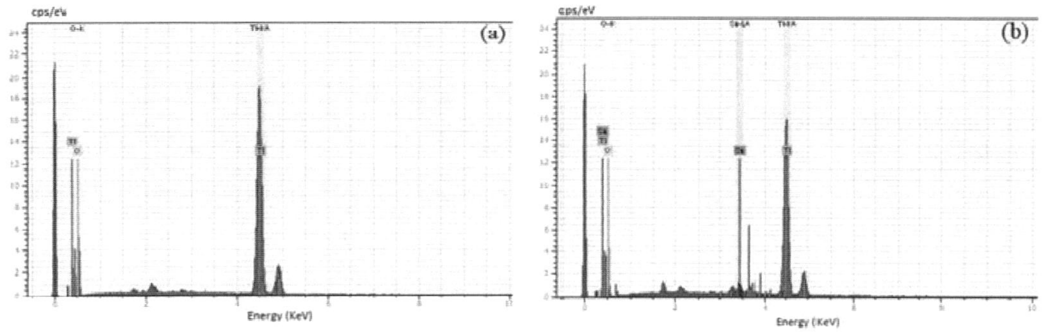

(a) (b)

Figure 3. The EDX spectra of (a) TiO_2 and (b) TiO_2-SnO_2 films

Table 1. The element compositions of the samples which are obtained from the EDX spectra

Ingredient number (at%)	TiO_2 film	TiO_2-SnO_2 film
Ti	67.24	71.06
O	32.76	27.66
Sn	0	1.28

The EDX spectra represents the element analysis of TiO_2 film and SnO_2 doped TiO_2 film. The EDX analysis of the TiO_2 film (figure 3a) confirms the presence of Ti and O. Figure 3b shows the existence of the SnO_2 dopants into TiO_2 lattice. From table 1 can be deduced that the Sn was presented around 1.28% in the TiO_2-SnO_2 film, and the percentage of O in the composites was decreased compared to TiO_2 film. The condition proved that the doping of SnO_2 has been successful.

4. Conclusions

In summary, the proposed work was a comparison the structural and morphological of the TiO_2-SnO_2 nanocomposite and the synthesized of TiO_2. The XRD pattern of TiO_2 anatase with SnO_2 besides showing the better crystallinity is also showing the stronger formation when compared to the anatase TiO_2. Additionally, the EDX analysis reveals the existence of the SnO_2 dopants into the TiO_2 lattice. The surface of the doped TiO_2 film showed smooth and rather homogeneous aggregate.

References

[1] Ye M, Wen X, Wang M, Iocozzia J, Zhang N, Lin C and Lin Z 2015 *Materials Today* **18** 155-62
[2] Liu J, Gu Y, Qin Z and Zhang Y 2014 *Int. Conf. on Materials for Renewable Energy and Environment (Chengdu)* vol 1 (IEEE) pp 66-9
[3] Hu B, Tang Q, He B, Lin L and Chen H 2014 *J. Power Sources* **267** 445-51
[4] Kontos A I, Kontos A G, Tsoukleris D S, Bernard M C, Spyrellis N and Falaras P 2008 *J. Mat. Processing Technol.* **196** 243-8
[5] Arunachalam A, Dhanapandian S and Manoharan C 2016 *Physica* E **76** 35-46
[6] Han D W, Heo J H, Kwak D J, Han C H and Sung Y M 2009 *JEET* **4** 93-7
[7] Falahatdoost S, Ara M H M, Shaban Z and Ghazyani N 2015 *Optical Materials* **47** 51-5
[8] Xu P, Tang Q, He B, Li Q and Chen H 2014 *Electrochimica Acta* **134** 281-6
[9] Sridhar D and Sriharan N 2014 *J. Nanoscience and Nanotechnol.* **2** 94-8

IOP Conf. Series: Materials Science and Engineering **188** (2017) 012062 doi:10.1088/1757-899X/188/1/012062

[10] Tian Z M, Yuan S L, He J H, Li P, Zhang S Q, Wang C H, Wang Y Q, Yin S Y and Liu L 2008 *J. Alloys and Compounds* **466** 26-30
[11] SharmilaDevi R, Venckatesh R and RajeshwariSivaraj 2014 *IJIRSET* **3** 15206-11
[12] Gurakar S, Serin T and Serin N 2014 *Adv. Mat. Lett.* **5** 309-14
[13] Stambolova I, Blaskov V, Vassilev S, Shipochka M and Dushkin C 2010 *J. Alloys and Compounds* **489** 257-61
[14] Qin Z, Huang Y, Liao Q, Zhang Z, Zhang X and Zhang Y 2012 *Materials Letters* **70** 177-80
[15] Kumar D A, Shyla J M and Xavier F P 2012 *Appl. Nanoscience* **2** 429-36
[16] Effendi M and Bilalodin 2012 *Prosiding Pertemuan Ilmiah XXVI HFI Jateng & DIY (Universitas Muhammadiyah Purworejo)* (HFI Jateng & DIY) pp 106-9